BUSINESS MATHEMATICS & STATISTICS

FOURTH EDITION

To my dad and in loving memory of my mom.

BUSINESS MATHEMATICS & STATISTICS

FOURTH EDITION

PETER WAXMAN

PRENTICE HALL

Sydney New York Toronto Mexico New Delhi
London Tokyo Singapore Rio de Janeiro

© 1998 by Prentice Hall Australia Pty Ltd

All rights reserved. No part of this publication may be reproduced, stored in a retrieval system, or transmitted in any form or by any means, electronic, mechanical, photocopying, recording, or otherwise, without written permission of the publisher.

Acquisitions Editor: Julie Catalano
Production Editor: Elizabeth Thomas
Cover and text design: Jack Jagtenberg
Typeset by Keyboard Wizards Pty Ltd, Allambie Heights, NSW

Printed in Australia by Ligare Pty Ltd, Riverwood, NSW

   2 3 4 5 02 01 00 99

ISBN 0 7248 0138 3

**National Library of Australia
Cataloguing-in-Publication Data**

Waxman, Peter
Business mathematics and statistics.

   4th ed.
   Includes index.
   ISBN 0 7248 0138 3

   1. Business mathematics. 2. Mathematical statistics.
   3. Business mathematics – Data processing. 4. Mathematical
   statistics – Data processing. I. Title.

650.01513

Prentice Hall of Australia Pty Ltd, *Sydney*
Prentice Hall, Inc., *Upper Saddle River*, *New Jersey*
Prentice Hall Canada, Inc., *Toronto*
Prentice Hall Hispanoamericana, *SA, Mexi*co
Prentice Hall of India Private Ltd, *New Delhi*
Prentice Hall International, Inc., *London*
Prentice Hall of Japan, Inc., *Tokyo*
Simon & Schuster (Asia) Pte Ltd
Editora Prentice Hall do Brasil Ltda, *Rio de Janeiro*

PRENTICE HALL

A division of Simon & Schuster

CONTENTS

Preface   ix
Acknowledgments   xii

Part 1 Business Mathematics   1

1. Basic mathematics   3
   1.1   Fractions   3
   1.2   Decimals   8
   1.3   Signed numbers   11
   1.4   Exponents   12
   1.5   Scientific notation   15
   1.6   Summary   17
   1.7   Multiple-choice questions   19
   1.8   Exercises   20
   Appendix 1A Significant figures   22

2. Percentages   25
   2.1   Conversion of percentages to fractions and decimals   26
   2.2   Using mathematical formulae to solve business problems   28
   2.3   Commission   32
   2.4   Discounts   38
   2.5   Taxation   46
   2.6   Profit and loss   53
   2.7   Summary   61
   2.8   Multiple-choice questions   63
   2.9   Exercises   64
   Appendix 2A Ratios and proportions   70

3. Simple interest   87
   3.1   Interest rates   87
   3.2   Calculation of simple interest   90
   3.3   Flat versus effective rate of interest   95
   3.4   Summary   98
   3.5   Multiple-choice questions   99
   3.6   Exercises   99

4. Compound interest   103
   4.1   Simple versus compound interest   103
   4.2   Present value   108
   4.3   Summary   112
   4.4   Multiple-choice questions   113
   4.5   Exercises   114

5. Annuities 117
- 5.1 Annuity defined 117
- 5.2 Future (accumulated) value of an ordinary annuity 118
- 5.3 Sinking fund (finding the periodic payment of annuity) 125
- 5.4 Present value of an ordinary annuity 128
- 5.5 Amortisation 132
- 5.6 Summary 135
- 5.7 Multiple-choice questions 136
- 5.8 Exercises 137

6. Depreciation 141
- 6.1 Key terms and principles of depreciation 142
- 6.2 Straight-line depreciation (prime cost) method 144
- 6.3 Reducing balance (diminishing value) method 148
- 6.4 Reducing balance depreciation schedule 150
- 6.5 Reducing balance method: the formula 153
- 6.6 Units-of-production method 157
- 6.7 Summary 160
- 6.8 Multiple-choice questions 161
- 6.9 Exercises 162

7. Line graphs 167
- 7.1 Graphing linear equations 167
- 7.2 Graphing simultaneous equations 176
- 7.3 Break-even analysis 183
- 7.4 Non-linear business-related graphs 188
- 7.5 Summary 191
- 7.6 Multiple-choice questions 191
- 7.7 Exercises 192

Part 2 Business Statistics 195

8. Introduction to statistics 197
- 8.1 Statistics 198
- 8.2 Data collection 203
- 8.3 Pitfalls in the use of statistics 207
- 8.4 Summary 221
- 8.5 Multiple-choice questions 224
- 8.6 Exercises 224

9. Visual presentation of business data 227
- 9.1 Why present information visually? 228
- 9.2 The range of visual presentations 231
- 9.3 Evaluation checklist for visual presentations 270
- 9.4 Summary 271

9.5 Multiple-choice questions 272
9.6 Exercises 272

10. Measurement of central tendency 279
10.1 Arithmetic mean 280
10.2 Median 292
10.3 Mode 300
10.4 Relationship between the mean, median and mode 305
10.5 Summary 307
10.6 Multiple-choice questions 309
10.7 Exercises 310
Appendix 10A Geometric mean 317

11. Dispersion 319
11.1 Range 321
11.2 Quartile deviation 323
11.3 Mean deviation 330
11.4 Standard deviation 338
11.5 Coefficient of variation 352
11.6 Summary 354
11.7 Multiple-choice questions 358
11.8 Exercises 358
Appendix 11A Coefficient of skewness 366

12. Sampling 369
12.1 Everyday observations of sampling 370
12.2 Types of sampling 373
12.3 Questionnaires 380
12.4 Summary 383
12.5 Multiple-choice questions 385
12.6 Exercises 386

13. Probability 391
13.1 Basic definitions and concepts 392
13.2 Probability rules and compound events 399
13.3 Conditional probability and dependent events 410
13.4 Bayes' Theorem for conditional probability 418
13.5 Summary 421
13.6 Multiple-choice questions 423
13.7 Exercises 424

14. Normal probability distribution curve 429
14.1 Area under the normal curve 431
14.2 Sampling distribution of the mean 446
14.3 Summary 455

14.4 Multiple-choice questions   456
14.5 Exercises   457
Appendix 14A Areas under the normal curve   461

15. Correlation and regression analysis   463
15.1 Scatter diagram   465
15.2 Correlation analysis   473
15.3 Rank correlation   480
15.4 Regression analysis   486
15.5 Standard error of estimate   496
15.6 Summary   500
15.7 Multiple-choice questions   503
15.8 Exercises   504

16. Index numbers   513
16.1 Simple index numbers   515
16.2 Simple aggregate (composite) index numbers   522
16.3 Weighted aggregate (or composite) index   527
16.4 Index number construction   542
16.5 Summary   547
16.6 Multiple-choice questions   553
16.7 Exercises   553

17. Time series   561
17.1 Time series components   562
17.2 Secular trend movements   567
17.3 Seasonal variation   594
17.4 Summary   611
17.5 Multiple-choice questions   613
17.6 Exercises   614

Appendix A Amount of compound interest   619
Appendix B Present value at compound interest   625
Appendix C Future value (amount) of an annuity   631
Appendix D Present value of an annuity   647
Appendix E Formulae used in the text   663
Appendix F Working with your calculator   673

Answers   681

Glossary   697

Index   707

PREFACE

In this age of information technology it is advisable for any educated individual to develop at least a basic level of understanding of mathematical and statistical tools necessary to make more informed daily decisions. One means of realising this objective is by attending classes and/or reading a textbook on the subject. However, your experience to date with maths and/or statistics may be one happily forgotten or suppressed into your subconscious. If so, I hate to be the one to inform you that as I speak (or write) and you read this, its applications continue to increase in areas of government, business, management, commerce, social science, physical science, medicine, education, agriculture and sport. Therefore, you really can't stall the inevitable.

Then why study from this book?
(a) Why not?
(b) Because it is a Christmas present;
(c) The teacher advised me to;
(d) Because I've heard such wonderful things about the book; no home should be without one.
(e) My parents made me do it;
(f) Peer pressure;
(g) All of the above.

Sorry, with over 800 questions, including 85 multiple-choice questions, I got carried away; the above question is not compulsory.

This book has not been written to change the course of your life (at least, not overtly) but it should relieve any negative feelings or pain associated with the word *maths* or *statistics*. This is no easy task, after your many years of conditioning. However, this text, when taken in proper doses, should alleviate any pain and at the same time help you cope and decipher the proliferation of statistical information invading our daily lives—you may even get to enjoy the subject!

How will this be done?
This is a tall order but the inclusion of the following innovations in this fourth edition should prove to be a recipe for successful mastery (and enjoyment) of the subject:
- This fourth edition is based on the National Module for Business Mathematics and Business Statistics.
- An introduction to the use of calculators is provided for those ill-at-ease in using these major time-savers.
- Learning outcomes are listed at the beginning of each chapter.
- A vignette (which is humorous and/or factual) opens most chapters.
- The chapter itself is presented in an informal, relaxed manner with up-to-date factual examples whenever possible.

Preface

- All major terms or concepts will be presented in bold lettering, indicating that it is defined in the glossary.
- Solutions are worked out for at least two examples for each new concept.
- Competency checks conclude each new section.
- Actual statistics are provided from publications such as the Australian Bureau of Statistics, the *Australian Financial Review*, the *Business Review Weekly* and the *Sydney Morning Herald*.
- A summary concludes each chapter.
- Five multiple-choice questions and 20 exercise questions are found at the conclusion of each chapter.
- Answers to approximately half of the questions requiring mathematical or statistical calculations can be found at the end of the text.
- Dashes of humour are sprinkled throughout the book.
- A glossary at the end of the book gives you a precise definition of all key terms used (and found in bold lettering) in the text.
- Formulae used in the text are presented separately to allow you immediate reference.
- An easily understood appendix at the end of the book on the use of calculators will make life simpler for those unfamiliar with more than just a few functions of the calculator.

In other words, this book has been written for you in such a manner as to minimise withdrawal symptoms, yet to maximise gain in the form of appreciation and understanding of key concepts influencing you and your organisation. The emphasis is on helping you become intelligent users of mathematics and statistics in the workplace and as consumers; therefore the importance of interpretation of data is stressed rather than mere number crunching or rote learning.

The fourth edition

Since the 1993 publication of the third edition of this book and subsequent reprints, I have been in constant contact with lecturers and students from all over Australia. The feedback has been gratefully received and ideas and advice seriously considered and integrated whenever practical and applicable.

The introduction of the National Module for Business Mathematics and Business Statistics rendered the subject matter covered in the previous edition excessive to your current needs. As a result, a fourth edition was needed to incorporate these changes and to allow the author to update examples and problems (both fact and fiction). Therefore, there are now 17 rather than the 26 chapters found in the third edition, the introduction of a new chapter, *Sampling*, and major revision to percentages, compound interest, annuities, introduction to statistics, index numbers, and correlation and regression analysis. The remaining chapters have all been altered and up-dated to ensure that this new edition caters specifically to your present requirements in a fashion which, hopefully, stimulates the senses.

The learning outcomes and assessment criteria per topic, as prescribed in the National Modules, have been presented at the beginning of each chapter, describing the material to be covered.

Preface

Use of this book

You should find that the chapters of this book correspond with the major topics prescribed by your teachers. In a few choice cases, the word 'Optional' has been placed next to those topics in certain chapters that are not part of the National Module per se but are sometimes covered by teachers, nevertheless, e.g., seasonal indexes in Chapter 17 on Time Series. Also, a number of chapters have appendices containing supplementary related information (such as ratios and proportions following Chapter 2 on Percentages).

Chapters 1–7 cover all material required for Business Mathematics while the remaining chapters 8–17 deal with Business Statistics. Since most of the examples, as well as the assigned questions, cover areas familiar to the reader as they deal with everyday commercial and/or household concerns, the relevance of each topic should enhance your understanding and retention of the relevant concepts.

You may find that some of the answers in the back of the text (particularly for chapters on Business Statistics) vary marginally with your own, a result of the derived answers for this text based on a floating decimal place of about 10 places. Therefore, as long as your answer approximates the answers given for the text and you understand the interpretation of the results, there is neither cause for alarm nor back-breaking attempts to get the identical answer to that found in the text.

A manual with the worked solutions to those problems that are not found in the 'Answers' section of the book is available to instructors.

Peter Waxman

ACKNOWLEDGMENTS

The most rewarding part of writing a text is to thank the many people who have made invaluable contributions to the improvement of the book over its many editions. I am, of course, indebted to Prentice Hall Australia, particularly Julie Catalano, Cheanne Yu and Liz Thomas for their continued confidence, support and advice in publishing the fourth edition within the space of 12 years.

I would also like to thank Peter Wheatley who provided insightful advice on improving the general text and whose meticulous eye surveyed all examples, questions and answers in the text in his technical review of the entire book.

I am grateful to Ken Stevenson for his key contribution in writing Chapter 13 on Probability; his wealth of experience in teaching mathematics is borne out in his well-structured, easily understood chapter.

Thanks are extended to Peter Mauthner for his major contribution to the first two editions of the book which helped distinguish it from the conventional mathematics and statistics text.

I would also like to express my thanks to Minas Poulos, Ian Joseph, Colin Granger and Kelly Chan for their comments and advice which were helpful in preparing this fourth edition.

The publication of a number of articles and the tables and graphs found throughout the text was made possible by the granting of permission by institutions (such as the Australian Government Publishing Service), organisations (Access Economics) and journalists (such as Max Walsh, Rowena Stretton and Ross Gittins) and I am very grateful to them all.

Most of the additions and deletions in this edition resulted from comments and recommendations made by reviewers from all States and I am indebted to you for your constructive criticism. I hope this edition does not disappoint you as I have tried, whenever possible, to take account of your suggestions.

Finally, it is a privilege to give thanks to those closest to me without whom this book would mean little. First, let me thank my dear father and my recently departed mother whose uncompromising love and support always helped me over the usual trials and tribulations experienced in life. My wife, Marianne, has stoically accepted her fate of being a weekend widow as she unselfishly has encouraged me to pursue my career as a lecturer, writer and researcher. However, the occasion of this book allows me the opportunity to express my heartfelt love and gratitude for allowing me the opportunity to have it all.

Peter Waxman

Part 1
BUSINESS MATHEMATICS

Small is beautiful, big is better, and other myths

By ROSS GITTINS
Economics Editor

HOW MUCH do we really know about what goes on in Australia's businesses? What do we know about the differences between small business and big business?

The surprising answer is: not all that much—until now.

Until now, we've had a few statistics for this and that, but mainly we've had to rely on our own experience—extensive or otherwise—and on vague impressions.

But now the Productivity Commission and the Industry Department have published a remarkably detailed study providing *A Portrait of Australian Business*.

It is based on a sample survey of nearly 9,000 firms, conducted in September 1995. The survey's results apply to virtually all non-agricultural private firms with employees.

Actually, it's the initial stage of Australia's first official "longitudinal" study of firms. That is, it will follow the same firms over five years—providing us with a wealth of information about how firms change (grow or die) over time.

The study is intended primarily to assist researchers and government policy-makers, but you and I can peek too.

The table summarises a tiny fraction of the information provided by the study's initial stage. Its columns divide up Australia's 393,000 firms by the number of people they employ.

The first column is for firms with one to four employees, the second for firms with five to nine employees, and so on, up to firms with 500 or more employees.

The final column shows the total or average for all firms. (It shows you just how misleading economy-wide averages can be.)

The first thing to note is the huge number of small firms we have. Fully 62 per cent of Australia's firms are "micro businesses" employing fewer than five people.

However, this 62 per cent of firms accounts for a mere 14 per cent of total private sector employment.

At the other end of the scale we have only 680 private firms employing more than 500 people. But these firms account for 22 per cent of total private sector employment.

Most of the other rows in the table show you how small business and big business differ on a host of characteristics.

Take foreign ownership. The survey's findings suggest that it's really only a relevant issue for big business. But almost a third of firms with 500 or more employees have at least 50 per cent foreign ownership.

(If I wanted to lie with statistics, however, I could tell you—quite

A PORTRAIT OF AUSTRALIAN BUSINESS

	Size of firm—maximum employees								
	4	9	19	49	99	199	499	500+	Total
Number of firms ('000s)	243	89	36	17	5	2	1	<1	393
				(per cent)					
Share of all firms	62	22	9	4	1	<1	<1	<1	100
Share of all employment	14	14	12	12	9	7	10	22	100
Majority foreign owned	<1	<1	3	5	13	20	28	32	1
Boss has uni degree	34	30	37	43	42	61	—70—		34
No union members	92	89	83	69	45	37	26	13	89
>75% unionists	3	3	4	5	8	10	18	20	3
Registered ent. agreement	2	2	2	4	11	18	32	54	2
Some individual contracts	47	40	38	39	50	50	51	47	44
Use "contracting out"	4	4	5	9	7	10	15	19	5
Introduced TQM	<1	<1	3	4	7	12	14	25	1
Introduced QA	2	5	10	18	24	33	17	21	5
No comparisons	86	79	72	64	67	51	42	38	82
Partic'n in govt programs	3	5	8	14	16	33	35	47	5
Firms exporting	2	4	8	11	20	33	34	46	4
Provide training	13	31	52	65	77	83	87	80	24

Source: 1995 Business Longitudinal Survey

truthfully—that only 1 per cent of Australian firms have majority foreign ownership. We have so many tiny firms that they swamp most national averages.)

The survey shows that one in three bosses (the survey calls them "major decision-makers") has a tertiary education. But the likelihood that the boss has a uni degree rises with the size of the firm.

The bosses of firms with 200 or more employees are twice as likely to be graduates as people running small businesses.

One thing the survey makes crystal clear is that unionism is very much a function of the size of the firm. The vast majority of firms with fewer than 50 workers have no union members at all whereas this is comparatively rare in big business.

But even in big firms it's clear that not everyone is in the union. Only one in five firms employing at least 500 workers has more the 75 per cent of them in a union.

You can see, too, that very few small and medium-sized firms have registered enterprise agreements. Bigness, unions and enterprise bargaining all go together.

On the other hand, individual contracts are common—for at least some employees—regardless of firms' size.

One surprising discovery from the survey is that various modern business practices which have received a lot of media publicity aren't, in fact, all that common. Take the practice of "contracting out" the provision of certain services rather than doing them in-house. This is often portrayed as one of the driving factors behind "downsizing".

But, according to the survey, contracting out is almost unknown in small business (perhaps the functions were never done in-house in the first place) and is used by fewer than one in five big businesses.

Then there's all the publicity about the spread of the enlightened management philosophy, TQM—total quality management. It's not happening in small business.

But 25 per cent of the biggest firms claim to have introduced TQM in the past three years. And, presumably, many big firms introduced it before then.

Even so, you can see that QA—quality assurance; a less philosophical, more standards driven approach—is spreading faster and wider than TQM.

I didn't include figures for the introduction of Just-In-Time inventory management in the graph—simply because almost no-one's introducing it. However, it's always possible many big firms introduced it years ago.

(Another thing I excluded was the proportion of businesses running at a loss in 1994-95. It was remarkably uniform across the size categories: one in five businesses incurred a loss in that year when the economy grew by 4 per cent.)

Have you heard all the chat in recent years about "benchmarking" to achieve "world best practice"? It involves making very detailed comparisons between the way your firm does things and the way the best-performing firm you can find does things.

The row labelled "No comparisons" shows you the proportion of firms in each size category that isn't doing any benchmarking—even of the most rough-and-ready kind.

Few small businesses bother to compare themselves with others. And a surprisingly high proportion of big businesses don't bother, either.

But, you may think, "I'll bet our small-business types aren't backward in coming forward when it comes to claiming handouts from the Government".

You may think it, but you'd be wrong. The survey finds that small business rarely participates in government programs, such as the tax concession for spending on research and development, the export market development grants, or even the National Industry Extension Service aimed at smaller businesses.

It turns out that what our naughty Treasurer calls "business welfare" is very much the preserve of big business.

Linked with this is the startling news of how many of our firms are engaged in exporting the goods or services they produce.

Clearly, small business doesn't export—and doesn't seek Government help in getting into it. On the other hand, it may surprise you to see that almost half our biggest firms *do* export. Overall, Australia exports 20 per cent of all it produces.

Yet another thing the survey makes clear is that we rely much more on big business than small to provide formal training for our workforce.

We live in an era when, where business is concerned, smallness is next to godliness. John Howard seems to worship the ground the small-business person walks on.

But the portrait the survey paints of small business isn't particulary flattering. Small business people are less educated, less willing to train their staff, less interested in applying modern management techniques, less responsive to government incentives and less export-oriented.

Big business doesn't win many plaudits, but it does a lot more to keep our economy moving forward.

Source: Ross Gittins, 'Small is beautiful, big is better, and other myths', *Sydney Morning Herald*, 22 March 1997. Reproduced with permission.

Chapter 1
BASIC MATHEMATICS

LEARNING OUTCOMES
After studying this chapter you should be able to:
- Perform calculations involving fractions
- Perform calculations involving decimals
- Perform calculations involving signed numbers
- Perform calculations involving exponents
- Perform calculations involving scientific notation.

It may be a number of years since you opened (or closed) a maths book, and studying maths and statistics may not be your idea of a good time, but now that you are enrolled in this subject, why not sit back, relax and enjoy the learning experience? This first chapter is a revision of the basics of mathematics, with subsequent chapters covering mathematical problems and concepts and ultimately progressing to statistics. You are not being groomed to be a statistician as much as a consumer or user of statistical information. To succeed in your new role, why not read and review this chapter thoroughly as it will make life much easier in later chapters?

1.1 Fractions

An arithmetic fraction is the division of one number by another; it is a part of a whole. In the fraction $\frac{3}{8}$, 3 (referred to as the numerator) is divided by 8 (called the denominator). This may be written as $3 \div 8$, $\frac{3}{8}$ or $3/8$. A number cannot be divided by zero (i.e. zero cannot be the denominator), as the result is undefined (e.g. $\frac{1}{0}$, $\frac{145}{0}$). However, if zero is divided by any other number (i.e. zero is the numerator), the fraction will always equal zero (e.g. $\frac{0}{2}$, $\frac{0}{115}$).

As illustrated above, fractions are commonly presented in $\frac{a}{b}$ form where a and b are whole numbers (**integers**). If the fraction is less than one (i.e. if the numerator is less than the denominator), it is known as a **proper fraction** (e.g. $\frac{4}{5}, \frac{2}{7}, \frac{2}{3}$). If the fraction is greater than one (i.e. if the numerator is larger than the denominator), it is known as an **improper fraction** (e.g. $\frac{7}{6}, \frac{13}{10}, \frac{12}{7}$). In this case, the fraction may be expressed as a **mixed number**— that is, a combination of a whole number and a proper fraction (e.g. $\frac{7}{6} = 1\frac{1}{6}, \frac{13}{10} = 1\frac{13}{10}, \frac{12}{7} = 1\frac{5}{7}$). In other words, to convert an improper fraction to a mixed number, divide the numerator by the denominator to the nearest whole number, and then divide the remainder by the denominator (e.g. $\frac{13}{10}$ = 1 plus 3 left over = $1\frac{3}{10}$). When converting a mixed number to an improper fraction, multiply the whole number by the denominator, add the numerator, then divide by the denominator. For example:

$$3\frac{7}{11} = \frac{(3 \times 11 + 7)}{11} = \frac{33 + 7}{11} = \frac{40}{11}$$

A *simple fraction* is one in which the numerator and the denominator are both whole numbers, such as $\frac{2}{3}, \frac{5}{7}, \frac{4}{9}$. A *complex fraction* is one in which the numerator, the denominator or both is a mixed number, such as:

$$\frac{5}{7\frac{1}{8}}, \quad \frac{4\frac{1}{2}}{6}, \quad \frac{2\frac{1}{3}}{1\frac{1}{2}}$$

Rules governing fractions

There are several rules that must be learned when dealing with fractions:

1. *Any fraction in which the denominator equals the numerator has a value of 1.* Hence:

    $$\frac{a}{a} = 1$$

    However, division by zero is a mathematically undefined operation. Hence:

    $$\frac{a}{0} = \text{undefined} \qquad \frac{0}{0} = \text{undefined}$$

2. *When multiplying or dividing the denominator and numerator by the same value, the fraction's value does not change.* Hence:

    $$\frac{a}{b} \times \frac{n}{n} = \frac{an}{bn} = \frac{a}{b}$$

    For example:

    (a) $\frac{3}{4} \times \frac{5}{5} = \frac{15}{20} = \frac{3}{4}$ $\qquad \frac{3}{4} \div \frac{5}{5} = \frac{15}{20} = \frac{3}{4}$

    (b) $\frac{2}{5} \times \frac{4}{4} = \frac{8}{20} = \frac{2}{5}$ $\qquad \frac{2}{5} \div \frac{4}{4} = \frac{8}{20} = \frac{2}{5}$

Chapter 1 – Basic mathematics

(c) $\dfrac{7}{8} \times \dfrac{1}{1} = \dfrac{7}{8} = \dfrac{7}{8}$ $\dfrac{7}{8} \div \dfrac{1}{1} = \dfrac{7}{8} = \dfrac{7}{8}$

3. *Always reduce a fraction to its lowest terms*—that is, divide both the numerator and the denominator by the largest divisor common to both the numerator and the denominator. This is referred to as *cancelling* and does not alter the value of the fraction. For example, to reduce $\dfrac{15}{20}$ to its lowest terms, 5 (the factor common to both parts of the fraction) is divided into 15 and 20 respectively; $\dfrac{15}{20}$ is reduced to $\dfrac{3}{4}$. There is no other common factor to further reduce the fraction.

4. When *adding or subtracting fractions*, the denominators of all the fractions must be the same. The numerators are then either added or subtracted (whatever the problem dictates), with the result then divided by the common denominator. For example:

    (a) $\dfrac{2}{11} + \dfrac{3}{11} = \dfrac{5}{11}$ (b) $\dfrac{2}{7} + \dfrac{4}{7} = \dfrac{6}{7}$

    (c) $\dfrac{6}{11} - \dfrac{4}{11} = \dfrac{2}{11}$ (d) $\dfrac{13}{15} - \dfrac{11}{15} = \dfrac{2}{15}$

5. When *adding or subtracting fractions* with different denominators, you must first find the lowest common denominator (LCD) for the fractions. For example, in the sum $\dfrac{1}{3} + \dfrac{2}{7}$ the lowest common denominator is 21. In fact, a common denominator is always found by multiplying the denominators. However, it can often be further reduced, as its multiplication does not necessarily result in the lowest common denominator.

    For example, to calculate $\dfrac{1}{8} + \dfrac{1}{6}$:

    $8 \times 6 = 48$

    Hence:

    $\dfrac{1}{8} = \dfrac{6}{48}$ $\dfrac{1}{6} = \dfrac{8}{48}$

    Therefore:

    $\dfrac{6}{48} + \dfrac{8}{48} = \dfrac{14}{48} = \dfrac{7}{24}$

    If it had been noted at the outset that the denominators of the fractions of $\dfrac{1}{8}$ and $\dfrac{1}{6}$ had a common factor of 24, the calculations could have been abbreviated to:

    $\dfrac{1}{8} = \dfrac{3}{24}$ $\dfrac{1}{6} = \dfrac{4}{24}$

    Hence:

    $\dfrac{3}{24} + \dfrac{4}{24} = \dfrac{7}{24}$

Business mathematics and statistics

Therefore, to calculate the above example of $\frac{1}{3} + \frac{2}{7}$, there is no smaller common denominator than 21; thus:

$$\frac{1}{3} = \frac{7}{21} \qquad \frac{2}{7} = \frac{6}{21}$$

Hence:

$$\frac{7}{21} + \frac{6}{21} = \frac{13}{21}$$

Note: To convert $\frac{1}{3}$ to a fraction with a denominator of 21, both the numerator and the denominator of $\frac{1}{3}$ are multiplied by 7 (i.e. $\frac{1}{3} \times \frac{7}{7} = \frac{7}{21}$). Similarly, to convert $\frac{2}{7}$ to a fraction with a denominator of 21, both the numerator and the denominator are multiplied by 3 (i.e. $\frac{2}{7} \times \frac{3}{3} = \frac{6}{21}$).

Problem: Calculate:

$$\frac{1}{3} + \frac{1}{6} - \frac{1}{12}$$

Solution: All numbers in the denominator are factors of 12. Therefore:

$$\frac{1}{3} = \frac{4}{12} \qquad \frac{1}{6} = \frac{2}{12} \qquad \frac{1}{12} = \frac{1}{12}$$

Hence:

$$\frac{4}{12} + \frac{2}{12} - \frac{1}{12} = \frac{5}{12}$$

6. When *adding mixed numbers*, first find the lowest common denominator for the fractions, add the fractions and then the whole numbers:

$$1\tfrac{7}{8} + 3\tfrac{1}{3} = 1\tfrac{21}{24} + 3\tfrac{8}{24} = 4\tfrac{29}{24} = 5\tfrac{5}{24}$$

7. *To multiply fractions*, multiply the numerators and divide this result by the number found by the multiplication of the denominators. Finally, reduce the result to its lowest terms. For example:

(a) $\frac{1}{2} \times \frac{2}{9} = \frac{2}{18} = \frac{1}{9}$ (LCD)

(b) $\frac{1}{11} \times \frac{3}{7} = \frac{3}{77}$

(c) $\frac{1}{3} \times \frac{4}{13} = \frac{4}{39}$

(d) $\frac{2}{3} \times \frac{7}{8} = \frac{14}{24} = \frac{7}{12}$

Chapter 1 – Basic mathematics

8. *To multiply fractions by a whole number*, multiply the numerator by the whole number, maintaining the same denominator. (The whole number can be thought of as a fraction, being divided by 1.) For example:

   (a) $\dfrac{1}{3} \times 7 = \dfrac{1 \times 7}{3} = \dfrac{7}{3} = 2\dfrac{1}{3}$

   (b) $\dfrac{2}{5} \times 8 = \dfrac{2 \times 8}{5} = \dfrac{16}{5} = 3\dfrac{1}{5}$

9. *To multiply mixed numbers*, you must first convert each mixed number (i.e. a whole number and a fraction) to an improper fraction (i.e. the numerator exceeds the denominator). To do this, multiply the whole number by the denominator, and add the numerator to that result. This is now the new numerator, with the denominator remaining the same as in the original mixed number.
   For example, to calculate $1\dfrac{5}{8} \times 3\dfrac{4}{7}$:

   $$1\dfrac{5}{8} = \dfrac{(1 \times 8 + 5)}{8} = \dfrac{13}{8}$$

   $$3\dfrac{4}{7} = \dfrac{(3 \times 7 + 4)}{7} = \dfrac{25}{7}$$

   Finally, rule 7 is applied:

   $$\dfrac{13}{8} \times \dfrac{25}{7} = \dfrac{325}{56} = 5\dfrac{45}{56}$$

10. *When dividing fractions*, invert (turn upside-down) the dividing fraction (the bottom fraction) and then multiply (as per rule 7). For example:

    (a) $\dfrac{\frac{7}{12}}{\frac{2}{3}} = \dfrac{7}{12} \times \dfrac{3}{2} = \dfrac{21}{24} = \dfrac{7}{8}$ or $\dfrac{7}{12} \div \dfrac{2}{3} = \dfrac{7}{12} \times \dfrac{3}{2} = \dfrac{7}{8}$

    (b) $\dfrac{\frac{3}{8}}{\frac{1}{2}} = \dfrac{3}{8} \times \dfrac{2}{1} = \dfrac{6}{8} = \dfrac{3}{4}$ or $\dfrac{3}{8} \div \dfrac{1}{2} = \dfrac{3}{8} \times \dfrac{2}{1} = \dfrac{6}{8} = \dfrac{3}{4}$

    (c) $\dfrac{\frac{1}{2}}{3} = \dfrac{1}{2} \times \dfrac{1}{3} = \dfrac{1}{6}$ or $\dfrac{1}{2} \div \dfrac{3}{1} = \dfrac{1}{2} \times \dfrac{1}{3} = \dfrac{1}{6}$

11. *When dividing a whole number by a fraction*, again invert the fraction and multiply. For example:

    $\dfrac{3}{\frac{1}{2}} = \dfrac{3}{1} \times \dfrac{2}{1} = 6$ or $3 \div \dfrac{1}{2} = 3 \times 2 = 6$

Business mathematics and statistics

1.1 Competency check

1. Calculate:

   (a) $\frac{1}{4} + \frac{1}{4}$ (b) $\frac{4}{7} - \frac{2}{7}$ (c) $\frac{1}{2} - \frac{0}{2}$ (d) $\frac{6}{7} \times 1$ (e) $\frac{3}{8} \div 1$

2. Add the following:

   (a) $\frac{3}{7} + \frac{2}{3}$ (b) $\frac{2}{5} + \frac{1}{6}$ (c) $4\frac{2}{5} + 7\frac{1}{8}$ (d) $1\frac{1}{3} + 2\frac{1}{5}$

3. Subtract the following:

   (a) $\frac{2}{3} - \frac{3}{7}$ (b) $\frac{2}{5} - \frac{1}{6}$ (c) $7\frac{1}{8} - 4\frac{2}{5}$ (d) $1\frac{1}{3} - 2\frac{1}{5}$

4. Multiply the following:

   (a) $\frac{3}{7} \times \frac{2}{3}$ (b) $\frac{2}{5} \times \frac{1}{6}$ (c) $2\frac{1}{2} \times \frac{3}{8}$ (d) $4\frac{2}{5} \times 7\frac{1}{8}$

5. Divide the following:

   (a) $\dfrac{\frac{3}{7}}{\frac{2}{3}}$ (b) $\dfrac{\frac{2}{5}}{\frac{1}{6}}$ (c) $\dfrac{4\frac{2}{5}}{7\frac{1}{8}}$ (d) $\dfrac{1\frac{1}{3}}{2\frac{1}{5}}$

6. In the past five days, Tom has worked overtime each day (in terms of hours): $2\frac{1}{2}$, $4\frac{2}{3}$, $1\frac{7}{8}$, $3\frac{1}{3}$ and $\frac{3}{4}$. What was Tom's total number of hours of overtime worked this past week?

1.2 Decimals

Our numerical system is based on the number 10, and therefore decimal fractions always have a power of 10 in the denominator (e.g. 10, 100, 1000). For example, the decimal equivalent of $\frac{1}{2}$ is 0.5 (i.e. five-tenths of one); likewise $\frac{13}{100}$ is 0.13 (i.e. thirteen hundredths of one). Similarly, $\frac{31}{1000}$ equals 0.031 and is read as thirty-one thousandths of one. Thus, all decimals are expressed as a number over the power of 10 (in the denominator). To convert a decimal to a fraction, divide the decimal result by the appropriate power of 10. For example:

$$0.1 = \frac{1}{10} \qquad 0.3 = \frac{3}{10} \qquad 0.875 = \frac{875}{1000} = \frac{7}{8}$$

Decimals can be thought of as fractions, the numerator of which has been divided by the denominator to arrive at the decimal equivalent. For example, the fraction $\frac{3}{4}$ indicates that 3 is divided by 4:

$$\frac{3}{4} = 4\overline{)3.00}^{\,0.75} = 0.75$$

The fraction $\frac{1}{9}$ indicates that 1 is divided by 9:

Chapter 1 – Basic mathematics

$$\frac{1}{9} = 9\overline{)1.000}^{\,0.111} = 0.111$$

Note: In the first example, the numerator is evaluated to two decimal places (i.e. 3.00); in the second example, the numerator is evaluated to three decimal places (i.e. 1.000).

The fraction $\frac{3}{4}$ is equivalent to 0.75 and is called a *terminating* decimal, as the denominator divides precisely into the numerator (i.e. there is no remainder). The fraction $\frac{1}{9}$ is called a *repeating* or *non-terminating* decimal (as a remainder exists). A non-terminating decimal is often designated with a bar over the repeating decimal—for example, $\frac{1}{6} = 0.1\overline{6}$.

Rules governing decimals
1. Addition and subtraction
When *adding or subtracting decimals*, ensure that all decimal points are aligned, tenths added to tenths, hundredths to hundredths, and so on. For example:

```
   0.13        0.1562       0.2
   0.105       0.1041       0.3
   0.213       0.0786       0.7
   0.3         0.4762       0.5
   ─────       ──────       ───
   0.748       0.8151       1.7
```

2. Multiplication and division
When *multiplying decimals*:
1. Multiply the two original numbers (disregard the decimal points).
2. Count the number of digits to the right of the decimal point in each number being multiplied.
3. In the answer (found in step 1), start at the extreme right and mark off to the left the number of decimal places found in step 2.

For example, 1.372×0.57 becomes:

```
      1.372
    × 0.57
    ───────
      9604
      6860
    ───────
    0.78204
```

Note: 1.372 has three decimal places to the right of the decimal point, and 0.57 has two; hence, five decimal places must be counted to the left from the last number in the answer.

Problem: 4.042×0.3702

Business mathematics and statistics

Solution:
```
    4.042
  × 0.3702
  ────────
    8084
    0000
   28294
   12126
  ────────
  1.4963484
```

As there are seven places to the right of the decimal points in the two factors (numbers), i.e. .042 and .3702, we count off seven places to the left from the last number in our answer and place the decimal at that point.

When *dividing decimals*:
1. Move the decimal in the divisor (i.e. the number being divided by) to the right to make it a whole number.
2. Move the decimal in the dividend (i.e. the number being divided into) the same number of decimal places as in the divisor.
3. Place the decimal in the quotient above the decimal point in the dividend.

For example, 1.5 ÷ 0.17. Since the decimal in the divisor (0.17) must be moved two decimal places to make it a whole number (from 0.17 to 17), the decimal in the dividend (1.5) must also be moved two places (from 1.5 to 150); thus, 1.5 becomes 150 and the division becomes 150 ÷ 17 = 8.8235294 (to seven decimal places).

Problem: 1.75 ÷ 0.002

Solution: As the divisor, 0.002, is two-thousandths of one (having three decimal places), three decimal places are 'stepped over' to convert it to a whole number. Therefore, three decimal places must be counted to the right of the decimal point in the dividend, 1.75. Therefore, 1.75 becomes 1750 (a zero must be added to fill the position of three places to the right). Thus, 1750 ÷ 2 = 875.

1.2 Competency check

1. Convert the following fractions to their decimal equivalents (to three decimal places):

    (a) $\frac{1}{3}$ (b) $\frac{15}{1000}$ (c) $\frac{3}{8}$

2. Convert the following decimals to fractions:

    (a) 0.11   (b) 0.25   (c) 0.04

3. Add the following:

    (a) 0.13 + 0.217 + 0.04

Chapter 1 – Basic mathematics

   (b) 1.305 + 0.155 + 0.004 + 0.078

   (c) 7.713 + 2.007 + 1.0056 + 2.42 + 1.1

4. Subtract the following:

   (a) 7.2316 – 2.1723   (b) 0.3104 – 0.16   (c) 0.2 – 0.1542

5. Multiply the following:

   (a) 3.217 × 0.423   (b) 0.707 × 0.707   (c) 2.35 × 0.042

6. Divide the following:

   (a) 2.55 ÷ 0.02   (b) 710 ÷ 0.5   (c) 0.458 ÷ 0.42

1.3 Signed numbers

Rules governing signed numbers

1. When adding numbers of the same sign, find the sum of the numbers and use the sign common to all factors. For example:

   (a) 7 + 6 + 4 + 3 = 20
   (b) –3 + (–2) + (–17) + (–111) = –133

2. When adding numbers of different signs, find the sum of the positive numbers and then the negative numbers. Subtract the smaller result from the larger one, and give the answer with the sign of the larger number. For example:

   (a) 134 + 14 + 17 + (–1) + (–122) + (–3) + (–2) = 165 + (–128) = 37
   (b) 37 + 42 + 81 + 3 + (–89) + (–101) + (–27) = 163 + (–217) = –54

3. When subtracting a number, reverse the sign of the number being subtracted and add this figure to the other numbers. For example:

   (a) 7 + 5 + 3 – (–7) = 7 + 5 + 3 + 7 = 22
   (b) 6 + 2 + 4 – (+2) = 6 + 2 + 4 – 2 = 10

4. When multiplying or dividing like signs, the answer is always positive (as a negative times a negative is a positive, and a positive times a positive is a positive). For example:

   (a) 7 × 6 = 42   (b) 121 × 3 = 363
   (c) 18 ÷ 6 = 3   (d) 121 ÷ 3 = $40\frac{1}{3}$
   (e) –7 × (–6) = 42   (f) –121 × (–3) = 363
   (g) –18 ÷ (–6) = 3   (h) –121 ÷ (–3) = $40\frac{1}{3}$

5. When multiplying or dividing unlike signs, the answer is always negative. For example:

   (a) 7 × (–6) = –42   (b) 121 × (–3) = –363
   (c) 18 ÷ (–6) = –3   (d) –18 ÷ (+6) = –3

Note: When multiplying or dividing like or unlike signs, carry out the required mathematical operation and then apply rules 4 and 5 for treatment of the signs. For example:

$$-7 + 8(-2) + (-6)(3) + (-4)(-5) + 3(6)$$
$$= -7 + (-16) + (-18) + 20 + 18$$
$$= -41 + 38$$
$$= -3$$

1.3 Competency check

Calculate:
1. $2 + 3 + 7 + (-6)$
2. $8 + 2 + 14 - (-6)$
3. $-8 + (-31) + 9 - (-32) + (-71)$
4. $-8 - 75 + 3(-8) - 9(-94)$
5. $(13(-3) + 7(-1) - (-3)) \div (-5)$
6. $(-8) \times (-4)$

1.4 Exponents

When a quantity is multiplied by itself one or more times, a superscript number can be placed at the upper right-hand side of the quantity. The superscript number, called an **exponent** or **index**, indicates how many times the quantity is to be multiplied by itself. For example, 3×3 can be written as 3^2, as 3 is multiplied by itself twice; $3 \times 3 \times 3$ can be written as 3^3, as 3 is multiplied by itself three times; $3 \times 3 \times 3 \times 3$ can be written as 3^4, as 3 is multiplied by itself four times.

An exponent relieves the writer of having to write in longhand the actual multiplication (e.g. $15^{11} = 15 \times 15 \times 15 \times 15 \times 15 \times 15 \times 15 \times 15 \times 15 \times 15 \times 15$).

Rules governing positive exponents

1. When a quantity (a) is raised to a power (n), and is multiplied by the same quantity (a) to another power (m), add the exponents ($n + m$). That is:

$$a^n \times a^m = a^{n+m}$$

Therefore, if a quantity (e.g. 8) is raised to a power (e.g. 4) and is multiplied by the same quantity (e.g. 8) to another power (e.g. 5), then:

$$8^4 \times 8^5 = 8^{4+5}$$
$$= 8^9$$

instead of having to write in longhand:

$$8^4 \times 8^5 = (8 \times 8 \times 8 \times 8) \times (8 \times 8 \times 8 \times 8 \times 8)$$
$$= 8 \times 8 \times 8 \times 8 \times 8 \times 8 \times 8 \times 8 \times 8$$
$$= 8^9$$

Chapter 1 – Basic mathematics

2. Multiply the exponents when a number already exponentially raised (a^n) is multiplied by itself m number of times. That is:
$$(a^n)^m = a^{nm}$$

Therefore, if a quantity exponentially raised (e.g. 8^3) is multiplied by itself, the number indicated by the second exponent (e.g. 4) is multiplied by the first exponent (i.e. 3). Hence:
$$(8^3)^4 = 8^{3 \times 4} = 8^{12}$$

instead of writing:
$$(8^3)^4 = (8 \times 8 \times 8) \times (8 \times 8 \times 8) \times (8 \times 8 \times 8) \times (8 \times 8 \times 8)$$
$$= 8 \times 8 \times 8 \times 8 \times 8 \times 8 \times 8 \times 8 \times 8 \times 8 \times 8 \times 8$$
$$= 8^{12}$$

3. Subtract the exponents $(n - m)$ when dividing identical quantities (a). That is:
$$\frac{a^n}{a^m} = a^{n-m}$$

Therefore, if a quantity with an exponent (e.g. 11^5) is divided by the same quantity raised to another power (e.g. 11^3), then subtract the exponents:
$$\frac{11^5}{11^3} = 11^{5-3} = 11^2$$

rather than:
$$\frac{11^5}{11^3} = \frac{11 \times 11 \times 11 \times 11 \times 11}{11 \times 11 \times 11} = 11^2$$

Rules governing negative exponents

A negative exponent is equivalent to the reciprocal of that factor. Thus:
$$a^{-n} = \frac{1}{a^n}$$

For example:
$$7^{-1} = \frac{1}{7^1} \qquad 4^{-2} = \frac{1}{4^2} \qquad 3^{-4} = \frac{1}{3^4}$$

If, on the other hand, one is presented with the reciprocal quantity, the quantity is raised to a negative exponent. For example:
$$\frac{1}{8^3} = 8^{-3} \qquad \frac{1}{4^4} = 4^{-4} \qquad \frac{1}{10^2} = 10^{-2}$$

The same rules apply to multiplication and division of positive and negative exponents as well as to numbers incorporating both positive and negative exponents.

1. When dividing like bases, subtract exponents. For example:
$$\frac{8^7}{8^2} = 8^{7-2} = 8^5$$

Business mathematics and statistics

2. When multiplying like bases, add the exponents. For example:

   (a) $(8^{-7})(8^{-4}) = 8^{-7-4} = 8^{-11}$ or $\frac{1}{8^{11}}$

   (b) $(8^{-3})(8^{-3}) = 8^{-3-3} = 8^{-6}$ or $\frac{1}{8^6}$

3. Any number divided by itself equals 1. For example, $\frac{11}{11}$ equals 1, which can also be written as:

   $$11^1 \times 11^{-1}$$

   which equals 11^0. As $\frac{11}{11}$ equals 1 and $\frac{11}{11}$ equals 11^0, 11^0 must therefore equal 1. This provides us with a most important corollary—any number raised to the power of zero is always equal to 1. That is:

   $$a^0 = 1$$

   Since $a^n \times a^{-n} = \frac{a^n}{a^n} = 1$, and $a^n \times a^{-n} = a^0$, then $a^0 = 1$.

Fractional exponents

Up to now, we have been dealing with whole numbers as exponents. Fractional exponents are also common and signify a root of a quantity. The root of a number is that quantity which when multiplied by itself equals the number. For example, the square root of 9, which is written $\sqrt{9}$, equals 3. The square root of 9 can also be written as $9^{\frac{1}{2}}$ (i.e. a fractional exponent). Likewise, the cube root of 27 can be written as $\sqrt[3]{27}$ or $27^{\frac{1}{3}}$.

To convert a root to its exponential form, the root is first converted to its reciprocal and the quantity is then raised to that reciprocal power. That is:

$$\sqrt[n]{a} = a^{1/n}$$
$$(\sqrt[n]{a})^m = a^{m/n}$$

For example, simplifying $\sqrt[4]{81}$, we get $81^{1/4}$ or 3; and $\sqrt[8]{256}$ becomes $256^{1/8}$, i.e. 2.

Simplify: $(\sqrt[8]{16})^4$

Solution: $(\sqrt[8]{16})^4 = (16)^{\frac{4}{8}} = 16^{\frac{1}{2}}$ or $\sqrt{16} = 4$

Simplify: $(\sqrt{9})^2$

Solution: $(\sqrt{9})^2 = (9)^{\frac{2}{2}} = 9^1 = 9$

Simplify: $(\sqrt[3]{8})^6$

Solution: $(\sqrt[3]{8})^6 = 8^{\frac{6}{3}} = 8^2 = 64$

Simplify: $(\sqrt[3]{64})(\sqrt[5]{32})$

Solution: $(\sqrt[3]{64})(\sqrt[5]{32}) = (64)^{\frac{1}{3}}(32)^{\frac{1}{5}} = (4^3)^{\frac{1}{3}}(2^5)^{\frac{1}{5}} = 4^{\frac{3}{3}} \times 2^{\frac{5}{5}} = 8$

Simplify: $(\sqrt[4]{16})(\sqrt{75})$

Solution: $(\sqrt[4]{16})(\sqrt{75}) = (16)^{\frac{1}{4}}(25)^{\frac{1}{2}}(3)^{\frac{1}{2}} = (2^4)^{\frac{1}{4}}(5^2)^{\frac{1}{2}}(3)^{\frac{1}{2}}$

$= (2)(5)(3)^{\frac{1}{2}} = 10\sqrt{3}$

1.4 Competency check

1. What is an exponent?
2. What is the exponent of $7 \times 7 \times 7 \times 7 \times 7$?
3. What is 8^0 equal to?
4. Simplify:

    (a) $\dfrac{11^5 \times 11^{-2}}{11^2 \times 11^{-4}}$ (b) $\dfrac{(10^5)^{-3} \times (10^{-2})^{-2}}{(10^2)^3 \times (10^3)^3}$

5. Simplify:

    (a) $\dfrac{7^4}{7^2}$ (b) $\dfrac{2^{-3}}{2^{-2}}$ (c) $\dfrac{2^{-3}}{2^{-2}}$ (d) $\dfrac{4^{-3}}{4^2}$ (e) $\dfrac{8^0}{8^2}$

6. Simplify:

    (a) $\sqrt{64}$ (b) $(\sqrt[4]{256})^{\frac{1}{2}}$ (c) $(\sqrt[3]{8})^4$ (d) $(\sqrt{3^4})^2$

1.5 Scientific notation (or standard form)

You have probably found, when using your calculator to determine an answer, that if the answer was too large to fit on the screen or was smaller than a pre-set figure programmed into the calculator, it was given in scientific or standard form. What this basically means is that the answer provided is a shorthand equivalent of the answer you would have derived had there been enough digits on the screen.

Problem: Using your calculator (unless you want to work it out in your head), multiply:

$845\,321 \times 685\,421$

Solution: On a calculator with a 10-digit display, you will read 5.794007651 11. This is a shorthand way of saying that: $845\,321 \times 685\,421 = 5.794007651 \times 10^{11}$ with $5.794007651 \times 10^{11}$, equivalent to the basic numeral answer of $579\,400\,765\,100$.

Therefore, the **scientific** or **standard notation** represents a number multiplied by the power of 10. Thus $5.794007651 \times 10^{11}$ becomes 5.79×10^{11}. (It is usual to express numbers in scientific notation using three significant figures, because the numbers are so large or small.)

Business mathematics and statistics

To convert any number to scientific notation, the number must be written as the product of a number from 1 to under 10 followed by a decimal and taken to the power of 10, with the power determined by the number of decimal places to the right of the decimal point. For example, note the notation of the following:

- $86 = 8.6 \times 10^1$ because there is one digit after the decimal point.
- $112 = 1.12 \times 10^2$ because there are two digits after the decimal point.
- $1967 = 1.967 \times 10^3$ because there are three digits after the decimal point.
- $15\,865 = 1.5865 \times 10^4$ because there are four digits after the decimal point.
- $250\,000 = 2.5 \times 10^5$ because there are five digits after the decimal point.

What about using scientific notation for small numbers? Once again, the number is written as a number from 1 to under 10 followed by a decimal, but this time the number is divided by the power of 10. This is done by counting the number of places to the left of the decimal point, which will indicate to which power of 10 it must be divided. For example, 0.0046 would be written as 4.6×10^{-3}. Remember that a minus sign is needed when an expression requires a division by the power of 10.

Problem: Express the following in scientific notation form:

(a) 0.0386   (b) 0.000073   (c) 0.0009

Solution:
(a) $0.0386 = 3.86 \times 10^{-2}$. Because the decimal point was moved two places to the right of the number 3, the power of ten to which the number is divided must be 2 (hence –2).
(b) $0.000073 = 7.3 \times 10^{-5}$. Because the decimal point was moved five places to the right of the number 7, the power of ten by which the number is divided must be 5 (hence –5).
(c) $0.0009 = 9 \times 10^{-4}$. Because the decimal point was moved four places to the right of the number 9, the power of ten by which the number is divided must be 4 (hence –4).

Finally, to convert to ordinary numerals from the scientific notation form when taken to the power of 10, the number of places to the right of the decimal point must equal the power to which 10 was taken. For example, 5.864×10^7 indicates that there are seven places or digits after the number 5, $5.864 \times 10^7 = 58\,640\,000$. As there are three numbers after the 5, four zeros have to be added to reach the 7th power.

1.5 Competency check

1. When is the use of scientific notation helpful?
2. Express the following in scientific notation form:

 (a) 8   (b) 11 995   (c) 899   (d) 876 442 921

3. Express the following in scientific notation form:

 (a) 0.00875   (b) 0.00096   (c) 0.006   (d) .0000987

Chapter 1 – Basic mathematics

4. Express the following as ordinary numerals:

    (a) 3.27×10^3 (b) 3.672×10^6 (c) 8.20704×10^6

5. Express the following as ordinary numerals:

    (a) 3.27×10^{-3} (b) 3.672×10^{-6} (c) 8.20704×10^{-5}

1.6 Summary

Fractions

A fraction is the division of one number (the numerator) by another (the denominator), and may be written as $\frac{a}{b}$, $a \div b$, or a/b. A proper fraction is one in which the numerator is less than the denominator; hence, the fraction is less than 1 (e.g. $\frac{13}{15}$, $\frac{17}{19}$, $\frac{2}{3}$). An improper fraction is one in which the numerator exceeds the denominator; hence, the fraction is greater than 1 (e.g. $\frac{18}{14}$, $\frac{3}{2}$, $\frac{6}{5}$). When an improper fraction is converted to a whole number plus a fraction (e.g. $\frac{18}{14} = 1\frac{2}{7}$, $\frac{3}{2} = 1\frac{1}{2}$, $\frac{6}{5} = 1\frac{1}{5}$), it is called a mixed number.

Several rules govern fractions. These are:

1. Any fraction in which the denominator equals the numerator has a value of 1. That is:

$$\frac{a}{a} = 1$$

However, division by zero is a mathematically undefined operation. Hence:

$$\frac{a}{0} = \text{undefined} \qquad \frac{0}{0} = \text{undefined}$$

2. When multiplying or dividing both the numerator and the denominator of a fraction by the same number, the value of the fraction does not change. Hence:

$$\frac{a}{b} \times \frac{n}{n} = \frac{an}{bn} = \frac{a}{b}$$

3. A fraction should always be reduced to its lowest terms by dividing both the numerator and the denominator by any factor common to both.
4. When adding or subtracting fractions, the numerators of all the fractions must be the same.
5. When adding or subtracting fractions with different denominators, you must first find the lowest common denominator.
6. When adding mixed numbers, first find the lowest common denominator for the fractions, then add the fractions and then the whole numbers.
7. To multiply fractions, multiply the numerators and divide that result by the result of the multiplication of the denominators.

8. To multiply fractions by a whole number, multiply the numerator by the whole number, maintaining the same denominator.
9. To multiply mixed numbers, convert the mixed numbers to improper fractions and apply rule 7.
10. When dividing fractions, invert the divisor and multiply.
11. When dividing a whole number by a fraction, invert the fraction and multiply.

Decimals

A fraction can be converted to decimal form by dividing the numerator by the denominator. The decimal is always expressed in terms of the power of 10. For example, $\frac{1}{5} = 0.2$ indicates that the fraction one-fifth is equal to two-tenths of one unit, while $\frac{3}{8} = 0.375$ indicates that three-eighths is equivalent to three hundred and seventy-five thousandths of one unit.

Signed numbers

The following rules apply when dealing with signed numbers:

1. When adding numbers of the same sign, find the sum of the numbers and use the sign common to all factors:

    $\boxed{+} + \boxed{+} = \boxed{+}$ $\quad\quad$ $\boxed{-} + \boxed{-} = \boxed{-}$

2. When adding numbers of different signs, find the sum of the positive and negative signs, subtract the smaller result from the larger one, and designate the answer with the sign of the larger number.
3. When subtracting a negative number, change the sign of the number being subtracted and add this figure to the other factors.
4. When multiplying or dividing like signs, the answer is always positive:

    $\boxed{+} \times \boxed{+} = \boxed{+}$ $\quad$ $\boxed{+} \div \boxed{+} = \boxed{+}$ $\quad$ $\boxed{-} \times \boxed{-} = \boxed{+}$ $\quad$ $\boxed{-} \div \boxed{-} = \boxed{+}$

5. When multiplying or dividing unlike signs, the answer is always negative:

    $\boxed{+} \times \boxed{-} = \boxed{-}$ $\quad$ $\boxed{-} \times \boxed{+} = \boxed{-}$ $\quad$ $\boxed{+} \div \boxed{-} = \boxed{-}$ $\quad$ $\boxed{-} \div \boxed{+} = \boxed{-}$

Exponents

The following rules govern exponents:

1. $a^n \times a^m = a^{n+m}$
2. $(a^n)^m = a^{nm}$
3. $\dfrac{a^n}{a^m} = a^{n-m}$
4. $a^{-n} = \dfrac{1}{a^n}$
5. $\sqrt[n]{a} = a^{1/n}$
6. $(\sqrt[n]{a})^m = a^{m/n}$

Chapter 1 – Basic mathematics

Scientific notation

When confronted with excessively large or small numbers, the use of scientific notation provides a shorthand method of expressing the number. To rewrite an ordinary numeral in scientific notation form (sometimes referred to as standard form), the number is written as the product between a number from 1 to under 10 and a power of 10. For example, 5.465×10^{11} indicates there are 11 places to the right of the decimal place; hence, 5.465×10^{11} is equivalent to 546 500 000 000. Similarly, 0.0092 when expressed in scientific notation is 9.2×10^{-3} since one must count three digits to place the decimal point after the number 9.

1.7 Multiple-choice questions

1. The value of $6\frac{3}{8}$ when converted to an improper fraction is:

    (a) $\frac{51}{8}$

    (b) $\frac{18}{8}$

    (c) $\frac{8}{18}$

    (d) $\frac{9}{8}$

    (e) None of the above.

2. The value of $117 \times (-3) - 1$ is:

    (a) 351   (b) 350   (c) –352   (d) 113   (e) 121

3. The value of $\dfrac{6^4 \times (6^{-3})^{-2}}{(6^2)^{-5} \times 6^{-2}}$ is:

    (a) 6^{-14} (b) 6^{-10} (c) 6^{-21} (d) 6^{22} (e) 6^{-26}

4. The value of $(\sqrt{16})^3 (\sqrt[3]{64})^{\frac{1}{2}}$ is:

    (a) 64
    (b) 128
    (c) 1024
    (d) 4
    (e) None of the above.

5. The scientific notation for 204.05 is:

    (a) 2.0405×10^4 (b) 2.0405×10^5
    (c) 2.0405×10^2 (d) 2.0405×10^3
    (e) 20.405×10^3

Business mathematics and statistics

1.8 Exercises

1. Convert the following mixed numbers to improper fractions:
   (a) $1\frac{5}{8}$ (b) $7\frac{1}{11}$ (c) $8\frac{2}{3}$ (d) $11\frac{11}{15}$ (e) $9\frac{2}{7}$

2. Convert the following improper fractions to mixed numbers:
   (a) $\frac{13}{12}$ (b) $\frac{3}{2}$ (c) $\frac{19}{16}$ (d) $\frac{135}{25}$ (e) $\frac{49}{32}$

3. Calculate:
   (a) $\frac{3}{7} + \frac{2}{7} + \frac{5}{7} - \frac{4}{7}$
   (b) $\frac{1}{8} - \frac{2}{8} - \frac{5}{8} + \frac{3}{8}$
   (c) $\frac{3}{7} + \frac{1}{2} - \frac{1}{4}$
   (d) $\frac{1}{3} + \frac{1}{2} - \frac{1}{8}$

4. Calculate:
   (a) $\frac{2}{5} \times \frac{3}{8} \times \frac{1}{2}$
   (b) $1\frac{3}{7} \times \frac{4}{8} \times \frac{2}{3}$
   (c) $\frac{7}{9} \times \frac{2}{3}$

5. Calculate:
   (a) $\frac{\frac{1}{2}}{\frac{1}{2}}$
   (b) $\frac{\frac{3}{1}}{4}$
   (c) $\frac{2\frac{5}{8}}{1\frac{1}{4}}$

6. What is the decimal equivalent of the following (to three decimal places)?
   (a) $\frac{3}{7}$ (b) $\frac{2}{9}$ (c) $\frac{1}{3}$ (d) $\frac{7}{62}$ (e) $\frac{9}{17}$ (f) $\frac{1}{7}$

7. Calculate:
   (a) $0.13 + 0.15 + 1.07 + 11.002$
   (b) 1.05×0.23
   (c) 1.002×1.15
   (d) $1.01 \div 0.01$

8. Calculate:
   (a) $2 + 7 + (-3) + 1$
   (b) $11 + (-3) + (-4) + (-7)$
   (c) $1 + 3 + (-6) + (-7)$

9. Calculate:
   (a) $(8 + (-2)) \times ((-6) + (-3)) \times (-1)$
   (b) $(18 \times 4 + (-3)) \times ((-2) + (-5)) \div (-11)$
   (c) $(-6 + (-3) + (-6)) \times (-2) \div (-3)$

10. Calculate:
    (a) $\frac{2^2}{2^4}$
    (b) 4^{-5}
    (c) $(2^3)^4$
    (d) $\frac{6^{-2}}{4^3}$

11. Calculate:
    (a) $\frac{\sqrt{4}}{2^{-3}}$
    (b) $\frac{(\sqrt[3]{27})^4}{3^2}$

12. Evaluate:

    (a) $3^2 \times (3^2)^{-1} \times 4^{-2}$ (b) $2^{-4} \times 4^2 \times (5^2)^{-2}$

    (c) $(3^2)^3 \times 3^{-2} \times (3^{-2})^3$

13. Write in its simplest form:

    (a) $(\sqrt[4]{162})(\sqrt{128})$ (b) $(\sqrt[4]{48})(\sqrt[3]{125})$

14. Robyn just had her 4-wheel drive filled up and it took $45\frac{3}{5}$ litres. If it normally travels $7\frac{1}{5}$ kilometres per litre, how far should Robyn be able to travel before running out of petrol?

15. To drive from Melbourne to Sydney takes approximately $9\frac{1}{2}$ hours. If Albury is considered one-third of the way to Sydney, how many more hours of driving can we expect until we reach Sydney?

16. Sydney residential property prices escalated by $\frac{3}{4}$ from 1987 to 1988. If a Sydney house sold for \$125 000 in 1987, what would be the expected selling price in 1988?

17. Brisbane's median residential house price rose by $1\frac{1}{5}$ from 1985 to 1995. If the median house price was \$60 000 in 1985, what was the median price in 1995?

18. Three tilers are tiling the outside entrance of a new commercial building. Charles has completed $\frac{1}{6}$ of the job, James $\frac{3}{8}$ and Alan $\frac{1}{3}$. What fraction of the job remains?

19. What is the scientific notation for the following?

    (a) 1   (b) 44   (c) 123.49

20. What is the scientific notation for the following?

    (a) .0155   (b) .0075   (c) .0002478

APPENDIX 1A
Significant figures

Sometimes questions arise about the degree of accuracy of figures with reference to sigfificant digits and rounding off of a number(s). For your convenience, a brief explanation follows.

1A.1 Significant figures

When a number is written, a significant figure (or digit) is one which has been measured (or is measurable). For example, if you travel to work by car and measure the distance, you may claim it to be 12 kilometres. This means that the distance travelled has been determined to the nearest kilometre, with 12 kilometres representing two significant figures—that is, 1 and 2. Had you stated that the distance was 12.7 kilometres, this would indicate that the distance had been measured to an accuracy of tenths of a kilometre with three significant figures (1, 2 and 7). Finally, if the distance travelled was 12.80 kilometres, and it is assumed that the zero (representing hundredths of a kilometre) was actually measured, four significant figures would be represented (1, 2, 8 and 0).

If the distance was measured by an instrument capable of reading only to tenths of a kilometre, it would be incorrect to record the zero because it cannot be considered a significant digit, even though it does not affect the magnitude of the reading.

The final digit may be thought of as guaranteeing the reliability and accuracy of the preceding digits. For example, 12.7 indicates that 12 is certainly correct and .7 is the result to the nearest tenth. Therefore, .7 implies that the exact value lies above .65 and below .75, although it may well be .7. Likewise, the first three figures of 12.80 are reliable (1, 2 and 8), while the last digit (0) may have been rounded off to the nearest hundredth.

What about zeros?

1. Note that the zero in 12.80 was considered significant because it appeared after a decimal.
2. If a zero(s) appears as the first digit(s) of a number, these are not significant because their sole purpose is the placement of the decimal point. For example, 0.075 displays two significant digits (7, 5) and .00750, three (7, 5 and 0).
3. If a zero appears between non-zero digits, it is a significant digit. For example, 12.704 has five significant digits.
4. Zeros found at the end of a whole number may or may not be significant, depending on the degree of rounding off. For example, if Joe quotes his weekly wages as $600, does this mean it is $600 exactly? The answer is usually 'no', since most salaries include cents as well. If Joe's annual salary is $31 130, his weekly wage is $598.65. Joe has rounded up the actual weekly wage to one significant figure. Therefore, unless we are informed that zeros found at the end of a whole number are actually measured or measurable, we assume that they are not significant figures.

1A.2 Rounding off

A calculation often results in more decimal places than is necessary, requiring a 'rounding off' to the appropriate number of decimal places. The simple rule to remember is that if the number subsequent to the required cut-off number is greater than 5, increase the number to its immediate left by one (the cut-off digit); if it is less than 5, the cut-off digit number is left unchanged.

For example, 0.345671 to the nearest tenth would be 0.3 (as the subsequent digit 4 is less than 5); to the nearest hundredth, 0.35 (as the digit subsequent to 4 is 5); to the nearest thousandth, 0.346; to the nearest ten-thousandth, 0.3457; and to the nearest hundred-thousandth, 0.34567. The more decimal places required, the more precise is the answer. For example, if the Consumer Price Index (CPI) for the quarter was 0.021 and wages were indexed to the CPI, wages would increase by 0.021 (or 2.1%). If your wage was $800 per week, a 2.1% (0.21) increase would add $16.80 to your weekly pay packet. However, if the CPI was rounded off to the nearest hundredth, the increase would be 0.02 (2%) of $800, or $16; thus, 80 cents per week, or $41.60 p.a., would be lost. Similarly, if income tax accounted for $\frac{1}{3}$ of your $800 weekly wage, your tax would be $240 if $\frac{1}{3}$ was carried to one decimal place (0.3). However, if carried to two decimal places ($800 × 0.33), your weekly tax would be $264, to three places, $266.40, and so on. For this reason, the tax scales are always terminating decimals so that no argument arises over which place to carry the decimal.

In the case of a number ending with a decimal half (e.g. 17.5), increase the number immediately preceding it by one if it is an odd number (thus, 17.5 becomes 18). If it is an even number (e.g. 18.5), leave it as that even number (thus 18.5 becomes 18). This rule eliminates over-rounding which would occur if all halves (0.5) resulted in increasing the results by one. Note the difference between (a) rounding up and (b) following the correct procedure:

	(a)	(b)
33.5	34	34
44.5	45	44
36.5	37	36
11.5	12	12
12.5	13	12
17.5	18	18
111.5	112	112
184.5	185	184
2.5	3	2
454.5	459	454

Problem: Round off 22.4786 to the nearest thousandth (i.e. three decimal places).

Solution: The third digit to the right of the decimal is 8 and since the subsequent digit is 6, which is greater than 5, round up the 8 thousandths digit to 9 thousandths. Hence the answer is 22.479.

Problem: Write 380.457 to three significant digits.

Solution: The first significant digit is 3, the second is 8, with the third being 0. Since the next digit is less than 5 (it is 4), the third digit, 0, remains as it is. Thus 380 is the answer.

Problem: Write 0.00706805 to three significant digits.

Solution: The first significant digit is 7, with 0 as the second and 6 as the third. Six (6) is followed by 8 and since 8 is greater than 5, the third significant digit, 6, is raised by 1 to give an answer of 0.00707.

Chapter 2
PERCENTAGES

LEARNING OUTCOMES

After studying this chapter you should be able to:
- Convert percentages to fractions and decimals
- Manipulate mathematical formulae to solve business problems
- Calculate commission (including brokerage)
- Calculate discounts (cash, trade, multiple or chain)
- Calculate tax (company, personal income, sales, stamp duty)
- Calculate profit and loss (mark-up and mark-down).

Over the past decade, inflation in Australia has varied from 1.0% to 8.0%, unemployment from 6.2% to 11.0%, annual wage rises from 2.0% to 6.8%, and population growth from 0.96% to 1.71% p.a.

Diet-associated diseases are claimed to be responsible for approximately 60% of Australian deaths. The average Australian diet consists of 16% protein, 40% fat and 44% carbohydrate. About 40% of Australians do not eat citrus fruit, and as many as 80% do not eat sufficient quantities of fruit and vegetables. In fact, one survey found that 73.6% of adult Australians have iron intakes below the recommended allowance. Of the $602 the average Australian household spends on goods and services each week, $111 is spent on food, $17 on alcohol, $14 on fruit, $13 on bakery products and cereals, and $30 on takeaway food.[1]

On the subject of diet, or perhaps the need for one, in 1994 nearly half of Australian men and one in three Australian women aged over 18 were overweight or obese. The proportion of overweight or obese adults increased from 32% at age 18–34 to 49% at age 55 and over.[2] Lack of regular exercise doesn't help matters, with about 35% of Australians not engaging in any deliberate exercise for sport or recreation.[3] However, in spite of the increasing levels of overweight and little progress on levels of exercise and blood cholesterol, Australian death rates among 20–69-year-olds from cardiovascular disease are now 70% lower than they were 25 years ago. Even among those aged 70 and over, the chance of dying from cardiovascular disease at a particular age is now about 50% less than for their counterparts in the late 1960s.[4] Thus, medical advances have assisted where Australians' diets have not!

The Australian male not only outperforms his female counterpart in dietary habits, (or should it be dietary hobbies?) but also excels in other areas. In New South Wales in 1994, 80% of suicides were men; more than 67% of motor vehicle accident fatalities were male; and 93% of people admitted to government alcohol treatment programs were male. Each year, 70% of those aged under 65 who die of heart problems are male. And it is claimed that two-thirds of marriage break-ups are initiated by women. Interestingly, the age group of 80 and over accounted for 50% of all female deaths but only 29% of all male deaths.[5] In the light of these statistics, it is surprising that there are any men left!

By 1996, the assets of the world's 358 billionaires exceeded the combined incomes of countries with 45% of the world's population.[6] Speaking of billions of dollars, gambling turnover in Australia in 1995 totalled $61 billion, up 25% nationally, up 52% in Victoria and 15% in New South Wales over the previous year. Nationally, Australians are spending approximately 2.83% of their household disposable income on gambling, which may drive some to drink.[7]

One obvious conclusion we can draw from all of the above is that inflation should be kept low so that we can afford more fruit which, in turn, will help keep us healthier and allow us to think more clearly—allowing us to make our fortunes. True, but in addition to this all-important message, the common denominator in the preceding paragraphs is per cent (%). 'Per cent' is derived from the Latin *per centum*, meaning by the hundred. Per cent indicates the relationship between the number preceding the per cent symbol and 100; in other words, when expressing numbers in percentage form, we are examining fractions with 100 in the denominator.

2.1 Conversion of percentages to fractions and decimals

According to the latest published census, 15% of families in Australia are headed by a single parent; that is, 15 families out of every 100 families are single-parent families (15/100). As we have seen, a percentage is easily converted to a fraction, and vice versa. To convert any percentage to a fraction, divide the number preceding the per cent symbol by 100 and drop the per cent sign. Always reduce the fraction to its lowest terms. For example, if there were a 25% customs duty on video recorders imported into Australia, this could be notated as 25/100, which equals 5/20 or 1/4. One-quarter can also be written in decimal form as 0.25 (see Chapter 1). Therefore, 25% equals 1/4, which equals 0.25.

Similarly, when converting a fraction to a percentage, multiply the fraction by 100 and place the per cent symbol after the answer. For example:

(a) $\frac{1}{2} = \frac{1}{2} \times 100\% = 50\%$ 　　(b) $\frac{2}{5} = \frac{2}{5} \times 100\% = 40\%$

(c) $\frac{2}{7} = \frac{2}{7} \times 100\% = 28.6\%$ 　　(d) $\frac{3}{10} = \frac{3}{10} \times 100\% = 30\%$

(e) $\frac{1}{8} = \frac{1}{8} \times 100\% = 12.5\%$ (f) $\frac{1}{5} = \frac{1}{5} \times 100\% = 20\%$

If converting a percentage to a decimal, move the decimal point from the last digit of the whole number two digits to the left of the per cent sign and drop the per cent symbol. For example:

(a) 83% = 0.83 (b) 77.5% = 0.775 (c) 5% = 0.05
(d) 0.08% = 0.0008 (e) 24% = 0.24 (f) 2.4% = 0.024
(g) 0.24% = 0.0024 (h) 417.876% = 4.17876

When converting a decimal to a percentage, move the decimal point two digits to the right and affix the per cent symbol (you are actually multiplying the number by 100%). For example:

(a) .5 = .50 = 50% (b) .05 = 5% (c) 1.45 = 145%
(d) .0025 = 0.25%

Normally, when dealing with rates (i.e. rates defined in percentages), the rate is calculated as a decimal number and then converted to a percentage. For example, by June 1995, US companies and residents had invested a total of A$88.6 billion in Australia, while total foreign investment was $400.9 billion.[8] To find the percentage of foreign investment in Australia from the United States, determine the **ratio** (a ratio being the relationship expressed in fraction form between two numbers of like units of measurement—in this case, dollar exports). The ratio or fraction is 88.6/400.9, which is equivalent to 0.2210. To determine the percentage, move the decimal point two digits to the right. Thus, 22.1% of foreign investment in Australia up to June 1995 came from the United States.

Problem: Japanese investment in Australia totalled $50.9 billion by June 1995. What percentage of foreign investment in Australia came from Japan?

Solution: $\frac{50.9}{400.9} = 0.1270 = 12.7\%$

2.1 Competency check

1. How do you convert a percentage to a decimal?

2. Convert the following decimal fractions to percentage form:

   (a) 0.53 (b) 0.04 (c) 1.75 (d) 0.375 (e) 0.002
   (f) 0.0002 (g) 0.3175

3. Convert the following percentages to decimals:

   (a) 15% (b) 1% (c) 17% (d) $24\frac{1}{2}\%$ (e) $568\frac{1}{5}\%$
   (f) $84\frac{1}{3}\%$

Business mathematics and statistics

4. Convert the following fractions to percentages:

   (a) $\frac{3}{8}$ (b) $\frac{13}{18}$ (c) $\frac{1}{6}$ (d) $\frac{19}{12}$ (e) $\frac{2}{3}$

5. Convert the following percentages to fractions:

   (a) 15%   (b) 245%   (c) 0.5%   (d) 6.5%   (e) 185%

2.2 Using mathematical formulae to solve business problems

Percentage refers to the value of any per cent of a given number. In every percentage problem, three numbers are involved: the base (the entire quantity), the rate (the per cent that is to be taken of the base) and the percentage (the value or amount).

We will use the symbols B, R and V to denote base, rate and value respectively, with their relationship to one another expressed as:

$$\text{Value } (V) = \text{Base } (B) \times \text{Rate } (R)$$

Provided that two of any three of these factors are known, the third can be found by substitution—that is, by substituting the known values into the equation and solving for the unknown.

Three basic questions which are often asked when dealing with percentages are:

1. A base number (i.e. the number to which other numbers are being compared) and a percentage or rate are given, and you are required to find the missing link. For example, rural exports accounted for 28.3% of Australia's 1995/96 exports.[9] If total exports were $75.2 billion, what was the dollar value of rural exports?

   The dollar value (V) of rural exports = total exports × rural's portion of total exports. Thus:

   $V = B \times R$
     $= \$75.2 \text{ billion} \times 28.3\%$
     $= \$75.2 \text{ billion} \times 0.283$
     $= \$21.28 \text{ billion}$

   Therefore, $21.28 billion worth of rural products were exported from Australia for the year 1995/96, which is equal to 28.3% of Australia's exports for that year.

2. A base number and value may be given, and you must determine the rate. For example, in 1995, of the 3 607 330 overseas visitors who arrived in Australia, 72 030 intended to stay for 12 months or more (but not permanently). What per cent of overseas visitors who arrived in 1995 intended to stay for a year or more? As we noted above, $V = B \times R$; thus, we are given V (72 030) and B (3 607 330) and must therefore find R. Hence:

   $V = B \times R$
   $72\,030 = 3\,607\,330 \times R$

Chapter 2 – Percentages

To find R, each side of the equation must be divided by B (3 607 330). Hence:

$$\frac{V}{B} = \frac{B \times R}{B}$$

$$\frac{V}{B} = R$$

$$\frac{72\,030}{3\,607\,330} = R$$

$$0.01997 = R$$

$$2\% = R$$

Therefore, when determining the percentage of one number compared with a base number, you must first construct a ratio or fraction with the base number in the denominator. Then calculate the decimal equivalent and convert it to a percentage (by moving two decimal places to the right and adding a per cent symbol).

Determination of rates is also required when you are asked to calculate changes from one period to another. For example, as of 30 June 1986 there were 326 registered trade unions in Australia compared with 157 on 30 June 1994. What was the percentage change (rate) in the number of trade unions between 1986 and 1994? The rate is the difference in the number of trade unions over the period, divided by the base year number—in this case, 326.

1986 (base year) value	326
1990 value	−157
Change over the period	169
$\frac{\text{Change }(V)}{1986 \text{ value }(B)}$	$\frac{-169}{326}$
	$-0.518 = -51.8\%$

In other words, there was a 51.8% decline in the number of trade unions between 1986 and 1994.

3. The percentage (R) and value (V) are given, but the base number must be determined. For example, in 1995/96, personal income taxation revenue to the Australian government totalled $60.4 billion, which was equivalent to approximately 51.9% of total taxation revenue. How much taxation was collected in Australia for the year 1995/96?

   To solve:

   $$V = B \times R$$

   $$\$60.4 \text{ billion} = B \times 51.9\%$$

   $$\$60.4 \text{ billion} = 0.519B$$

   $$\frac{\$60.4 \text{ billion}}{0.519} = \frac{0.519B}{0.519}$$

$$\frac{\$60.4 \text{ billion}}{0.519} = B$$

$$\$116.38 \text{ billion} = B$$

Total Federal Government taxation collections were $116.4 billion in 1995/96, of which 51.9% (or $60.4 billion) was received in the form of personal income tax.

Problem: If sales are 10% down on last year's figures, and this year's results total $4.5 million, find last year's results.

Solution: Since this year's results are 10% below last year's, it means that this year's results of $4.5 million are equal to 90% of last year's results. Thus, as $B \times R = V$, 90% (R) of last year's results (B) is $4.5 million ($V$). Hence:

$$B \times R = V$$
$$B \times 90\% = \$4.5 \text{ million}$$
$$0.90B = \$4.5 \text{ million}$$
$$B = \frac{\$4.5 \text{ million}}{0.9}$$
$$B = \$5 \text{ million}$$

Therefore, last year's results totalled $5 million. To check the answer, multiply the calculated answer of $5 million for last year by 90% (this year's result compared to the previous year's). This gives us $4.5 million for this year's result. A common mistake is to calculate 10% of $4.5 million and to add this amount (i.e. $450 000) to this year's result; hence $4.95 million would be last year's sales. However, this answer is incorrect. The correct procedure is as explained above.

Problem: A recently released motor vehicle, La Bombe, retails for $37 500, approximately 16% more than its major competitors. What is the price of a comparable motor vehicle from one of La Bombe's competitors?

Solution: Since La Bombe's selling price is 16% greater than its competitors, that means its price is 116% of theirs. Therefore, solving for B, which is the average price of a competitor's vehicle, we have:

$$V = B \times R$$
$$\$37\,500 = B \times R$$
$$\$37\,500 = B \times 1.16$$
$$\frac{\$37\,500}{1.16} = \frac{1.16B}{1.16}$$
$$\$32\,328 = B$$

To check the answer, assume that we are given the average price of a motor vehicle as $32 328 but are informed that La Bombe's price is 16% higher. To find La Bombe's selling price:

$$V = B \times R$$
$$= \$32\,328 \times 1.16$$
$$= \$37\,500$$

Problem: An antique purchased three years ago is said to increase in value by 20% p.a. What is the value of the article today if it cost $400 three years ago?

Solution: As the article was purchased three years ago, the 20% gained was three times compounded. That is:

- *First year:* $400 × 0.20 = $80, thus worth $400 + $80 = $480.
- *Second year:* $480 × 0.20 = $96, thus worth $480 + $96 = $576.
- *Third year:* $576 × 0.20 = $115.20, thus worth $576 + $115.20 = $691.20.

Hence, the value of the antique today is $691.20. Had you merely added the 20% three times (which is a common mistake) and taken 60% of $400, the total value of the antique would have been $640. Why underprice your article?

2.2 Competency check

1. When tenants decided to vacate their premises, they expected to receive their entire $500 bond; instead, they received only $365. What per cent of their bond was kept by the landlord?

2. Bill Gates' estimated personal wealth reached about US$18 billion in 1995, compared with US$15.3 billion for Warren Buffet, US$10.6 billion for Li Ka-shing and US$7 billion for Tan Yo.[10] What per cent (to the nearest per cent) of Gates' wealth did each of the other billionaires realise?

3. In 1995, 304 875 US visitors graced Australian shores, a rise of 5.2% on the previous year. How many US tourists (to the nearest whole number) had arrived in 1994?

4. Rocky put a deposit of $5250 on a secondhand 4-wheel drive. If the balance is 85% of the selling price, what was the selling price of the vehicle?

5. A Japanese company sold a commercial property at 31 Bligh Street in Sydney's CBD in 1995 for $3 million—80% less than the price paid for it in 1989. How much was the commercial property sold for in 1989?

Too many people operate under the mistaken notion that the need to understand the basics of finance should be relegated to those whose professions encompass the world of high finance. This could not be further from the truth. Anyone who has ever worked on a commission basis, received bonuses on the basis of sales, received discounts for purchases, deposited money with a building society or bank, borrowed funds for a personal loan or mortgage, paid premiums for insurance, or taken the occasional plunge on the horses should be aware of the methods of determining the appropriateness of the decision. (For example, should I accept a retainer plus commission, or a straight salary? Should I borrow at 9.25% flat or 15% reducible?) In other words, a minimum level of understanding of business mathematics is necessary for any worker/consumer who wants to improve his or her business or financial decision-making ability.

2.3 Commission

Many sales representatives receive income that is dependent on the level of sales generated. This is referred to as a **commission** and is usually calculated as a percentage of the sales value. For example, if an individual operating on a straight commission basis sells merchandise to the value of $10 000 on a commission of 7% of sales, the commission would be $700. That is:

$$\text{Commission } (C) = \text{Sales } (S) \times \text{Commission rate } (R)$$
$$C = S \times R$$
$$= \$10\,000 \times 7\%$$
$$= \$10\,000 \times 0.07$$
$$= \$700$$

Sales representatives often receive a retainer plus commission. If you were given the option of working on a straight commission basis or a retainer plus commission basis, you would need to forecast the average sales and thereby determine which of the two alternatives is more financially advantageous. Let's look in on Norm, who must decide between taking a straight commission of 12% or a retainer of $1200 per month plus a commission of 5% on sales. On average monthly sales of $35 000 working under commission only, Norm would receive:

$$\text{Commission } (C) = \text{Sales } (S) \times \text{Commission rate } (R)$$
$$C = S \times R$$
$$= \$35\,000 \times 12\%$$
$$= \$4200$$

On a retainer plus commission basis, he would receive:

$$\text{Earnings } (E) = \text{Sales } (S) \times \text{Commission rate } (R) + \text{Retainer}$$
$$E = S \times R + \text{Retainer}$$
$$= \$35\,000 \times 5\% + \$1200$$
$$= \$1750 + \$1200$$
$$= \$2950$$

In this particular instance, the straight commission would be more favourable than the retainer plus commission.

By applying algebra, the value of sales can be determined if the amount as well as the rate of commission are known.

Problem: A car sales representative received $3250 for a month's commission. If he received a straight commission of 5% on sales, what was the value of the month's sales?

Solution: To solve for S:

$$C = S \times R$$
$$\frac{C}{R} = S$$
$$\frac{\$3250}{0.05} = S$$
$$\$65\,000 = S$$

Therefore, sales of $65 000 at a 5% commission rate provided an income for the month of $3250. You can check the answer by solving the C in the equation $C = S \times R$, substituting $65 000 for S and 5% for R:

$$C = S \times R$$
$$= \$65\,000 \times 5\%$$
$$= \$3250$$

Problem: The Waite Fur Nuting Company's sales representative received a commission of $510 for sales of $2040. What was the rate of commission?

Solution: To solve for R:

$$C = S \times R$$
$$\frac{C}{S} = R$$
$$\frac{\$510}{\$2040} = R$$
$$0.25 = R$$

To convert a decimal to a percentage, move the decimal point two places to the right—that is, multiply by 100 and insert the per cent symbol; therefore, R equals 25%. Therefore, the rate of commission paid to the sales representative was 25%. Check your answer.

As an incentive to improve performance, some sales representatives receive graduated commissions—that is, the percentage commission received increases as sales increase.

Business mathematics and statistics

Problem: A sales representative receives 3.5% on sales up to $25 000, 5% on the next $25 000, and 7.5% on sales over $50 000 per month. What commission was earned if sales for the month amounted to $71 000?

Solution: To solve for C:

Commission on first $25 000 = $S \times R$
$\qquad\qquad\qquad\qquad\quad = \$25\,000 \times 3.5\% \;=\; \$\;875$
Commission on next $25 000 = \$25\,000 \times 5\% \;\;=\; \$1250$
Commission on final $21 000 = \$21\,000 \times 7.5\% = \;\$1575$

Total commission $\qquad\qquad\qquad\qquad\quad = \$3700$

Check the answer.

Problem: A sales representative operates on a graduated commission basis with 5% received on the first $10 000 sales, 7.5% on the next $10 000 and 10% thereafter. What did sales amount to if the sales representative received $2500 in commission?

Solution: To solve for S:

Commission on first $10 000 = $S \times R$
$\qquad\qquad\qquad\qquad\quad = \$10\,000 \times 5\% = \$500$
Commission on next $10 000 = \$10\,000 \times 7.5\% \;= \$750$
Commission on further sales $= S \times 10\%$

Therefore:

$$\$2500 = \$500 + \$750 + 0.1S$$
$$\$2500 - \$500 - \$750 = 0.1S$$
$$\$1250 = 0.1S$$
$$\frac{\$1250}{0.1} = S$$
$$\$12\,500 = S$$

Therefore, total sales equalled $32 500 ($10 000 + $10 000 + $12 500). Check the answer.

Problem: Since mid-1993, real estate agent fees in New South Wales have been deregulated. The fee is now a matter of negotiation between the vendor and the agent, with a range in 1997 of between 1.5% and 3.5%. Up to 1993, the fee was determined according to a set scale which was as follows: 5% on the first $15 000, 3% on the next $45 000, 2.5% on the subsequent $40 000 and 2% thereafter.

(a) Up to 1993, how much could a real estate agent expect to earn from the sale of a one-bedroom unit at $97 500 and two houses sold at $435 000 and $216 250?

Chapter 2 – Percentages

(b) If you as an agent had a choice between this scale or a negotiated 2.5%, which of these two fee structures would you prefer?

Solution:
(a) *To solve for C:*
- *Unit at $97 500:*

  Commission on first $15 000 = $S \times R$

  $\qquad\qquad\qquad\qquad\qquad\quad$ = $15 000 \times 5\%$ $\quad$ = $ 750.00

  Commission on next $45 000 = $45 000 \times 3\%$ $\quad$ = $1350.00

  Commission on next $37 500 = $37 500 \times 2.5\%$ $\quad$ = $ 937.50

  Total commission $\qquad\qquad\qquad\qquad\qquad$ = $3037.50

- *House at $435 000:*

  Commission on first $15 000 $\quad$ = $ 15 000 \times 5\%$ $\quad$ = $ 750

  Commission on next $45 000 $\quad$ = $ 45 000 \times 3\%$ $\quad$ = $1350

  Commission on next $40 000 $\quad$ = $ 40 000 \times 2.5\%$ $\quad$ = $1000

  Commission on next $335 000 = $335 000 \times 2\%$ $\quad$ = $6700

  Total commission $\qquad\qquad\qquad\qquad\qquad$ = $9800

- *House at $216 250:*

  Commission on first $15 000 $\quad$ = $ 15 000 \times 5\%$ $\quad$ = $ 750

  Commission on next $45 000 $\quad$ = $ 45 000 \times 3\%$ $\quad$ = $1350

  Commission on next $40 000 $\quad$ = $ 40 000 \times 2.5\%$ $\quad$ = $1000

  Commission on next $116 250 = $116 250 \times 2\%$ $\quad$ = $2325

  Total commission $\qquad\qquad\qquad\qquad\qquad$ = $5425

Therefore, from the sale of the three properties, the real estate agent can expect to earn $18 262.50 ($3037.50 + $9800 + $5425).

(b) To solve for *C* at 2.5% flat commission:

- *Unit at $97 500:*

  Commission (C) = Sales (S) × Commission rate (R)

  $\qquad C = S \times R$

  $\qquad\quad = \$97\,500 \times 2.5\%$

  $\qquad\quad = \$2437.50$

- *House at $435 000:*

  Commission (C) = Sales (S) × Commission rate (R)

  $\qquad C = S \times R$

  $\qquad\quad = \$435\,000 \times 2.5\%$

  $\qquad\quad = \$10\,875$

- *House at $216 250:*

  Commission (C) = Sales (S) × Commission rate (R)

  $\qquad C = S \times R$

  $\qquad\quad = \$216\,250 \times 2.5\%$

  $\qquad\quad = \$5406.25$

Business mathematics and statistics

Therefore, from the sale of the three properties, the real estate agent working on a flat 2.5% commission rate would earn $18 718.75 ($2437.50 + $10 875 + $5406.25).

Of the two fee structures, the flat fee is preferable in this particular case. You may have noted that it was the higher priced property ($435 000) which clinched the higher income, since 2.5% was applied to the entire amount while only 2% above $100 000 was applied under the old scale of fees.

Brokerage

Those interested in investing in securities ('securities' refers to the type of investment offered by a company or authority, e.g. shares, debentures, bonds) quoted on the Australian Stock Exchange must use a member organisation of the Exchange to act on their behalf in the purchase and sale of shares (also referred to as securities). A charge, called **brokerage**, is made by the stockbroker for this service. Approximately 15.5% of the Australian adult population owns shares in public companies listed on the Australian Stock Exchange (listed companies), either directly or through managed equity funds.[11]

Brokerage rates are commonly charged on a scale per order per class of stock, such as: $25 plus 2% on the first $15 000, plus 1.5% for $15 000 to $50 000, plus 1% in excess of $50 000, subject to a minimum charge of $80, although generic brokerage firms are now charging as low as $50 minimum.

There are a variety of other brokers similarly charging fees (usually on a declining scale basis), including business brokers who provide advice and expertise on the purchase or sale of a large variety of businesses, and insurance brokers.

Problem: In September 1996, BHP shares were selling at $17.40 a share. What would be the total brokerage payable if you purchased 1000 shares?

Solution: The purchase of 1000 BHP shares at $17.40 a share will cost $17 400 plus brokerage. Applying the scale cited above:

$$
\begin{aligned}
\text{Fixed fee} &= & & \$25 \text{ fee} \\
\text{Commission on first } \$15\,000 &= S \times R & & \\
&= \$15\,000 \times 2\% & &= \$300 \\
\text{Commission on } \$15\,000 \text{ to } \$50\,000 & & & \\
&= \$2400\ (\$17\,400 - \$15\,000) \times 1.5\% &&= \underline{36} \\
& & & \$361
\end{aligned}
$$

Therefore, a brokerage fee of $361 is owing.

Problem: Calculate the cost of a share portfolio if the brokerage payable was $2250.

Solution: To determine the total investment, solve for S:

$$
\begin{aligned}
\text{Commission on first } \$15\,000 &= S \times R \\
&= \$15\,000 \times 2\% \\
&= \$300
\end{aligned}
$$

Commission on $15\,000$ to $50\,000$ = $35\,000 \times 1.5\%$
= $525
Commission on further sales = $S \times 1\%$

Therefore, for the first $50\,000 investment the accrued brokerage was $300 + $525 plus the $25 fixed fee—that is, we have:

$$\$2250 = \$300 + \$525 + \$25 \text{ (fixed fee)} + 0.01S$$
$$\$2250 - \$300 - \$525 - \$25 = 0.01S$$
$$\$1400 = 0.01S$$
$$\frac{\$1400}{0.01} = \frac{0.01S}{0.01}$$
$$\$140\,000 = S$$

Thus, if a brokerage of $2250 was payable, $190\,000 in shares was purchased (i.e. $140\,000 + $50\,000). Remember, you can check the answer by working backwards—that is, calculate the brokerage for an investment portfolio of $190\,000.

Problem: If the shares you were interested in are selling for $2.50 each, how many would you have to purchase in order to owe more than the minimum brokerage fee of $80?

Solution: As $25 is a fixed fee, it indicates that the minimum brokerage of $80 needs to have an additional $55 commission payable. Therefore, to find how many shares one must purchase to pay above the minimum, solve for S:

$$C = S \times R$$
$$\$55 = S \times 2\%$$
$$= S \times .02$$
$$= .02S$$
$$\frac{\$55}{.02} = \frac{.02S}{.02}$$
$$\$2750 = S$$

Therefore, the shares must cost more than $2750 to have to pay above the $80 minimum brokerage fee. At $2.50 a share, if more than 1100 shares ($2750/$2.50 = 1100) are bought, a higher brokerage fee applies.

2.3 Competency check

1. Enrique is working on a flat commission basis of 3.75% of sales. If Enrique's sales for the month totalled $98\,000, how much has he earned for the month?

2. A wage of $680 per week is equivalent to what rate of commission (to two decimal places) if the commission is calculated on annual sales of $318\,710? Assume equal sales per week.

3. In 1997, Mr A. Korn received an income of $41 600, which was based on a retainer plus a commission on sales. If his commission was 5.5% on all sales and he sold $360 545 worth of shrubbery, what was his weekly retainer?

4. Ms A. Drenalin receives a retainer of $300 per week plus commission on all sales exceeding $3000 per week. Over the past 10 weeks, Ms Drenalin has achieved sales totalling $143 500, with her earnings over the same period totalling $7550 (including retainer). What was the rate of commission on sales exceeding $3000 per week? (Assume equal sales during each of the 10 weeks.)

5. Calculate the brokerage due if you just spent $37 400 on shares. (Apply the scale presented in the above brokerage example.)

2.4 Discounts

A reduction in the list (or original) price of a product is known as a **discount**, with the net price being the final price to the purchaser after all discounts have been deducted from the original list price.

Discounts are often given to induce a purchase, for buying in bulk, to clear slow-moving merchandise and to encourage early payment. We are bombarded daily by advertisements offering percentage discounts or absolute dollar savings, whichever appears the more impressive. (For example, on a major appliance, a saving of $100 sounds more enticing than a 10% saving.)

To determine the discount rate, you may use one of two methods. The first calculates the discount rate by dividing the amount of the discount by the original price. That is:

$$\text{Discount rate } (R) = \frac{\text{Discount } (D)}{\text{List price } (LP)}$$

For example, a colour television normally retails for $650 but has been reduced to $575. To find the percentage discount, find the amount which the purchaser would save as a result of the discount—that is, $650 minus $575 equals $75. This savings, or amount of discount, is then divided by the original selling price. Therefore, we have:

$$R = \frac{D}{LP}$$
$$= \frac{\$75}{\$650}$$
$$= 11.5\%$$

Thus the colour television was discounted by 11.5%. You can check the answer by taking 11.5% of $650 and subtracting the result ($75) from $650 to arrive at the discounted price of $575. An easier way of confirming the answer is to realise that the discount rate of 11.5% indicates that the discounted price is 88.5% of the original selling price. Therefore, $650 × 88.5% = $575 (to the nearest dollar).

Chapter 2 – Percentages

A second means of determining the discount rate is to divide the discounted price by the original price, convert to a per cent and subtract this result from 100% to find the discount rate. Looking at the above example:

$$\text{Discount rate} = 100\% - \frac{(\text{Discounted price} \times 100)}{\text{Original price}}$$

$$= 100\% - \left(\frac{575 \times 100}{650}\right)$$

$$= 100\% - 88.5\%$$

$$= 11.5\%$$

Therefore, the television is selling for 88.5% of what it originally sold for; hence, an 11.5% discount (100% − 88.5%).

Problem: A deluxe knife block is advertised at $17.55, reduced from $21.95. What is the percentage discount?

Solution: To determine the discount rate, you must first calculate the discount *(D)*, which in this case is equal to $4.40 ($21.95 (*LP*) − $17.55 (*DP*)). You then substitute into the formula:

$$R = \frac{D}{LP}$$

$$= \frac{\$4.40}{\$21.95}$$

$$= 0.2005$$

$$= 20.05\%$$

The answer may also be derived by:

$$\text{Discount rate} = 100\% - \frac{(\text{Discounted price} \times 100)}{\text{Original price}}$$

$$R = 100\% - \frac{(17.55 \times 100)}{21.95}$$

$$= 100\% - 79.95\%$$

$$= 20.05\%$$

Problem: A department store is selling a heater for $153, after a 10% discount. What was the original selling price (*LP*)?

Solution: The $153 is equal to 90% of the original selling price (as there was a 10% discount). Therefore:

Business mathematics and statistics

$$90\% \text{ of } LP = \$153$$
$$0.9LP = \$153$$
$$LP = \frac{\$153}{0.9}$$
$$= \$170$$

Therefore, a discount of 10% on a list price of $170 results in a discounted sale price of $153.

Note: Always check your answer by substituting and solving. For example, in the above problem, a list price of $170 discounted at 10% results in a reduction of $17; hence, a discounted sale price of $153.

Problem: A vacuum cleaner has been marked down to $195 after being discounted by 25%. Calculate:

(a) the original list price; and
(b) the amount by which it has been discounted.

Solution:
(a) Because a discount of 25% was applied to the vacuum cleaner, the marked-down price of $195 must be equal to 75% of the list price. Therefore:

$$\$195 = .75LP$$
$$\frac{\$195}{.75} = LP$$
$$\$260 = LP$$

(b) Discounted amount = List price − Marked-down price
$$= \$260 - \$195$$
$$= \$65$$

You may also derive your answer by letting:

MD = marked-down price
LP = original list price
DR = discount rate
$MD = LP - LP\,(DR)$
$\$195 = LP - LP\,(.25)$
$\$195 = .75LP$
$\$260 = LP$

To check your answer, calculate the discount rate as follows:

$$DR = \frac{D}{LP}$$
$$= \frac{65}{260}$$
$$= .25 \text{ or } 25\%$$

Cash discounts

Cash discounts are often given as an inducement to the purchaser to pay the bill by cash or cheque within a specific discount period.

Problem: The terms of payment to a particular retailer are a discount of 3.75% for payment within 7 days and 2.5% for payment within 30 days. If Max buys footwear for his shop to the value of $25 000, how much should he make out his cheque for if he pays:

(a) within 7 days?
(b) within 30 days?

Solution:
(a) Payment within 7 days:

$$\text{Discount } (D) = \text{List price } (LP) \times \text{Discount rate } (R)$$
$$= \$25\,000 \times 3.75\%$$
$$= \$937.50$$

Therefore, the cheque should be made out for $24 062.50 ($25 000 − $937.50).

(b) Payment within 30 days:

$$D = LP \times R$$
$$= \$25\,000 \times 2.5\%$$
$$= \$625$$

Therefore, the cheque should be made out for $24 375 ($25 000 − $625).

An even faster way of determining the discounted price Max would have to pay in (a) is to recognise that a discount of 3.75% for payment within 7 days indicates Max will be paying 96.25% of the original price (i.e. 100% − 3.75%). Therefore, the cheque should be made out in the amount of:

$$\text{Value} = \text{Base} \times \text{Rate}$$
$$V = \$25\,000 \times 96.25\%$$
$$= \$24\,062.50$$

and in (b):

$$\text{Value} = \text{Base} \times \text{Rate}$$
$$V = \$25\,000 \times 97.5\%$$
$$= \$24\,375.00$$

Businesses take advantage of such cash discounts as often as possible, even if they have to borrow to do so. For example, for Max to save the $937.50 in (a), he had to make the payment within 7 days. If he had used his overdraft facility with his bank and borrowed the entire $24 062.50 for the same period (7 days) at a rate of, say, 15%, the interest over the week period would have been:

Business mathematics and statistics

$$\text{Interest } (I) = \text{Principal } (P) \times \text{Rate } (R) \times \text{Time } (T)$$
$$I = P \times R \times T$$
$$= \$24\,062.50 \times .15 \times \tfrac{1}{52}$$
$$= 69.41$$

Therefore, the interest expense of $69.41 incurred over the 7-day period is more than compensated for by the $937.50 savings through early payment. You may argue that Max could have waited until the invoiced payment was due (normally 30 to 60 days), but in that event the discounted savings would have been forgone (up to $937.50).

Trade discounts

Discounts are often offered by manufacturers to wholesalers, wholesalers to retailers and retailers to certain customers. For example, tradespeople normally receive 10–25% discount on the list price of materials used in their profession. To determine the amount of **trade discount**, the list price is multiplied by the rate of discount.

Problem: A tiler is eligible for a 15% trade discount on all purchases from a particular hardware shop. If $240 was the total list price of the goods purchased, what is the net price to the tiler?

Solution:
Amount of trade discount (D) = List price $(LP) \times$ Trade discount rate (R).
$$D = LP \times R$$
$$= \$240 \times 0.15$$
$$= \$36$$

Therefore, the purchases cost the tiler $204 ($240 – $36).

Again, since a trade discount of 15% indicates that the tradesperson is paying 85% of the list price:

$$\text{Discounted price} = LP \times 85\%$$
$$= \$240 \times .85$$
$$= \$204$$

Thus, the tiler paid $204 for the items.

Problem: With a trade discount of 15%, a mechanic paid $83.30 for materials. What would the materials have cost without the trade discount?

Solution: $\$83.30 = 85\%$ of LP
$$\$83.30 = .85\,LP$$
$$\frac{\$83.30}{0.85} = LP$$
$$\$98 = LP$$

Therefore, without the trade discount, the materials would have cost $98. Check the answer.

Chain discounts

Sometimes more than one discount is given in order to clear stock and generate more sales and/or cash payments. For example, it is not uncommon to receive both a trade discount and a discount for payment in cash. Discount houses that have already 'slashed' prices by $x\%$ often slash them by a further $y\%$ to clear slow-moving stock and to generate an improved cash flow. This is referred to as a **chain** (or **multiple**) **discount**.

Problem: Mr S. Hammer, a carpenter, receives a trade discount of 20% and a cash discount of 3% at the local hardware shop. He bought materials, the list price for which totalled $4532.50.

(a) Find the invoice price and the actual amount paid in cash for the materials.
(b) What percentage of the list price did Mr Hammer save?

Solution:
(a)
$$D = LP \times R$$
$$= \$4532.50 \times 20\%$$
$$= \$906.50$$

$$\text{Invoice price } (IP) = LP - D$$
$$= \$4532.50 - \$906.50$$
$$= \$3626$$

Thus, the invoice price was $3626. To determine the cash discount:

$$D = LP \times R$$
$$= \$3626 \times 3\%$$
$$= \$108.78$$

$$\text{Amount paid } (AP) = LP - D$$
$$= \$3626 - \$108.78$$
$$= \$3517.22$$

Thus, the actual amount paid by Mr Hammer was $3517.22.

(b) To determine Mr Hammer's percentage savings, first find the total discount.

$$\text{Total discount } (TD) = LP - AP$$
$$= \$4532.50 - \$3517.22$$
$$= \$1015.28$$

Next, find the total discount as a percentage of the list price.

$$R = \frac{TD}{LP}$$
$$= \frac{\$1015.28}{\$4532.50}$$
$$= 0.224$$
$$= 22.4\%$$

Business mathematics and statistics

The actual amount of $3517.22 paid by Mr Hammer could have been found by a simpler, faster shorthand method which we can refer to as the combined discount rate. To find the result of multiple or chain discounts, apply the following formula:

Combined discount rate $(CDR) = 1 - (100\% - $ 1st discount$) \times$
$(100\% - $ 2nd discount$) \times \ldots (100\% - n$th discount$)$

In the above example, we have:

$$\begin{aligned}
CDR &= 1 - (100\% - 20\%)(100\% - 3\%) \\
&= 1 - (80\%)(97\%) \\
&= 1 - .776 \\
&= .224 \\
&= 22.4\%
\end{aligned}$$

In other words, Mr Hammer pays 77.6% of the original list price (77.6% of $4532.50 = $3517.22)—that is, he has received a 22.4% discount (22.4% of $4532.50 = $1015.28). (Note how the number of calculations has been dramatically reduced, thus minimising potential errors.)

Try one more example to reinforce your understanding of chain discounts.

Problem: The end of season has seen a local retailer mark down all items by 30%, with a further 10% trade discount and a 5% cash discount. If Mr Win Doe, a mechanic, is eligible for the trade discount, what will he pay in cash for items at a list price of $450?

$$\begin{aligned}
CDR &= 1 - (100\% - 30\%)(100\% - 10\%)(100\% - 5\%) \\
&= 1 - (70\%)(90\%)(95\%) \\
&= 1 - (.7)(.9)(.95) \\
&= 1 - (.5985) \\
&= .4015 \\
&= 40.15\%
\end{aligned}$$

Solution: The selling price to Mr Win Doe is 59.85% of the list price; hence:

$$\begin{aligned}
SP &= \$450 \times .5985 \\
&= \$269.32
\end{aligned}$$

Therefore, Mr Win Doe must pay 59.85% of the list price of $450, or $269.32. In other words, the combined discount rate of 40.15% results in a saving of $180.68 (40.15% of $450, or $450 − $269.32).

Problem: A local discount shop advertises all merchandise at 20% off the recommended retail price. A further discount has been applied to the shop's last refrigerator. The refrigerator originally retailed for $975 *(LP)*; after the second discount it is advertised at $663. What was the second discount rate?

Solution: First, find the discounted price after the 20% discount was applied to the list price of $975:

$$D = LP \times R$$
$$= \$975 \times 20\%$$
$$= \$195$$
$$\text{Discounted price } (DP) = LP - D$$
$$= \$975 - \$195$$
$$= \$780$$

As the second discount resulted in a final sale price of $663, an additional discount of $117 ($780 − $663) must have been applied. To determine the second discount rate, you must substitute into the equation:

$$D = \text{Selling price } (SP) \times R$$
$$\$117 = \$780 \times R$$
$$\frac{\$117}{\$780} = R$$
$$0.15 = R$$
$$15\% = R$$

Thus, the chain discount rates were 20% and 15%. Check the answer.

Once again, the shorthand method can be applied. After the first discount of 20% off the list price, the merchandise sells for 80% of the list price; thus:

$$\text{Selling price} = (100\% - 20\%)\,LP$$
$$= (80\%)\,LP$$
$$= (.80)\,\$975$$
$$= \$780$$

As the second discount resulted in a final price of $663, to find the additional discount rate, use the formula:

$$\text{Second discount rate} = \frac{\text{Discounted price}}{\text{Selling price (after the first discount)}}$$
$$= \frac{\$663}{\$780}$$
$$= .85$$
$$= 85\%$$

Since the second discount resulted in the refrigerator being purchased for 85% of the selling price after the first discount, this indicates that there was a further 15% discount. To check the answer, use the combined discount rate equivalent:

$$CDR = 1 - (100\% - 20\%)(100\% - 15\%)$$
$$= 1 - (.8)(.85)$$
$$= 1 - (.68)$$
$$= .32$$

The final selling price was 68% of the original list price—that is, $975 × .68 = $663—which corresponds with the advertised amount.

Business mathematics and statistics

2.4 Competency check

1. A colour television's marked price is $480, but a discount of $72 is given on that price.
   (a) What is the percentage discount?
   (b) What is the discounted selling price?

2. One of the major department-store chains receives a discount on recommended list prices of 20% on all purchases from Underwater Basketweavers Inc., while a small speciality shop receives a 5% discount from the same source. How much more does it cost the owner of the speciality shop to buy merchandise with a list price totalling $5450?

3. An artist pays $1255 for materials after receiving a trade discount of 15% and a further cash discount of 5%. What was the list price of the purchased equipment (to the nearest dollar)?

4. A long-time customer has just spent $2465 in a hardware shop after receiving the regular 15% discount reserved for all customers, as well as a further 10% for his long-time patronage and 3.5% discount for payment in cash. What would have been the list price for the goods if no discount had been provided (to the nearest dollar)?

5. Last week a sports shop, Run For Your Life, offered a discount of 20% on running shoes, and this week it advertised a further 15% discount. The discounts save the purchaser $45 in total.
   (a) What was the original list price of the running shoes (to the nearest dollar)?
   (b) What was the discounted price (to the nearest dollar)?

2.5 Taxation

Company tax is a flat or proportional tax applied at the rate of 36% irrespective of income. It was reduced to 33% (from 39%) in the early 1990s in order to be competitive with company tax rates in other Asia-Pacific countries. However, the company tax rate was increased in 1995 to 36% as the Federal Government tried to raise more taxation revenue in an attempt to reduce the Budget deficit.

Problem: How much tax is payable for a company with a taxable earning of $2 500 000?

Solution: Companies pay 36% tax irrespective of the level of income, with this particular company liable as follows:

$$\begin{aligned} \text{Company tax payable} &= \text{Income} \times 36\% \\ &= \$2\,500\,000 \times 0.36 \\ &= \$900\,000 \end{aligned}$$

Chapter 2 – Percentages

Problem: How much company tax should B4UCIM Gone Pty Ltd pay on a taxable income of $16 300?

Solution: Company tax payable = Income × 36%
= $16 300 × 0.36
= $5868

Personal income tax

All Australian workers who earn over a certain amount of income (called the threshold, currently $5400) must pay **personal income tax**. The Australian income tax regime is similar to that of the income tax structure of the industrialised world, in that it is progressive—that is, as an individual's income increases, the percentage of tax paid also rises. For example, according to Table 2.1, someone earning $48 000 would pay a marginal tax rate of 43% since this income can be found in the $38 000 to $50 000 income bracket. The amount of tax paid in this income range would have to be added to the tax calculated through the first two marginal tax rates and accompanying income brackets.

(*Note:* It should be remembered that a Medicare levy of 1.7% (incorporating the 'gun levy') is applied as well (from 1997), with an additional 1% for those high income earners not subscribing to private health insurance.)

Table 2.1 *Personal income tax rates in Australia*

Income bracket ($)	Marginal rate (%)
0–$5400	nil
5400–20 700	20
20 700–38 000	34
38 000–50 000	43
More than 50 000	47

Problem:
(a) Calculate the amount of tax that Christopher has to pay on his taxable income of $26 400.
(b) Calculate Alan's tax on an income of $45 000.
(c) Calculate Kerry's tax bill if she earns $64 000.

Solution:
(a) Christopher's income of $26 400 is in the third income bracket. To determine the amount of tax he has to pay, we must work through the first three income brackets—that is, there is no tax payable on the first $5400 earned but 20% tax is payable on $15 300 (i.e. the second tax bracket, $5400 to $20 700). Therefore, $3060 tax is payable on incomes of $20 700 (i.e. $15 300 × .20 = $3060).

Since Christopher earned $26 400, the remaining balance of his income, $5700 (i.e. $26 400 – $20 700), is found in the third tax bracket at a marginal rate of 34%; thus, an additional tax of $1938 ($5700 × 34%) must be added to the $3060 to derive the total income tax payable by Christopher.

Total tax payable on $26 400 = $3060 (from the second income bracket at marginal rate of 20%) + $1938 = $4998.

(b) Alan's income tax on earnings of $45 000 can be determined by recognising that a marginal rate of 43% applies to amounts of $38 000 up to $50 000, within which bracket Alan's income falls. However, prior to getting to this point, we must calculate the tax payable up to $38 000. We already know that up to $20 700, there is a tax of $3060, but we must calculate the tax payable in the third tax bracket ($20 700 – $38 000) and then, finally, the fourth tax bracket ($38 000 – $50 000). Therefore, we have:

Tax payable on
$$\begin{aligned}
\$45\,000 &= \$3060 + 34\% \text{ of } (\$38\,000 - \$20\,700) + 43\% \text{ of } (\$45\,000 - \$38\,000) \\
&= \$3060 + (34\% \times \$17\,300) + (43\% \times \$7000) \\
&= \$3060 + (0.34 \times \$17\,300) + (0.43 \times \$7000) \\
&= \$3060 + \$5882 + 3010 \\
&= \$11\,952
\end{aligned}$$

(c) Kerry's earnings of $64 000 take her into the highest tax bracket of 47%, for those earning above $50 000. We have already calculated tax payable on incomes up to $38 000, so we need only determine the tax payable on the income above that.

Tax payable on
$$\begin{aligned}
\$64\,000 &= \$3060 + \$5882 + 43\% \text{ of } (\$50\,000 - \$38\,000) + 47\% \text{ of } (\$64\,000 - \$50\,000) \\
&= \$3060 + \$5882 + (43\% \times \$12\,000) + (47\% \times \$14\,000) \\
&= \$3060 + \$5882 + (0.43 \times \$12\,000) + (0.47 \times \$14\,000) \\
&= \$3060 + \$5882 + \$5160 + \$6580 \\
&= \$20\,682
\end{aligned}$$

If you took the average tax rate paid by each of these three individuals, you would find that the rate increased as their incomes rose—that is, Christopher on $26 400 (i.e. $4998 ÷ $26 400) paid 18.9% income tax, while Alan paid 26.6% on $45 000 in earnings and Kerry paid 32.3% tax on her income.

Sales tax

Sales tax is

> ... a once only tax on goods which are manufactured in Australia or imported into Australia. Some goods, eg., most food and clothing, are always exempt from sales tax. Some goods are exempt if they are used in a particular way or by particular people or organisations. Goods to be exported from Australia are also exempt from sales tax provided they are not used before being exported. The tax is intended to fall once

on the sale of goods immediately before the sale to the end user or final consumer. Often this will occur when the last wholesale sale occurs, i.e., the sale by the wholesaler to a retailer. Sales tax may be payable on other transactions involving, for example, leases and rentals, or when goods are given away free of charge or are used as demonstration stock. The tax may also be payable by manufacturers or wholesalers if they use the goods for their own purposes instead of selling them.[12]

Sales tax is therefore a percentage increase in the wholesale price of an article, which is ultimately added to the retail price. The appropriate rate of tax depends on the classification of the goods involved. The *Sales Tax Act 1992* contains seven Schedules that identify the goods that are exempt from tax (Schedule 1). It also classifies other goods in order to determine the rate of tax to be imposed on those goods. For example:[13]

- *Schedule 1 (tax-exempt goods):* includes drugs and medicines, books and magazines, food and drink for human consumption.
- *Schedule 2 (12% tax):* includes household goods, maps and related goods, and certain confectionery, flavoured milk and certain fruit juice products.
- *Schedule 4* (22% tax):* includes motor vehicles under $36 995, sporting goods, toys, stationery, handbags, leather/imitation leather goods, cosmetics and toiletries.
- *Schedule 5 (32% tax):* luxury goods, including fur skins, jewellery, watches, clocks, cameras, televisions, videotape recorders, and record and compact disc players.
- *Schedule 6 (45% tax):* luxury motor vehicles (if wholesale price exceeds $36 995).
- *Schedule 7 (26% tax):* alcoholic wine, cider, etc.

* There is no Schedule 3.

Problem: Video recorders attract a sales tax rate of 32%. If the wholesale price of a video recorder is $400, how much will the sales tax amount to?

Solution: Amount of sales tax = Wholesale price × Sales tax rate
= $400 × 32%
= $128

Therefore, the Federal Government would receive $128 in sales tax for a video recorder wholesaling for $400.

Problem: Chocolate confectionery attracts a 12% sales tax. How much would a chocolate bar retailing for $1.20 cost us cocoa-addicted sufferers if there were neither a sales tax of 12% nor the 40 cents profit margin applied by the retailer?

Solution: Let x = wholesale price of chocolate.
Retail price = $1x + .12x + 40$
$120 = 1.12x + 40$
$120 - 40 = 1.12x$
$80 = 1.12x$
$\dfrac{80}{1.12} = \dfrac{1.12x}{1.12}$
$71.43 = x$

Therefore, the chocolate bar's wholesale price was 71 cents. You can check the answer by starting with the wholesale price and subsequently calculating 12% of that price (which is the sale tax) and adding this amount (8.52 cents), along with the 40 cents profit margin, to the 71 cents to get $1.1952 or $1.20 (and the chocolate bar).

Problem: Nova Office Furnishings has just bought office equipment from a wholesaler at a total cost of $1950, including $325 in sales tax. What is the rate of sales tax applied to office equipment?

Solution: The wholesale tax of $325 must be subtracted from the $1950 retail price in order to determine the wholesale price, i.e. $1950 − $325 = $1625. Therefore:

$$\text{Amount of sales tax} = \text{Wholesale price} \times \text{Sales tax rate}$$
$$\$325 = \$1625 \times x$$
$$= \$1625x$$
$$\frac{\$325}{\$1625} = x$$
$$.20 = x$$

A sales tax of 20% applied on the wholesale price of $1625 results in a retail price of $1950.

Let's now turn to a more complicated problem, one which we are all exposed to—that is, the impact of not only sales tax but also tariffs and profit margins.

Problem: Passenger vehicles attract a sales tax of 45% if the vehicle wholesales for over $36 995, and 22% sales tax if less. The tax is progressive—22% applies up to $36 995, and then increases to 45%. If a passenger vehicle's wholesale price is $42 000 and a profit margin of 20% is applied on the wholesale price (inclusive of sales tax):

(a) What is the final selling price of this vehicle?
(b) What percentage of the wholesale price is the final selling price?

Solution:
(a) Sales tax on the first $36 995 is 22%. Therefore:

$$\text{Amount of sales tax} = \text{Wholesale price} \times \text{Sales tax rate}$$
$$= \$36\,995 \times 22\%$$
$$= \$8138.90$$

Sales tax on remaining $5005 (i.e. $42 000 − $36 995):

$$\text{Additional sales tax} = \$5005 \times 45\%$$
$$= \$2252.25$$

Therefore, total sales tax = $10391.15

The profit margin of 20% is applied to the $52 391 (i.e. $42 000 + $10 391). Hence:

$$\text{Profit} = \$52\,391 \times 20\%$$
$$= \$10\,478.20$$

To determine the selling price, we must add the costs and profit margin.

$$\text{Final price} = \text{Wholesale price} + \text{Sales tax} + \text{Profit}$$
$$= \$42\,000 + \$10\,391 + \$10\,478$$
$$= \$62\,869$$

Therefore, the final selling price of the vehicle is $62 869.

(b) To determine what per cent the final selling price is of the wholesale price, i.e.:

$$\frac{\text{Final selling price} - \text{Wholesale price}}{\text{Wholesale price}} = \frac{\$62\,869 - \$42\,000}{\$42\,000} = \frac{\$20\,869}{\$42\,000}$$
$$= 0.497 = 49.7\%$$

The owner of this new vehicle will not take much comfort in knowing that the selling price is just short of 50% (49.7% to be precise) above the wholesale price. Is it any wonder that cars are so pricey in Australia! And we haven't even introduced the 22.5% tariff on imported cars which applies to approximately 50% of the new cars registered each year in Australia which are imported.

Stamp duty

Each State imposes duties on certain legal transactions and commercial and financial instruments. For instance, there is a **stamp duty** on the transfer of property ownership, and on the transfer as well as registration of motor vehicle ownership. For example, in New South Wales the stamp duty for the registration of a motor vehicle is 3% for every $100 (and any fractional part of $100) of the value of the vehicle. Therefore, to register a car valued at $22 000 the stamp duty would amount to $660 ($22 000 × 0.03).

The stamp duty on a mortgage registered in New South Wales is $5.50 for the first $15 000 and $4 per $1000 thereafter. Therefore, a mortgage of $100 000 will entail a stamp duty of $345.50. That is:

$$\text{Stamp duty} = \$5.50 + \frac{(\$100\,000 - \$15\,000) \times \$4}{\$1000}$$
$$= \$5.50 + \frac{\$85\,000 \times \$4}{1000}$$
$$= \$5.50 + \$340$$
$$= \$345.50$$

Finally, there is a stamp duty based on the purchase price of property, be it vacant land or an existing home. Table 2.2 gives a schedule of the rates applying in New South Wales.

Business mathematics and statistics

Table 2.2 *Stamp duty on the purchase price of property in New South Wales*

Purchase price ($)	Stamp duty
0–14 000	$1.25 per $100 ($10 min.)
14 001–30 000	plus $1.50 per $100 over $14 000
30 001–80 000	plus $1.75 per $100 over $30 000
80 001–300 000	plus $3.50 per $100 over $80 000
300 001–1 000 000	plus $4.50 per $100 over $300 000
more than 1 000 000	plus $5.50 per $100 over $1 000 000

Source: New South Wales Office of State Revenue, September 1996.

Problem: According to the Real Estate Institute of New South Wales, in April 1997 the median price of a house sold in Sydney was $225 300. How much would the purchaser have to allocate for the stamp duty?

Solution: Calculate the stamp duty for each of the first three scales and $145 300 for the fourth scale (i.e. $225 300 – $80 000). Thus:

$$0-\$14\ 000: \quad \frac{\$14\ 000}{\$100} \times \$1.25 = \$175$$

$$\$14\ 001-\$30\ 000: \quad \frac{\$16\ 000}{\$100} \times \$1.50 = \$240$$

$$\$30\ 001-\$80\ 000: \quad \frac{\$50\ 000}{\$100} \times \$1.75 = \$875$$

$$\$80\ 001-\$225\ 300: \quad \frac{\$145\ 300}{\$100} \times \$3.50 = \frac{\$5085.50}{\$6375.50}$$

Therefore, a stamp duty of $6375.50 must be paid to the NSW state government for purchasing a Sydney property for $225 300.

2.5 Competency check

1. What is the difference in the tax structure between personal income tax and company tax?

2. In Australia, television sets attract a 32% sales tax. If a retailer applies a 20% profit margin, how much does the set retail for if the wholesale price (excluding sales tax) is $280 (to the nearest dollar)?

3. A carpet retails for $50 per square metre after the retailer has applied a 12% sales tax and a 30% profit margin. What was the cost of the carpet per square metre to the retailer?

Chapter 2 – Percentages

4. Calculate the income tax payable (to the nearest dollar) for those on the adult male average annual income of $39 775 (February 1996).[14]

5. Calculate the stamp duty on a unit sold for $165 000 if the schedule in Table 2.2 applies.

. . . .

2.6 Profit and loss

No-one starts a business venture without the immediate aim of covering all costs and the ultimate aim of realising a profit. **Profit** can be defined as the remainder of funds available to the proprietor after deducting all costs and expenses from total sales revenue. A profit is made by selling a product at a price exceeding the total cost of the merchandise to the seller or vendor. The difference between the cost and selling price of the item is known as the **mark-up** or **profit margin**. It is normally expressed as a percentage of the cost price or list price (i.e. retail price). Therefore, the appropriate equations would be:

$$\text{Selling (or List) price } (SP) - \text{Cost } (C) = \text{Mark-up } (M)$$
$$\text{Selling price } (SP) = \text{Cost } (C) + \text{Mark-up } (M)$$

This formula should be used irrespective of the mark-up being applied to the cost or list price.

Problem: Giftware often carries a mark-up of 50% on cost. If the cost of a vase to a retailer is $84, what will the giftware retail for?

Solution: $SP = C + M$
$= \$84 + 50\% \text{ of } \$84$
$= \$84 + .5\,(\$84)$
$= \$84 + \$42$
$= \$126$

Problem: A refrigerator ticketed at $1150 has been marked up on cost by 30%. What was the cost price to the retailer?

Solution: Since the 30% mark-up is on cost, substitute $.3C$ into the equation for mark-up. Thus:

$$SP = C + M$$
$$\$1150 = C + .3C$$
$$\$1150 = 1.3C$$
$$\frac{\$1150}{1.3} = \frac{1.3C}{1.3}$$
$$\$884.62 = C$$

Therefore, the cost of the refrigerator to the retailer was $884.62.

Check the answer:

$$SP = C + M$$
$$= \$884.62 + .3(\$884.62)$$
$$= \$884.62 + \$265.38$$
$$= \$1150$$

Problem: A computer shop bought a new range of hobby computers for $750 each and is retailing them for $999. Calculate the percentage mark-up on cost.

Solution:
$$SP = C + M$$
$$\$999 = \$750 + M$$
$$\$999 - \$750 = M$$
$$\$249 = M$$

Since the mark-up is on cost,

$$\text{Mark–up rate} = \frac{\$249}{\$750}$$
$$= .332 \text{ or } 33.2\%$$

A faster way of calculating the mark-up rate would be to divide the selling price by the cost price and convert the answer to a percentage. Since the base (the cost price in this case) is always equal to 1 (or 100%), this indicates the mark-up was 33.2% (1.332 – 1.00).

$$\text{Mark–up} = \frac{\$999}{\$750} - 1.00$$
$$= 1.332 - 1.00$$
$$= .332 \text{ or } 33.2\%$$

Problem: The sale of yoyos exhibits cyclical variations, with 1996, for example, witnessing a boom in sales after a number of lean years. I would like to carry the top-of-the-range yoyo, which I have noted is selling for $9.50 in most shops. Since I normally apply a mark-up of 45% on cost, up to how much should I be prepared to pay for the yoyos?

Solution:
$$SP = C + M$$
$$\$9.50 = C + .45C$$
$$= 1.45C$$
$$\frac{\$9.50}{1.45} = \frac{1.45C}{1.45}$$
$$\$6.55 = C$$

I should be prepared to pay $6.55 for top-of-the-range yoyos.

Chapter 2 – Percentages

To check the answer, substitute $6.55 for cost and solve for the selling price.

$$SP = C + M$$
$$= \$6.55 + \$6.55(.45)$$
$$= \$6.55 + \$2.95$$
$$= \$9.50$$

Problem: A panelbeater marks up all materials by 40% based on the selling price. What would the panelbeater charge for a bonnet if it cost him $950?

Solution: Since the mark-up is on the selling price, $.4SP$ would be substituted for M in the equation; hence:

$$SP = C + M$$
$$= \$950 + .4SP$$
$$SP - .4SP = \$950$$
$$\frac{.6SP}{.6} = \frac{\$950}{.6}$$
$$SP = \$1583.33$$

The panelbeater would charge $1583.33 if he marks up by 40% on selling price the bonnet he bought for $950.

To check the answer, substitute $1583.33 and solve for either C or M. If we solve for C, we have:

$$SP = C + M$$
$$\$1583.33 = C + .4(\$1583.33)$$
$$\$1583.33 = C + \$633.33$$
$$\$1583.33 - \$633.33 = C$$
$$\$950.00 = C$$

Problem: The cost to the retailer of a tennis racquet that sold for $165 was $110.

(a) Find the profit as a percentage based on cost.
(b) Find the profit as a percentage based on selling price.

Solution:
(a) Rate of profit based on cost price:

$$SP = C + M$$
$$\$165 = \$110 + M$$
$$\$55 = M$$

Since mark-up was $55 and profit was based on cost, divide the $55 by the cost of $110. Thus:

$$\text{Percentage mark-up on cost} = \frac{\text{Mark-up}}{\text{Cost}}$$

$$= \frac{\$55}{\$110}$$

$$= .5 \text{ or } 50\%$$

Therefore, the profit of $55 was marked up 50% on the cost price of $110. Once again, the faster way of calculating the percentage mark-up is to divide the selling price by the cost price. As the cost price is the base, hence equal to 1 or 100%, you must subtract 1 from 1.5 to determine the mark-up on cost, as follows:

$$\text{Percentage mark-up on cost} = \frac{\text{Selling price}}{\text{Cost}} - 1$$

$$= \frac{\$165}{\$110} - 1$$

$$= 1.5 - 1$$

$$= .5 \text{ or } 50\%$$

(b) Rate of profit on selling price:

The mark-up was calculated to be $55. Hence:

$$\text{Percentage mark-up rate on selling price} = \frac{\text{Mark-up}}{\text{Selling price}}$$

$$= \frac{\$55}{\$165}$$

$$= .333 \text{ or } 33.3\%$$

Note: To determine the percentage mark-up on selling price, if given the mark-up rate on cost price, divide the cost mark-up rate by 100% plus the cost mark-up rate. Therefore, in the above example:

$$\text{Mark-up rate on selling price} = \frac{\text{Mark-up rate on cost}}{100\% + \text{Mark-up rate on cost}}$$

$$= \frac{50}{150}$$

$$= 33.3\%$$

If the mark-up on selling price is known and you want to determine the mark-up on cost, divide the mark-up on selling price by 100% minus the selling price mark-up rate. Thus, we have:

Chapter 2 – Percentages

$$\text{Mark-up rate on cost} = \frac{\text{Mark-up rate on selling price}}{100\% - \text{Mark-up rate on selling price}}$$

$$= \frac{33.3\%}{100\% - 33.3\%}$$

$$= \frac{33.3\%}{66.7\%}$$

$$= 50\%$$

Problem: A fountain pen cost $12 and was sold for $15.
(a) What is the mark-up rate on cost?
(b) What is the mark-up rate on selling price?

Solution:
(a) To determine the mark-up rate on cost, first calculate the mark-up:

$$M = SP - C$$
$$= \$15 - \$12$$
$$= \$3$$

Now substitute into the equation:

$$\text{Mark-up rate} = \frac{\text{Mark-up}}{\text{Cost}}$$

$$= \frac{\$3}{\$12}$$

$$= .25 \text{ or } 25\%$$

(b) Mark-up rate on selling price:

$$\text{Mark-up rate} = \frac{\text{Mark-up}}{\text{Selling price}}$$

$$= \frac{\$3}{\$15}$$

$$= .20 \text{ or } 20\%$$

Problem: The cost of a garment to a boutique was $400.
(a) If a profit margin of 17% was made on the selling price, find the selling price.
(b) If a loss of 2.5% was made on the selling price, find the selling price.

Solution:
(a)
$$SP = C + M$$
$$SP = \$400 + .17SP$$
$$SP - .17SP = \$400$$
$$.83SP = \$400$$
$$SP = \frac{\$400}{.83}$$
$$= \$481.93$$

The garment sold for $481.93 after application of a 17% profit margin to the $400 cost of the garment to the boutique.

Check the answer:

$$\text{Mark-up } (M) = \frac{\text{Selling price } (SP) - \text{Cost } (C)}{\text{Cost } (C)}$$

$$M = \frac{SP - C}{C}$$

$$= \frac{\$481.93 - \$400}{\$481.93}$$

$$= \frac{\$81.93}{\$481.93}$$

$$= 17\%$$

(b)
$$SP = C + M$$

$$SP = \$400 + (-0.025)SP$$

$$SP = \$400 - 0.025 SP$$

$$1.025 SP = \$400$$

$$SP = \frac{\$400}{1.025}$$

$$SP = \$390.24$$

A selling price of $390.24 on a garment costing $400 results in a loss of 2.5% on the selling price.

Check the answer:

$$M = SP - C$$
$$= \$390.24 - \$400$$
$$= -\$9.76$$

$$\text{Mark-up rate} = \frac{M}{SP}$$

$$= \frac{-\$9.76}{\$390.24}$$

$$= -0.025$$

$$= -2.5\%$$

Problem: The local shop marks up all goods by 25% on cost, but because of slow-moving stock it has been forced to offer a discount of 10% (a mark-down) on all merchandise.

(a) What would an article now retail for if it cost the proprietor $50?
(b) What was the final mark-up rate on cost following the sale of the discounted item?

Solution:

(a)
$$SP = C + M$$
$$= \$50 + \$50(.25)$$
$$= \$50 + \$12.50$$
$$= \$62.50$$
$$\text{Discount of } 10\% = \$62.50 \times .10$$
$$= \$6.25$$
$$\text{Final selling price} = \text{Selling price} - \text{Discount}$$
$$= \$62.50 - \$6.25$$
$$= \$56.25$$

The final selling price is $56.25.

A faster way of calculating this would be to remember the principle in deriving a single percentage from multiple percentages. In this case, we have:

$$\text{Final selling price} = \$50(1.25)(.9)$$
$$= \$50(1.125)$$
$$= \$56.25$$

(b)
$$\text{Mark-up rate} = \frac{\text{Sale price} - \text{Cost}}{\text{Cost}}$$
$$= \frac{\$56.25 - \$50}{\$50}$$
$$= \frac{\$6.25}{\$50}$$
$$= 0.125$$
$$= 12.5\%$$

The mark-up rate on cost is 12.5%.

Some firms operate at a loss, as costs and expenses are not covered by the revenue generated from sales. Some firms purposely sell products at a price below cost with the aim of attracting customers who will buy more non-discounted items. Products that are purposely sold below cost to increase customer traffic are referred to as loss leaders. However, a product is often sold at a loss to clear costly space and/or to generate cash flow.

Problem: An electrical appliance was sold for $667 after being reduced by $33\frac{1}{3}\%$ on the selling price. The sales representative received a commission of 7.5% on the sale, and the delivery cost of $10 was borne by the shop. The appliance originally cost the retail firm $625.

(a) What was the original selling or list price?
(b) What was the net profit/loss?

Solution:

(a) $667 = 66.7\%$ of selling price (as a 33.33% discount)

$$\$667 = 0.667 SP$$
$$\frac{\$667}{0.667} = SP$$
$$\$1000 = SP$$

(b)   Cost of appliance = \$625.00
      7.5% commission = \$50.02 (7.5% of \$667)
        Delivery charge = <u>\$10.00</u>
              Total cost = \$685.02
           Profit/loss = Discounted price – Total cost
                    = \$667 – \$685.02
                    = –\$18.02

Therefore, the net loss was \$18.02.

Problem: Booksellers often apply a 40% mark-up on the selling price of textbooks.
(a) If the retail price for a textbook is \$50, what was the cost price?
(b) What was the mark-up rate on cost?

Solution:

(a) $SP = C + M$
           $SP = C + .40 SP$
          $\$50 = C + .40(\$50)$
      $\$50 - \$20 = C$
            $\$30 = C$

Therefore, the textbook cost the bookseller \$30.

(b)   The mark-up rate on cost $= \dfrac{\text{Mark-up}}{\text{Cost}}$

                                   $= \dfrac{\$20}{\$30}$

                                   $= .667$ or 66.7%

There is a 66.7% mark-up on cost by the bookseller.

2.6 Competency check

1. In a recent sale, tracksuits were selling for \$49.99, providing the shop owner with a 5% return on cost. How much had each tracksuit cost the shop owner?

2. Charles bought 50 coffee tables for his shop at a cost of \$45 each. However, in transit from the wharf, three of the tables were damaged beyond repair

and Charles had no insurance coverage. How much should Charles charge per table in order to realise a profit margin of 35% on cost?

3. A boutique has discounted all items by 30% on the list price.
   (a) If an outfit sold for $140 at the discounted price, what was the original list price?
   (b) If the original mark-up was 40% on the list price, what was the cost price of the outfit to the boutique?

4. Mr Mobler bought a leather lounge suite for his furniture shop. It cost him $3750, and he is selling it at a list price of $6999. What is the percentage mark-up on cost (to one decimal place)?

5. In question 4, what is the percentage mark-up on selling price (to one decimal place)?

2.7 Summary

A per cent indicates the relationship between the number preceding the per cent symbol and 100. Thus, 40% is forty one-hundredths (i.e. $\frac{40}{100}$). To convert a percentage to a fraction, divide the number preceding the per cent sign by 100 and then reduce the result to its lowest terms. For example, $80\% = \frac{80}{100} = \frac{8}{10} = \frac{4}{5}$.

When converting a percentage to a decimal equivalent, place the decimal point two places to the left of the last digit of the whole number preceding the per cent symbol. For example, to convert 17% to a decimal, move the decimal point two places to the left of the 7; hence, 17% equals 0.17. To convert 23.25% to a decimal, moving the decimal point two places to the left of 3 results in 0.2325. To convert a decimal to a percentage, move the decimal point two places to the right and affix the per cent symbol.

When determining the percentage given the base and rate, the relationship is expressed as $V = B \times R$. Should the base or rate be the unknown factor, solve for each by transposing—that is, $B = V \div R$ and $R = V \div B$.

Commission

Commission is income that is received as a result of selling a product or service. It is normally based on a percentage of the sales value. Some sales representatives work on a retainer plus commission basis—that is, they receive a weekly wage and a percentage fee based on the sales volume (based on all sales over a certain allocated amount).

Brokerage is the commission received by a broker (stock or business broker, for example) for undertaking a business transaction on the client's behalf.

Discounts

In order to increase the likelihood of selling an item, a reduction in the price (known as a discount) is given. To determine the amount of discount received, multiply the selling (list) price by the discount rate; to determine the list price (or retail price), add the discount and sale price; to determine the discount rate, divide the discount amount by the original selling price.

Cash discounts are discounts given as an inducement for early payment by cash or cheque.

Trade discounts are often offered by manufacturers to wholesalers, wholesalers to retailers, or retailers to certain customers.

Chain discounts—that is, more than one discount on an item or service, with each subsequent discount calculated on the previously discounted price—are often applied to clear stock and/or to increase cash flow. For example, if discounts of 20% and 15% are given, the product is sold for 68% of its original selling price.

$$\text{Discount } (D) = \text{Selling price} \times \text{Discount rate}$$
$$\text{Let } x = \text{Selling price}$$
$$\text{First discount} = x \times 20\%$$
$$= 0.2x$$
$$\text{Discounted price} = 1x - 0.2x = 0.8x$$
$$\text{Second discount} = \text{Selling price} \times \text{Discount rate}$$
$$= 0.8x \times 15\%$$
$$= 0.12x$$
$$\text{Final sale price} = \text{Selling price} - \text{discount}$$
$$= 0.8x - 0.12x$$
$$= 0.68x$$

A faster means of arriving at the same answer is to determine the combined discount rate (*CDR*) which is defined as:

$$CDR = 1 - (100\% - \text{1st discount})(100\% - \text{2nd discount}) \ldots (100\% - n\text{th discount})$$

Thus, we have:

$$CDR = 1 - (100\% - 20\%)(100\% - 15\%)$$

Convert percentages to decimals:

$$= 1 - (.8)(.85)$$
$$= 1 - .68$$
$$= .32 \text{ or } 32\%$$

In other words, the product is discounted by 32%, therefore selling at 68% of the original selling price.

Taxation

Company tax

Company tax is the tax applied on a company's profits after all expenses have been taken into account. The current company tax is a flat tax of 36%, which means that all registered companies pay the same percentage tax rate irrespective of their taxable income.

Personal income tax

Australia's personal income tax regime is progressive—that is, as an individual's income increases, the percentage tax paid also rises. There is also a threshold ($5400) below which no personal income tax is payable.

Sales tax
Sales tax is a tax added to the wholesale price of goods. It is calculated as a percentage of the wholesale price and is ultimately added to the selling price. Sales tax is a vital source of revenue to the Federal Government (providing approximately 10% of revenue).

Stamp duty
A stamp duty is a duty imposed at state level on certain instruments and commercial and financial transactions, such as property, motor vehicle and insurance transactions.

Profit and loss
The ultimate aim of a business is to make a profit—that is, when sales revenue exceeds expenditure. The difference between the cost and the selling price (or list price) is referred to as the mark-up or profit margin. It is normally expressed as a percentage of the cost or selling price.

2.8 Multiple-choice questions

1. The stamp duty on a sale of property is approximately 2.5% of the selling price. If a house sold for $260 000, how much did the state government receive in stamp duty revenue from the sale?

    (a) $6500 (b) $65 000 (c) $650 (d) $65

2. As of 30 June 1994, Japanese investment in Australia was 357% greater than Australian investment in Japan. If Japanese investment in Australia amounted to $48.9 billion, how much had Australians invested in Japan?

    (a) $13.7 billion
    (b) $10.7 billion
    (c) $174.6 billion
    (d) $137 billion
    (e) None of the above.

3. If you are working on a commission basis only, earning 4.25% on all sales, and just received a cheque for $3476, what were your month's sales?

    (a) $362 373 (b) $147 730 (c) $333 429 (d) $81 788

4. An electric fan was reduced to $51 after a 15% discount had been applied. What was the original list price of the fan?

    (a) $58.65 (b) $95.35 (c) $60.00 (d) $43.35

5. Referring to Table 2.1, the personal income tax payable on an annual income of $32 000 is:

    (a) $6902 (b) $8942 (c) $10 880 (d) $6400

2.9 Exercises

1. Determine:

    (a) 70% of 400 (b) 13.5% of 51 (c) 2% of 20
    (d) 0.2% of 20 (e) 115% of 18 (f) 49% of 94

2. Calculate the following:

    (a) 16 is what per cent of 64? (b) 9 is what per cent of 81?
    (c) 13.5 is what per cent of 25? (d) 8 is what per cent of 25?
    (e) 12 is what per cent of 144?

3. Calculate the following:

    (a) What per cent of 80 is 40? (b) What per cent of 1250 is 25?
    (c) What per cent of 8 is 16? (d) What per cent of 90 is 45?
    (e) What per cent of 0.5 is 2?

4. Calculate the following:

    (a) 18 is 40% of what number? (b) 75 is 30% of what number?
    (c) 0.5 is 20% of what number? (d) 148 is 250% of what number?

5. The total gold medal tally since the first Olympics in 1896,[15] and including the 1996 Atlanta games, is as follows:

    USA = 790; USSR = 440; Germany = 250; Great Britain = 178; France =160 and Australia = 80.

    What percentage of the US gold medal tally do each of the countries have (to the nearest per cent)?

6. In a local real estate agency, commission on rentals makes up 13% of total weekly income. If rental income is $948 per week, what is the agency's weekly income?

7. In June 1996, of the Australian part-time labour force, 544 000 were men and 1 538 700 were women.[16] What percentage of the total part-time labour force were men and what percentage were women?

8. In 1995, of 49 million cars sold worldwide, the top four manufacturers were GM which sold 8.56 million, Ford 6.6 million, Toyota 4.56 million and Volkswagen 3.56 million.[17] What share of total car sales did each of the four manufacturers attain?

9. In Australia, the median age at first marriage increased from 23.4 years in 1971 to 27.2 years in 1994 for bridegrooms and from 21.1 years to 25.1 years for brides. What was the percentage rise for grooms versus brides in waiting to tie the knot? Should they have waited still longer, since in 1995 there were 44 divorces for every 100 marriages?[18]

10. The Marshfield family bought a home for $274 000, paying a 10% deposit with the remainder to be paid over the 20-year term of the loan. How much was the balance after the deposit was paid?

11. Each section in a department store is given a target for the year, with Jack's area targeted for an increase of 25% over last year's result. If last year's sales were $1.25 million, what was Jack's targeted sum?

12. In 1997 the Hoati Hotel's annual wage bill was $20 million, which was 35% of its annual expenditure. What was the Hoati's expenditure for 1997?

13. According to the 1996/97 Federal Budget, Federal Government expenditure was forecast to rise by 2.4% to $129.7 billion. What was the Federal Government expenditure in the 1995/96 financial year?

14. A 1996 Reserve Bank estimate of private sector wealth in Australia found that dwellings accounted for 59% (or $1040 billion), business assets 28%, consumer durables and motor vehicles 6%, private sector holdings of government securities 6%, and notes and coins in circulation 1%.[19] Determine the dollar amount (to the nearest billion dollars) for each category of wealth.

15. Australian foreign investment in New Zealand[20] totalled $8.5 billion, or 6.5% of Australian overseas investment at 30 June 1994. What was the total Australian overseas investment (to the nearest billion dollars)?

16. World trade in 1994 of $5405 billion was an increase of 13% on the previous year. What was world trade (to the nearest billion dollars) in 1993?

17. Skygarden Shopping Centre in Sydney was valued in late 1995 at $120 million,[21] 40% of the offer received in 1989. How much was the 1989 offer?

18. A multinational firm found that its Australian subsidiary contributed 19% of its total worldwide sales and 8% of its total worldwide profit. If the subsidiary's sales and profit results were $112.5 million and $2.6 million respectively, find the multinational's worldwide sales and profit.

19. Sonny Blue works for Greystanes Inc. Last month he received a commission of $837.59 for sales of $9854. What was the rate of commission paid?

20. Ms Di Mone earned $32 000 last year as a sales representative for an importer of custom-made jewellery. She operates on the basis of a $175 per week retainer plus 5% commission on all sales. How much did Ms Di Mone generate in sales last year?

21. Egbert Everidge works on a graduated commission basis, receiving 5% on monthly sales up to $10 000, 7.5% on the subsequent $10 000 and 10% thereafter. If 'Eggie' (as his close friends call him) scrambles home with $2875 in commission for the month, what were his sales for that month?

22. Jack B. Nimble works for Jack B. Quik's candle factory and has been offered the option of working on a monthly retainer plus commission basis ($500 plus 7%), or on a straight commission basis (at 11%). Jack B. Nimble's monthly sales average is $23 000. Which option should he choose?

23. Mr F. Disk sells computer hardware. He has been offered a choice of straight commission of 5% on all sales, or a retainer of $275 per week plus 4% on sales

Business mathematics and statistics

exceeding $4000 per week. If projected sales for the year total $875 000, which arrangement is the more favourable for Mr Disk? Assume equal sales per week.

24. A real estate agent sold a house and received a commission of $4505. If rates of 5% applied on the first $15 000, 3% on the next $45 000, 2.5% on the next $40 000 and 2% thereafter, how much was the house sold for?

25. A top screen star on contract to a major film studio receives royalties of 3.5% on all film and video sales plus a weekly wage. Last year the celebrity earned $4 486 250. Total film and video sales for that year were $68 750 000. How much did the celebrity receive per week (to the nearest dollar):

    (a) in royalties? (b) as wages?

26. An author received $4800 in royalties for a book. If the author's royalty is equal to 15% of the publisher's selling price to the bookseller, what was the sales value of the total number of copies sold?

27. A solar calculator was advertised as being reduced from $12.50 to $8.75. What was the rate of discount?

28. A suit is advertised at a sale price of $297 after 45% is slashed off the regular price. What was the regular price?

29. A painter, who is eligible for a 20% trade discount, bought materials with a total list price of $750. How much did the materials cost the painter?

30. A local discount shop offering a 25% discount on all items is now offering a further 15%, as well as a 2.5% cash discount. The Espinosa family paid $745 in cash for a lounge suite. What was the original list price (i.e. if no discounts had been applied)?

31. Ms Ruby Quartz bought a diamond ring originally selling for $18 000 but discounted three times—by 20%, 15% and 25%. What was the discounted selling price to Ruby for the diamond ring?

32. A discount of 5% is given for payment within seven days and 2.5% for payment within 14 days. If the invoice amount totalled $17 845, how much should the buyer make out the cheque for if:

    (a) the account is paid within five days?
    (b) the account is paid on the 12th day?

33. Referring to Table 2.1, how much personal income tax is payable if:

    (a) $4890 is earned?
    (b) $14 700 is earned?
    (c) $55 000 is earned?

34. Company tax is applied at a flat rate of 36%. How much company tax must be paid if the company's taxable income is:

    (a) $4890? (b) $14 700? (c) $55 000?

Chapter 2 – Percentages

35. Next time you are in the market for toilet pans and seats, remember that the Federal Government wants a piece of the action—that is, there is a 12% sales tax on these necessities. If the top-of-the-range toilet seat wholesales for $175, how much sales tax will you have to pay (up-front)?

36. That new motorbike you are considering buying is hit with a 22% sales tax and currently retails for $9750. Assuming that there is also a 20% profit margin, what is the wholesale price of the bike (to the nearest dollar)? How much is the sales tax? The profit margin?

37. A new laptop IBM compatible retails for $3184 after a sales tax of $484 has been applied to the wholesale price and a $500 profit margin added. What is the rate of sales tax applied to computers (to the nearest dollar)?

38. (a) Motor vehicle parts attract a 22% sales tax. How much sales tax will the Federal Government receive from your purchase of motor parts worth $450 at the wholesale price?

    (b) You have just paid the retail price of $600 for a lawnmower which was hit by a 22% sales tax and a profit margin of $75. What was the wholesale price of the lawnmower (to the nearest dollar)?

    (c) You have just bought yourself a new camera for $780, which includes the 32% sales tax and a $185 profit margin. What was the wholesale price of the camera, and what was the amount of sales tax (to the nearest dollar)?

39. In Victoria, the state government applies the following stamp duty fees to mortgages: $4.00 up to $8000, plus 70 cents/$200 up to $10 000, plus 80 cents/$200 above $10 000. If the value of a particular residence is $72 000 and a mortgage for 80% of the value is taken, calculate the stamp duty the Victorian government can expect to receive from this transaction.

40. Orrefors crystal is advertised at 35% savings on the normal selling price of $75. What is the discount price for the crystal?

41. Norm always works on a 40% margin on the selling price. If the cost of the item is $235, what will be the selling price (to the nearest dollar)?

42. All goods in a particular shop are marked up at 25% on the list price.

    (a) What is the cost to the proprietor of an article with a list price of $30?

    (b) What is the percentage mark-up on cost?

43. A retailer applies a profit margin of 50% on all goods but has now been forced to discount all goods by 25%. What is the retailer's percentage profit on cost after the discount?

44. All candles in a particular gift shop are priced at $4.25 after a 25% mark-up on cost. What did the candles cost the proprietor?

45. All menswear has been marked down by 40%, with suits now selling for $225.

Business mathematics and statistics

Calculate:
(a) the original selling price of the suits;
(b) the amount by which the suits have been discounted.

46. A bookseller bought 1000 copies of a particular title at a cost of $7.50 each. As they did not sell well, a discount of $1.38 a book was given, which resulted in a profit of 15%.

    (a) What per cent discount was given?
    (b) What was the book's original list price?

47. A local shop applies a profit margin on costs of 45% on all merchandise. A sale was announced three weeks ago with across-the-board discounts of 25%. Yesterday an additional discount of 20% was announced for all remaining stock.

    (a) If an item was sold today for $25, what was the cost price to the shop owner?
    (b) What was the original list price?
    (c) What was the original profit-margin rate based on list price (assuming that the list price calculated in (b) applied)?

48. A particular style of running shoe cost the sports shop owner $55 a pair.

    (a) If the shop owner hopes to make 40% on the selling price, what will be the selling price?
    (b) What is the mark-up rate on cost?

49. A computer was marked up by 40% on cost, and subsequently a 20% discount was given. If the discounted selling price was $3750, what was:

    (a) the original cost of the computer?
    (b) the mark-up rate on the discounted selling price (to one decimal place)?

50. Last summer a retailer applied a 40% mark-up on cost to all toys. Now, in order to clear the stock, the retailer has discounted all toys by 25% off the list price. What was the cost to the retailer of a toy discounted down to $37.50?

Notes

1. Kim Sweetman, 'Takeaway food eats up budget', *Sunday Telegraph*, 2 March 1996, p. 17.
2. National Heart Foundation of Australia, *Heart and Stroke Facts* (1996), p. 32.
3. Ibid.
4. Ibid, p. 3.
5. Australian Bureau of Statistics, *Demography New South Wales 1994* (Cat. No. 3311.1).
6. Geoff Barker, 'World Bank challenge as the poor get poorer', *Australian Financial Review*, 19 June 1996, p. 23.
7. Andrew Darby, 'A good bet', *Sydney Morning Herald*, 11 June 1996, p. 8.
8. Reserve Bank of Australia, *Australian Economic Statistics 1949–50 to 1994–95*, June 1996.
9. Australian Bureau of Statistics, *Balance of Payments, Australia* (Cat. No. 5301.0).

Chapter 2 – Percentages

10. 'Asians move up Forbes list', *Australian Financial Review*, 7 February 1996, p. 12.
11. Australian Stock Exchange, *Your Guide to Investment*, p. 3. This is an excellent introduction to the whys and wherefores of such investment.
12. Australian Taxation Office, *Sales Tax: How it affects you*, September 1995.
13. CCH Australia Limited, *Australian Master Tax Guide*, pp. 1381–2.
14. Australian Bureau of Statistics, *Average Weekly Earnings, States and Australia* (Cat. No. 6302.0).
15. Alistair Adams-Smith, 'The feats to beat', *Sydney Morning Herald*, 17 July 1996, p. 44.
16. Australian Bureau of Statistics, *The Labour Force* (Cat. No. 6202.0).
17. *Australian Financial Review*, 14 June 1996, p. 11.
18. Australian Bureau of Statistics, *Australian Demographic Statistics* (Cat. No. 3101.0).
19. Commonwealth Treasury of Australia, *Economic Roundup*, Summer 1996, p. 48.
20. Stephen Ellis, 'ABS figures map out the investment routes', *Australian Financial Review*, 14 July 1995, p. 28.
21. Kathy MacDermott, '$600m sale of problem holdings', *Australian Financial Review*, 7 December 1995, p. 31.

APPENDIX 2A
Ratios and proportions

2A.1 Ratios

A ratio can be thought of as a description of the relationship between two or more numbers. It is used throughout our daily lives in areas as diverse as business, sports, demographic and crime statistics, cookbooks, maps and pharmaceutical drugs. For example, in 1995, Australia had 1137 cinema screens for a population of 18 192 000, or 1 screen per 16 000 people,[1] as well as 14.6 cafes and restaurants per 100 000 people.[2] In the great Olympic medal chase since 1896, Finland is the leader with 1 medal (includes gold, silver and bronze) per 18 000 population, just ahead of Sweden at 1 per 20 000, with Australia at 1 per 71 000 and the United States at 1 for 136 000 of population.[3] In Australia there are 99.2 males per 100 females and 2.3 persons per household. In New South Wales and Victoria respectively, the murder rate was 1.72 and 1.38 per 100 000 head of population; assault, 619.1 and 351.1 per 100 000; break, enter and steal, 2178.5 and 1575.4 per 100 000; and motor vehicle theft, 761.85 and 649.9 per 100 000 population.[4] These statistics are all ratios and may also be written in the following manner, for example:

Restaurants and cafes:

14.6/100 000   14.6:100 000 $\dfrac{14.6}{100\ 000}$

Cinema screens:

1/16 000   1:16 000 $\dfrac{1}{16\ 000}$

Demographic statistics illustrate the use of ratios, as these statistics are normally expressed on a per 1000 population basis (from now on, 'per' will be written as /). For example, in Australia in 1994 there were 2.7 divorces/1000 population versus 6.2 marriages/1000 population.[5] In the same year the crude birth rate in Australia was 14.5/1000 population versus a mortality rate of 7.1/1000 population.[6] In the same year in Australia, 161/100 000 males died of cardiovascular diseases compared with 181/100 000 from cancer.[7] A total of 21 000/100 000 of the world's population are chronically undernourished; consequently, over 35 000 people die daily from starvation. The infant mortality rate per 1000 births is 5.8/1000 in Australia, 8.3/1000 in New Zealand, 4.5/1000 in Japan and 75/1000 in Indonesia.[8]

Why are statistics commonly presented in such a fashion? The above statistics could be converted to a per head basis: 0.0027 divorces per head, 0.0062 marriages per head, 0.01447 births per head, etc. However, the statistics are more easily understood when the numerator ratio is expressed as a mixed number rather than as a decimal. It should be realised, however, that ratios can be written as fractions, decimals or

Chapter 2 – Percentages

percentages and, as such, the same rules apply. (For example, see the rules governing fractions in Chapter 1, Section 1.1.)

Ratios are also used to express rates—for example, 60 kilometres/hour, 2.3 children/family and 12 kilometres/litre of petrol. In each of these cases the denominator, or the expression to the right of the '/', is understood to be one unit—that is, 60 kilometres/1 hour, 2.3 children/1 family and 12 kilometres/1 litre.

When confronted with two quantities of different units, it is sometimes possible to express them in the same unit of measurement. For example, if you had to compare 5 centimetres with 1 metre in the form of a ratio, since 100 centimetres equals 1 metre, the ratio would be 5/100, 1/20, 1:20, or 0.05 to 1.

As has been seen, a measure must be compared with a standard to have any meaning. For example, the fact that Christine ran the 14 kilometres of the City to Surf Fun Run in 50 minutes has little meaning unless it is compared with the average female running time. If the average time for a female to complete the course was 90 minutes, then the ratio of Christine's time to the average time was 50/90, 5/9, 5:9 or 1:1.8. As the distance completed was identical, this means that Christine finished the race in 5/9 the time of the average female runner; and that for every 5 metres the average female ran, Christine ran 9 metres.

Problem: Value Industries has a number of divisions, whose 1997 sales performance and corresponding profits are indicated in Table 2A.1.

(a) What is the ratio of each division's sales to the engineering division's sales?
(b) What is the ratio of profits to sales for each division?
(c) What is the ratio of total profits to total sales?

Table 2A.1 *Value Industries: sales and profits for 1997*

Division	Sales ($ million)	Profits ($ million)
Wines and spirits	24	1.5
Sportswear	18	1.4
Graphics	28	2.9
Engineering	42	2.1

Source: Hypothetical.

Solution:
(a) Since we are comparing each division with the engineering division, the sales result for engineering will always be in the denominator. Therefore, we have:

$$\text{Wines and spirits} \quad \frac{24}{42} = \frac{4}{7}$$

$$\text{Sportswear} \quad \frac{18}{42} = \frac{3}{7}$$

$$\text{Graphics} \quad \frac{28}{42} = \frac{2}{3}$$

Business mathematics and statistics

Note that in each case the ratio has been expressed as a fraction in its lowest terms.

(b) To calculate the ratio of profits to sales, the profit result is placed in the numerator and the sales result in the denominator. Always place the subject being compared (in this case, sales) in the denominator. Thus:

Wines and spirits $\quad \dfrac{1.5}{24.0} = 0.0625 = 6.25:100$

Sportswear $\quad \dfrac{1.4}{18.0} = 0.0777 = 7.77:100$

Graphics $\quad \dfrac{2.9}{28.0} = 0.1036 = 10.36:100$

Engineering $\quad \dfrac{2.1}{42.0} = 0.0500 = 5.00:100$

A percentage may be thought of as a ratio, as it expresses the relationship between a number and 100. Therefore, you could have converted the above decimals to percentages. A common method of expressing a ratio is to express the number being compared to as 100, as this may make the ratio clearer to the reader (particularly if the ratio is not an even number).

In the above example, Wines and spirits realised profits of $6.25 for every $100 sales, with Sportswear at $7.77 profit for each $100 sales, Graphics at $10.36 profit for every $100 sales and Engineering at $5 per $100 sales.

(c)
$$\dfrac{\text{Total profits}}{\text{Total sales}} = \dfrac{1.5 + 1.4 + 2.9 + 2.1}{24.0 + 18.0 + 28.0 + 42.0} = \dfrac{7.9}{112.0} = .0705 \text{ or about } 7.1:100$$

In other words, for every $100 worth of sales, there was a profit of approximately $7.

Problem: In 1997 a full-page colour advertisement in a reputable Australian weekly magazine cost $22 080 for one issue. If 3300 inquiries resulted, what was:

(a) the cost per inquiry?
(b) the number of inquiries per advertising dollar?
(c) the cost per printed copy if circulation is 1 202 526 (as claimed by the *Australian Women's Weekly*)?

Solution:

(a) $\quad \dfrac{\text{Cost}}{\text{Number of inquiries}} = \dfrac{\$22\,080}{3300} = \$6.69/\text{inquiry}$

(b) $\dfrac{\text{Number of inquiries}}{\text{Cost}} = \dfrac{3300}{\$22\,080} = 0.1495/\$1$

However, 14.95 responses per $100 spent (or 149.5 responses per $1000 spent) is more easily understood than 0.1495 inquiries per $1.

(c) $\dfrac{\text{Cost}}{\text{Circulation}} = \dfrac{\$22\,080}{1\,202\,526} = 0.0184 = 1.84$ cents/copy

Problem: The James and Kelly brothers formed a partnership and opened up a gun shop. Of the four participants, Jesse and Frank James contributed $\frac{1}{4}$ and $\frac{1}{5}$ to the partnership, with Ned putting in $\frac{3}{10}$ and Gene the remainder. If profits are to be shared in the same ratio as their original outlay, how will profits of $1 095 000 be allocated?

Solution: First, find Gene's share of the partnership:

$\frac{1}{4} + \frac{1}{5} + \frac{3}{10} + x = 1$

$\frac{5}{20} + \frac{4}{20} + \frac{6}{20} + x = 1$

$x = 1 - \frac{15}{20}$

$x = \frac{5}{20} = \frac{1}{4}$

Next, multiply each partner's respective ratio by the amount of profit earned:

Jesse $= \frac{1}{4} \times \$1\,095\,000 = \$\ \ 273\,750$

Frank $= \frac{1}{5} \times \$1\,095\,000 = \$\ \ 219\,000$

Ned $\ \ = \frac{3}{10} \times \$1\,095\,000 = \$\ \ 328\,500$

Gene $= \frac{1}{4} \times \$1\,095\,000 = \underline{\$\ \ 273\,750}$

$\hspace{5cm} \$1\,095\,000$

Problem: Mr T. Leaves, Mr A. Tom and Ms B. Quik decided to form a partnership, with each contributing $45 000. Mr T. Leaves is the full-time manager of the business, Mr A. Tom is his part-time assistant, and Ms B. Quik is a silent partner. Their agreed division of profits is in the ratio of 10:4:2 respectively. Interest is paid on their original capital outlay at the market rate of 10.5% p.a., and Mr Leaves' salary is $28 000 p.a. If Mr Leaves' income for the year was $57 725 (including salary, interest and profit-sharing):

(a) what was the partnership's total profit?
(b) how much did Mr A. Tom and Ms B. Quik each receive in income for the year?

Business mathematics and statistics

Solution:
(a) First, determine Mr Leaves' profit-sharing income:

$$\text{Total income} = \text{Salary} + 10.5\% \text{ interest on capital outlay} + \text{Profit-sharing income } (x)$$

$$\$57\,725 = \$28\,000 + \$4725 + x$$
$$\$57\,725 = \$32\,725 + x$$
$$\$57\,725 - \$32\,725 = x$$
$$\$25\,000 = x$$

Therefore, Mr Leaves' share of the profit was $25 000.

Now determine what share of the total profit accrued to Mr Leaves. Find the ratios of the agreed sharing of profits—that is, 10:4:2 indicates that there are 16 equal parts to the partnership (10 + 4 + 2 = 16). Hence:

$$\begin{aligned}
\text{Mr T. Leaves' share at } & 10/16 = 5/8 \\
\text{Mr A. Tom's share at } & 4/16 = 2/8 \\
\text{Ms B. Quik's share at } & \underline{2/16 = 1/8} \\
& 16/16 \quad 8/8
\end{aligned}$$

As Mr T. Leaves' share of total profits is 5/8 and he received $25 000, if x = total profit, then:

$$\frac{5}{8}x = \$\,25\,000$$
$$5x = \$200\,000$$
$$x = \$\,40\,000$$

(b) Mr A. Tom's share is:

$$\frac{2}{8} \times \$40\,000 = \$10\,000$$

Therefore:

$$\begin{aligned}
\text{Mr A. Tom's income} &= \text{Profit share} + \text{Interest} \\
&= \$10\,000 + \$4725 \\
&= \$14\,725
\end{aligned}$$

Ms B. Quik's share is:

$$\frac{1}{8} \times \$40\,000 = \$5000$$

Therefore:

$$\begin{aligned}
\text{Ms B. Quik's income} &= \text{Profit share} + \text{Interest} \\
&= \$5000 + \$4725 \\
&= \$9725
\end{aligned}$$

Financial ratios

The financial position of a company is of major concern not only to the owner and/or management but also to parties outside the organisation itself, such as creditors, suppliers and the government. Investment and lending decisions by owners, shareholders and creditors are influenced to a major extent by the financial information available in the form of the profit and loss statement, the balance sheet and the funds statement.

When financial statements are evaluated, one often analyses the relationship between one financial variable and another (e.g. that of profit to sales). This relationship is referred to as a *financial ratio* or *percentage*. When financial ratios over several periods are compared with a prescribed standard, the degree to which the results vary can be determined. Therefore, ratios are often used in evaluating financial statements, as these statements, although containing significant accounting information, do not always immediately highlight vital relationships. For example, ratios can be used to assess the profitability, efficiency, liquidity and/or solvency of a business, thus permitting a better understanding of the financial status and performance of the business. This section emphasises the diversity of the application and the usefulness of ratios.

To evaluate a firm's financial position, the first items that are needed are the firm's profit and loss statement and its **balance sheet**. Examples are presented in Tables 2A.2 and 2A.3.

Table 2A.2 *The Alliance Group Pty Ltd profit and loss statement (year ended 30 June 1997)*

	$	$
Sales		1 500 000
Less: Cost of goods sold		875 000
Gross profit		625 000
Less: Total operating expenses		
Marketing	204 000	
Administration	185 000	
Finance	24 500	413 500
Less: Company tax		40 000
Net profit		171 500

Source: Hypothetical.

Profitability ratios

Return on total assets

The return on assets is the net profit earned in relation to the total resources used by the firm in realising that profit. It is defined as:

$$\text{Return on total assets} = \frac{\text{Net profit}}{\text{Total assets}}$$

Table 2A.3 The Alliance Group Pty Ltd balance sheet (year ended 30 June 1997)

Assets ($)			Equities ($)		
Current assets			Current liabilities		
Cash	14 000		Trade creditors	102 000	
Trade debtors	85 000		Company tax		
Stock	55 000		payable	38 000	
Bills receivable	9 000		Bank overdraft	52 000	
Total current assets		163 000	Total current liabilities		192 000
Investments		185 000	Long-term debt		70 500
Property, plant and equipment			Total liabilities		262 500
Land	85 000		Shareholders' equity		
Buildings	90 000		80 000 shares	181 500	
Plant and equipment	59 000	234 000	Retained earnings	138 000	319 500
Total assets		582 000	Total equities		582 000

Source: Hypothetical.

Therefore, in the Alliance group's case:

$$\text{Return on total assets} = \frac{\$171\ 500}{\$582\ 000} \quad \begin{array}{l}\text{(from profit and loss statement)}\\\text{(from balance sheet)}\end{array}$$

$$= 0.295$$

Therefore, a net profit of 29.5 cents was earned for every dollar used by the company. Evaluation of this result with the firm's performance and other business proportions can be very helpful in determining which investment has the greatest stability and likelihood of continued success.

Earnings per share

Earnings per share, which are often quoted in the media, are a widely used measure of financial performance and are often compared with those of previous years in order to determine trends. The ratio is defined as:

$$\text{Earnings per share} = \frac{\text{Net profit after tax}}{\text{Issued shares}} \quad \begin{array}{l}\text{(from profit and loss statement)}\\\text{(from balance sheet)}\end{array}$$

$$= \frac{\$171\ 500}{80\ 000}$$

$$= \$2.14/\text{share}$$

Return on sales

The return on sales is a comparison of either gross or net profit and total sales (sometimes referred to as gross or net profit margin):

$$\text{Gross profit on sales} = \frac{\text{Gross profit}}{\text{Sales}}$$
$$= \frac{\$625\,000}{\$1\,500\,000} \quad \begin{array}{l}\text{(from profit and loss statement)}\\ \text{(from profit and loss statement)}\end{array}$$
$$= 0.4166$$

Therefore, for every dollar received in sales, a gross profit of 41.66 cents was earned. This ratio provides a basis for evaluation and comparison over time, since the use of the absolute dollar terms of gross profit becomes less meaningful over time unless related to sales.

$$\text{Net profit on sales} = \frac{\text{Net profit}}{\text{Sales}}$$
$$= \frac{\$171\,500}{\$1\,500\,000} \quad \begin{array}{l}\text{(from profit and loss statement)}\\ \text{(from profit and loss statement)}\end{array}$$
$$= 0.1143$$

Therefore, for every dollar received in sales, a net profit of 11.43 cents was earned.

As net profit is determined by the deduction of all operating expenses and company tax from gross profit, it indicates the relative efficiency of the business. The use of both gross and net profit enables the businessperson to assess how, for example, expenses have been controlled in relation to sales. For instance, if expenses are increasing at a faster rate than sales, then a decline in net profit on sales will occur, although the profit ratios can increase or remain constant.

Asset turnover
Companies such as Coles, Woolworths and Franklins work with high turnover and low profit margins to cover all operating expenses and to yield a satisfactory return on investment. Companies such as Tiffany's and Rolls-Royce may operate with comparatively low turnover and high profit margins. The asset turnover ratio supplements the previously explained profitability ratios to account for the differences in business efficiency rather than margin on sales.

$$\text{Asset turnover} = \frac{\text{Sales}}{\text{Total assets}} \quad \begin{array}{l}\text{(from profit and loss statement)}\\ \text{(from balance sheet)}\end{array}$$
$$= \frac{\$1\,500\,000}{\$582\,000}$$
$$= \$2.58$$

Therefore, $1 of assets supported every $2.58 of sales.

Wherever possible, profitability ratios should be used in conjunction with one another in evaluating current and past periods because movements in one ratio can then be analysed in relation to movements in another, thus permitting a more conclusive appraisal of the firm's profit performance.

Business mathematics and statistics

Efficiency ratios

The efficiency of a company is often assessed by evaluating the turnover of stock, days' supply of stock on hand, debtor turnover, and the average collection period per debtor.

Stock turnover

Stock turnover is the number of times the average stock has been sold over a period (or the number of times the inventory has been turned over). It is determined by dividing the cost of goods sold by the average stock. Average stock is the average of the opening and closing stock. That is:

$$\text{Average stock} = \frac{\text{Opening stock} + \text{Closing stock}}{2}$$

Assume that the Alliance group's opening stock on 1 July 1996 was $62 000. Therefore:

$$\begin{aligned}
\text{Stock turnover} &= \frac{\text{Cost of goods sold}}{\text{Average stock}} \quad \begin{array}{l}\text{(from profit and loss statement)}\\ \text{(from balance sheet)}\end{array}\\
&= \frac{\$875\,000}{\frac{\$62\,000 + \$55\,000}{2}}\\
&= \frac{\$875\,000}{\$58\,500}\\
&= 14.96
\end{aligned}$$

Therefore, the turnover of the average inventory was 14.96 times per annum.

The stock turnover for companies such as Coles, Woolworths and Franklins will far exceed that of speciality shops (e.g. whitegoods, pianos, jewellery). High stock turnover can indicate either that stock and funds are being used efficiently or that not enough stock is on hand to satisfy immediate demand. A low rate can likewise indicate a large amount of investment stock, a slow-moving line, depressed business conditions or managerial inefficiency. However, rates of inventory turnover depend to a large extent on the business itself—for example, a milk bar has a high turnover, while an art dealer has a low turnover.

Supply of stock on hand

The number of days' stock on hand is calculated by:

$$\text{Days' stock on hand} = \frac{365}{\text{Stock turnover}}$$

In the Alliance group's case, therefore:

$$\begin{aligned}
\text{Days' stock on hand} &= \frac{365}{14.96}\\
&= 24.4
\end{aligned}$$

Businesses with fast-moving lines (e.g. perishables, low-cost items) have a high stock turnover rate and, consequently, the stock is held over for fewer days than in firms with slow-moving lines (e.g. high-priced furniture, whitegoods).

As inventory is money tied up, and as money is expensive (if borrowed, the cost is interest; if using working capital—that is, internally generated funds, the cost is the amount that could have been earned elsewhere) a firm must determine what level of inventory is necessary to ensure that consumer demand is met with as little delay as possible. The efficiency of Japanese car manufacturers, for example, can be explained, in part, by exceedingly low levels of inventory of components. The Japanese use the 'kanban', or 'just-in-time', system of inventory control. This system obliges suppliers to deliver parts exactly when they are needed. Hewlett-Packard has been using such a system since 1982 when there was an average of 22 days of inventory on hand versus the present average of one day.

Debtor turnover
Debtor turnover should be used with the average collection period per debtor to determine the effectiveness of the credit policy of the firm. If the average collection period exceeds the normal credit period in the industry, then either the firm is using a liberal credit policy to stimulate sales revenue or the policing of its credit policy is unsatisfactory.

$$\text{Debtor turnover} = \frac{\text{Credit sales}}{\text{Debtors}}$$

Assume that the Alliance group's credit sales were $1 000 000. Therefore:

$$\text{Debtor turnover} = \frac{\$1\,000\,000}{\$85\,000} \quad \text{(from balance sheet)}$$
$$= 11.76$$

The average collection period per debtor can now be determined:

$$\text{Average collection period} = \frac{365}{\text{Debtor turnover}}$$
$$= \frac{365}{11.76}$$
$$= 31.04$$

In other words, the average debt-collection period is approximately 31 days. Should this period be outside the normal credit terms, then a more stringent credit policy must be enforced.

Once again, it is always advisable to examine all efficiency ratios so that an informed decision can be made.

Liquidity ratios

Liquidity ratios measure the ability of the firm to meet its current monetary commitments by the use of its current assets. The most commonly used liquidity ratios are current working capital ratios and acid test ratios.

Current working capital ratios

Current working capital ratios are probably the most popular indicator of near-term liquidity. The ratio is defined as current assets divided by current liabilities:

$$\text{Current ratio} = \frac{\text{Current assets}}{\text{Current liabilities}} \quad \begin{array}{l}\text{(from profit and loss statement)} \\ \text{(from balance sheet)}\end{array}$$

In the Alliance group's case, therefore:

$$\text{Current ratio} = \frac{\$163\,000}{\$192\,000}$$
$$= 0.85$$

Thus, for every $1 in current liabilities owed, the firm has 85 cents in current assets to meet such commitments. Naturally, the higher the current ratio, the more able the firm is to meet its obligations. Whether the Alliance group's current ratio of 0.85 is adequate depends on the variability of the firm's cash flow, and the extent to which the current liability is likely to be presented for settlement. If the firm's working capital ratio is low, there might not be sufficient funds available to operate at a satisfactory level. If, on the other hand, the working capital ratio is high, it could mean that short-term funds are not being used profitably, perhaps because of excessive holdings of cash or financial assets yielding low returns.

Acid test ratios

Current assets include stock (or inventory) which may not be convertible to ready cash at short notice. Therefore, in the Alliance group's case, if the short-term obligations were called for payment within a month, $55 000 worth of stock would have to be sold fairly quickly although there would be no guarantee of it being sold or of selling it for $55 000.

The Alliance group has a bank overdraft of $52 000 which, although a current liability, would not normally have to be repaid in the immediate future since overdrafts are customarily provided over an indefinite period. Thus, in order to eliminate these defects of the current ratio, the acid test or quick or liquid ratio (as it is sometimes referred to) is applied:

$$\text{Acid test ratio} = \frac{\text{Current assets} - \text{Stock (as of 30/6/1997)}}{\text{Current liabilities} - \text{Bank overdraft}}$$
$$= \frac{\$163\,000 - \$55\,000}{\$192\,000 - \$52\,000}$$
$$= \frac{\$108\,000}{\$140\,000}$$
$$= 0.77$$

This result could be evaluated in terms of previous years' results. If, for example, the acid test ratio of 0.77 for 1997 was far less than the 1996 result, this could reflect a large increase in inventory or a reduced overdraft, with the former being perhaps a cause for concern. Thus, the acid test ratio indicates the ability of the firm to meet its immediate financial obligations from its own liquid resources.

Chapter 2 – Percentages

2A.2 Proportions

A **proportion** is a statement that two ratios are equal—for example, 2/7 = 6/21. A proportion can also express the comparison of a part to a whole and is usually represented in fractional form. Each number in a proportion is called a term, with the first and fourth terms (in the above example, 2 and 21 respectively) referred to as the extremes and the second and third terms (7 and 6 respectively) referred to as the means. In any proportion, the product of the means equals the product of the extremes. This property permits the ratios to be checked (in the above case, $2 \times 21 = 6 \times 7 = 42$; hence the proportion is correct) and enables a missing term within a proportion to be determined.

Proportions indicating two ratios are equal

Problem: If examinations normally result in a failure rate of 5 per 100 students, how many failures can be expected if 3400 students sit for an examination?

Solution: $\dfrac{5}{100} = \dfrac{x}{3400}$

Students are represented in the numerators of both fractions. As the average ratio of 5:100 is repeated when the examination is taken by 3400 students, the two ratios are equal.

Let x equal the number of failures. As the product of the means equals the product of the extremes in a proportion:

$$100x = 5(3400)$$
$$100x = 17\,000$$
$$x = \dfrac{17\,000}{100}$$
$$x = 170$$

To check, substitute 170 for x:

$$\dfrac{5}{100} = \dfrac{170}{3400}$$
$$5(3400) = 170(100)$$
$$17\,000 = 17\,000$$

or both 5/100 and 170/3400 can be reduced to 1/20.

Problem: Lot 45 is a block of land that is 48 metres long by 20 metres wide. If a scale drawing is made of the lot, with 1 centimetre equalling 10 metres, find the length and width of the scaled drawing.

Solution: Since the scale is 1 centimetre = 10 metres, and the actual width of the lot is 20 metres, to determine the width of the land on the drawing let w = width. Hence:

$$\frac{1}{10} = \frac{w}{20}$$

As the product of the means equals the product of the extremes:

$$10w = 1(20)$$
$$w = \frac{20}{10}$$
$$w = 2$$

Thus, the width of the land, as seen on the scaled drawing, is 2 centimetres. To determine the length, let l = length. Hence:

$$\frac{1}{10} = \frac{l}{48}$$
$$10l = 1(48)$$
$$l = \frac{48}{10}$$
$$l = 4.8$$

Thus the length of the land, as seen on the scaled drawing, is 4.8 centimetres.

Proportions expressing comparison of a part to a whole

So far we have examined proportions that are expressed as ratios equal to one another. Proportions are also used to indicate the relationship of a part to a whole, normally presented in the form of a fraction. For example, a tennis player may receive 4/7 of his or her income from tournaments, 2/7 from sponsorships and 1/7 from exhibition matches (i.e. 4/7 + 2/7 + 1/7 = 7/7 = 1).

Problem: In 1997 a company recorded sales of $171 million. The allocation of these sales dollars is shown in Table 2A.4. Calculate the proportion of each sales dollar that is spent on each category listed.

Table 2A.4 *Revenue split-up for 1997*

Category	($ millions)
Materials	106
Staff	30
Taxes	22
Depreciation	5
Interest	1
Profit	7

Source: Hypothetical.

Solution: To determine the proportion of each sales dollar allocated, divide each expenditure area by the sales result of $171. Therefore:

Table 2A.5 *Revenue split-up per sales dollar*

Category	Revenue split-up	Per sales dollar
Materials	$\dfrac{106}{171}$	0.62
Staff	$\dfrac{30}{171}$	0.18
Taxes	$\dfrac{22}{171}$	0.13
Depreciation	$\dfrac{5}{171}$	0.03
Interest	$\dfrac{1}{171}$	< 0.01
Profit	$\dfrac{7}{171}$	0.04

Materials consume 62 cents of every sales dollar, staff 18 cents, taxes 13 cents and depreciation 3 cents. Interest payments are negligible, and profits are 4 cents of every sales dollar.

Problem: Moe, Larry and Curly bought a business for $1.2 million, with the partners' contribution to the purchase price in the ratio of 5:8:11 respectively. As none of the three wanted to be seen as a stooge, they agreed that profits would be distributed in the same proportion as their investment contribution.

(a) How much did each partner contribute?
(b) How much could each expect to receive if a profit of $135 000 was realised at the end of the first year?

Solution:
(a) First, determine the total number of parts contributing to the $1.2 million—that is:

$$5 + 8 + 11 = 24$$

Now, we must calculate each partner's share of the whole (24) and multiply that by the investment amount ($1.2 million). Hence we have:

```
Moe   =  5/24 × $1 200 000 = $  250 000
Larry =  8/24 × $1 200 000 = $  400 000
Curly = 11/24 × $1 200 000 = $  550 000
            Total          = $1 200 000
```

Business mathematics and statistics

(b) To determine the partners' share of the profits, multiply their respective ratios by the profit amount:

$$\begin{aligned}
\text{Moe} &= 5/24 \times \$135\,000 = \$\ 28\,125 \\
\text{Larry} &= 8/24 \times \$135\,000 = \$\ 45\,000 \\
\text{Curly} &= 11/24 \times \$135\,000 = \underline{\$\ 61\,875} \\
&\hspace{3.2cm} \text{Total} = \$135\,000
\end{aligned}$$

Problem: Four individuals formed a partnership to manufacture titanium bicycles. The initial capital outlay was $202 500, with Ms D. Kline investing twice as much as Ms B. Shore, who invested 30% more than Mr G. Wizz, whose investment was double that of Mr Y. Nott. It was agreed that Mr G. Wizz would act as manager on a salary of $65 000 p.a. and that each partner would receive interest of 12.5% p.a. on invested capital, with these expenses met before any calculation of net profit or loss was made. The profit or loss would be paid on the basis of the proportion of each individual's original investment to the total investment of $202 500. The gross profit for the first year was $425 000 (excluding Mr G. Wizz's salary). How much did each partner receive?

Solution: First, determine how much each partner contributed. Therefore, let:

$$x = \text{Mr G. Wizz's contribution}$$
(by choosing Mr G. Wizz, it makes calculation easier)
$$x + 0.3x = \text{Ms B. Shore's contribution}$$
$$2(x + 0.3x) = \text{Ms D. Kline's contribution}$$
$$0.5x = \text{Mr Y. Nott's contribution}$$

Therefore:

$$x + x + 0.3x + 2(x + 0.3x) + 0.5x = \$202\,500$$
$$x + x + 0.3x + 2x + 0.6x + 0.5x = \$202\,500$$
$$5.4x = \$202\,500$$
$$x = \frac{\$202\,500}{5.4}$$
$$= \$37\,500$$

Therefore:

$$\begin{aligned}
\text{Mr G. Wizz's contribution} &= x &&= \$\ 37\,500 \\
\text{Ms B. Shore's contribution} &= x + 0.3x &&= \$\ 48\,750 \\
\text{Ms D. Kline's contribution} &= 2(x + 0.3x) &&= \$\ 97\,500 \\
\text{Mr Y. Nott's contribution} &= 0.5x &&= \underline{\$\ 18\,750} \\
&\text{Total contribution} &&= \$202\,500
\end{aligned}$$

Now that each partner's individual contribution is known, the interest received on their investment can be determined. Each partner received 12.5% p.a. Therefore:

Chapter 2 – Percentages

$$\text{Investment} \times 12.5\% = \text{Interest received}$$

Mr G. Wizz   = $37 500 × 12.5% = $ 4 687.50
Ms B. Shore = $48 750 × 12.5% = $ 6 093.75
Ms D. Kline = $97 500 × 12.5% = $12 187.50
Mr Y. Nott   = $18 750 × 12.5% = $ 2 343.75

              Total interest received = $25 312.50

The disbursement of profits should now be determined:

Profit = Gross profit − Mr G. Wizz's salary
      = $425 000 − $65 000
      = $360 000

The profit of $360 000 must be distributed according to the proportion originally invested by each person.

Table 2A.6 *Amounts invested as a proportion of original outlay*

Investor	Amount invested ($)	Proportion of original outlay ($)
Mr G.Wizz	37 500	37.5/202.5 = 0.1852
Ms B. Shore	48 750	48.75/202.5 = 0.2407
Ms D. Kline	97 500	97.5/202.5 = 0.4815
Mr Y. Nott	18 750	18.75/202.5 = 0.0926
	202 500	202.5/202.5 = 1.000

Each partner's profit can now be determined to the nearest $100:

Partner's profit = Total profit × Proportion of original outlay
Mr G. Wizz   = $360 000 × 0.1852
                = $66 700
Ms B. Shore = $360 000 × 0.2407
                = $86 700
Ms D. Kline = $360 000 × 0.4815
                = $173 300
Mr Y. Nott   = $360 000 × 0.0926
                = $33 300

($66 700 + $86 700 + $173 300 + $33 300 = $360 000)

Finally, the year's earnings for each partner can be determined as follows:

	Salary	+	Interest	+ Profit share	= Total income
Mr G. Wizz =	$65 000	+	$ 4 687.50	+ $ 66 700	= $136 387.50
Ms B. Shore =	0	+	$ 6 093.75	+ $ 86 700	= $ 92 793.75
Ms D. Kline =	0	+	$12 187.50	+ $173 300	= $185 487.50
Mr Y. Nott =	0	+	$ 2 343.75	+ $ 33 300	= $ 35 643.75
Total =	$65 000	+	$25 312.50	+ $360 000	= $450 312.50

Business mathematics and statistics

2A.3 Summary

A ratio is the description of the relationship between two or more numbers, permitting a comparison between like or unlike measures. A ratio can be written as *a/b*, *a:b* or $\frac{a}{b}$. Ratios also allow more useful and accurate ways of understanding the relationship between whole numbers than, for example, decimals. For instance, recent statistics for the Sydney metropolitan area showed that one-seventh of families are single-parent families. Thus, one in every seven families is headed by a single parent (this is more easily understood than expressing the ratio 1/7 as a decimal 0.1428 and stating that there are 0.1428 single-parent families per family unit). Teachers often argue for smaller classes, saying that the ideal class size is 1:20 (teacher:students). If this ratio was presented in decimal form as 0.05 teachers/student, it would be difficult to put the ratio into its right perspective.

A proportion is a statement of equality between two ratios. A proportion may also express the comparison of a part to a whole, normally in fractional form. Each number in a proportion is referred to as a *term*, with the first and fourth terms called the *extremes*, and the second and third terms called the *means*. In any proportion, the product of the means equals the product of the extremes, which permits the determination of an unknown term when the other three are known.

Notes

1. Greg Maddox, 'The big picture', *Sydney Morning Herald*, 30 March 1996, p. 12s.
2. Cathy Bolt, 'The new-look food industry', *Australian Financial Review*, 4 April 1996, p. 42.
3. Alistair Adams-Smith, 'The feats to beat', *Sydney Morning Herald*, 17 July 1996, p. 44.
4. Greg Bearup, 'State's culpable driving deaths soar', *Sydney Morning Herald*, 20 July 1996, p. 8.
5. Australian Bureau of Statistics, *Year Book, Australia, 1996* (Cat. No. 1301.0).
6. Australian Bureau of Statistics, *Australian Social Trends, 1996* (Cat. No. 4102.0).
7. Ibid.
8. Australian Bureau of Statistics, *Year Book, Australia, 1996* (Cat. No. 1301.0).

Chapter 3
SIMPLE INTEREST

LEARNING OUTCOMES

After studying this chapter you should be able to:
- Understand the relationship between inflation and interest rates
- Describe the factors influencing the level of interest rates
- Perform calculations involving simple interest
- Manipulate the simple interest formula
- Distinguish between a flat and an effective rate of interest
- Estimate the effective rate of interest.

3.1 Interest rates

Interest is both the money paid for the use of money borrowed and the return to the depositor of funds for forgoing its use, and is stated as a percentage and/or dollar value. The amount actually borrowed or lent is referred to as the **principal**. The **nominal interest rate** is the rate of interest expressed in terms of current dollars (i.e. today's dollars), while the **real interest rate** is the rate of interest expressed in terms of inflation-adjusted value (i.e. the nominal rate less the rate of inflation). Rates of interest may vary on, for example, a business, housing or car loan. This can be attributed to varying degrees of risk on the respective loans, as well as the term (period of time) and size of the loans.

In Australia, interest rates have been highly volatile over the past decade. Interest rates are affected by several variables which, if understood, should assist you in assessing future interest-rate movements.

Inflation, measured by the general increase in the retail price level, has a direct bearing on interest rates. The higher the inflation rate, the higher the corresponding interest rate must be to attract prospective depositors. No-one wants an investment that is worth less in real terms at maturity than when originally deposited (i.e. a rate of return on the investment that is less than the rate of inflation). As inflation declines, there are greater prospects for reduced interest rates. For example, in 1992 interest rates in Australia hit new 30-year lows coinciding with marked reductions in inflation. However, as economic growth surged in 1994, the Reserve Bank, fearful of accompanying inflation, announced three interest rate increases within the space of four months. Then in 1996, when the March quarter inflation rate came in below expectations, talk immediately reverted to lower interest rates.

The size of the government's budget deficit or surplus influences the rate of interest. When the government has a budget deficit, it finances a large part of it by selling bonds and securities to the public and the financial sector; therefore, the larger the deficit, the greater the amount of bonds that must be sold. Hence, greater upward pressure is put on interest rates to attract a larger number of subscribers. If, on the other hand, the government has a budget surplus, this implies that the government will not have to raise as much money from selling bonds, especially if all or part of the surplus is used to retire previously incurred government debt. Thus, there would be a tendency for downward pressure on interest rates in the event of a budget surplus, with the size of the surplus influencing the degree of change in the interest rate level. This was clearly demonstrated in September 1987 when then Treasurer Keating announced that the 1987/88 federal budget would be balanced. Within half an hour of this announcement, one Australian bank announced a 0.5% cut in the home loan interest rate; the other banks followed in a matter of days.

Accordingly, competition between government, semi-government, corporate and consumer sectors for loan funds exerts pressure on interest rates; the greater the demand for funds by these sectors, the greater the pressure on the rates to escalate (the law of supply and demand).

The deregulation of Australia's financial system (i.e. reducing the degree of government intervention and permitting the market forces of supply and demand to operate) has included the floating of the Australian dollar, the entry of foreign banks, and interest rate levels determined by market forces. As a result, we find the Australian consumer and private enterprise being enticed to deposit and/or borrow funds from many financial intermediaries. For example, the entry of new non-banking financial institutions (such as Aussie Home Loans) into the home mortgage market foray has resulted in intense rivalry and banks' introduction of no-frills home loans at rates of up to 0.7% below the usual variable home loan interest rate. Therefore, the increased level of competition in the financial system assures the potential investor and/or borrower of competitive interest rates.

The level of deposits in financial institutions will also affect interest rates, as shown by the events subsequent to the October 1987 stockmarket crash. Many sharemarket investors became security-conscious and withdrew what funds remained in their share portfolio and deposited them in banks. This resulted in a dramatic increase in savings bank deposits, which contributed to a 1% cut in home loan interest rates shortly afterwards.

The value of the Australian dollar against other currencies (i.e. the foreign exchange rate) may also influence the level of interest rates. In November 1985, in an attempt to dampen domestic demand by increasing the cost of borrowing (i.e. raising interest rates), the Reserve Bank tightened monetary policy, hoping to attract at the same time foreign investment with the higher level of interest rates. Since the Australian dollar's exchange rate is determined by the market forces of supply and demand (as it was floated in December 1983), the increased foreign investment leads to an increased demand for the Australian dollar, hence an appreciation or increase in the value of the Australian dollar versus other currencies.

In November 1996 the Australian dollar nudged past the 81 US cents level, explained, in part, by the continued attractiveness of Australia's real interest rate level (i.e. the interest rate on offer discounted for the rate of inflation) when compared with other countries such as the United States, Germany and Japan. As a result, by mid-1996 the Australian dollar had reached five-year highs against the US greenback, with some economists predicting a breakthrough of the $0.85 barrier within 12 months.

Much foreign investment in Australia has been in the form of portfolio investment—that is, investment in interest-bearing deposits, company debentures and/or shares. The real interest rate has been the major attraction for both local and overseas investors.

Rates of change in wage levels have been a major concern of the Reserve Bank of Australia (RBA) as such changes ultimately impact on inflation (unless matched by corresponding increases in productivity) and, therefore, interest rates. In May 1997, the RBA reduced official interest rates when it was convinced that wage levels were only growing within the Bank's prescribed target range and, as such, would not fuel inflationary pressure.

A fall in interest rates, although welcomed by borrowers yet upsetting to lenders, is often accompanied by an increase in domestic demand, a portion of which spills over into increased imports. In fact, statistics suggest that in Australia for every 1% increase in domestic demand there is a corresponding 2–3% increase in import demand. As Australia already suffers from a huge deficit in its balance of payments, increased imports without a similar rise in exports will add to Australia's $190 billion net foreign debt. Therefore, upward pressure on interest rates was evident from mid-1988 as the Federal Government once again tightened monetary policy in an attempt to dampen domestic demand, and hence import demand. The success of this strategy, however, came at a high price—a recession (said to be the most severe economic downturn since the Depression) with an unemployment rate of over 11%).

Australia's balance of payments problem and corresponding rising foreign debt continues to plague our economy and to influence interest rate direction. For example, as our economy experiences economic growth, imports grow at a greater rate than domestic demand, which may well contribute to higher deficits in the current account (that part of the balance of payments dealing predominantly with the import and export of goods and services, interest payments going overseas for past debt, and dividends going overseas). Reductions in interest rates would be ill-conceived under these circumstances, since lower interest rates entice individuals and companies to borrow and, lo and behold, more is spent on imports. Therefore, although other indications may point to lower interest rates, a rising deficit in the current account may negate any such interest rate reduction.

Finally, in 1990, with the economy experiencing more than the 'soft landing we had to have', the Reserve Bank began to ease monetary policy, thereby reducing interest rates. This attempt to stimulate the economy occurred no less than 16 times from 1990 to 1993. This loosening of monetary policy, along with increasing federal budget deficits, helped to get the economy back on the track of economic growth (although there are many commentators who feel economic mismanagement prevented an earlier return with less trauma). By 1997, Australia had experienced the longest period of sustained economic growth since the end of the Second World War.

Thus, inflation and inflationary expectations, overseas interest rates, the government's budget, domestic demand, the state of the balance of payments, wage growth, the value of the Australian dollar, competition for funds, deposits, and the deregulation of the financial system all contribute to fluctuations in interest rates, which naturally affect the return on your investment (if you are the depositor) or the cost of borrowing.

3.1 Competency check

1. Briefly define what is meant by *interest rates*.
2. What is meant by the *principal* of a loan?
3. How would you explain the term *inflation*?
4. What is the effect of inflation on the value of your money?
5. What is the usual relationship between the inflation rate and the interest rate level?
6. How would you differentiate between nominal interest rates and real interest rates?

3.2 Calculation of simple interest

Simple interest is interest earned for the term of the loan or investment on the original principal only. For example, simple interest or a 'flat' rate of interest may be used for a short-term loan (e.g. a personal loan). The amount of simple interest (I) is calculated by considering the principal (P), the cost to you of the borrowed funds—that is, the rate of interest expressed as a percentage (R), and the length of time you will need to repay the loan (T). The time is expressed in terms of a year or part thereof.

Mathematically, simple interest is defined as:

$$\text{Dollar amount of interest} = \text{Principal} \times \text{Rate} \times \text{Time}$$
$$I = P \times R \times T$$

where
- I = total dollar amount of simple interest
- P = original amount borrowed or deposited (the principal)
- R = annual rate of interest (%)
- T = number of years or part thereof.

For example, to find the simple interest on $500 deposited at 9.5% p.a. for three years:

$$I = P \times R \times T$$
$$= \$500 \times 9.5\% \times 3$$
$$= \$142.50$$

Chapter 3 – Simple interest

Thus, $142.50 was earned over three years, or $47.50 was earned each year. Therefore, the 9.5% was applied only to the original principal of $500 for each year of the three years.

Problem: How much will Granny earn in simple interest if she deposits $10 000 for five years and three months at:

(a) 9.25% p.a.?
(b) 7.5% p.a.?

Solution:

(a) $I = P \times R \times T$
$ = \$10\,000 \times 9.25\% \times 5.25$
$ = \$4856.25$

(b) $I = P \times R \times T$
$ = \$10\,000 \times 7.5\% \times 5.25$
$ = \$3937.50$

(As expected, Granny must strive to get the highest interest rate possible; therefore, option (a) would be chosen.)

Sometimes you may have to solve for an unknown other than *I*. In this case, the algebraic rules of transposition must be applied:

Given $\quad I = P \times R \times T$

then, $\quad \dfrac{I}{PT} = R$

and $\quad \dfrac{I}{PR} = T$

and $\quad \dfrac{I}{RT} = P$

Problem:

(a) Dana has just received $400 in interest after having an account for four years in which she had deposited $2000. Calculate the simple interest rate she received.
(b) What simple rate of interest did Monique earn on her $2500 deposit of eight years ago if she has just collected interest of $1500?

Solution:

(a) $I = P \times R \times T$ (b) $I = P \times R \times T$

This time, we must determine the simple rate of interest (*R*); therefore, solve for *R*.

In (a) $I = 400$, $P = 2000$ and $T = 4$.   (b) $I = 1500$, $P = 2500$ and $T = 8$

Business mathematics and statistics

$$\frac{I}{PT} = R \qquad\qquad \frac{I}{PT} = R$$

$$\frac{400}{2000 \times 4} = R \qquad\qquad \frac{1500}{2500 \times 8} = R$$

$$\frac{400}{8000} = R \qquad\qquad \frac{1500}{20\,000} = R$$

$$0.05 = R \qquad\qquad 0.075 = R$$

$$5\% = R \qquad\qquad 7.5\% = R$$

Dana (a) received a simple rate of interest of 5%, while Monique (b) received a 7.5% simple interest rate.

Problem: How long must an investor leave her $7500 in an account in order to have a maturity value of $12 000 if the simple rate of interest is:

(a) 9% p.a.?
(b) 7% p.a.?

Solution: To solve for T, first solve for I:

Accumulated interest = Maturity value (i.e. Sum of interest + Principal) − Principal
= $12 000 − $7500
= $4500

Now solve for T:

(a) where $R = 9\%$ $\qquad\qquad$ (b) where $R = 7\%$

$\qquad\qquad I = P \times R \times T \qquad\qquad\qquad\qquad I = P \times R \times T$

$\qquad\qquad 4500 = 7500 \times 9\% \times T \qquad\qquad 4500 = 7500 \times 7\% \times T$

$$\frac{4500}{675} = T \qquad\qquad\qquad \frac{4500}{525} = T$$

$\qquad\qquad 6.67 \text{ years} = T \qquad\qquad\qquad 8.57 \text{ years} = T$

Therefore, after 6.67 years (6 years and 8 months), $7500 will amount to $12 000 at 9% p.a. simple interest, and at 7% simple interest it would take 8.57 years. You may check the answer by solving for the rate of interest. For example, let's check the first answer of 6.67 years:

$$I = P \times R \times T$$

$$\frac{I}{PT} = R$$

$$\frac{4500}{7500 \times 6.67} = R$$

$$\frac{4500}{50\,025} = R$$

$$0.0899 = R$$

$$9\% = R$$

You can also check that the answer for (b) is correct. The following example illustrates yet another way of solving the same problem.

If you are to determine the **maturity value** (S) of the principal, let:

$$S = \text{Principal } (P) + \text{Interest } (I)$$

As $I = PRT$, the I in the above equation of $P + I$ can be replaced by PRT. Hence:

$$S = P + PRT$$
$$= P(1 + RT)$$

S is therefore the maturity value of the principal and includes the principal and accumulated interest. To solve for P:

$$P = \frac{S}{1 + RT}$$

Problem: Laurie calculated that he will need $15 500 in five years in order to be clear of all debts by then. How much should he invest today to have a maturity value of $15 500 if the simple interest rate is:

(a) 10.25%? (b) 6.75%?

Solution:

(a) $P = \dfrac{S}{1 + RT}$

$= \dfrac{\$15\ 500}{1 + (10.25\% \times 5)}$

$= \dfrac{\$15\ 500}{1.5125}$

$= \$10\ 247.93$

(b) $P = \dfrac{S}{1 + RT}$

$= \dfrac{\$15\ 500}{1 + (6.75\% \times 5)}$

$= \dfrac{\$15\ 500}{1.3375}$

$= \$11\ 588.79$

As you would expect, Laurie needs a larger deposit the lower the level of interest rate applying.

Each of the answers to the problem cited above can be checked by substituting the calculated variable into the equation $I = PRT$ and solving for another variable. For example, in the above problem, you could solve for I in (a):

$$I = PRT$$
$$= \$10\ 247.93 \times 10.25\% \times 5$$
$$= \$5252.06$$

If this interest of $5252.06 is added to the principal of $10 247.93, we get $15 499.99, which corresponds to the amount listed in the original problem (a difference of 1 cent due to rounding).

Problem: Twelve years ago the Simpsons received an inheritance of $16 000 and decided to invest at the fixed simple interest rate of 14% p.a. The investment matures today.

(a) What is the maturity value of their investment?
(b) How much interest accumulated over the 12-year period?

Solution:

(a) $S = P(1 + RT)$
$= \$16\,000 (1 + (14\% \times 12))$
$= \$16\,000 (1 + 1.68)$
$= \$16\,000 (2.68)$
$= \$42\,880$

The maturity value was $42 880.

(b) Interest accumulated (I) = Maturity value (S) – Principal (P)
$= \$42\,880 - \$16\,000$
$= \$26\,880$

Problem: Charles is interested in doubling his money in eight years. If he has $5000 to invest, what simple rate of interest will ensure that he reaches his goal?

Solution: Charles wants to have $10 000 in eight years' time (he wants to double his $5000 investment), so we must solve for R. Therefore:

$$S = P(1 + RT)$$
$$S = P + PRT$$
$$\frac{S - P}{PT} = R$$
$$\frac{10\,000 - 5000}{5000 \times 8} = R$$
$$\frac{5000}{40\,000} = R$$
$$\frac{1}{8} = R$$
$$12.5\% = R$$

A simple interest rate of 12.5% p.a. will ensure that Charles can double his money in eight years.

Problem: How long will it take Granny to double her money if she deposits $9000 at a simple interest rate of 9.5% p.a.? Can she afford to wait so long?

Solution: Again, we use the maturity value formula (as Granny wants to double her $9000 investment) and we must solve for T.

Chapter 3 – Simple interest

$$S = P + PRT$$
$$\frac{S - P}{PR} = T$$
$$\frac{18\,000 - 9000}{9000 \times 9.5\%} = T$$
$$\frac{9000}{855} = T$$
$$10.53 \text{ years} = T$$

Another method can be used to calculate the time: since Granny hopes to earn interest of $9000, we have only to substitute into the formula $I = PRT$ and solve for T.

3.2 Competency check

1. What is meant by the *simple interest* on a loan?
2. What is the formula used to determine the total amount of simple interest received?
3. Determine the interest on a principal of $20 000 deposited for five years at 8.5% p.a. simple interest.
4. Find the simple interest rate which will result in a principal of $5000 amounting to $6450 after four years.
5. After what period of time will a principal of $1450 amount to $2374.38 at 8.5% p.a. simple interest?
6. What principal will amount to $18 177.50 after 7.25 years at 9% p.a. simple interest?

3.3 Flat versus effective rate of interest

In applying for a short-term loan—for example, a personal loan—a flat rate of interest is often cited. However, owing to increased recognition of the need for improved consumer protection, most States now make it mandatory to include the **effective rate of interest** in the final contract. This is because the flat rate of interest on a loan is at a far higher 'effective' rate of interest than the flat rate implies and, as such, it is deceptive for the prospective borrower.

For example, if you were to take out a personal loan of $5000 over five years at 10% p.a., the interest on the loan would be:

$$\begin{aligned} I &= PRT \\ &= \$5000 \times 10\% \times 5 \\ &= \$2500 \end{aligned}$$

As the principal of $5000 and interest of $2500 are to be paid over five years, the monthly repayments would be $125 ($7500 ÷ 60 months), of which $41.67 would be interest and $83.33 principal ($2500/60 = $41.67; $5000/60 = $83.33). At the end of the first year of the loan you will have paid back $1000 of the principal ($83.33 × 12), at the end of the second year $2000 of the principal ($83.33 × 24), and so on. However, the rate of interest of 10% applies to the principal of $5000 (or $500) for the duration of the loan. Therefore, at the beginning of the third year of the loan, for example, only $3000 of the principal is owed, yet you are paying interest of $500 on that $3000, which over the year is at an interest rate of 20%. (As $3000 is owed at the beginning of the third year and $2000 at the end of the third year, an average of $2500 is owed for the year with $500 in interest paid; the effective rate of interest is therefore $500/$2500 = 20%.) Table 3.1 shows the effective interest rate for each of the five years.

Table 3.1 *Effective rate of interest*

Year	Amount of principal owing at beginning of year ($)	Mean amount of principal owing during year ($)	Amount of interest p.a. ($)	Effective interest rate (%)
1	5000	4500	500	11.11
2	4000	3500	500	14.29
3	3000	2500	500	20.00
4	2000	1500	500	33.33
5	1000	500	500	100.00
6	0			

The overall effective rate indicates the actual rate of interest as if each repayment was made on the remaining loan balance. The effective rate of interest in Table 3.1 is almost double the stated 10% flat rate and, as indicated earlier, recognition of the need for consumer protection now makes it obligatory to include the effective rate of interest paid over the term of the loan in documentation signed by the borrower. To determine the approximate effective rate of interest on a designated flat rate of interest, use the following formula:

$$ER = \frac{2N \times R}{N + 1}$$

where ER = effective rate of interest
 N = number of repayments
 R = flat rate of interest.

In the example of the $5000 personal loan repaid monthly over five years (i.e. 60 repayments) at a 10% p.a. flat rate of interest, the approximate effective rate of interest is:

$$ER = \frac{2(60) \times 10\%}{60 + 1}$$
$$= \frac{1200\%}{61}$$
$$= 19.67\%$$

Therefore, the effective rate of interest of 19.67% is almost twice the cited 10% flat rate.

Problem: A $10 000 loan is taken out over five years at 8% p.a.

(a) Calculate the total simple interest on the loan.
(b) If the repayments are monthly, what is the effective rate of interest?

Solution:

(a) $I = PRT$
$= \$10\,000 \times 8\% \times 5$
$= \$4000$

The total simple interest is therefore $4000.

(b) $ER = \dfrac{2N \times R}{N + 1}$
$= \dfrac{2(60) \times 8\%}{60 + 1}$
$= \dfrac{960\%}{61}$
$= 15.7\%$

In other words, the 8% flat rate over five years is equivalent to an effective rate of 15.7%.

Problem: B-Ware Usury Pty Ltd have lent Granny $12 000 for seven years with monthly payments at an effective rate of 36% p.a. What is the equivalent flat rate of interest?

Solution: $ER = \dfrac{2N \times R}{N + 1}$

Since we wish to determine the simple rate of interest (R), solve for R in the above equation N is the number of payment periods, which is 84 (7 years × 12/year).

$R = \dfrac{ER \times (N + 1)}{2N}$
$= \dfrac{36(84 + 1)}{2(84)}$
$= \dfrac{36(85)}{168}$
$= \dfrac{3060}{168}$
$= 18.2\%$

Poor Granny has been taken for an effective ride, as the flat rate of 18.2% is equivalent to an approximate effective rate of 36%.

Business mathematics and statistics

3.3 Competency check

1. What is meant by a *flat rate of interest*?
2. How would you define the *effective rate of interest*?
3. Calculate the effective rate of interest (to one decimal place) for a three-year loan with monthly repayments at 10% p.a.
4. Calculate (a) the total simple interest and (b) the effective rate of interest (to one decimal place) if $15 000 is borrowed over five years at 9% p.a. with monthly repayments.
5. Calculate (a) the total simple interest and (b) effective rate of interest (to one decimal place) for $7500 borrowed over two years at 8.5% p.a. with monthly repayments.
6. Calculate the equivalent flat rate of interest (to the nearest percentage) for a personal loan to be repaid monthly over four years at an effective rate of 22%.

3.4 Summary

Interest is the price paid for the use of money and is stated as a percentage. The actual funds borrowed or lent is known as the principal. The varying interest rate levels for different types of loans can be explained by the degree of risk, as well as the size and term (usually in years) of the loan. Interest rate movements may be influenced by such factors as inflation and inflationary expectations, overseas interest rates, the value of the Australian dollar and the competition for funds.

Simple interest or a flat rate of interest is calculated on the principal of a loan. The formula for calculating simple interest is:

$$\text{Dollar amount of interest } (I) = \text{Principal } (P) \times \text{Rate } (R) \times \text{Time } (T)$$
$$I = PRT$$

One may manipulate the formula to solve for any of the four variables (i.e. I, P, R or T).

To determine the principal when the amount paid at maturity, the rate of interest, and the time over which the interest accrued are known, use the formula:

$$\text{Principal } (P) = \frac{\text{Sum of the principal and interest } (S)}{1 + (\text{Rate } (R) \times \text{Time } (T))}$$

$$P = \frac{S}{1 + RT}$$

The effective rate of interest is the rate of interest actually paid when a flat rate is cited. To determine the effective rate of return, use the formula:

$$ER = \frac{2(\text{Number of repayments }(N)) \times \text{Flat rate of interest }(R)}{\text{Number of repayments }(N) + 1}$$

$$ER = \frac{2N \times R}{N + 1}$$

The effective rate is just less than double the flat rate, and this can be explained, in part, by recognising that when calculating simple interest, the interest rate is always applied on your original principal (or loan) rather than on the balance that would exist if your repayments had been deducted from the original principal.

3.5 Multiple-choice questions

1. The effective rate of interest is approximately how many times the flat rate of interest?

   (a) The rates are the same for both.   (b) double
   (c) triple   (d) five times

2. If inflation is 5% and the nominal interest rate you receive for your deposit is 9%, you are realising a real return of approximately:

   (a) 14% on your money   (b) 5% on your money
   (c) 4% on your money   (d) None of the above.

3. A baker has to repay a $50 000 loan (a fair bit of dough in anyone's language) monthly over five years at a simple rate of interest of 7.5% p.a. The total simple interest on the loan was:

   (a) $187 500   (b) $18 750   (c) $1875   (d) $68 750

4. Mr I.M. Payscent is anxiously awaiting the time when his $5625 deposit receiving 8% p.a. simple interest will reach a maturity value of $10 000. How long must Mr. Payscent wait (to one decimal place)?

   (a) 5 years   (b) 5.5 years   (c) 9.7 years   (d) 12 years

5. J.P. Getty wanted to ensure he was well looked after at retirement and calculated that he needed $500 000 at age 65 years to live comfortably thereafter. How much (to the nearest dollar) should Mr Getty invest today if he retires in seven years and a simple interest rate of 8.5% p.a. applies?

   (a) $313 480   (b) $297 500   (c) $186 520   (d) $202 500

3.6 Exercises

1. What is meant by the *nominal interest rate*?

2. How does the nominal rate of interest differ from the real rate of interest?

3. What happens to the purchasing power of your deposit if the interest rate you receive on your deposit is less than the prevailing inflation rate?

Business mathematics and statistics

4. There is no single factor which explains today's interest rate level. What do you think have been the more prominent factors affecting interest rates over the past year?

5. Do you believe it should be compulsory for all flat rates to be converted to effective rates of interest? Briefly discuss your answer.

6. (i) What is meant by the *maturity value*?
   (ii) What is the formula for determining the maturity value of the principal?

7. What is the formula for the effective rate of interest?

8. Determine the amount of simple interest for the following:

   (a) $10 000 principal at 10% p.a. over 10 years;
   (b) $5000 principal at 8% p.a. over 292 days (assume 365 days in a year);
   (c) $6500 principal at 9.25% p.a. over 2 years and 3 months;
   (d) $2450 principal at 7.5% p.a. over 5 years.

9. Find the simple interest rate which will result in:

   (a) a principal of $3000 amounting to $3450 after 3 years;
   (b) a principal of $6000 amounting to $7125 after 2 years and 6 months;
   (c) a principal of $17 500 amounting to $30 976 after 7 years;
   (d) a principal of $24 250 amounting to $40 158 after 6 years and 146 days.

10. After what period of time will:

    (a) a principal of $4250 amount to $5100 at 8% p.a. simple interest?
    (b) a principal of $1000 amount to $1625 at 12.5% p.a. simple interest?
    (c) a principal of $115 000 amount to $143 980 at 10.5% p.a. simple interest?
    (d) a principal of $35 000 amount to $46 025 at 9% p.a. simple interest?

11. What principal will amount to:

    (a) $18 125 after 6 years at 7.5% p.a. simple interest?
    (b) $7793.75 after 2.6 years at 9.5% p.a. simple interest?
    (c) $14 301 after 73 days at 10.75% p.a. simple interest?
    (d) $9500 after 11 years and 3 months at 8% p.a. simple interest?

12. What is the effective rate of interest on a loan (to one decimal place):

    (a) over 5 years at 9% p.a. simple interest, repaid monthly?
    (b) over 3 years at 9% p.a. simple interest, repaid monthly?
    (c) over 1 year at 9% p.a. simple interest, repaid monthly?
    (d) over 4.5 years at 9% p.a. simple interest, repaid monthly?
    (e) over 5 years at 9% p.a. simple interest, repaid quarterly?
    (f) over 5 years at 9% p.a. simple interest, repaid twice yearly?

13. Christine bought a stereo system for $1750 with a deposit of $250. The residual is to be paid over 15 months at a flat rate of interest of 12% p.a.

    (a) What is Christine's monthly repayment?
    (b) What is the effective rate of interest?

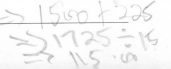

Chapter 3 – Simple interest

14. A video-cassette recorder is advertised at $650. For customers buying it on terms, it is 10% down with repayments of $26 per month over three years.
    (a) What is the flat rate of interest?
    (b) How much interest is paid?
    (c) What is the effective rate of interest?
    (d) What is the total cost of the video-cassette recorder to the customer?

15. Monique, Dana and Robyn invested $8000, $12 000 and $14 000 respectively. The total amount returned a simple interest of 8% p.a. for a period of four years. Calculate the total amount (i.e. interest plus principal) each receives after the investment period.

16. How long would it take $10 000 invested at 10.25% p.a. flat to earn $3485 in simple interest?

17. If $25 000 is invested at 9.75% p.a. simple interest for 10 years, how much interest will have accumulated at the end of the 10 years?

18. How long will it take for $2500 to amount to $3616.25 at 11.75% p.a. simple interest?

19. The Forgutin family bought a car on terms. Its cash price was $19 500 and the Forgutins paid a deposit of 10%. It is to be paid off monthly over a five-year period, with monthly repayments of $511.88.
    (a) What is the total amount of interest payable over the term of the loan?
    (b) What is the simple or flat rate of interest?
    (c) What is the effective rate of interest?

20. Calculate the equivalent flat rate of interest (to the nearest percentage) for a personal loan to be repaid monthly over five and a half years at an effective rate of 16%.

Chapter 4
COMPOUND INTEREST

LEARNING OUTCOMES
After studying this chapter you should be able to:
• Define compound interest
• Distinguish between simple and compound interest
• Compare the calculations of simple and compound interest
• Calculate compound interest (i.e. future value of a deposit)
• Use the Amount at compound interest table
• Solve problems involving transposing compound interest formulae.

4.1 Simple versus compound interest

Most individuals who have a savings or investment account with a financial institution leave their deposits in the institution for a period of time, expecting to accrue an increased amount of interest as more time elapses. If the deposits are made in an account carrying a flat rate of interest calculated on the original deposit for the entire duration of the account, a **simple interest** has been applied to this account—that is, the interest received for each period has been based solely on the original deposit or principal.

All accounts at financial institutions, in fact, accrue interest on the principal, plus interest received from previous periods, and the sum of the original principal plus total interest earned is called the **accumulated value**. The interest is said to be compounded, since interest is paid on interest earned from a previous period(s). Therefore, as you would expect, a **compound rate of interest** applying to a deposit generates a greater interest amount than an identical simple rate of interest applying to that same deposit. This is a direct result of the compound rate of interest applying to the principal plus accumulated interest, versus the simple interest which earns interest only on the original principal.

The terms 'future value' and 'accumulated value' are often used to refer to the sum of the principal plus the interest expected at the end of a period(s), while the difference between this accumulated value and the original principal is accordingly referred to as the compound interest accrued. In other words, the **future value** of a

Business mathematics and statistics

principal or deposit is the total amount in the account at the conclusion of the term and includes the original deposit (or principal), as well as all the interest that has accumulated over the term of the deposit.

Calculation of simple interest and compound interest compared

To compare and contrast the calculations of simple and compound interest, it is best to examine the application of an identical rate of interest to both simple and compound interest applying on a principal.

Problem: Mr I. Sighte deposited $10 000 in an account carrying an interest rate of 10% p.a. What amount would Mr I. Sighte have earned after four years if:

(a) He received simple interest?
(b) He received compound interest?

Solution:

(a) The formula for simple interest (which you no doubt recall) is:

Interest amount (I) = Principal (P) × Rate of interest (R) × Period of time (T)

$$\begin{aligned} I &= P \times R \times T \\ &= \$10\,000 \times 10\% \times 4 \\ &= \$10\,000 \times .10 \times 4 \\ &= \$4000 \end{aligned}$$

Therefore, a total of $4000 was received in interest for the $10 000 deposit left over four years. As seen in Table 4.1, $1000 of simple interest was received at the end of each of the four years.

Table 4.1 *Simple interest and compound interest of 10% p.a. on a principal of $10 000 for four years*

Year	(a) Simple	(b) Compound
1	$10 000 (*P*) × 0.10 (*R*)	$10 000 (*P*) × 0.10 (*R*)
2	$1 000 (*I*) $10 000 (*P*) × 0.10 (*R*)	$1 000 (*I*) $11 000 (*P*) × 0.10 (*R*)
3	$1 000 (*I*) $10 000 (*P*) × 0.10 (*R*)	$1 100 (*I*) $12 100 (*P*) × 0.10 (*R*)
4	$1 000 (*I*) $10 000 (*P*) × 0.10 (*R*) $1000 (*I*)	$1 210 (*I*) $13 310 (*P*) × 0.10 (*R*) $1331 (*I*)

(b) To determine the compound interest received over the four-year period, the interest received at the end of each year must be added to the original principal. Therefore, at the end of the first year, for example, the $10 000 deposit has generated $1000 in interest (i.e. 10% of $10 000) which results in an accumulated value of $11 000 at the start of the second year. The 10% interest now applies to this $11 000 (see Table 4.1).

At the end of the second year, another $1100 interest has been generated as a result of the 10% applying to the $11 000. In effect, to determine the compound interest in Table 4.1, the principal of $10 000 was multiplied by 1.1 four times. Hence, the 10% interest compounded annually was equal to:

$$\$10\,000(1.1)(1.1)(1.1)(1.1) \text{ or } \$10\,000(1.1)^4$$

If Mr I. Sighte had deposited the funds for eight years, the compound interest would have been calculated as:

$$\$10\,000(1.1)(1.1)(1.1)(1.1)(1.1)(1.1)(1.1)(1.1) \text{ or } \$10\,000(1.1)^8.$$

Thus, from the above example, we can extract an appropriate formula for deriving the compound interest without going through this lengthy process of calculating the interest at the conclusion of each period. The formula for compound interest is:

$$S = P(1+i)^n$$

where S = sum of the principal and the compounded interest,
P = principal
i = interest rate per period
n = number of periods.

Therefore, to calculate the interest earned on Mr I. Sighte's $10 000 over a period of four years at 10% p.a.:

$$\begin{aligned} S &= \$10\,000(1 + 0.1)^4 \\ &= \$10\,000(1.4641) \\ &= \$14\,641 \end{aligned}$$

Therefore, interest received was $4641 (i.e. $14 641 − $10 000). You may note that the same answer is derived in Table 4.1 by adding the annual interest accrued (i.e. $1000 + $1100 + $1210 + $1331).

Similarly, to determine the interest received on the $10 000 deposit over eight years at 10% p.a., use the formula:

$$S = P(1+i)^n$$

Therefore:

$$\begin{aligned} S &= \$10\,000(1 + 0.1)^8 \\ &= \$10\,000(2.14358881) \\ &= \$21\,435.89 \end{aligned}$$

Note: Interest received was $11 435.89 (i.e. $21 435.89 − $10 000).

Interest is normally compounded monthly by financial institutions, although an increasing number of institutions are compounding deposits on a daily basis. When calculating compound interest for a period of less than a year, the interest rate (i) and time periods (n) in the formula must be adjusted, as the interest rate is quoted on an annual basis and must be altered to allow for the period of time in question. For example, an interest rate of 10% p.a. compounded on a quarterly basis results in an interest rate of 2.5% per quarter (i.e. 10/4), while a rate of 10% p.a. calculated on a monthly basis results in an interest rate of 0.833% (e.g. 10/12%) per month. The number of periods must also be adjusted. If, for example, interest is compounded quarterly over eight years, there are 32 periods under consideration (4×8), while another deposit compounded monthly over three years has 36 periods (12×3).

Problem: Mr I. Sighte deposited his $10 000 in a bank for four years, paying 10% p.a. interest compounded semi-annually. How much interest did Mr I. Sighte earn?

Solution: First, calculate the interest rate per period and the number of interest rate periods. As 10% is the annual rate for two periods per year, the interest rate per period must be 5% ($10\% \div 2 = 5\%$). The number of interest rate periods is 8 (4 years $\times$ 2 periods per year). To determine the amount of compound interest, substitute into the formula:

$$S = P(1 + i)^n$$
$$= \$10\,000(1.05)^8$$
$$= \$10\,000(1.05)(1.05)(1.05)(1.05)(1.05)(1.05)(1.05)(1.05)$$
$$= \$10\,000(1.4774554)$$
$$= \$14\,774.55$$

As the amount of compound interest is the accumulated value of the principal and accrued interest, the interest received is $4774.55 ($14 774.55 − $10 000). Note that the interest earned semi-annually is higher than that earned annually (i.e. $4641). Therefore, Mr I. Sighte would have earned:

(a) $4000 ($1000 + $1000 + $1000 + $1000) on 10% p.a. simple interest;
(b) $4641 ($1000 + $1100 + $1210 + $1331) on 10% p.a. compounded annually;
(c) $4774.55 on 10% p.a. interest compounded twice yearly.

Therefore, as expected, at a given principal, rate of interest and term, simple interest generates the least amount of accumulated interest, while the greater the number of times the interest rate is compounded annually, the larger the interest amount received.

Amount at Compound Interest Table

The larger the number of periods considered, the more tedious the calculation. For example, to calculate the future value of $15 000 invested over 20 years at 8% p.a. compounded quarterly, a total of 80 periods (20 years $\times$ 4 quarters per year) would have to be considered. Note that the rate is 2% per period (i.e. 8% p.a. with interest compounded quarterly, or 8/4% or 2% interest received per quarter). That is:

$$S = P(1 + i)^n$$
$$= \$15\,000(1 + 0.02)^{80}$$

Imagine having to multiply 1.02 by 1.02 eighty times. Do not despair! There is a time-saver available—an Amount at Compound Interest Table (see Appendix A at the back of the book) which gives the values for $(1 + i)^n$. The value is extracted from the table and substituted into the equation $S = P(1 + i)^n$.

Therefore, to find the future value of the $15 000 deposit, locate 2% (the interest rate) at the top of the table and read down the left-hand side to locate the appropriate number of periods, 80. As there are only 60 periods contained within the table and we need the 80th period, apply the law of exponents. That is:

$$(1 + 0.02)^{80} = (1.02)^{40} \times (1.02)^{40}$$
$$= (2.20803966) \times (2.20803966)$$
$$= 4.87543914$$

Note that you can use any combination of exponents to get the same answer. For example:

$$(1 + 0.02)^{80} = (1.02)^{30} \times (1.02)^{50}$$
$$= (1.81136158) \times (2.69158803),$$
$$= 4.87543915$$

Hence:

$$S = P(1 + i)^n$$
$$= \$15\,000(1 + 0.02)^{80}$$
$$= \$15\,000(4.8754391)$$
$$= \$73\,131.59$$

The original principal of $15 000 accrued compound interest of $58 131.59 over the 20-year period (i.e. the accumulated value of $73 131.59 minus the original principal of $15 000).

Problem: Determine the future value of $5000 invested at 12% p.a. interest compounded monthly over five years.

Solution: Interest rate per period = 12%/12 = 1% per period
Number of periods = 5 years × 12 periods/year = 60

Therefore:

$$S = P(1 + i)^n$$
$$= \$5000(1 + 0.01)^{60}$$

The Amount of Compound Interest Table indicates that 1.01 for 60 periods (i.e. $(1 + 0.01)^{60}$) is equivalent to 1.81669670. Hence:

$S = \$5000(1.81669670)$
$= \$9083.48$

Therefore, the $5000 deposited for five years accumulated in value to $9083.48 with $4083.48 as the compound interest received.

Some of you may already own financial calculators which allow you to calculate future and present values, since such calculators have built-in functions incorporating the compound interest formula. However, it is important for you to understand the logic behind such calculations.

4.1 Competency check

1. (a) If you had a choice between depositing your principal in accounts exhibiting identical rates of interest, would you choose a simple or compound interest generating account? Briefly explain your answer.
   (b) What is meant by the *future* or *accumulated value*?

2. What is the interest rate per period if a deposit attracts 12% p.a. interest compounded:

   (a) semi-annually? (b) quarterly?
   (c) monthly? (d) daily?

3. How many interest rate periods are there if a deposit is compounded monthly over:

   (a) two years? (b) three years and six months?
   (c) six months? (d) five years and 11 months?

4. $7500 is deposited in an account for three years at a rate of 9% p.a. Calculate the interest received (to the nearest dollar) in:

   (a) a simple interest bearing account;
   (b) a compound interest bearing account.

5. Determine the accumulated value and compound interest for the following:

   (a) $1000 at 9% p.a. compounded monthly over three years;
   (b) $25 000 at 10% p.a. compounded quarterly over two years;
   (c) $50 at 12% p.a. compounded semi-annually over 10 years;
   (d) $275 at 8% p.a. compounded quarterly over three years.

4.2 Present value

Have you ever wondered how much you would have to deposit today in an account at a certain rate of interest to receive a predetermined accumulated amount at a future date? If, for example, you hope to have a deposit of $20 000 for a home within five years, you must ascertain how much must be deposited today at the appropriate interest

Chapter 4 – Compound interest

rate (e.g. 10% p.a. interest compounded quarterly) if it is to realise $20 000 in five years.

The amount which you must invest today to realise a predetermined total in five years is referred to as the *present value*. Using the compound interest formula $S = P(1 + i)^n$, P is the present value of the funds which must be deposited over n periods at i interest rate per period in order to realise the total amount of S. To solve the present value (P) algebraically:

$$S = P(1 + i)^n$$
$$\frac{S}{(1 + i)^n} = P$$
$$S(1 + i)^{-n} = P$$

To determine the present value, examine the Present Value at Compound Interest Table (Appendix B at the back of the book) which provides the present-value factor (i.e. $(1 + i)^{-n}$). Then solve for P in the formula $P = S(1 + i)^{-n}$.

Therefore, to find the present value of $20 000 needed in five years at 10% p.a. interest compounded quarterly:

$$\text{Rate of interest per period} = 10\%/4 = 2.5\%$$
$$\text{Number of periods} = 5 \text{ years} \times 4 \text{ periods/year} = 20$$

Therefore:

$$P = S(1 + i)^{-n}$$
$$= \$20\,000(1 + 0.025)^{-20}$$

The value of $(1 + 0.025)^{-20}$ in the Present Value at Compound Interest Table is .61027094. Hence:

$$P = \$20\,000(.61027094)$$
$$= \$12\,205.42$$

In other words, $12 205.42 invested today (the present value) at 10% p.a. interest will result in an accumulated or future value of $20 000 in five years.

To check the answer, find the accumulated value (S) in the equation if $P = \$12\,205.42$, $i = 2.5\%$ and $n = 20$.

$$S = P(1 + i)^n$$
$$= \$12\,205.42(1 + 0.025)^{20}$$

The value which corresponds to $(1 + 0.025)^{20}$ in the Amount at Compound Interest Table is 1.63861644. Hence:

$$S = \$12\,205.42(1.63861644)$$
$$= \$20\,000$$

As $20 000 is the targeted future value, the answer is correct.

Problem: How much must Mr A. Musse invest in an account receiving 8% p.a. interest compounded quarterly in order to accumulate $15 000 in three years?

Business mathematics and statistics

Solution: Rate of interest per period = 8%/4 = 2%
Number of periods = 3 years × 4 periods/year = 12

Therefore:

$$P = S(1 + i)^{-n}$$
$$= \$15\,000(1 + 0.02)^{-12}$$

Looking up the Present Value at Compound Interest Table:

$$P = \$15\,000(.78849318)$$
$$= \$11\,827.40$$

Therefore, $11 827.40 invested at 8% p.a. interest compounded quarterly will result in interest of $3172.60 and, therefore, the desired accumulated amount of $15 000.

Check the answer:

$$S = P(1 + i)^n$$
$$= \$11\,827.40(1 + 0.02)^{12}$$

Looking up the Amount at Compound Interest Table:

$$S = \$11\,827.40(1.26824179)$$
$$= \$15\,000$$

Calculation of the interest rate

We have used the equation $P = S(1 + i)^n$ to solve for S and for P, but what if you wanted to determine the rate of interest you would need in order to realise a certain future value on the basis of today's investment? Look no further than the next line, where we again use our compound interest equation but now solve for I:

$$S = P(1 + i)^n$$

Divide both sides by P:

$$\frac{S}{P} = \frac{P(1 + i)^n}{P}$$

Take the nth root of both sides:

$$\sqrt[n]{\frac{S}{P}} = \sqrt[n]{(1 + i)^n}$$

$$\sqrt[n]{\frac{S}{P}} = 1 + i$$

To solve for i, transpose the number 1 from the right side to the left side of the equation.

$$\sqrt[n]{\frac{S}{P}} - 1 = i$$

Problem: What interest rate compounded quarterly will result in a principal of $10 000 realising a future value of $13 469 in five years?

Solution: Interest rate per period = i
Number of periods = 5 years × 4 periods/year = 20
Principal (P) = $10 000
Accumulated sum = $13 469

Therefore:

$$i = \sqrt[n]{\frac{S}{P}} - 1$$

$$= \sqrt[20]{\frac{13\,469}{10\,000}} - 1$$

$$= (1.0150017) - 1$$

$$= 0.015$$

Therefore, an interest rate of 1.5% per quarter or 6% p.a. will result in an investment of $10 000 realising an accumulated value of $13 469 in five years. Remember that you can always check your answer by substituting 1.5% for i in the equation and solving for either S or P.

Let's try another problem for good measure.

Problem: You are interested in accumulating a total of $50 000 in 10 years based on an investment today of $20 000. What rate of interest will you need to obtain in order to realise your aim if interest is compounded monthly?

Solution: Interest rate per period = i
Number of periods = 10 years × 12 periods/year = 120
Principal = $20 000
Accumulated sum = $50 000

$$r = \sqrt[n]{\frac{S}{P}} - 1$$

$$= \sqrt[120]{\frac{50\,000}{20\,000}} - 1$$

$$= (1.00766498) - 1$$

$$= .00766498$$

An annual rate of interest of approximately 9.2% p.a. compounded monthly (i.e. .00766498 per month × 12) will ensure that the $20 000 invested today will have a future value of $50 000 in 10 years.

The distinction between the future or accumulated value and the present value is that the former asks what value will occur in the future after earning compound interest on the known principal, while the latter, the present value, asks what is today's

Business mathematics and statistics

value (the unknown principal or deposit) which results in a certain (known) sum in the future. In other words, to determine the future or accumulated value the principal deposited must be known and you must find the future value or accumulated value of the deposit plus accrued interest. To determine the present value, the future or accumulated value is known and you must find the principal necessary to realise such a future value. As previously mentioned, solutions to all of the above calculations can also be undertaken with the use of financial calculators.

4.2 Competency check

1. (a) What is meant by *present value*?
   (b) Briefly explain the difference between the future or accumulated value and the present value.

2. To calculate the present value of a future amount, the compound interest formula can be applied—that is, $S = P(1 + i)^n$. However, now we are solving for P instead of S. Solve for P.

3. Find the present values of the following future values (to the nearest dollar):

   (a) $5000 for five years at 8% p.a. interest compounded semi-annually;
   (b) $1500 for two years at 12% p.a. interest compounded quarterly;
   (c) $250 for eight years at 12% p.a. interest compounded monthly;
   (d) $250 000 for nine years at 18% p.a. interest compounded monthly.

4. If you are given the future and present value of the deposit as well as the term, but must discover the rate of interest, how would you solve for the interest rate (i)?

5. Determine the interest rate (to one decimal place) needed for:

   (a) a principal of $2000 to reach an accumulated value of $3000 in four years if the interest is compounded monthly;
   (b) a principal of $7500 to have an accumulated value of $15 000 in seven years if the interest is compounded quarterly;
   (c) a principal of $100 000 to have an accumulated value of $250 000 in 12 years if the interest is compounded semi-annually.

4.3 Summary

The simple interest rate involves applying the rate of interest to the original principal (P) only—that is, the interest received during previous periods is not added to the principal (P) but instead the interest rate over all periods is calculated solely on the value of the original principal (P).

Normally, however, when depositing funds in a financial institution, you receive compound interest on that deposit—that is, the interest plus the principal generates further interest (unlike simple interest where interest is earned only on the principal). The total amount you will receive at some future date is (logically) referred to as the future or accumulated value. To determine the future or accumulated value, use the formula:

$$S = P(1 + i)^n$$

where
S = sum of the principal and the compounded interest
P = principal
i = interest rate per period
n = number of periods.

In other words, the application of the identical principal, rate of interest and term to a simple rate of interest compared to a compound rate of interest will generate a lower amount in the former since interest does not accrue on interest.

If interest is compounded more often than once a year, the stated per annum rate of interest is called the nominal rate, and i and n must be adjusted. For example, if the rate of interest is 12% p.a. compounded quarterly for three years, then:

i = 12%/4 periods/year = 3%
n = 3 years × 4 quarters/year = 12

To determine the present value of a future value (i.e. the amount that must be invested today to realise a certain sum in the future), use the formula and solve for P:

$$S = P(1 + i)^n$$
$$\frac{S}{(1 + i)^n} = P$$
$$S(1 + i)^{-n} = P$$

If all variables but the rate of interest are given, you must solve for i in the compound interest formula:

$$S = P(1 + i)^n \qquad \text{Therefore:} \qquad i = \sqrt[n]{\frac{S}{P}} - 1$$

4.4 Multiple-choice questions

1. Another term for *future value* is:

   (a) present value;
   (b) principal;
   (c) accumulated value;
   (d) compound interest.

2. How many interest rate periods are there if a deposit is compounded monthly over 7.75 years?

   (a) 93 (b) 91.75 (c) 84.75 (d) 96

3. What is the interest rate per period if a deposit attracts 10.5% p.a. compounded monthly?

   (a) 0.0875   (b) 0.875   (c) 0.00875   (d) 8.75

4. What is the simple interest (i) and compound interest (ii) received for a $2000 principal at 8% p.a. over four years?

   (a) (i) $640   (ii) $2721
   (b) (i) $2640   (ii) $2721
   (c) (i) $6400   (ii) $721
   (d) (i) $640   (ii) $721

5. Calculate the accumulated value and compound interest (to the nearest dollar) for $10 000 invested at 12% compounded quarterly for three years.

   (a) Accumulated value = $14 049.   Compound interest = $4049.
   (b) Accumulated value = $14 258.   Compound interest = $4258.
   (c) Accumulated value = $14 185.   Compound interest = $4185.
   (d) Accumulated value = $16 010.   Compound interest = $6010.

6. What is the present value (to the nearest dollar) of the accumulated value of $15 000 earning 10% p.a. interest compounded quarterly over 10 years?

   (a) $5586   (b) $5783   (c) $11 718   (d) $7151

4.5 Exercises

1. Why is compound interest preferable to simple interest on a deposit?
2. Why would an understanding of the concept and mechanics of *accumulated value* be of interest to an investor?
3. Cite an example where the concept of present value would be of interest to a soon-to-be retiree?
4. Calculate the simple and compound interest that accrues on a $50 000 investment earning 8.25% p.a. over 11 years.
5. Calculate the future value (to the nearest dollar) of $5000 invested at 12% p.a. interest compounded semi-annually over five years.
6. Calculate the future value (to the nearest dollar) of $100 000 invested at 9% p.a. interest compounded quarterly over four years.
7. If the rate of inflation is 5% p.a. each year for the next 15 years, what value will $1 value today have in 15 years?
8. A $10 000 investment attracts interest of 10% p.a. compounded quarterly over a 10-year period. How much interest will have accrued at the end of that period?
9. Which of the following investment choices is the more financially favourable:

   (a) 12% p.a. simple interest on $10 000 over five years?
   (b) 10% p.a. interest compounded quarterly on $10 000 over five years?

Chapter 4 – Compound interest

10. The brothers Des and Mal Grimm had $8000 and $9000 respectively in savings. Des deposited his funds for three years in a fixed-term deposit account attracting 12% p.a. interest compounded quarterly. Mal found a return of 16% p.a. compounded quarterly from a less reputable financial institution. After three years they pooled their resources (the accumulated value of their respective deposits) and deposited the money with an institution offering 12% p.a. compounded monthly for five years. How much did the Grimm brothers have in total at the end of the eight-year period (to the nearest dollar)?

11. What principal will you have to deposit (to the nearest dollar) if you need $10 000 in five years invested at an interest rate of 12% p.a. compounded semi-annually?

12. How much money (to the nearest dollar) will you receive in 4.5 years if you invest $2000 at 9% p.a. interest compounded quarterly?

13. Determine the interest rate needed for:

    (a) a principal of $25 000 to have an accumulated value of $40 000 in 3.5 years if the interest is compounded monthly;
    (b) a principal of $50 000 to have an accumulated value of $100 000 in five years if interest is compounded semi-annually;
    (c) a principal of $35 000 to have an accumulated value of $40 000 in two years if interest is compounded quarterly.

14. Anna's parents have decided to send her to Switzerland in four years to study chocolate-making, as both her parents have a sweet tooth. They want to have $15 000 at that time and have two investment options. Lay-Low Finance Company offers 14% p.a. interest compounded semi-annually, while Fly-Bye-Nite Inc. offers 12% p.a. interest compounded quarterly. What principal (to the nearest dollar) must be deposited today with each of the two companies to realise $15 000 in four years? Which of the two choices is the more attractive?

15. What principal (to the nearest dollar) would you need to invest to receive:

    (a) $15 000 at 8% p.a. interest over five years compounded quarterly?
    (b) $15 000 at 8% p.a. interest over five years compounded monthly?

16. If you invested $4500, how much interest (to the nearest dollar) would you receive if:

    (a) interest of 9% p.a. compounded semi-annually was earned over 4 years?
    (b) interest of 12% p.a. compounded quarterly was earned over 7.5 years?

17. Mr D. Kline started a savings account for his daughter 18 years ago so that she would receive $25 000 on her 18th birthday. What was the original principal (to the nearest dollar) if a 10% p.a. rate of interest compounded quarterly applied at the time of the deposit?

18. Mr B. Ware deposited $15 000 at 9% p.a. interest compounded monthly in C.U. Laiter Finance Company. Three years later to the day, Mr B. Ware deposited a further $6000 in the account for another two years at the same rate of interest.

How much (to the nearest dollar) will Mr B. Ware have earned in interest at the end of the five years?

19. Mr G. Wizz deposited $5000 on 1 January 1992 with Y.B. Heer Finance Company, receiving interest of 8% p.a. compounded quarterly. On 1 January 1994, interest rates were increased to 9% p.a. How much did Mr G. Wizz receive in interest on 1 January 1997?

20. Ms M. Bags inherited $1 500 000 on her 11th birthday but will not have access to the funds until her 21st birthday. If the funds are invested by her trustee, how much (to the nearest $1000) can Ms Bags expect to receive in total on her 21st birthday if the interest rate is 12.75% p.a. compounded:

    (a) monthly?   (b) daily?

Chapter 5
ANNUITIES

LEARNING OUTCOMES
After studying this chapter you should be able to:
- Define an annuity
- Describe the different types of annuities in the marketplace
- Apply future value annuities through problem-solving
- Calculate the periodic payment of a future value annuity (sinking fund)
- Apply present value annuities through problem-solving
- Calculate the periodic payment of a present value annuity (amortisation).

5.1 Annuity defined

Approximately 70% of Australians either own or are paying off their own home. To many Australians, home ownership is as Australian as meat pies and Vegemite (the latter, by the way, is now owned by the American firm, Kraft). Taking out a mortgage to finance the purchase of a home is, by far, the single largest commitment for most Australians. A home mortgage, a personal loan, insurance premiums, or even Austudy, involves a periodic payment (monthly, quarterly, semi-annually or annually) of equal amounts of money over a period of time. These periodic payments are referred to as annuities.

Annuities may be classified as simple annuities, or general annuities. In the case of simple annuities, the compounding of interest occurs on the same date as the annuity payments, while with general annuities the payments are made either more or less frequently than interest is compounded.

We will be examining only simple annuities which can be classified into four types: ordinary annuity, annuity due, deferred annuity and perpetuity. Each periodic payment made at the outset of the equal intervals of time is known as *annuity due* (e.g. life insurance premium), while a payment made at the end of each period is an *ordinary or simple annuity* (e.g. house mortgage). If the payments of an annuity don't

Business mathematics and statistics

begin until a number of compounding periods have elapsed, the annuity is referred to as a *deferred annuity*. Finally, when the annuity continues forever, it is a *perpetuity*. This chapter will explain in detail the operations of ordinary annuities, focusing on:

- **Future value annuities**, which refers to the accumulation of interest (compounded) on periodic payments or deposits in an account over a period of time—that is, the value at the end of the last period. For example, if you deposit $100 every three months in an account receiving interest of 10% p.a. compounded quarterly, the amount you will receive in five years is the future value of your annuity; in other words, the amount due at the end of the term.
- The **present value of an annuity**, on the other hand, starts with an amount (e.g. the principal of a loan—the value of the annuity at the beginning of the term) which accrues interest and is reduced by periodic payments by the borrower, with the subsequent interest charge calculated on the current balance. These annuity payments continue until such time as the current balance is zero (i.e. the loan is paid off). In other words, the present value is the value at the beginning of the first period.

5.1 Competency check

1. What is meant by an *annuity*?
2. Simple annuities can be classified into four types. Nominate and define two of these and provide an appropriate example of each.
3. During the course of a working month, you can expect to be exposed to a variety of ordinary annuities. List two and briefly explain how each operates.
4. What is meant by the *future value of an annuity* and why may understanding its calculation come in handy?
5. When would an understanding of the calculation of the *present value of an annuity* be useful?

5.2 Future (accumulated) value of an ordinary annuity

If you deposited $1000 (the periodic payment) in an account on a quarterly basis (payment interval) earning 12% p.a. interest compounded quarterly for two years (term), the problem would be to calculate the accumulated or future value of your annuity.

To determine the future value, first find the balance of the account at the end of each period (e.g. quarter). As 12% p.a. is compounded quarterly, this is equivalent to 3% per quarter. The quarterly payments are shown in Table 5.1. The interest on the balance (column 3) is calculated by applying the 3% quarterly interest rate to the balance at the beginning of the particular quarter in question. For example, at the beginning of the fourth quarter, there was an opening balance of $3090.90. Applying the 3% quarterly interest rate results

Chapter 5 – Annuities

in interest of $92.73, which when added to the opening balance and the quarterly payment of $1000 produces a balance at the end of the quarter of $4183.63.

Table 5.1 *Quarterly payments of $1000 at 12% p.a. interest compounded quarterly*

Quarter	Balance at beginning of quarter ($)	Interest on balance ($)	Quarterly payment ($)	Total at end of quarter ($)
1st	0	0	1000	1000.00
2nd	1000	30	1000	2030.00
3rd	2030	60.90	1000	3090.90
4th	3090.90	92.73	1000	4183.63
5th	4183.63	125.51	1000	5309.14
6th	5309.14	159.27	1000	6468.41
7th	6468.41	194.05	1000	7662.46
8th	7662.46	229.87	1000	8892.33

Therefore, $892.33 would have been earned in interest from investing $1000 in an annuity on a quarterly basis for two years.

When considering the future value of a periodic payment earning compound interest (an annuity), think of each periodic payment earning the compounded interest over the term of the loan, with the summation of all the payments and their corresponding compounded interest equal to the future value of the annuity. For example, re-examining the above problem:

- The first $1000 deposited is in the account for the seven remaining quarters.
- The second $1000 deposited is in the account for the six remaining quarters.
- The third $1000 deposited is in the account for the five remaining quarters.
- The fourth $1000 deposited is in the account for the four remaining quarters.
- The fifth $1000 deposited is in the account for the three remaining quarters.
- The sixth $1000 deposited is in the account for the two remaining quarters.
- The seventh $1000 deposited is in the account for the one remaining quarter.
- The eighth $1000 deposited is in the account for the zero remaining quarters.

The future value can be determined by using the Amount at compound interest table (Appendix A at the back of the book). In the above problem, 3% interest is earned during each quarter. Therefore:

```
1st   $1000(1.22987387) = $1229.87
2nd   $1000(1.19405230) = $1194.05
3rd   $1000(1.15927407) = $1159.27
4th   $1000(1.12550881) = $1125.51
5th   $1000(1.09272700) = $1092.73
6th   $1000(1.06090000) = $1060.90
7th   $1000(1.03000000) = $1030.00
8th   $1000(1.00000000) = $1000.00
                          $8892.33
```

Business mathematics and statistics

This method results in the same answer but with less work and, hence, reduced risk of error.

However, instead of having to plough through the tedious workings above, you can apply the following formula to determine the future value of an annuity:

$$S = p \frac{(1+i)^n - 1}{i}$$

where S = future value (i.e. total deposit and accrued interest)
p = periodic payment
i = interest rate per period
n = number of periods.

A shorthand method of writing this formula is sometimes used:

$$S = p s_{\overline{n}|i} \quad \text{where} \quad s_{\overline{n}|i} = p \frac{(1+i)^n - 1}{i}$$

To solve the above problem, we know that p = $1000, i = 3% (12% ÷ 4) and n = 8. Therefore:

$$S = \$1000 \frac{(1 + 0.03)^8 - 1}{0.03}$$

In determining the future and/or present value of an annuity, you may use the tables in the appendix or a calculator which has an exponential function y^x. For those lacking the necessary function keys in their calculators, reference is made within the text to the appropriate tables to be applied.

To determine the value of $(1 + 0.03)^8$, you may use your calculator or the Amount at Compound Interest Table. Thus:

$$S = \$1000 \frac{(1.26677008) - 1}{0.03}$$

$$= \$1000 \frac{(0.26677008)}{0.03}$$

$$= \$1000(8.892336)$$

$$= \$8892.34$$

Problem: Find the amount of an annuity of deposit of $50 per month for 10 years at 12% p.a. interest compounded monthly.

Solution:

$$S = p \frac{(1+i)^n - 1}{i}$$

As p = $50, i = 1% (12% ÷ 12) and n = 120 (12 × 10),

$$S = \$50 \frac{(1 + 0.01)^{120} - 1}{0.01}$$

As the Amount at Compound Interest Table contains information up to 60 periods only, the law of exponents (indices) must be applied. That is:

$$(1 + 0.01)^{120} = (1 + 0.01)^{60} \times (1 + 0.01)^{60}$$

Looking up the Amount at Compound Interest Table for 60 periods at 1% period:

$$(1 + 0.01)^{120} = (1.8166967)(1.8166967)$$
$$= 3.3003869$$

Now substitute 3.3003869 for $(1 + i)^n$ in the formula:

$$S = p \frac{(1 + i)^n - 1}{i}$$

$$S = \$50 \frac{(3.3003869) - 1}{0.01}$$

$$= \$50 \frac{(2.3003869)}{0.01}$$

$$= \$50(230.03869)$$

$$= \$11\,501.93$$

Imagine if you tried to solve this problem using the method first illustrated!

Future Value (amount) of an Annuity Table

To further minimise time and effort in solving future value annuity problems, the Future Value (amount) of an Annuity Table can be used (Appendix C at the back of the book). This table is used in the same way as the Amount at Compound Interest Table. For example, the $1000 deposited quarterly over two years earning 12% p.a. interest compounded quarterly results in an interest rate per period of 3% for a total of eight periods (i.e. 4 quarters × 2 years = 8 quarters or periods). Corresponding to eight periods of 3% per period is 8.89233605, which is the ordinary annuity of $1 per period. As $1000 was invested per period, the future value of the annuity is $1000(8.89233605) = $8892.34, which corresponds to our previous calculations, although this time with much less time and effort.

Problem: The Skalskys have started a savings account for their recently born daughter, Monique. If the Skalskys deposit $25 a quarter for 18 years, earning 10% p.a. interest compounded quarterly, how much will Monique have at the age of 18?

Solution: Interest rate per period = 10%/4 = 2.5%
Number of periods = 4 × 18 = 72

As 72 periods at 2.5% interest per period corresponds to 196.68912249 in the Future Value (amount) of an Annuity Table:

$$S = \$25(196.68912249)$$
$$= \$4917.23$$

In other words, $25 deposited quarterly over 18 years earning 10% p.a. interest compounded quarterly will give Monique $4917.23. Her parents will have to invest $1800 ($25/quarter × 72 quarters) to realise the $4917.23 (hence, interest received would be $3117.23).

Problem: Mr B. Dazzle has been offered two fairly attractive contracts by opposing publishers for the rights to his impressive autobiography. Lystin & Dumpin has offered Mr B. Dazzle $1 000 000 outright, while Uppiti Dumpiti Inc. has offered royalties of $130 000 every year for the next 15 years. If the current return is 12% p.a. interest compounded annually, which of the two contracts should be accepted? (Assume that if Mr B. Dazzle accepted the $1 000 000 outright, he would deposit it in the 12% p.a. compounded annually interest-bearing account.)

Solution: *Alternative 1:* If Mr B. Dazzle accepted Lystin & Dumpin's offer, he would deposit the $1 000 000 at 12% p.a. interest compounded annually for 15 years. Therefore, the number of periods is 15 and the interest rate per period is 12%. As 15 periods at 12% interest per period corresponds to 5.47356576 in the Amount at Compound Interest Table:

$$\begin{aligned} S &= p(1+i)^n \\ &= \$1\,000\,000(1+.12)^{15} \\ &= \$1\,000\,000(5.47356576) \\ &= \$5\,473\,565.76 \end{aligned}$$

Alternative 2: If Mr B. Dazzle accepted Uppiti Dumpiti's offer, he would deposit $130 000 every year for the next 15 years at 12% p.a. interest compounded annually. As 15 periods at 12% interest per period corresponds to 37.27971466 in the Future Value (amount) of an Annuity Table:

$$\begin{aligned} S &= p\frac{(1+i)^n - 1}{i} \\ &= \$130\,000\,\frac{(1+.12)^{15} - 1}{.12} \\ &= \$130\,000\,(37.27971466) \\ &= \$4\,846\,362.91 \end{aligned}$$

Therefore, Alternative 1 would return Mr B. Dazzle $627 202.85 more than Alternative 2, even though the latter entails Uppiti Dumpiti paying $1 950 000 over the duration of the contract ($130 000 × 15 years) versus the $1 000 000 paid by Lystin & Dumpin.

Problem: Find the future value of an annuity of $100 deposited quarterly over eight years at 10% p.a. interest compounded quarterly.

Solution: $S = p\dfrac{(1+i)^n - 1}{i}$

As $p = \$100$, $i = 2.5\%$ ($10\% \div 4$) and $n = 32$ (4×8),

$$S = \frac{\$100(1 + 0.025)^{32} - 1}{0.025}$$

The Future Value (amount) of an Annuity Table (Appendix C) indicates that corresponding to 32 periods at 2.5% interest per period is 48.15027751. That is:

$$\frac{(1 + 0.025)^{32} - 1}{0.025} = 48.15027751$$

Substituting into the formula:

$$S = p\frac{(1 + i)^n - 1}{i}$$
$$= \$100(48.15027751)$$
$$= \$4815.03$$

Problem: A unit sold recently is to be paid off over 10 years at monthly repayments of $950. If the $950 monthly payment is instead placed in a 12% p.a. compounded monthly interest-bearing deposit, what would be the value of all these deposits in 10 years?

Solution: $S = p\frac{(1 + i)^n - 1}{i}$

As $p = \$950$, $i = 1\%$ ($12\% \div 12$) and $n = 120$ (10×12),

$$S = \$950\frac{(1 + 0.01)^{120} - 1}{0.01}$$
$$= \$950\frac{(3.30038689) - 1}{0.01}$$
$$= \$950\,(230.038689)$$
$$= \$218\,536.75$$

Therefore, deposits totalled $114 000 ($950 × 120). Since the money was earning 12% p.a. compounded monthly, the value of all the deposits over the 10-year period was $218 536.75, with $104 536.75 in accrued interest (i.e. $218 536.75 − $114 000).

Problem: Ms B. Leave deposits $350 at the end of each quarter in an account receiving 10% p.a. interest compounded quarterly. If she deposits this amount quarterly for six years, how much will be in the account five years after the last deposit has been made? (Assume the 10% p.a. interest rate applies and remains constant over the entire 11-year period.)

Solution: $S = p\dfrac{(1+i)^n - 1}{i}$

As $p = \$350$, $i = 2.5\%$ ($10\% \div 4$) and $n = 24$ (4×6),

$$S = \$350\,\dfrac{(1 + 0.025)^{24} - 1}{0.025}$$

Looking up the Future Value (amount) of an Annuity Table (24 periods at 2.5% interest per period):

$S = \$350(32.34903798)$
$= \$11\,322.16$

Thus, after six years the future value of $350 deposited quarterly at 10% p.a. interest compounded quarterly is $11 322.16, comprising deposits of $8400 ($350 × 24) and accrued interest of $2922.16 ($11 322.16 − $8400).

The $11 322.16 is to remain in the account for another five years after the last deposit attracting interest of 10% p.a. compounded quarterly; hence the compound interest formula must be applied:

$$S = P(1 + i)^n$$

As $P = \$11\,322.16$, $r = 2.5\%$ ($10\% \div 4$) and $n = 20$ (4×5),

$S = \$11\,322.16(1 + 0.025)^{20}$
$= \$11\,322.16(1.63861644)$
$= \$18\,552.68$

Therefore, there will be $18 522.68 in the account five years after the last deposit is made.

5.2 Competency check

1. When would calculation of the future value of an annuity be useful for you?

2. Find the future value of an annuity if $50 is deposited monthly in an account for 10 years at 9% p.a. interest compounded monthly.

3. Find the future value of an annuity if $300 is deposited semi-annually in an account for 10 years at 9% p.a. interest compounded semi-annually.

4. Find the future value of an annuity if $150 is deposited quarterly in an account for 10 years at 9% p.a. interest compounded quarterly.

5. Assume that the future value calculated in question 4 is transferred to another account for an extra seven years attracting the same rate of interest of 9% p.a. compounded quarterly. How much will be in the account after the 17 years?

Note: Questions 2 to 4 involve deposits of $600 p.a.; observe the varying results.

Chapter 5 – Annuities

5.3 Sinking fund (finding the periodic payment of annuity)

Have you ever decided to save for a special purpose, such as an overseas trip, and set yourself a target figure, and then wondered how much you had to put away each month in order to reach that targeted amount (realising that each deposit would earn interest compounded for the duration of the account)? This is referred to as a **sinking fund**: the required future amount is known, but the periodic payments necessary to realise this future sum (i.e. what must be deposited or sunk into the account) must be determined. Well, your problem is over (at least the problem of how to calculate how much to save periodically). To determine your annuity payments, apply the formula for future value of an annuity and solve for p:

$$S = p \frac{(1+i)^n - 1}{i}$$

Therefore, to solve for p:

$$p = \frac{S}{\frac{(1+i)^n - 1}{i}} \quad \text{or} \quad p = \frac{S}{s_{\overline{n}|i}}$$

Thus, the formula can be thought of as:

$$\text{Periodic payments} = \frac{\text{Future value}}{s_{\overline{n}|i}}$$

Problem: Carol plans to go on a working holiday in three years and needs $5000 for her trip. How much should she deposit in an account every quarter to have $5000 in three years if the account attracts 8% p.a. interest compounded quarterly?

Solution: $S = p \dfrac{(1+i)^n - 1}{i}$

As $i = 2\%$ (8% ÷ 4), $n = 12$ (4 × 3) and $S = \$5000$,

$$\$5000 = p \frac{(1 + 0.02)^{12} - 1}{0.02}$$

Looking up the Future Value (amount) of an Annuity Table (12 periods at 2% interest per period), $\dfrac{(1.02)^{12} - 1}{0.02} = 13.41208973$

$$\$5000 = p\,(13.41208973)$$

$$\frac{\$5000}{13.41208973} = p \quad \text{(solving for } p\text{)}$$

$$\$372.80 = p$$

In other words, Carol must deposit $372.80 every quarter to have $5000 in three years. Check the answer by substituting into the formula:

$$S = p \frac{(1+i)^n - 1}{i}$$

where $p = \$372.80$
$i = 2\%$
$n = 12$.

Therefore, as 13.41208973 corresponds to 2% for 12 periods in the Future Value (amount) of an Annuity Table:

$S = \$372.80(13.41208973)$
$= \$5000$

Problem: Peter must repay a debt of $12 000 (including interest) in five years. If deposits in an account attract 10% p.a. interest compounded semi-annually, how much should Peter deposit every six months to ensure an accumulation of $12 000 in five years?

Solution: $S = p \dfrac{(1+i)^n - 1}{i}$

As $S = \$12\,000$, $i = 5\%$ ($10\% \div 2$) and $n = 10$ (2×5),

$$\$12\,000 = p \frac{(1 + 0.05)^{10} - 1}{0.05}$$

$\$12\,000 = p(12.57789254)$

$\dfrac{\$12\,000}{12.57789254} = p$

$\$954.05 = p$

Thus, $954.05 must be deposited semi-annually to accumulate $12 000. Peter must deposit a total of $9540.50; hence, the interest would be $2459.50. To check the answer, substitute into the formula:

$$S = p \frac{(1+i)^n - 1}{i}$$

where $p = \$954.05$
$i = 5\%$
$n = 10$.

Therefore:

$$S = \$954.05 \frac{(1 + 0.05)^{10} - 1}{0.05}$$

As 12.57789254 corresponds to 5% over 10 periods in the Future Value (amount) of an Annuity Table,

$$S = \$954.05(12.57789254)$$
$$= \$11\,999.94 \text{ (difference due to rounding of } i\text{)}$$

Problem: The body corporate of strata plan 007 believes a sinking fund of $100 000 is necessary in four years. If there is currently a zero balance in 007's sinking fund, how much must strata owners deposit in aggregate into an account each quarter to raise these funds in four years if the account attracts 8.5% p.a. interest compounded quarterly?

Solution: $S = p\dfrac{(1+i)^n - 1}{i}$

As $S = \$100\,000$, $i = 2.125\%$ (8.5% ÷ 4) and $n = 16$ (4 × 4),

$$\$100\,000 = p\dfrac{(1.02125)^{16} - 1}{0.02125}$$

$$\$100\,000 = p\dfrac{(1.399951895) - 1}{0.02125}$$

$$\$100\,000 = p(18.82126563)$$

$$\dfrac{\$100\,000}{18.82126563} = p$$

$$\$5313.14 = p$$

Therefore, $5313.14 must be deposited quarterly over the four years ($85 010.24 in total—i.e. $5313.14 × 16) to accumulate a total of $100 000 at the end of the three-year period. Interest of $14 989.76 accrued over the period.

Problem: How much should be deposited monthly into a 9% p.a. monthly interest-bearing account in order to accumulate $50 000 in five years?

Solution: $S = p\dfrac{(1+i)^n - 1}{i}$

As $S = \$50\,000$, $p = ?$, $i = 0.75\%$ (9% ÷ 12) and $n = 60$ (5 × 12),

$$\$50\,000 = p\dfrac{(1 + .0075)^{60} - 1}{0.0075}$$

$$\$50\,000 = p\dfrac{(1.565681027) - 1}{0.0075}$$

$$\$50\,000 = p(75.42413693)$$

$$\dfrac{\$50\,000}{75.42413693} = p$$

$$\$662.92 = p$$

Thus, approximately $663 a month would have to be deposited at 9% p.a. to accumulate $50 000 in five years.

5.3 Competency check

1. What is meant by a *sinking fund*?
2. Give an example of when you personally have or could have used this concept.
3. Mr I. Soare plans to have paid off the mortgage on his home in four years. The bank has informed Mr I. Soare that a debt of $24 000 (including interest) will be due in four years. If deposits in an account attract 8% p.a. interest compounded quarterly, how much should Mr I. Soare deposit every quarter to ensure an accumulation of $24 000 in four years?
4. Mike and Pat would like to send their son, Ben, to a private university in 10 years. According to their calculations, the tuition and living away expenses would total approximately $58 000. How much must Mike and Pat deposit each month over the next 10 years to accumulate $58 000 with the rate of interest received on deposits at 12% p.a. compounded monthly?
5. How much must they deposit each month if the rate of interest in question 4 is 9% compounded monthly?

5.4 Present value of an ordinary annuity

Would you like to retire when you turn 45 on $7000 per quarter for the rest of your life? If so, how much would you have to deposit at the age of 45 at, say, 9% p.a. interest compounded quarterly so that you could withdraw $7000 per quarter until, say, the age of 75? To answer this question, the lump sum (or present value of the lump sum) which will permit you to withdraw $7000 per quarter for 30 years (75–45) must be determined. In other words, we are determining the amount of principal which must be deposited today which will allow payments of a certain amount ($7000) to be made at the end of each period (e.g. quarter), with interest calculated and accruing to the current balance until such time as the current balance is zero (e.g. in 30 years).

The formula for solving the present value is:

$$PV = p \frac{1 - (1 + i)^{-n}}{i}$$

where PV = present value
 i = interest rate per period
 n = number of periods
 p = periodic payments.

A shorthand method of writing this formula is:

$$PV = P \, pv_{\overline{n}|i} \quad \text{where} \quad pv_{\overline{n}|i} = \frac{1-(1+i)^{-n}}{i}$$

In the above example, if the lump sum was deposited in an account earning 9% p.a. interest compounded quarterly, then $i = 2.25\%$ ($9\% \div 4$), $n = 120$ (4×30) and $p = \$7000$. Therefore:

$$PV = p\frac{1-(1+i)^{-n}}{i}$$

$$= \$7000 \, \frac{1-(1+0.0225)^{-120}}{0.0225}$$

You may use the Present Value of an Annuity Table (Appendix D at the back of the book) which provides values for

$$\frac{1-(1+i)^{-n}}{i}$$

As 41.36679266 corresponds to 2.25% over 120 periods in the Present Value of an Annuity Table, i.e. $\frac{1-(1+0.0225)^{-120}}{0.0225}$

$$PV = \$7000 \, (41.36679266)$$
$$= \$289\,567.55$$

Therefore, $289\,568 deposited at the age of 45 enables withdrawals of $7000 per quarter for 30 years from an account earning 9% p.a. interest compounded quarterly. Note that the withdrawals ($7000 × 120) amount to $840\,000; hence, $550\,432 in interest accrues over the period of the account.

Problem: Nicholas is going overseas for three years and wants to withdraw $2000 every three months while abroad. Upon leaving Australia, how much must he deposit in an account paying 10% p.a. interest compounded quarterly?

Solution: $PV = p\frac{1-(1+i)^{-n}}{i}$

As $p = \$2000$, $i = 2.5\%$ ($10\% \div 4$) and $n = 12$ (3×4),

$$PV = \$2000 \, \frac{1-(1+0.025)^{-12}}{0.025}$$

If you don't have an exponential function on your calculator, use the Present Value of an Annuity Table (Appendix D), according to which 10.2577646 corresponds to 2.5% over 12 periods. That is:

$$\frac{1-(1+0.025)^{-12}}{0.025} \quad \text{is equivalent to } 10.2577646.$$

Therefore:

$$PV = \$2000(10.2577646)$$
$$= \$20\,515.53$$

Thus, $20 515.53 deposited by Nicholas when leaving Australia will permit him to withdraw $2000 every three months for three years (12 withdrawals of $2000 each), a total of $24 000. Therefore, the interest accrued over the term of his withdrawals is $3484.47 (i.e. $24 000 − $20 515.53).

Problem: Mr C. Daar wants to spend five years researching a new book on the timber industry. He calculated that he needs $1600 a month to live on over the five years. How much must Mr C. Daar deposit today in an account earning 9% p.a. interest compounded monthly in order to withdraw $1600 a month over the next five years? (Use the Present Value of an Annuity Table.)

Solution: $$PV = p\frac{1-(1+i)^{-n}}{i}$$

As $p = \$1600$, $i = 0.75\%$ (9% ÷ 12) and $n = 60$ (12 × 5),

$$PV = \$1600\frac{1-(1+0.0075)^{-60}}{0.0075}$$

As 48.17337352 corresponds to 0.0075% over 60 periods in the Present Value of an Annuity Table (Appendix D),

$$PV = \$1600(48.17337352)$$
$$= \$77\,077.40$$

Therefore, a deposit of $77 077.39 will allow Mr C. Daar living expenses of $1600 per month over five years. In other words, $77 077.39 deposited at 9% p.a. interest compounded monthly will earn interest of $18 922.61 over the term (60 × $1600 − $77 077.39 = $18 922.61).

Problem: If a cottage sells for $135 000 down and repayments of $1610.75 per month over 25 years, what is the equivalent cash price (to the nearest dollar) if the interest rate is 6% p.a. compounded monthly?

Solution: The cash price of the cottage is the present value of the 300 repayments (12 × 25) added to the $135 000 down-payment. Therefore, the present value of the annuity is:

$$PV = p\frac{1-(1+i)^{-n}}{i}$$

As $p = \$1610.75$, $i = 0.5\%$ (6% ÷ 12) and $n = 300$ (12 × 25),

$$PV = \$1610.75 \frac{1-(1+0.005)^{-300}}{0.005}$$

$$= \$1610.75 \frac{1-(0.22396568)}{0.005}$$

$$= \$1610.75 \frac{(0.77603432)}{0.005}$$

$$= \$1610.75(155.206864)$$

$$= \$249\,999.46$$

Therefore, the present value of the annuity is equal to approximately $250 000, which when added to the initial deposit of $135 000 results in the equivalent cash price of $385 000.

Problem: Mr D. Kaye's scheduled debt repayment was for $500 per quarter for the next five years. However, he was delinquent in his first eight payments, forcing the lender to demand that a single payment be made at the end of the ninth quarter cancelling the entire debt. How much must the single payment be if money is worth 8% p.a. interest compounded quarterly?

Solution: The first eight payments plus the ninth form a future value annuity, while the remaining 11 payments form a present value annuity. Thus:

$$\text{Single payment} = p\frac{(1+i)^n - 1}{i} + p\frac{1-(1+i)^{-n}}{i}$$

As $p = \$500$, $i = 2\%$ $(8\% \div 4)$ and $n = 9$ and 11,

$$\text{Single payment} = \$500 \frac{(1+0.02)^9 - 1}{0.02} + \$500 \frac{1-(1+0.02)^{-11}}{0.02}$$

$$= \$500 \frac{(1.19509257) - 1}{0.02} + \$500 \frac{1-(0.804263039)}{0.02}$$

$$= \$500(9.75462843) + \$500(9.78684805)$$

$$= \$4877.31 + \$4893.42$$

$$= \$9770.73$$

5.4 Competency check

1. Distinguish between the usage of the present value of an annuity and the future value of an annuity.

2. Find, on the Present Value of an Annuity Table, the equivalent of $\frac{1-(1+i)^{-n}}{i}$ for:

    (a) 12% p.a. compounded monthly over 10 years;

(b) 9% p.a. compounded quarterly over 8 years;
(c) 6% p.a. compounded semi-annually for 15 years.

3. How much must you deposit today (to the nearest dollar) at 10% p.a. interest compounded semi-annually to be able to withdraw $2500 semi-annually (or an annuity of $5000 p.a.) over each of the next six years?

4. Norm's parents want to help Norm while he does his four-year apprenticeship in underwater basketweaving. They would like him to be able to withdraw $900 every three months for the duration of his apprenticeship. How much must Norm's parents deposit today in an account bearing interest of 10% p.a. compounded quarterly?

5. Dana wants to determine how much to deposit today in order to meet loan repayments of $288 per month for the next 15 months. If the current interest rate is 8% p.a. compounded monthly, what lump sum should Dana deposit (to the nearest dollar)?

. . . .

5.5 Amortisation

When you are repaying a debt—that is, principal and interest—the debt is **amortised** (i.e. paid off) by making equal periodic payments. Each periodic payment covers the interest on the balance of the principal and results in a decline in the principal so that by the end of the term of the loan, after the last repayment, there is nothing owing.

The Present Value of an Annuity Table (Appendix D) may again be used to determine the periodic payments necessary to pay off a debt. The same table can be used to determine the remaining debt after x repayments have been made.

Problem:
(a) Determine the quarterly repayments on a $70 000 mortgage at 10% p.a. interest compounded quarterly over 25 years.
(b) Calculate the total interest paid over the 25-year loan period.

Solution:
(a) To solve for p, use the formula:

$$PV = p \frac{1 - (1+i)^{-n}}{i}$$

As $PV = \$70\,000$, $i = 2.5\%$ ($10\% \div 4$) and $n = 100$ (25×4),

$$\$70\,000 = p \frac{1 - (1 + 0.025)^{-100}}{0.025}$$

As 36.61410526 corresponds to 2.5% over 100 periods in the Present Value of an Annuity Table:

Chapter 5 – Annuities

$$\$70\,000 = p(36.61410526)$$
$$\frac{\$70\,000}{36.61410526} = p$$
$$\$1911.83 = p$$

Therefore, the repayments would be $1911.83 per quarter.

(b) Total repayments = Quarterly repayments × 100 (25 years × 4 quarters)
   = $1911.83 × 100 = $191 183

   Interest paid = Total repayments − Principal
   = $191 183 − $70 000
   = $121 183

Problem: What is the balance outstanding at the end of the ninth year of the $70 000 mortgage described in the above problem?

Solution: As nine years of the loan have passed, there are 16 years remaining. To determine the balance outstanding, apply the formula:

$$PV = p\frac{1-(1+i)^{-n}}{i}$$

As $p = \$1911.83$, $i = 2.5\%$ (10% ÷ 4) and $n = 64$ (16 × 4),

$$PV = \$1911.83\frac{1-(1+0.025)^{-64}}{0.025}$$

Looking up the Present Value of an Annuity Table (Appendix D):

$$\frac{1-(1+i)^{-n}}{i} = \frac{1-(1+0.025)^{-64}}{0.025} = 31.76369148$$

$$PV = \$1911.83(31.76369148)$$
$$= \$60\,726.78$$

Therefore, at the end of the ninth year, $60 726.78 is owing on the 25-year loan.

Notes:
(a) As repayments have been made for nine years (9 × 4 = 36), there are 64 payments (16 × 4) remaining.
(b) Over the nine years (36% of the loan's term), payments totalling $68 825.88 have been made, yet the principal owing was reduced from $70 000 to $60 726.78. In other words, the principal has been reduced by $9273.22; hence, $59 552.66 ($68 825.88 − $9273.22) of the repayments so far are interest payments. In fact, although nine years, or approximately 36%, of the term of the loan have passed, only 13.25% of the principal has been repaid. The early repayments of a mortgage are predominantly comprised of interest payments.

Business mathematics and statistics

Problem: Mr D. Fault borrowed $80 000 over 15 years at 9% p.a. interest compounded monthly. After three years, interest rates climbed to 12% p.a. compounded monthly and Mr D. Fault's repayments were increased to allow him to complete the repayments by the end of the 15-year term.

(a) How much did Mr D. Fault owe at the end of the first three years?
(b) What were his monthly repayments for the remaining 12 years?

Solution:
(a) First, determine Mr D. Fault's monthly repayments during the first three years.

As $PV = \$50\,000$, $i = 0.75\%$ ($9\% \div 12$) and $n = 180$ (15×12),

$$PV = p\,\frac{1-(1+i)^{-n}}{i}$$

$$\$80\,000 = p\,\frac{1-(1+0.0075)^{-180}}{0.0075}$$

$$\$80\,000 = p(98.59340884)$$

$$\frac{\$80\,000}{98.59340884} = p$$

$$\$811.41 = p$$

Now find the amount owing after three years' payments. As three years' (36 periods) payments have been made, there are 12 years' (144 periods) payments remaining. Therefore, $p = \$811.41$, $i = 0.75\%$ ($9\% \div 12$) and $n = 144$ (12×12):

$$PV = \$811.41\,\frac{1-(1+0.0075)^{-144}}{0.0075}$$

$$= \$811.41(87.87109195)$$

$$= \$71\,299.48$$

Therefore, Mr D. Fault owed $71 299.48 at the end of the first three years. Repayments of $29 210.76 ($811.41 × 36) were made over the three years, but the present value of the loan was reduced by only $8700.52 (i.e. the original loan of $80 000 minus the amount outstanding of $71 299.48 equals $8700.52).

(b) To find Mr D. Fault's quarterly repayments for the remaining 12 years, use the formula:

$$PV = p\,\frac{1-(1+i)^{-n}}{i}$$

As $PV = \$71\,299.48$, $i = 1\%$ ($12\% \div 12$) and $n = 144$ (12×12),

$$\$71\,299.48 = p\,\frac{1-(1+0.01)^{-144}}{0.01}$$

Looking up the Present value of an annuity table and applying the law of exponents:

$$\$71\,299.48 = (76.13715747)\,p$$

$$\frac{\$71\,299.48}{76.13715747} = p$$

$$\$936.46 = p$$

Therefore, Mr D. Fault's monthly repayments over the remaining 12 years were $936.46.

Note: If the $80 000 loan is taken for the full 15 year term, it would cost in total (i.e. principal and interest) $164 061 ($29 210.76 when the interest rate was 9% p.a. for three years, and $134 850.24 when the interest rate was 12% p.a. for the remaining 12 years).

5.5 Competency check

1. What is meant by the term *amortisation*?

2. Determine the monthly repayments on a $75 000 mortgage at 12% p.a. interest over 20 years.

3. What would be the balance outstanding (to the nearest dollar) at the end of the fifth year of the $75 000 mortgage described in problem 2?

4. Ms G. Wilkins borrowed $85 000 over 25 years at 13.5% p.a. interest compounded monthly. At the end of the seventh year, interest rates climbed to 15% p.a. compounded monthly and Ms G. Wilkins' repayments were increased to allow her to complete the repayments by the end of the 25-year term. How much did Ms G. Wilkins owe at the end of the first seven years?

5. In question 4, what were Ms G. Wilkins' monthly repayments for the remaining 18 years?

5.6 Summary

Annuities are periodic payments of equal amounts made over a period of time (e.g. loan repayments and payments of life insurance premiums).

The future or accumulated value of an annuity is the accumulation of compound interest on periodic payments or deposits in an account over a period of time. It indicates how much you will receive—for example, in 10 years—if you deposit $50 every month in an account receiving 7% p.a. interest compounded monthly. Thus, the periodic payment is known and the future or accumulated value of these periodic payments and interest must be determined. The following formula is used to determine the future value of an ordinary annuity:

Business mathematics and statistics

$$S = p\frac{(1+i)^n - 1}{i}$$

where S = future value (i.e. total deposits plus accrued interest)
p = periodic payments
i = interest rate per period
n = number of periods.

This formula can also be written as:

$$S = ps_{\overline{n}|i} \qquad \text{where} \qquad s_{\overline{n}|i} = \frac{(1+i)^n - 1}{i}$$

To determine the periodic payments (referred to as the sinking fund), p must be solved for:

$$p = \frac{S}{\frac{(1+i)^n - 1}{i}} \qquad \text{or} \qquad p = \frac{S}{s_{\overline{n}|i}}$$

The present value of an annuity is today's value of a future amount. It indicates the sum that must be invested today to accumulate a particular amount at a future time. It is used for the amortisation of, for example, home loans and is defined by the formula:

$$PV = p\frac{1 - (1+i)^{-n}}{i}$$

where PV = present value
p = periodic payments
i = interest rate per period
n = number of periods.

The formula can also be written as:

$$PV = pv_{\overline{n}|i} \qquad \text{where} \qquad pv_{\overline{n}|i} = \frac{1 - (1+i)^{-n}}{i}$$

5.7 Multiple-choice questions

1. An ordinary or simple annuity is one where each periodic payment is made:
   (a) at the end of each period;
   (b) at the beginning of each period;
   (c) at some later date;
   (d) forever.

2. If Anna deposits $100 a month into a savings account realising 6% p.a. compounded monthly, how much will she have accumulated after five years?
   (a) 6977   (b) $13 488   (c) $3299   (d) $10 253

3. Ma and Pa Kettle have to start saving for their baby Kettle's wedding reception. They need to have $25 000 in seven years' time. If current interest rates of 8.6% p.a. compounded quarterly apply over the term, how much must Ma and Pa Kettle deposit every three months to make sure their pride and joy's reception is fit for a Kettle (answer to the nearest dollar)?

   (a) $382
   (b) $660
   (c) $538
   (d) $745

4. At the motor vehicle auction, Harley purchased a car for $17 500 but had to borrow $15 000 over four years at 9% p.a. compounded quarterly. What will be the quarterly repayments?

   (a) $1472.25
   (b) $1126.75
   (c) $1314.54
   (d) None of the above.

5. I want to study full-time for the next four years but need $1000 a month to live on. How much will I need to deposit in a lump sum in an account paying 9% p.a. interest compounded monthly in order to be able to withdraw $1000 each month over the next four years (to the nearest dollar)?

   (a) $41 541
   (b) $37 521
   (c) $40 185
   (d) $32 397

5.8 Exercises

1. Calculate the future values of the annuities listed in the following table (to the nearest dollar).

Payment	Term	Interest rate p.a. (%)	Compounded
(a) $5000/quarter	4 years	10.5	quarterly
(b) $300/quarter	12 years	9.0	quarterly
(c) $1500/6 months	6 years	8.5	semi-annually
(d) $6750/annum	7 years	10.5	annually
(e) $200/month	3 years	12.0	monthly

2. Mr A. Cute contributed $15 per month to an endowment policy which provided him with $4420 after 10 years. If, instead, he had deposited the $15 per month in an account bearing 12% p.a. interest compounded monthly over the 10 years, how much better or worse off would he have been?

3. Find the future value of an annuity of $250 deposited monthly for each month for five years if the interest rate is 6% p.a. compounded monthly.

Business mathematics and statistics

4. Find the future value of an annuity of $1000 deposited each quarter for three years if the interest rate is 10% p.a. compounded quarterly.

5. Find the present value of a series of quarterly payments (to the nearest dollar) of $5000 for six years if money attracts 12% p.a. interest compounded quarterly.

6. Find the present value of payments of $5000 received at the end of each of the next 10 years if the compound interest rate is 12% p.a.

7. Ms D. Meanour agreed to buy her partner's share for $120 000 in six years. How much must she deposit in an account every six months if the account attracts 10% p.a. interest compounded semi-annually?

8. Pat owes $50 000 which must be repaid in seven years. If deposits attract 7% p.a. interest compounded quarterly, how much should Pat deposit every three months to accumulate the $50 000 in seven years?

9. Ms B. Head deposited $1000 at the end of each year for 10 years in an account earning 12% p.a. interest compounded annually. If she left the funds in the account for another five years after the last deposit, how much (to the nearest dollar) would she receive at the end of this time?

10. A commercial property recently sold at auction to Mr B. Ware for $500 000 deposit and $10 000 per month for the next seven years. If money attracts interest of 10% p.a. compounded monthly, what is the equivalent cash value?

11. What are the annual repayments on a $100 000 mortgage at 8% p.a. compounded annually over 20 years?

12. Mr Abacus reckoned that in three years the average price of a new motor vehicle would be $30 000. If he hoped to save that money over the three years by depositing monthly payments into an 8% p.a. (compounded monthly) interest-bearing account, what should Mr Abacus deposit monthly (to the nearest dollar)?

13. Ms I. Ran has decided to save for a holiday in the Sahara Desert in three years. The cost of the package deal is $3000, which includes accommodation at Tent City and fly-drive (driving by camel). How much should Ms I. Ran deposit every month in an account which pays 6% p.a. interest compounded monthly to have the $3000 in three years?

14. Mike Tieson is unhappy with available superannuation packages and decides to deposit $400 per month in a 9% interest-bearing account (compounded monthly) for the remainder of his working life of 35 years.

    (a) How much (to the nearest dollar) will Mike have accumulated at the end of the 35 years?
    (b) How much of that will be accrued interest?

15. Mr D. Mist bought a new car, putting $8000 down and paying off the rest at $250 a month for six years. If current interest rates are at 9.5% p.a. compounded monthly, what is the equivalent cash price (to the nearest dollar) for the car?

Chapter 5 – Annuities

16. David has borrowed $35 000 at an interest rate of 11.5% p.a. compounded monthly. What are the monthly repayments if the loan is taken over 20 years?

17. (a) From question 16, what are the monthly repayments if the $35 000 loan is taken over 15 years?
    (b) How much extra interest does David have to pay if he takes the loan over 20 years compared with 15 years?

18. If David takes the $35 000 loan over 15 years, what is the balance outstanding after five years?

19. If 16.5 years have passed on a 25-year $80 000 mortgage at 12% p.a. interest compounded monthly:
    (a) What is the monthly repayment?
    (b) What is the balance outstanding (to the nearest dollar)?

20. Mr T. Bone took out a $1 million loan on a steakhouse over a 10-year period at an interest rate of 15% p.a. compounded monthly. After 3.5 years, interest rates climbed to 18% p.a. compounded monthly, with Mr T. Bone making no bones about his beef at the increased repayments necessary to allow him to complete the repayments by the end of the 10-year term.
    (a) What were Mr T. Bone's monthly repayments at the beginning of the 10-year period?
    (b) How much (to the nearest dollar) did Mr T. Bone owe at the end of the first 3.5 years?
    (c) What were his monthly repayments for the remaining 6.5 years?

Chapter 6
DEPRECIATION*

LEARNING OUTCOMES
After studying this chapter you should be able to:
- Define the key terms and principles of depreciation
- Calculate the total depreciation, annual depreciation and rate of depreciation of items with and without residual value
- Prepare a depreciation schedule using the straight-line method
- Describe the reducing balance method
- Prepare a depreciation schedule using the reducing balance method
- Calculate the current written-down value of an asset, using the reducing balance method
- Calculate the depreciation rate using the units-of-production method.

Every business owns a variety of assets. **Assets** are, put simply, anything a business or an individual owns, and may be conveniently categorised as either current assets or non-current assets. Bank deposits, bank bills, receivables and other assets that can be converted into cash, sold or consumed within the space of one year exemplify the former, while other assets, such as buildings, plant and equipment, usually displaying a life exceeding 12 months, are referred to as non-current assets or sometimes as fixed assets.

You will find that, without exception, all businesses invest in plant and equipment, although obviously the extent of such investment will vary according to the type of industry. For example, the mining industry is a capital-intensive industry which relies heavily on plant and equipment and relatively few workers. Clothing manufacture and the travel and tourism industry on the other hand, with their dependence on a large volume of workers, are labour-intensive.

---

* Sincere thanks to Colin Granger for his invaluable advice in the preparation of this chapter.

Over time, businesses' fixed assets such as plant and equipment deteriorate because of age, use and obsolescence, and the share of the machinery's total service potential that has expired during that period is referred to as **depreciation**. According to Professor Carrick Martin, depreciation describes the writing off of a productive asset over its active life, with each asset 'consisting of a bundle of services'. As these services are progressively used, the asset gradually expires and becomes an expense during that period of use.[1]

Usually, the active life of plant and equipment exceeds a period of one year, and therefore the cost of the equipment cannot be completely written off in the year of purchase. Instead, for taxation purposes a business spreads the capital expenditure over the assumed or useful life of the equipment—that is, the equipment is depreciated (with the estimated life of capital items recommended by the Australian Taxation Office). However, any item purchased after 1 July 1991 at an initial cost of $300 or less or with an effective life of less than three years is entitled to 100% depreciation in the year of purchase.

In review, depreciation can be seen as measuring the use or consumption of productive resources during a given period of time, with the business firm permitted to allocate the cost of using the asset (i.e. depreciation) against the benefits received. Depreciation charges can also be seen as imparting improved quality of income information to shareholders and/or directors alike if the depreciation changes.

6.1 Key terms and principles of depreciation

Businesses should always attempt to keep accurate annual records of the amounts that their assets have depreciated, with the annual depreciation reflecting the share of the particular asset's total life that has expired during the accounting period. This will not only assist shareholders and any potential investors in assessing the financial statements, but it will also allow management to plan for eventual replacement of the asset once its useful life has been realised (and it has been fully depreciated).

As a fixed asset is depreciated on an annual basis over the term of its life, the **total depreciation** or **accumulated depreciation** of an asset at any point in time is the cumulative sum of all depreciation since its purchase.

What happens if you purchase a fixed asset during the financial year rather than at the beginning of the financial year? Since you do not have use of the asset over the entire year, you can only deduct a portion of the annual depreciation charge during that first financial year of ownership—that is, that portion of the year since you acquired the asset. For example, if an asset was bought in April of the financial year (assume the financial year is 1 July to 30 June), then you have had use of the asset for three months of the financial year and can therefore claim one-quarter of the annual depreciation expense (i.e. 3/12 = 1/4) that first year. Likewise, if the purchase had been made in February, 5/12ths of the annual depreciation charge would have been claimed.

The current value of an asset—that is, the unexpired cost or 'remaining bundle of services'—at a particular point in time is found by subtracting the accumulated depreciation from its original cost (or purchase price), and is said to reflect the **book value** (or **closing written-down value**) of an item. Therefore:

Book value = Original cost of asset − Accumulated depreciation

In order to determine the annual depreciation amount—in other words, that portion of an asset's lifetime service potential that has been consumed during the year—three variables must be known, these being the **acquisition cost** of the item, its **estimated life** and its **residual value**. At times, the latter two variables may be difficult to estimate, as predictions of future events must be made. However, in many cases, it is possible to obtain an estimate of the life of the asset from the manufacturer of the plant and equipment or, in the case of a building, from the architect, construction engineer or the like. In other cases, a best estimate must be made based on past experience, advice from insurance companies or, indeed, trade associations. Occasionally the residual value—that is, the estimated value of the asset at the end of its useful life—is also known at the time of purchase. If, for example, you leased a motor vehicle over, say, four years, you will be told at the outset the residual value of the vehicle in four years' time. However, in many other instances, when you have to estimate the residual value in terms of trade-in value, scrap value or sales value, precision is lost since you must estimate the residual value at the time of purchase if you are to accurately determine the annual depreciation amount.

In order to determine the total depreciable amount of an asset, you must therefore subtract the residual value from the original cost. That is:

Acquisition cost − Residual value = Total depreciable amount

or:

Residual value = Acquisition cost − Total depreciable amount

or:

Acquisition cost = Total depreciable amount + Residual value

There are four traditional methods of depreciating non-current or fixed assets: the *straight-line method*, the *units-of-production method*, the *reducing balance method* and the *sum-of-the digits method*. Although the last two named methods are accelerated methods (i.e. the depreciation amounts are larger during the early years of the asset's life), only the straight-line and reducing balance methods are acceptable to the Australian Taxation Office. The straight-line method is generally the preferred depreciation method used by corporations in reporting to shareholders, while the reducing balance method is the preferred method for tax purposes.[2]

In summary, the amount of depreciation will depend on the type and cost of the asset, the length of its estimated useful life, any residual (or scrap) value at the end of its useful life, and the depreciation method selected. Finally, it should also be understood that the depreciation does not represent the replacement value of the asset but is, instead, the amount of the acquisition cost allocated over the total useful life of an asset.

Business mathematics and statistics

6.1 Competency check

1. Distinguish between a current asset and a non-current asset, giving examples of each.
2. (a) What is meant by *depreciation*?
   (b) Why is it necessary for all businesses to allow for depreciation charges?
3. What is the difference between *accumulated depreciation* and *annual depreciation*?
4. How is the book value of an asset calculated?
5. What is meant by *residual value of an asset*?

6.2 Straight-line depreciation (prime-cost) method

The **straight-line method** deducts a fixed amount of the expenditure every year for the duration of the plant's and/or equipment's life and, as suggested earlier, it is often the preferred option in corporate reporting to shareholders.

To determine the annual amount to be depreciated using the straight-line depreciation method:

1. Subtract the residual value from the acquisition cost in order to determine the total depreciation over the asset's useful life.
2. Divide this total depreciation amount by the number of years of its estimated useful life.

$$\text{Depreciation p.a.} = \frac{\text{Acquisition cost} - \text{Residual value}}{\text{Estimated life}}$$

To calculate the straight-line depreciation rate, you can either divide the estimated life of the asset into 100 or divide the annual depreciation amount by the total depreciable amount.

$$\text{Depreciation rate} = \frac{100}{\text{Estimated life}} \quad \text{or} \quad \frac{\text{Annual depreciation}}{\text{Total depreciation}} \times 100$$

Finally, a depreciation schedule can be constructed which illustrates how the asset's depreciation would be displayed on a balance sheet with examples provided below.

Problem: A photocopier was purchased for $10 500 with an expected useful life of five years with no residual value. Find the annual depreciation expense.

Chapter 6 – Depreciation

Solution:

$$\text{Depreciation p.a.} = \frac{\text{Cost} - \text{Residual value}}{\text{Life}}$$

$$= \frac{\$10\,500 - 0}{5}$$

$$= \frac{\$10\,500}{5}$$

$$= \$2100$$

Problem: A company car was purchased at a cost of $34 900. The useful life of the car is four years with a residual of $7500.
(a) Find the annual depreciation charge.
(b) Find the annual straight-line depreciation rate.

Solution:

(a) $\quad \text{Depreciation p.a.} = \dfrac{\text{Cost} - \text{Residual value}}{\text{Estimated life}}$

$$= \frac{\$34\,900 - \$7500}{4}$$

$$= \frac{\$27\,400}{4}$$

$$= \$6850$$

Therefore, the depreciation amount of $27 400 will be allocated evenly over the four-year period at $6850 p.a.

(b) $\quad \text{Annual depreciation rate} = \dfrac{100}{\text{Life}} \quad$ or $\quad \dfrac{\text{Annual depreciation}}{\text{Total depreciation}}$

$$= \frac{100}{4} \quad \text{or} \quad \frac{\$6850}{\$27\,400} \times 100\%$$

$$= 25\%$$

Problem: A delivery truck was bought seven years ago at a cost of $56 800. Its estimated life is eight years, and a residual value of $8400 is expected.[3]
(a) What is the annual depreciation charge?
(b) What is the book value (i.e. the written-down value after depreciation has been deducted) of the truck at the end of the fifth year?

Solution:

(a) $\quad \text{Depreciation p.a.} = \dfrac{\text{Cost} - \text{Residual value}}{\text{Estimated life}}$

$$= \frac{\$56\,800 - \$8400}{8}$$

$$= \frac{\$48\,400}{8}$$

$$= \$6050$$

Thus, the annual depreciation charge is $6050.

(b) The firm has written off $30 250 of the cost of the truck over the five-year period (i.e. $6050 × 5 = $30 250). Hence, the book value at the end of the fifth year would be $26 550 (i.e. $56 800 − $30 250).

Problem: Assume that office furniture is depreciated by the straight-line method over five years. If the equipment originally cost $25 000 and there is no residual value:

(a) What is the depreciation amount per annum?
(b) Compile a five-year depreciation schedule.

Solution:

(a) $$\text{Depreciation p.a.} = \frac{\text{Cost} - \text{Residual value}}{\text{Estimated life}}$$

$$= \frac{\$25\,000 - 0}{5}$$

$$= \frac{\$25\,000}{5}$$

$$= \$5000$$

(b) **Table 6.1** *Five-year depreciation schedule: furniture*

Year	Original cost ($)	Opening written-down value ($)	Depreciation ($)	Closing written-down value ($)
1	25 000		5 000	20 000
2		20 000	5 000	15 000
3		15 000	5 000	10 000
4		10 000	5 000	5 000
5		5 000	5 000	0

In other words, after the first year, the furniture had a closing written-down value (book value) of $20 000, while at the end of the third year its book value was $10 000. Therefore, by the start of the fourth year of use, the furniture had already depreciated by $15 000 and had two more useful years of life. Also, note that at the conclusion of the fifth year, the written-down value is zero. This means that no future depreciation is permitted as the equipment has been completely written off.

Problem: Computer systems can be depreciated by the straight-line system over three years under current Australian taxation rules on self-assessment. If the computer originally cost $31 800 and there is a residual value of $4500:

(a) What is the annual depreciation expense?
(b) What is the depreciation rate?
(c) Complete a three-year depreciation schedule.

Solution:

(a) Depreciation p.a. = $\dfrac{\text{Cost} - \text{Residual value}}{\text{Life}}$

$= \dfrac{\$31\,800 - \$4500}{3}$

$= \dfrac{\$27\,300}{3}$

$= \$9100$

(b) Depreciation rate $= \dfrac{100}{\text{Life}}$ or $\dfrac{\text{Annual depreciation}}{\text{Total depreciation}}$

$= \dfrac{100}{3}$ or $\dfrac{\$9100}{\$27\,300} \times 100\%$

$= 33\%$

(c) **Table 6.2** *Three-year depreciation schedule: computer system*

Year	Original cost ($)	Opening written-down value ($)	Depreciation ($)	Closing written-down value ($)
1	31 800		9 100	22 700
2		22 700	9 100	13 600
3		13 600	9 100	4 500

In this example, note that after the conclusion of the third year, there is a written-down (book value) of $4500—that is, this is its residual value, since the estimated life of three years has elapsed.

6.2 Competency check

1. (a) How is the annual depreciation expense determined using the straight-line depreciation method?
   (b) What is the advantage(s) of using this method?

2. For the following items, find the total depreciation expense, the annual depreciation amount and the rate of depreciation using the straight-line depreciation method.

Business mathematics and statistics

	Acquisition cost ($)	Residual value ($)	Estimated life	Total depreciation ($)	Annual rate of depreciation (%)
(a)	7 500	1 425	5 years		
(b)	176 000	17 400	20 years		
(c)	13 000	0	4 years		
(d)	36 900	8 500	10 years		

3. Mr Sole had a security system installed in his seafood restaurant at a cost of $22 500. It has an estimated life of 10 years, with a residual value of $1500.

   (a) What is the annual depreciation amount?
   (b) What was the book value of the security system at the end of the fourth year?

4. A new taxi has just been bought for $28 000. Construct a depreciation schedule over the life of the taxi if it is depreciated at the rate of 25% p.a. using the straight-line depreciation method and has a residual value of $5000.

5. A stereo system costing $48 000 was recently installed in a reception centre and has an estimated life of 12 years with an estimated residual value of $4800. Assuming the straight-line depreciation method is used:

   (a) What is the annual depreciation expense?
   (b) What is the annual depreciation rate?
   (c) Construct a depreciation schedule for the first three years of its estimated life.
   (d) What is the book value at the end of the third year?

6. An industrial sewing machine originally costing $16 400 was expected to have a useful life of 12 years with a residual value of $2000.

   (a) What was the annual depreciation expense?
   (b) What is the annual depreciation rate?
   (c) How much would the industrial machine have depreciated after four years?

6.3 Reducing balance (diminishing value) method

Another method of depreciation for fixed assets is the **reducing balance method**. According to this approach, the annual depreciation charge declines over the estimated life of the asset; in other words, the asset will decline more in the earlier years on the basis that the benefits of the service provided by the asset are greater in the earlier years and therefore the depreciation expense should reflect this. As the reducing balance method assumes that the depreciation is highest at the outset of the purchase, it is

Chapter 6 – Depreciation

therefore referred to as an **accelerated depreciation method** and is often the preferred option for corporate reporting to the Australian Taxation Office.

To calculate the depreciation in the first year using this method, the depreciation rate is applied to the original cost of the asset. The second and subsequent years' depreciation is calculated by applying this same depreciation rate on the opening written-down value (book value) at the beginning of that year (rather than on the original cost as per the straight-line method).

If you need to calculate the purchase price, the book value (for any nominated period) or the depreciation rate, use the formula:

$$A = P(1-i)^n$$

where P = original purchase price
 i = interest rate per period expressed in decimal form
 n = number of periods
 A = book value (i.e. the written-down value of the item after depreciation has been deducted).

For your information, the Australian Taxation Office determines the reducing balance depreciation rate by taking 1.5 times the straight-line rate (e.g. 7.5% straight-line depreciation is equivalent to 11.25% reducing balance depreciation).

6.3 Competency check

1. What is meant by *accelerated depreciation*?
2. How does the reducing balance method of calculating the annual depreciation expense differ from the straight-line method?
3. Under what circumstances would management consider the reducing balance method appropriate for calculating the depreciation expense?
4. Why is the reducing balance method the preferred option for management when reporting to the Australian Taxation Office?
5. Determine the reducing balance depreciation rate for taxation purposes if the straight-line rate is:
   (a) 15% for video recorders;
   (b) 10% for refrigerated fruit-juice dispensers;
   (c) 20% for poker machines.
6. Calculate the straight-line rate and the reducing balance rate (which is 1.5 times the straight-line rate) for each of the following periods:
   (a) 4 years   (b) 8 years   (c) 10 years   (d) 20 years
7. If you were asked to find the purchase price (P) of an asset but were informed of the rate of depreciation (i), the number of periods (n) and its current book value (A), how would you solve for P?

6.4 Reducing balance depreciation schedule

Just as a depreciation schedule was constructed for the straight-line method of depreciation, we may create a similar schedule for presenting the reducing balance method. However, in the latter case, you will find that the depreciation amount will not be a constant figure over the duration of its useful life (as it was in Tables 6.1 and 6.2); instead, it will be greatest in the first year and decline with each successive period. You may recall that this is a result of the belief that the asset generates its greatest benefit to the owner at the beginning of its life and, accordingly, the depreciation expenses should be greatest at that time as well.

For example, a 22.5% depreciation rate (using the reducing balance rate) applies to a car used for business purposes. Depreciation for the first four years on a vehicle that originally cost $34 000 would be as shown in Table 6.3. (Note that only whole dollars are used.)

Table 6.3 *Reducing balance method: motor vehicle*

Year	Original cost ($)	Opening written-down value ($)	Depreciation (22.5%) ($)	Closing written-down value ($)
1	34 000		7 650	26 350
2		26 350	5 929	20 421
3		20 421	4 595	15 826
4		15 826	3 561	12 265
5		12 265	2 760	9 505

The depreciation amount is determined as follows:

1. Multiply the opening written-down value by the depreciation rate of 22.5% except for year 1, when the original cost ($34 000) is multiplied by the 22.5% depreciation rate.
2. Deduct the depreciation of the first year ($7650) from the original cost ($34 000) to determine the closing written-down value (or book value) at the end of year 1 (i.e. $34 000 − $7650 = $26 350).
3. Transpose year 2's opening written-down value from the previous year's closing written-down value (i.e. $26 350) and multiply by the rate of 22.5% to get $5929, which is the depreciation expense for year 2.
4. This leaves $20 421 for the closing written-down value for the same year (i.e. $26 350 − $5929 = $20 421) as well as the opening book value for year 3.

Problem: The restaurateur, Pierre Pomme Frites has requested his accountant to prepare a depreciation schedule for the $165 000 spent on renovations for his El Berato cafe. Construct a depreciation schedule for the renovations for the first four years using the reducing balance method at a depreciation rate of 30%.

Chapter 6 – Depreciation

Solution:

Table 6.4 *Reducing balance method: renovations*

Year	Original cost ($)	Opening written-down value ($)	Depreciation (30%) ($)	Closing written-down value ($)
1	165 000		49 500	115 500
2		115 500	34 650	80 850
3		80 850	24 255	56 595
4		56 595	16 979	39 616

Note that the first year experienced the highest depreciation expense, with each successive year illustrating reduced charges.

Problem: Office equipment costing $30 000 can be depreciated by using either the straight-line or the reducing balance method. If the useful life of the equipment is 10 years and there is no residual value, construct a five-year depreciation schedule for:

(a) straight-line depreciation;
(b) reducing balance depreciation.
(c) Which of the two methods would you recommend to the manager for tax reporting purposes? Why?

Solution:

(a) *Straight-line depreciation method*

$$\text{Depreciation amount} = \frac{\text{Cost}}{\text{Life}}$$
$$= \frac{30\,000}{10}$$
$$= \$3000$$

Thus, depreciation of $3000 p.a. for 10 years (i.e. 10% p.a. straight-line depreciation) will be allowed as a tax deduction.

Table 6.5 *Straight-line method: office equipment*

Year	Original cost ($)	Opening written-down value ($)	Depreciation (10%) ($)	Closing written-down value ($)
1	30 000		3 000	27 000
2		27 000	3 000	24 000
3		24 000	3 000	21 000
4		21 000	3 000	18 000
5		18 000	3 000	15 000

(b) *Reducing balance method*

The reducing balance depreciation rate is 1.5 times the straight-line rate, and as the latter is 10% in this example (i.e. 100/10 or 3000/30 000 × 100), then a reducing balance depreciation rate of 15% applies.

Table 6.6 *Reducing balance method: office equipment*

Year	Original cost ($)	Opening written-down value ($)	Depreciation (15%) ($)	Closing written-down value ($)
1	30 000		4 500	25 500
2		25 500	3 825	21 675
3		21 675	3 251	18 424
4		18 424	2 764	15 660
5		15 660	2 349	13 311

(c) The depreciated amounts are higher for the first three years when applying the reducing balance method and, as such, the company reduces its taxable income by a greater amount than by using the straight-line depreciation method. From the fourth year onward, the straight-line depreciation method's annual depreciation expense exceeds the reducing balance, but since businesses prefer larger tax write-offs sooner rather than later, the reducing balance method would be more favoured for taxation purposes.

The straight-line depreciation method with its lower depreciation charges during the first three years permits management to report better net profit results to its owners or shareholders, while the higher reducing balance depreciation expenses in those same years permit management to claim higher expenses when calculating their tax liability.

6.4 *Competency check*

1. Why is it that in spite of the depreciation rate remaining constant, the depreciation amounts decline each year when using the reducing balance method of depreciation?

2. As a court judge you are permitted to claim a depreciation rate of 20% when using the straight-line method and 30% when applying the reducing balance method (i.e. reducing balance rate = 1.5 times the straight-line rate) for your judge's robe. What is the written-down value of the robe at the end of the second year if it originally cost $750 using:

    (a) the straight-line method?
    (b) the reducing balance method?

3. Applying the reducing balance method, find the first year's depreciation of a stereo system which originally cost $1600 with a useful life of five years.

4. Determine the written-down value of fittings at the end of the third year if they cost $5000 and they can be depreciated at an annual rate of 15% (reducing balance method).

5. An office building cost $5 million and had an estimated life of 40 years. Prepare a depreciation schedule for the first three years of its life using the reducing balance method. (Remember that the depreciation rate using this method is equivalent to 1.5 times the straight-line rate.)

6. Construct a three-year depreciation schedule for a conference table and chairs originally costing $8500 and depreciated at an annual rate of 15% (reducing balance method).

6.5 Reducing balance method: the formula

You may find that the depreciation rate is not given, or you could be asked to determine an asset's projected book value or even its purchase price. At such times, the formula below will be a time-saver, as well as reduce the likelihood of error in calculation since it will entail only a few steps to derive the correct answer.

$$A = P(1 - i)^n$$

where P = original purchase price
 i = interest rate per period expressed in decimal form
 n = number of periods
 A = book value (i.e. the written-down value of the item after depreciation has been deducted).

In order to ensure that you understand and apply this formula to real-life problems, examples of solving for each variable of the formula follow.

Problem: Applying the reducing balance method, determine the current written-down value (book value) of machinery purchased six years ago for $145 000 if it is depreciated at 15% p.a.

Solution: $A = P(1 - i)^n$

As P = $145 000, i = 15% and n = 6,
A = $145 000(1 − 0.15)^6$
  = $145 000(0.85)^6$
  = $54 686.68

In other words, the machinery, originally costing $145 000, has a book value of $54 687 at the end of six years of use.

Note: The exponential result was taken to five decimal places. Should you use a different number of decimal places, the final depreciation charge will vary marginally.

Problem: Applying the same method, calculate the book value of a motor vehicle at the end of the third year if it originally cost $39 300 and is depreciated at 22.5% p.a. (reducing balance method).

Solution: $A = P(1 - i)^n$

As $P = \$39\,300$, $i = 22.5\%$ and $n = 3$,
$A = \$39\,300(1 - 0.225)^3$
$= \$39\,300(0.775)^3$
$= \$18\,293.54$

Therefore, the $39 300 car has been depreciated to a value of $18 294 after three years.

Just to make life a bit more complicated, try the next problem.

Problem: A printing company purchased a printing press costing $84 500, with a reducing balance depreciation rate of 30%.

(a) Find its book value after seven years.
(b) What was the depreciation charge for the seventh year?

Solution:
(a) This part simply requires the substitution of the appropriate figures into the formula below:

$A = P(1 - i)^n$
$= \$84\,500(1 - 0.3)^7$
$= \$84\,500(0.70)^7$
$= \$6958.94$

The printing press, which originally cost $84 500, has a written-down value of $6959.

(b) To calculate the depreciation charge for the seventh year without having to construct a depreciation schedule, calculate the book value at the end of the sixth year and subtract this result from the book value we calculated for the seventh year (i.e. $6959). Thus:

$A = P(1 - i)^n$
$= \$84\,500(1 - 0.3)^6$
$= \$84\,500\,(0.7)^6$
$= \$9941.34$

If the book value at the end of the sixth year was $9941 and the seventh year $6959, then the depreciation charge is the difference between the two years in book values. That is:

Depreciation charge for seventh year = $9941 − $6959
= $2982

Chapter 6 – Depreciation

What happens if you are required to determine the original cost of the asset? Once again, it will simply be a matter of substituting all the figures into the given formula except for *P* which you will solve for.

Problem: A computer purchased three years ago has a book value of $5110. If a reducing balance rate of depreciation of 30% applies, what was the original cost?

Solution: To solve for *P*:

$$A = P(1 - i)^n$$

As $A = \$5110$, $i = 0.3$ and $n = 3$,

$$\$5110 = P(1 - 0.3)^3$$
$$\$5110 = P(0.7)^3$$
$$\$5110 = P(0.343)$$
$$\frac{\$5110}{0.343} = P$$
$$\$14\,897.96 = P$$

Therefore, the computer's original cost was $14 898.

You can check the answer by, for example, substituting $14 898 for *P* in the formula and solving for, say, the book value *A* at the end of the third year. Therefore:

$$A = P(1 - i)^n$$
$$= \$14\,898\,(1 - 0.3)^3$$
$$= \$14\,898\,(0.7)^3$$
$$= \$14\,898(0.343)$$
$$= \$5110.01$$

As the calculated book value of $5110 is the same as the given book value, the original cost of $14 898 must be correct.

Before concluding this particular section, you should also be aware of how to calculate the reducing balance depreciation rate (if not provided with the straight-line rate).

Problem: A video camera purchased for business purposes four years ago at a cost of $1499 is now worth $541. If the reducing balance method is applied, what was the depreciation rate?

Solution: To solve for *i*:

$$A = P(1 - i)^n$$

As $A = \$541$, $P = \$1499$ and $n = 4$,

$$\$541 = \$1499(1 - i)^4$$

$$\frac{\$541}{\$1499} = (1 - i)^4$$

$$\sqrt[4]{\frac{\$541}{\$1499}} = 1 - i$$

$$0.77508 = 1 - i$$

$$i = 1 - 0.77508$$

$$= 0.22492$$

$$= 22.5\%$$

To check the answer, substitute 22.5% for i and solve for A:

$$A = P(1 - i)^n$$
$$= \$1499(1 - 0.225)^4$$
$$= \$1499(0.775)^4$$
$$= \$1499(0.36075)$$
$$= \$540.76$$
$$= \$541$$

6.5 Competency check

1. What is the formula when solving for an unknown when using the reducing balance method?

2. Why bother using this formula for, say, finding the book value of an asset after its fourth year if you could just as well prepare a depreciation schedule to discover that same answer?

3. (a) For parts (i) to (iii) find the first year's depreciation using the reducing balance method.

Table 6.7 Calculating depreciation using the reducing balance method

	Original cost ($)	Residual value ($)	Rate of depreciation (%)	First year's depreciation ($)
(i)	4000	500	15	
(ii)	280	0	11.25	
(iii)	1350	200	20	

(b) Calculate the third year's depreciation for (i) to (iii) using the reducing balance formula.

4. Find the book value of the following items after six years using the reducing balance formula.

	Asset x	Asset y
Original cost	$17 000	$5 500
Depreciation rate	7.5%	15%

5. Ms Dye O'Hare opened a beauty salon five years ago with equipment costing $46 000. If a reducing balance rate of depreciation of 30% is applicable, what is the book value at the end of the fifth year?

6. Aircraft used predominantly for agricultural spraying or dusting are eligible for a 37.5% depreciation allowance (reducing balance method). If an aircraft originally cost $375 000:

   (a) What will be the book value at the end of the fourth year?
   (b) What was the depreciation charge for the fourth year?

6.6 Units-of-production method

The **units-of-production depreciation method** is based on the notion that an asset's life reflects the physical usage of the asset. In fact, this method can be thought of as very similar to the straight-line depreciation method except that rather than being based on estimated life in terms of years, the depreciation of an asset is based on usage. Therefore, the depreciation amount calculated each year will depend on the degree of usage of the asset. In other words, if the asset is heavily used, it is heavily depreciated; should it be infrequently used, then the depreciation expense for that period will be considerably less.

To determine the depreciation expense using the units-of-production method, you must first determine the depreciation cost per unit as follows:

1. Subtract the residual value from the acquisition cost in order to determine the total depreciation of the asset.
2. Divide the total depreciation (determined in the first step) by the estimated number of units used, as this results in the depreciation per unit of production.
3. Multiply the units of production of each year by the depreciation cost per unit (derived in previous step) to arrive at the depreciation cost for each year.

$$\text{Depreciation expense per unit} = \frac{\text{Acquisition cost} - \text{Residual value}}{\text{Estimated units of use}}$$

For example, if a lawn mower is purchased for $760 with a residual value of $100 and an estimated engine life of 600 hours, then the depreciation charge on a hourly basis would be:

$$\text{Depreciation per hour} = \frac{\$760 - \$100}{600}$$

$$= \frac{\$660}{600}$$

$$= \$1.10/\text{hour}$$

Business mathematics and statistics

Therefore, if the lawn mower is used for 150 hours the first year, 170 the second, and 48 the third, the depreciation charges would be:

First year = 150 hours × $1.10/hour = $165
Second year = 170 hours × $1.10/hour = $187
Third year = 48 hours × $1.10/hour = $52.80 or $53

Problem: A refrigerated milk truck cost $64 500 and has an estimated useful life of 60 000 kilometres and a trade-in value of $7500. If the milk truck was driven 18 000 kilometres this year, how much should the milk vendor deduct as a depreciation expense if applying the units-of-production method?

Solution:

$$\text{Depreciation per kilometre} = \frac{\text{Acquisition cost} - \text{Residual value}}{\text{Estimated total kilometres usage}}$$

$$= \frac{\$64\ 500 - \$7500}{60\ 000}$$

$$= \frac{\$57\ 000}{60\ 000}$$

$$= \$0.95$$

Since the cost per kilometre is 95 cents, the depreciation expense for 18 000 kilometres is:

Year's depreciation expense = Depreciation per unit × Number of units
= $0.95 × 18 000
= $17 100

Problem: A photocopier was purchased for $7500 and had a residual value of $250 as well as an estimated useful life of 250 000 copies. Table 6.8 shows the number of copies made over the six years of use. Determine the depreciation expense for each year.

Table 6.8 *Units-of-production method: photocopier*

Year	Units copied
1	35 259
2	39 354
3	48 765
4	59 805
5	36 202
6	42 423

Chapter 6 – Depreciation

Solution: First, determine the depreciation cost per copy.

$$\text{Depreciation cost per copy} = \frac{\text{Acquisition cost} - \text{Residual value}}{\text{Total estimated number of copies}}$$

$$= \frac{\$7500 - \$250}{250\,000}$$

$$= \frac{\$7250}{250\,000}$$

$$= 0.029$$

Therefore, each photocopy cost 2.9 cents. Applying this unit charge to the number of copies produced gives the results shown in Table 6.9:

Table 6.9 *Units-of-production method: photocopier*

Year (col. 1)	Units copied (col. 2)	Depreciation for year ($) (col. 3) (col. 2 × 0.029)	Total depreciation ($) (col. 4)	Book value ($) (col. 5) ($7500 − col. 4)
				7 500
1	35 259	1 023	1 023	6 477
2	39 354	1 141	2 164	5 336
3	48 765	1 414	3 578	3 922
4	59 805	1 734	5 312	2 188
5	36 202	1 050	6 362	1 138
6	42 423	888*	7 250	250

Note: *Year 6's depreciation was adjusted to $888 instead of a calculated $1230 (42 423 × 0.029) in order for the book value not to fall below the residual value of $250. This revision can be explained by remembering that the life of the photocopier was estimated at the time of purchase, but due to the actual life of the photocopier exceeding the forecast life an adjustment of the depreciation for the year (column 3) was necessary.

The units-of-production method may be considered by some to be more precise than the straight-line method in determining the annual depreciation expenses, as it reflects the actual physical usage of the asset. However, the problem is that while the straight-line method is very easy to calculate, the units-of-production approach requires regular data collection which is an additional expense to business not usually relished.

Business mathematics and statistics

6.6 Competency check

1. What forms the basis of the units-of-production depreciation method?
2. How does this method differ from the straight-line depreciation method?
3. What is a possible advantage of the units-of-production method when compared to the straight-line depreciation method?
4. How is the depreciation per unit of usage determined?
5. Tread-Softly tyres cost $450 each and have an estimated life of 65 000 kilometres and no residual value.
   (a) What is the depreciation expense per kilometre per tyre?
   (b) If 20 500 kilometres were travelled on the tyres in 1997, what is the depreciation expense for that year (using the units-of-production method)?

6.7 Summary

Fixed assets such as buildings, motor vehicles, or plant and equipment used for income-producing purposes normally have effective lives of greater than one year. In other words, these assets are expected to be in use for more than just one year and, as a result, the original cost of the asset cannot be written off completely in the year of purchase but can be depreciated over its effective life (which is defined by the Australian Taxation Office as 'the estimated life . . . over which the item can be expected to be used for income producing purposes'). The effective life is determined by the asset's physical deterioration and obsolescence. As wear and tear or obsolescence takes hold, the asset increasingly loses its ability to create benefits for the business.

Two methods of depreciation are commonly used, as these are considered acceptable by the Australian Taxation Office: the straight-line (or prime-cost) method and the reducing balance (the diminishing value) method.

Under the straight-line depreciation method, a constant depreciation amount is deducted every year for the estimated life of the item. The formula for calculating the amount of depreciation per annum is:

$$\text{Depreciation p.a.} = \frac{\text{Acquisition cost} - \text{Residual value}}{\text{Estimated life}}$$

The depreciation rate can be found by dividing the estimated number of years of useful life of the asset into 100, or by dividing the total depreciable amount by the annual depreciation amount.

The reducing balance method, on the other hand, is based on the assumption that as the future benefits of an asset to a business are usually greatest at the beginning of its use, the depreciation expense should, accordingly, be greatest at this time. The annual depreciation charge is based on a fixed rate of depreciation applied to the book value (i.e. the written-down value). The fixed rate is applied at the beginning of each year, and the first year's depreciated amount is found by multiplying the fixed depreciation rate by the original cost of the article. The formula used for the reducing balance method is:

$$A = P(1-i)^n$$

where A = written-down or book value at the end of period
       P = original purchase price
       i = interest rate per period expressed in decimal form
       n = number of periods.

The reducing balance depreciation rate is 1.5 times that of the straight-line rate and is therefore referred to as accelerated depreciation. The straight-line method is used by most companies when reporting to their shareholders, while the reducing balance method is favoured when management reports to the Australian Taxation Office.

The units-of-production approach to depreciation reflects the actual usage of an asset rather than assuming there is a constant depreciation amount or rate each year. The depreciation amount for each unit used is found using the following formula:

$$\text{Depreciation rate} = \frac{\text{Acquisition cost} - \text{Residual value}}{\text{Total estimated number of units}}$$

Its calculation is very similar to that of the straight-line depreciation approach (with straight-line depreciation divided by the estimated life in years instead of by units of life), but its major drawback is the time and energy required by management in record-keeping and data collection to determine the annual usage. This contrasts markedly with the simple calculations when using the straight-line method. Furthermore, often when applying either of these two approaches the depreciation charges do not differ sufficiently to warrant the additional expense incurred in using the units-of-production approach.

Finally, it must be understood that the annual depreciation expense does not reflect the current cost of replacement of the asset but, instead, allows for the historical cost of a fixed asset to be spread over the term of its effective or estimated life.

6.8 Multiple-choice questions

1. Which of the following is not considered an asset?

    (a) plant and equipment;
    (b) bank deposits;
    (c) business loan;
    (d) summer home.

2. To determine the annual depreciation amount, the three variables that must be considered are:
    (a) accumulated depreciation, estimated life and residential value;
    (b) straight-line, unit-of-production and reducing balance;
    (c) accelerated depreciation, book value and estimated life;
    (d) acquisition cost, estimated life and residual value.

3. When converting the depreciation rate from straight-line to reducing balance, one must:

   (a) multiply the rate of the former (i.e. straight-line) by 1.5;
   (b) divide the rate of the former by 1.5;
   (c) multiply the rate of the former by 2;
   (d) keep the rate constant when using either method.

4. Tread-Softly tyres cost $450 each and have an estimated life of 35 000 kilometres and no residual value. What is the depreciation expense per kilometre per tyre (using the units-of-production method)?

   (a) 1.29 cents/kilometre (b) 0.0128 cents/kilometre
   (c) 7.7 cents/kilometre   (d) 0.78 cents/kilometre

5. After six years of use, Ms R. Tide's laundromat of 10 washing machines costing $62 800 in total is now worth $13 600. What is the appropriate depreciation rate if the reducing balance method is applied?

   (a) 22.5%   (b) 15%   (c) 13%   (d) 20%

6.9 Exercises

1. Why is it difficult, at times, to determine the residual value of an asset?

2. What is the residual value of an asset if it cost $2500 at the time of purchase and the total depreciable amount is $2050?

3. Find the original cost of an asset if the total depreciable amount is $5500 and it has a residual value of $300.

4. What are the three variables that must be known in order to calculate the annual depreciation expense?

5. Your manager has requested that you calculate the current book value of each of the fixed assets listed below.

Equipment	Cost of equipment ($)	Accumulated depreciation ($)
(a)	250 000	207 500
(b)	84 750	73 875
(c)	46 990	40 090
(d)	164 250	81 500

6. A motorcycle costing $6000 is expected to have a trade-in value of $2000 after four years. What is the annual depreciation amount if calculated on a straight-line basis?

7. A 10-speed bicycle bought for $750 in 1992 was sold five years later for $250. Calculate the annual depreciation amount, using the straight-line method.

Chapter 6 – Depreciation

8. The book value of a straight-line depreciated six-year-old machine is $18 000. If the machine originally cost $54 000, what is its useful life, assuming that there is no residual value?

9. A stereo system was recently on sale for $1096. Its estimated useful life is eight years, with no residual value at the end of that time. What would be the book value (to the nearest dollar) after three years by the straight-line method?

10. A dishwasher originally costing Alex $915 was traded in for $195 after six years.

    (a) What was the annual depreciation expense (using the straight-line method)?
    (b) What was the annual depreciation rate?

11. Ms I. O'Dine, a pharmacist, operates a chemist shop which she opened five years ago. The fittings' current written-down value (book value) is $25 500. If the depreciation rate (using the straight-line depreciation method) is 10%:

    (a) How much did the fittings originally cost Ms I. O'Dine if there is no expected residual value,
    (b) What is the annual depreciation expense?
    (c) What would be the annual depreciation expense if there had been a residual value of $5500 on the fittings?

12. As Mr Bedford's accountant, construct the depreciation schedule using the straight-line method for a bulk-tipper bought for $36 000 with an estimated life of five years and a residual value of $8500.

13. A 15% reducing balance depreciation rate applies to camels used in business ventures. If the original cost of a herd of camels (not sure if fleet discounts apply in this case) was $25 000, how much will the herd be worth after four years?

14. Taxis are depreciated at 37.5% p.a. (reducing balance). If a taxi's book value at the end of three years is $7350, what was the original cost of the taxi (to the nearest $1000)?

15. Using the reducing balance formula, determine the written-down value (book value) of plant and equipment (to the nearest dollar) at the end of the:

    (a) third year if the original cost was $80 000 at 22.5% depreciation p.a.;
    (b) fifth year if the original cost was $18 250 at 30% depreciation p.a.;
    (c) eighth year if the original cost was $41 750 at 37.5% depreciation p.a.;
    (d) second year if the original cost was $32 150 at 30% depreciation p.a.

16. Prepare a five-year depreciation schedule for poker machines just purchased at the local RSL for $80 000 and depreciated at a rate of 30% p.a. (reducing balance method).

17. At what annual rate (using the reducing balance method and to the nearest per cent) have a conference table and chairs been depreciated if they are worth $3325 after five years and originally cost $7500?

18. A hotel had new carpet laid seven years ago at a cost of $256 000. The carpet's book value is now $21 083. What rate of depreciation applies if it is calculated by the reducing balance method?

19. A computer bought seven years ago for $4500 is now worth $371 (book value).

    (a) What is the rate of depreciation (reducing balance)?
    (b) What was the depreciation charge for the seventh year?

20. Aircraft used predominantly for agricultural spraying or dusting are eligible for a 37.5% depreciation allowance (reducing balance). If an aeroplane originally cost $340 000:

    (a) What will be the book value at the end of seven years?
    (b) What was the depreciation charge for the seventh year?
    (c) Construct a depreciation schedule for the first six years.

21. Poker machines costing $145 000 were bought five years ago. Prepare a five-year depreciation schedule if:

    (a) the poker machines are depreciated at the rate of 20% p.a. (straight-line method), with no residual value;
    (b) the poker machines are depreciated at the rate of 30% p.a. (reducing balance method).

22. Calculate the depreciation for the units of equipment used the following.

	Original cost ($)	Residual value ($)	Estimated life	Units used
(a)	2 760	0	800 hrs	151
(b)	19 000	2 000	4 000 hrs	1 440
(c)	6 375	0	300 000 units	56 000
(d)	115 000	14 500	150 000 km	39 500

23. A taxi cost $28 100 new, with a residual value of $5000 and an estimated life of 220 000 kilometres. If the taxi was driven 70 400 kilometres in its first year, what was the depreciation expense that year if the units-of-production approach was applied?

Chapter 6 – Depreciation

Notes

1. Carrick Martin, *An Introduction to Accounting* (Sydney: McGraw-Hill Book Co., 1987), p. 150.
2. Plant acquired after 26 February 1992 has the following general depreciation rates, applying with the new schedule not applying to 'works of art' or passenger vehicles:

Years in effective life	Annual depreciation percentage	
	Diminishing value	Prime cost
Less than 3	Immediate	Immediate
3 to less than 5	60	40
5 to less than 6 2/3	40	27
6 2/3 to less than 10	30	20
10 to less than 13	25	17
13 to less than 30	20	13
30 and over	10	7

Source: Australian Taxation Office, *Tax Pack 1996*, p. 52.

Also, from 1 July 1991, taxpayers have been permitted to self-assess the effective life of their depreciable plants acquired after 12 March 1991, having regard to their particular circumstances of use. Alternatively, taxpayers may elect to use the Commissioner's published determination of effective years (Australian Taxation Office, *Tax Pack 1996*, p. 52.)

3. For further explanation of residual value and appropriate calculation, refer to Horngren, Harrison, Best, Fraser and Izan, *Accounting* 2nd edition (Sydney: Prentice Hall Australia, 1997).

Chapter 7
LINE GRAPHS

LEARNING OUTCOMES
After studying this chapter you should be able to:
- Plot an ordered pair of data of a given linear function to form a straight-line graph
- Interpret linear function (straight-line) graphs
- Solve simple simultaneous equations using graphs
- Apply the principle of simultaneous equations to break-even analysis
- Examine non-linear business-related graphs.

To simplify a collection of data, a visual presentation of the collected data is often desirable. Such presentations vary from line graphs to pie charts, with a large variety requiring the plotting of such information on appropriate graph paper. Furthermore, much of the data we collect, if presented graphically, approaches a straight line. Therefore, an understanding of how to graph statistical information, as well as how to apply it to business decisions (such as break-even analysis), could well prove invaluable in the workplace.

7.1 Graphing linear equations

In most disciplines, you will find that observations of data are presented in the form of **ordered pairs**, such as in the rate of change in inflation and average weekly earnings; advertising expenditure and sales turnover; and the growth in property values versus stock market growth, Gross Domestic Product (GDP) and the Federal Budget deficit, to name but a few.

The first coordinate (numbers in *ordered pairs* are referred to as **coordinates**) within each set of parentheses corresponds to the x value, or **abscissa**; the second coordinate is the **ordinate**, or y value.

Note: 'Order' implies that (x, y) is different from (y, x). See below.

Business mathematics and statistics

For example, Table 7.1 shows the exports and imports of Australia's merchandise trade from 1989/90 to 1994/95. How may these be presented as ordered pairs of observations?

Table 7.1 *Australia's exports and imports, 1989/90 to 1995/96*

Year	Exports ($ billion)	Imports ($ billion)
1989/90	49	51
1990/91	52	49
1991/92	55	51
1992/93	60	59
1993/94	64	64
1994/95	66	75
1995/96	75	77

Source: Australian Bureau of Statistics, *Balance of Payments,* Australia (Cat. No. 5302.0).

To convert these results into ordered pairs, just select the exports and imports for each year. In other words, (49, 51), (52, 49), (55, 51), (60, 59), (64, 64), (66, 75) and (75, 77) correspond to the results presented for the period 1989/90 to 1995/96.

The visual presentation of such data may be shown on a line graph which is constructed by using two perpendicular axes by convention (with the x values on the horizontal axis and the y values on the vertical axis), with the point of intersection known as the point of **origin** and labelled 0.

Figure 7.1 illustrates the two axes of x and y. Observe how to the right of the origin (0) there are positive ascending values on the x axis, while to the left of the origin there are negative ascending values. Similarly, upwards from the origin (0) positive ascending values are found for y, and downwards from the origin are negative values. These two scales on the x and y axes are known as **Cartesian coordinates**.

The visual presentation of data (such as the export and import results) may be undertaken by means of plotting each of the ordered pairs on these scales—that is, the ordered x and y value represents a single point on the line graph. Therefore, (49, 51) would be one point and (52, 49) would be another, and so on.

Figure 7.1 *The x and y axes*

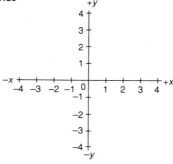

Chapter 7 – Line graphs

To plot data, count the number of graduations horizontally or vertically from the point of origin corresponding to the data. For example, when presented with the points A(–1, 3) and B(4, –5) the first coordinate within each set of parentheses corresponds to the x value while the second coordinate is the y value (see Figure 7.2).

Figure 7.2 *Plot of A(–1, 3) and B(4, –5)*

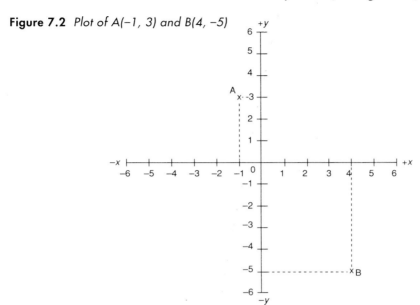

Point (–1, 3) means that the first number, the x value, is –1, so on the x axis go one graduation to the left of the point of origin. The second number, the y value, is 3, so go three graduations up from –1 (the x value). Point (4, –5) indicate 4 on the abscissa and –5 on the ordinate, so we count four graduations along the x axis (as the x value is positive, it is marked off to the right of the point of origin) and down vertically five graduations (corresponding to a y value of –5).

Problem: Plot A(3, 4), B(3, –5) and C(–2, –2).

Solution: See Figure 7.3.

Problem: Plot the trading results for Australia over the 1989/90 to 1995/96 period—that is, A(49, 51), B(52, 49), C(55, 51), D(60, 59), E(64, 64), F(66, 75), G(75, 77).

Solution: See Figure 7.4.

How, then, do you plot an equation such as $y = 4x + 7$? All you have to do is give hypothetical values to x and solve for y. For example, in $y = 4x + 7$, make up a table of values for x, solve for y and then plot.

Figure 7.3 Plot of A(3, 4), B(3, −5) and C(−2, −2)

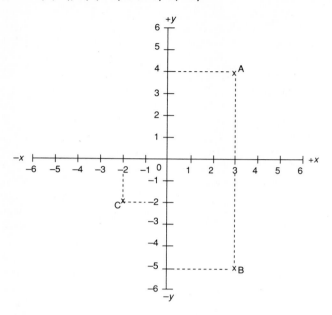

Figure 7.4 Australia's exports and imports, 1989/90 to 1995/96

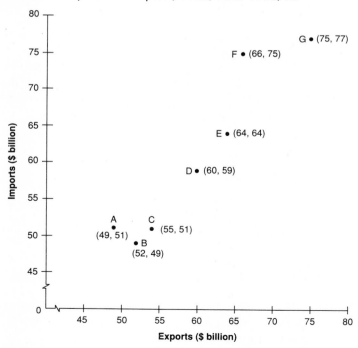

Chapter 7 – Line graphs

	A	B	C	D
x	0	2	−1	−5
y	7	15	3	−13

Plot these points and observe that when connecting these a straight line is the result (see Figure 7.5). Although a straight line can be drawn by using only two points, it is recommended to use at least three points as a check.

An equation in the form of $y = mx + b$ (or $y = a + bx$) is known as a linear equation because when the equation is represented graphically, it is depicted in the form of a straight line (as in Figure 7.5). The gradient or slope of a line—that is, 'm' in the equation—is defined as the change in the ordinate (the y coordinate) divided by the change in the abscissa (the x coordinate) between two points on the line. Thus, the slope indicates how steeply the line slants and is the coefficient of x in the equation. The y intercept—that is, that point where the straight line intersects the y axis—is the value for b.

Figure 7.5 *Graph of line: y = 4x + 7*

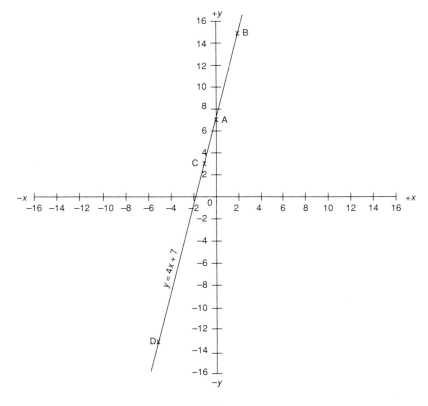

Business mathematics and statistics

In the equation $y = 4x + 7$, the slope, or the coefficient of x, is 4 (indicating that there are four graduations in the y coordinate for each graduation in the x coordinate). This can be confirmed by selecting any two points used in plotting the data—for example, $A(0, 7)$ and $B(2, 15)$—and determining the slope:

$$\frac{\text{The change in } y}{\text{The change in } x} \quad \frac{15-7}{2-0} = \frac{8}{2} = 4$$

The y intercept, b, is +7 and this too can be confirmed by letting $x = 0$ in the equation (since at the y intercept the x coordinate necessarily equals zero). Thus:

$$y = 4(0) + 7$$
$$y = 7$$

Hence the coordinates at the y intercept in this example are (0, 7).

Problem: Plot $y = 3x - 2$. Determine the slope and y intercept.

Solution: Construct a table of values for x and solve for y.

	A	B	C	D
x	0	2	4	−1
y	−2	4	10	−5

Figure 7.6 illustrates the graphing of $y = 3x - 2$ with a slope of 3 and a y intercept of −2.

Problem: The director of a company earns double the salary of his assistant. Graph this word statement. What are the slope and the y intercept?

Solution: To convert this word statement into an equation, if we let the manager's salary equal y, it is equal to double the assistant manager's salary ($2x$). Therefore, the equation would read:

$$y = 2x$$

To plot this, choose any value for x and calculate the resulting y value. For example, if we let $x = -2, 0, 2$ and 4, we have:

	A	B	C	D
x	−2	0	2	4
y	−4	0	4	8

The graphic representation, therefore, of $y = 2x$ is as illustrated in Figure 7.7.

Figure 7.6 Graph of line: y = 3x − 2

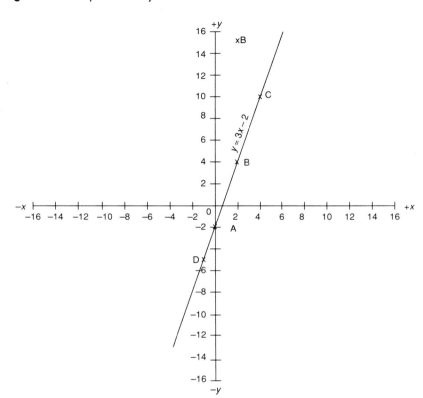

From this, we can determine the manager's salary when told the assistant's wage and vice versa. The slope is 2, indicating that for every graduation change in *x* (the assistant's salary) there is twice the graduation in the director's salary. There is no *y* intercept as there is no value for *b*. In fact, the plotted line intersects the point of origin. If we solve for *y* by letting *x* equal zero, *y* will also equal zero, thus confirming that the straight line intersects the point of origin (0, 0).

Problem: Graph 3*y* = 12. Determine the slope and the *y* intercept.

Solution: Note that there is no *x* value, and hence no slope. Instead, there exists only *b* or the *y* intercept. Solving for *y* in the equation 3*y* = 12, *y* = 4. Therefore, to plot this equation draw a straight line parallel to the *x* axis through (0, 4). (See Figure 7.8.)

Business mathematics and statistics

Figure 7.7 Graph of line: y = 2x

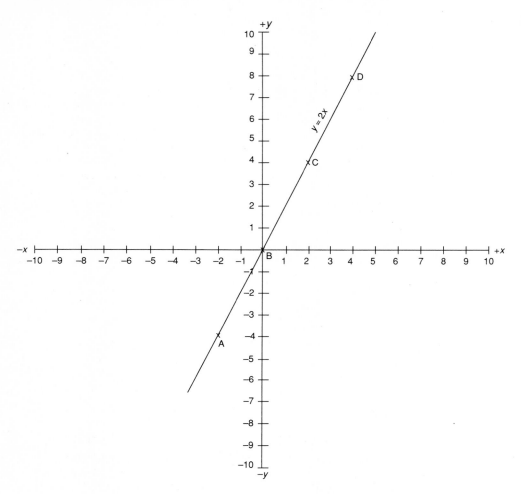

Figure 7.8 *Graph of line: 3y = 12*

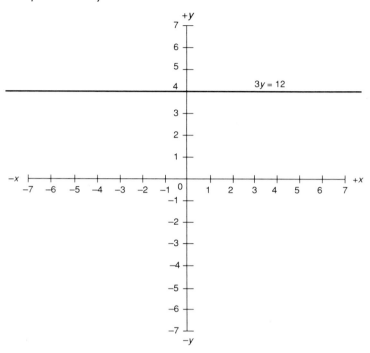

7.1 Competency check

1. Observations of data can be presented in the form of ordered pairs such as sales turnover and profit, inflation rate and interest rates, wage growth rate and unemployment rate, etc. Can you think of another two ordered pairs of observations which have not already been presented in this chapter?

2. Plot and clearly label each of the following points on Cartesian scales:

    (a) (5, 1)   (b) (−3, 5)   (c) (4, −4)   (d) (−4, −6)

3. The construction of a line graph by using two perpendicular axes results in a point of intersection of the x and the y scales. What is this point of intersection referred to as?

4. What is the slope and the y intercept for each of the following?

    (a) $x = 7$ (b) $y = 5x$ (c) $x = y$ (d) $2y = 6x - 8$

5. Plot $3y = 2x$.

Business mathematics and statistics

7.2 Graphing simultaneous equations

Just as single equations may be graphed, **simultaneous equations** may be solved by using the graphic method. Each equation is individually plotted with the equation's point of intersection indicating the coordinates satisfying both equations. It should be understood that some simultaneous equations may have no solution—that is, they do not intersect. An example is parallel lines (as they have an identical slope).

The precision is sometimes lost when extracting the points of intersection from the graph, when compared with determining the x and y values by using an algebraic method.

Problem: Solve for x and y using the graphic method:

$$x + 2y = 5 \quad (1)$$
$$2x - y = 5 \quad (2)$$

Figure 7.9 Intersection of graphs: $x + 2y = 5$ and $2x - y = 5$

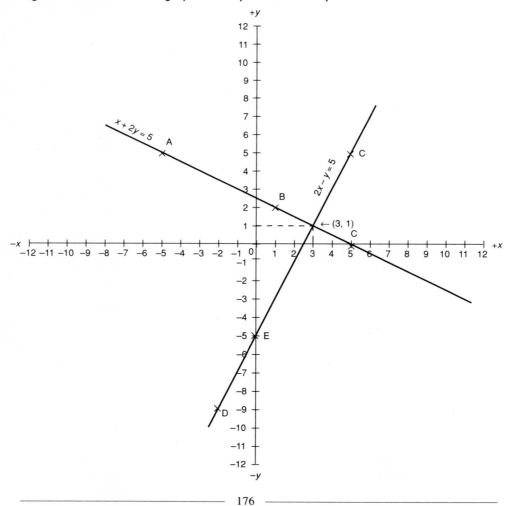

Chapter 7 – Line graphs

Solution: Solve for y in equation (1). Make up a table of values and plot $y = \frac{-x}{2} + \frac{5}{2}$. For example, let $x = -5$, 1 and 5. Remember, you may pick any values for x, but it is a good idea to examine at least one positive and one negative value.

	A	B	C
x	-5	1	5
y	5	2	0

Solve for y in equation (2). Make up a table of values and plot $y = 2x - 5$.

	D	E	F
x	-2	0	5
y	-9	-5	5

The result is shown in Figure 7.9, with the point of intersection (3, 1) providing the solution to the simultaneous equation. You can check this answer by solving for x and y in the simultaneous equation. That is:

$$x + 2y = 5 \quad (1)$$
$$2x - y = 5 \quad (2)$$

If we solve for x, we must eliminate y. To do so, there must be equal coefficients of y for both equations, resulting in their sum equalling zero. In this particular example, if we multiply the top equation (1) by 1 and the lower equation (2) by +2, we have:

$$x + 2y = 5 \quad (3)$$
$$2x(2) - y(2) = 5(2) \quad (4)$$

Add equations (3) and (4):

$$x + 2y = 5$$
$$4x - 2y = 10$$
$$5x = 15$$
$$x = 3$$

Substituting $x = 3$ into the equation $x + 2y = 5$, we have:

$$3 + 2y = 5$$
$$2y = 5 - 3$$
$$2y = 2$$
$$y = 1$$

Therefore, the coordinates are $x = 3$, $y = 1$, or (3, 1), the same result we derived using the graphic method in Figure 7.9.

Business mathematics and statistics

Problem: Solve for x and y using the graphic method.

$$2x + 6y = 44 \quad (1)$$
$$6x - 2y = 52 \quad (2)$$

Solution: Solve for x in equation (1). Make up a table of values and plot the following:

First, $2x = 44 - 6y$ can be simplified to $x = 22 - 3y$.
Now, solve for x in the equation $x = 22 - 3y$.

	A	B	C
x	28	22	10
y	−2	0	4

Next, solve for y in equation (2). Make up a table of values and plot the following:

You can simplify $6x - 52 = 2y$ to $3x - 26 = y$.

	D	E	F
x	4	7	11
y	−14	−5	7

The result is illustrated in Figure 7.10, with the point of intersection (10, 4) providing the solution to the simultaneous equation.

Graphic representation of word problems

Word problems which can be translated into simultaneous equations may be solved by using not only the elimination and substitution method but also the graphic method. For example, to determine the length and width of a rectangular block of land with a perimeter of 110 metres and length 5 metres less than three times the width, simultaneous equations can be created which represent this information:

$$2l + 2w = 110 \quad \text{or}$$
$$l + w = 55 \quad (1)$$
$$l = 3w - 5 \quad (2)$$

Make up a table of values for equation (1) and plot the following:

	A	B	C
w	−10	0	10
l	65	55	45

Chapter 7 – Line graphs

Make up a table of values for equation (2) and plot the following (see Figure 7.11):

	D	E	F
w	−1	0	16
l	−8	−5	43

Figure 7.10 *Intersection of graphs: 2x + 6y = 44 and 6x − 2y = 52*

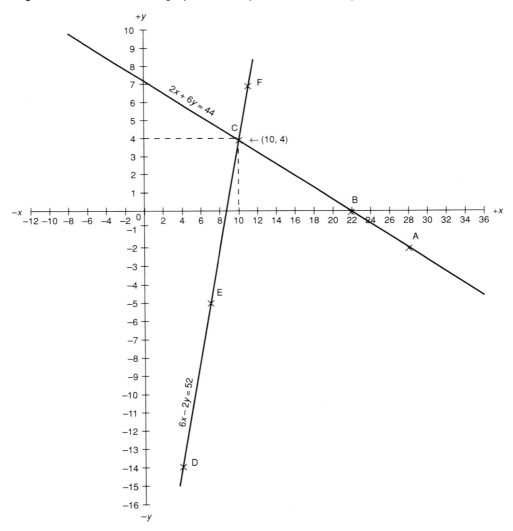

Business mathematics and statistics

Figure 7.11 *Intersection of graphs: 2l + 2w = 110 and l = 3w − 5*

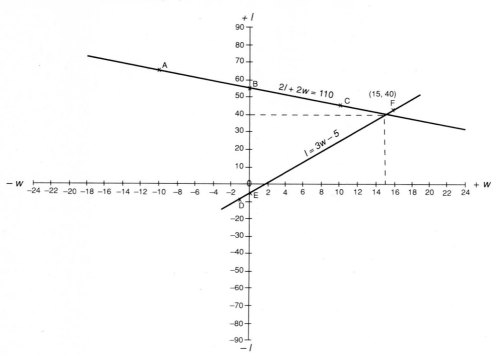

The point of intersection (15, 40) of the two lines satisfies the two equations, but notice when graphing the difficulty in determining the corresponding coordinates without a loss in precision.

Problem: Drive'Em Crazy, car importers, received two deliveries (price does not include Australian tariffs and sales tax): one contained three Mazda MX-6s and two Ford Mustangs for $79 000, and the other contained two MX-6s and four Mustangs for $98 000. How much did the importer pay for each Mazda MX-6 and each Mustang?

Solution:
  Let x = Cost of the Mazda MX-6
  Let y = Cost of the Mustang

The first delivery would therefore be represented by: $3x + 2y = \$79\,000$ (1)
The second delivery would be represented by: $2x + 4y = \$98\,000$ (2)
Make up a table of values (in $000s) for equation (1).

	A	B	C
x	11	1	21
y	23	38	8

Chapter 7 – Line graphs

A table of values for equation (2) must also be created:

	A'	B'	C'
x	1	9	19
y	24	20	15

Plotting this in Figure 7.12, we find $x = \$15(000)$ and $y = \$17(000)$. (Believe it or not, these are the approximate prices for these vehicles sold secondhand—one to two years old—in the United States, excluding sales tax).

Figure 7.12 *Intersection of the graphs: $3x + 2y = \$79\,000$ and $2x + 4y = \$98\,000$*

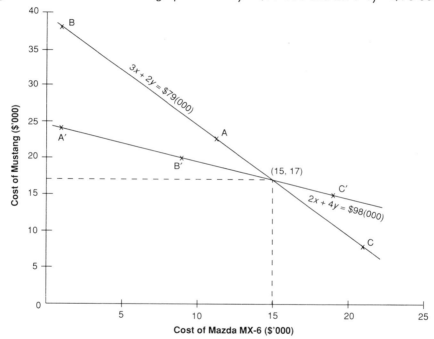

Source: The Oregonian, 5 October 1996.

The faster way of solving for the simultaneous equations would have been to solve for either x or y. For example:

$3x + 2y = \$79\,000$ (1)
$2x + 4y = \$98\,000$ (2)

To solve for x, the coefficient of y must be identical for both equations except for the sign so that the sum equals zero. Therefore, equation (1) is multiplied by 2, hence:

$2(3x) + 2(2y) = 2(\$79\,000)$ or
$6x + 4y = \$158\,000$ (3)

and equation (2) is multiplied by −1, hence:

$$-1(2x) + (-1)(4y) = (-1)(\$98\,000) \quad \text{or}$$
$$-2x - 4y = -\$98\,000 \quad (4)$$

Equations (3) and (4) are added together, resulting in:

$$6x + 4y = \$158\,000$$
$$\underline{-2x - 4y = -\$98\,000}$$
$$4x = \$60\,000$$
$$x = \$15\,000$$

To determine the y value, $15 000 is substituted for x in either of the original equations (1) or (2). Thus:

$$3x + 2y = \$79\,000 \quad (1)$$
$$3(\$15\,000) + 2y = \$79\,000$$
$$\$45\,000 + 2y = \$79\,000$$
$$2y = \$34\,000$$
$$y = \$17\,000$$

To check the answers, substitute the x and y answers into equation (2):

$$2x + 4y = \$98\,000 \quad (2)$$
$$2(\$15\,000) + 4(\$17\,000) = \$98\,000$$
$$\$30\,000 + \$68\,000 = \$98\,000$$
$$\$98\,000 = \$98\,000$$

7.2 Competency check

1. Will the graphing of simultaneous equations always result in a solution—that is, an intersection of the two equations? Briefly explain your response.

2. You may determine the solution to a simultaneous equation by algebraically solving for x or y or by graphing the two equations. Which of these two methods is the less precise? Why?

3. Solve using the graphic method (and check algebraically):

   $4x + 3y = 23$
   $3x - y = 14$

4. Solve using the graphic method (and check algebraically):

   $3x + 2y = 5$
   $x - y = 5$

5. Three months ago, the tennis pro shop ordered 15 tennis racquets and five pairs of tennis shoes for $1400 and last week ordered another eight racquets

Chapter 7 – Line graphs

and three pairs of tennis shoes for $765. Using the graphic method, determine the cost to the proprietor of a tennis racquet and a pair of tennis shoes.

. . . .

7.3 Break-even analysis

The principals of most business ventures have as their prime objective the realisation of profits—that is, where the total revenue received exceeds the total expenditure incurred in realising that revenue.

Total costs in the short term are said to be fixed and variable, with the **fixed costs** incurred irrespective of the level of production and exemplified by leasing charges, insurance premiums and depreciation expenses. In other words, fixed cost remains constant as it does not vary with output levels. **Variable cost**, on the other hand, increases with the level of production and includes payments for materials, labour, fuel and transport services. Therefore:

$$\text{Total cost} = \text{Fixed cost} + \text{Variable cost}$$

If a company's total revenue (TR)—that is, the price times the quantity sold—equals its fixed and variable costs (i.e. its total costs), it is said to be breaking even, hence the term **break-even analysis**. If, however, the revenue is not enough to meet the total cost, the firm is operating at a loss. Therefore, the break-even point is where total revenue = total costs.

To determine the break-even point—the number of units that must be produced and sold to cover all costs—we must discover the point where total costs equal total revenue. If we assume that the variable cost per unit remains constant and that all production is sold, we can solve this problem both algebraically and graphically, as follows.

Problem: Luckless Inc. has a total fixed cost of $100 000 p.a., with a constant variable cost of $5 per unit. If the product is marketed to sell for $9 per unit, how many units must be produced and sold for the firm to cover costs (i.e. what is the break-even point)?

Solution: Since the break-even point occurs where total revenue = total cost, we must substitute the values given in the example into the equation $TR = TC$. Thus, if we let x = number of units to be produced and sold, we have:

$$\text{Total revenue} = \text{Price} \times \text{Quantity sold}$$
$$TR = 9x$$

and

$$\text{Total cost} = \text{Fixed cost} + \text{Variable cost}$$
$$TC = 100\,000 + 5x$$

Therefore:

$$9x = 100\,000 + 5x$$
$$4x = 100\,000$$
$$x = 25\,000$$

Hence, if 25 000 units are produced and sold:

$$TC = \$100\,000 + 5x$$
$$= \$100\,000 + \$5(25\,000)$$
$$= \$100\,000 + \$125\,000$$
$$= \$225\,000$$

and

$$TR = 9x$$
$$= \$9(25\,000)$$
$$= \$225\,000$$

then:

$$\text{Profit} = \text{Total revenue} - \text{Total cost}$$
$$= \$225\,000 - \$225\,000$$
$$= 0$$

Therefore, at 25 000 units, the total revenue equals the total cost, thus resulting in a break-even situation for the principals of the company.

If, however, 27 000 units are sold:

$$TC = \$100\,000 + \$5(27\,000)$$
$$= \$100\,000 + \$135\,000$$
$$= \$235\,000$$

and

$$TR = \$9(27\,000)$$
$$= \$243\,000$$

then:

$$\text{Profit} = \text{Total revenue} - \text{Total cost}$$
$$= \$243\,000 - \$235\,000$$
$$= \$8000$$

If, on the other hand, only 20 000 units were sold:

$$TC = \$100\,000 + \$5(20\,000)$$
$$= \$100\,000 + \$100\,000$$
$$= \$200\,000$$

and

$$TR = \$9(20\,000)$$
$$= \$180\,000$$

Chapter 7 – Line graphs

then:

    Profit = Total revenue – Total cost
          = \$180 000 – \$200 000
   Loss = \$20 000

You should have noted that both the total cost and total revenue equations are linear, thereby permitting the graphic solution to the problem. Simply plot each equation on the graph as shown in Figure 7.13. To discover the break-even point, draw a perpendicular line from the point of intersection of the two lines to the x-axis.

Finally, to determine the resultant profitability when producing at any point other than the point of intersection (at 25 000), draw a perpendicular line from the nominated point to where it intersects both the *TC* and *TR* lines. The vertical gap between the *TR* and *TC* lines will indicate the profit or loss realised. Profitability at production levels of 20 000 and 27 000 units is indicated by means of perpendiculars, confirming the algebraic solutions calculated above.

Figure 7.13 *Break-even analysis*

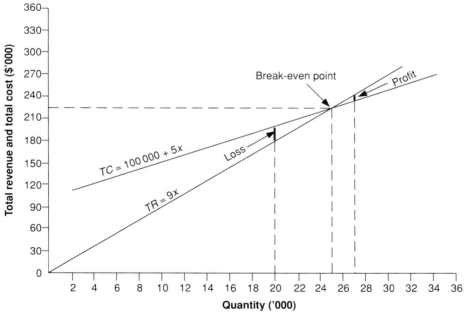

Problem: A restaurateur estimates that her fixed costs per week are \$4800, with a constant variable cost per meal of \$7. If the average meal per customer is \$19, algebraically determine and graphically illustrate:

(a) how many customers are needed by the proprietor to break even on her culinary escapade.

(b) how many meals would have to be sold to realise a profit of $1500 per week.

Solution:
(a) Let x = number of meals per week that must be prepared and sold.

Then:
$$\text{Total revenue} = 19x$$

and
$$\text{Total cost} = \text{Fixed cost} + \text{Variable cost}$$
$$TC = 4800 + 7x$$

Since the break-even point is where $TR = TC$, then:
$$19x = 4800 + 7x$$
$$19x - 7x = 4800$$
$$12x = 4800$$
$$x = 400$$

Therefore, 400 customers are needed for the proprietor to break even. You should check the answer by substituting into $TR = TC$. Hence:
$$TR = 19x \text{ or } \$19(400)$$
$$= \$7600$$

and
$$TC = \$4800 + 7x$$
$$= \$4800 + 7(400)$$
$$= \$7600$$

As $TR = TC$, 400 is, indeed, the break-even point. The graphic representation is found in Figure 7.14.

(b) Since a profit of $1500 is desired, this means that $TR - TC = \$1500$.
Therefore, since $TR = 19x$ and $TC = \$4800 + 7x$,
$$19x - (4800 + 7x) = 1500$$
$$19x - 4800 - 7x = 1500$$
$$12x = 4800 + 1500$$
$$12x = 6300$$
$$x = 525$$

Thus, 525 customers per week need to be served for a profit of $1500 to be realised. The graphic representation of this can be found in Figure 7.14. You may check the answer:

$$TR = 19x = \$19(525) = \$9975$$
minus $\quad TC = \$4800 + 7x = \$4800 + 7(525) = \$8475$
$$\overline{\$1500}$$

Chapter 7 – Line graphs

Figure 7.14 *Break-even analysis for restaurateur*

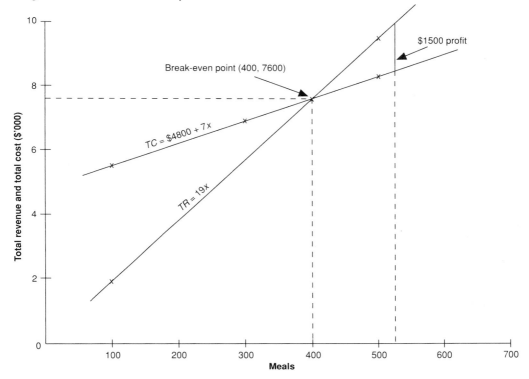

Source: Hypothetical.

7.3 Competency check

1. In determining the total expenditures of a business over the short run, there are two cost classifications that must be accounted for. What are these? Distinguish between the two.

2. How would you define the *break-even point*?

3. Algebraically determine and check graphically the break-even point for a company with fixed costs of $800 300 p.a. with a constant variable cost per unit of $6.45 if the product is sold for $14 per unit.

4. If, for the firm cited in problem 3, a profit of $105 700 is desired for the year, how many units must be sold?

5. Graphically illustrate the point where the profit of $105 700 is made (as calculated in question 4).

7.4 Non-linear business-related graphs

Unfortunately, the graphing of actual business statistics such as sales or profitability doesn't always exhibit changes which can be conveniently described or portrayed by a straight line. In other words, these are referred to as **non-linear graphs** and they are derived from equations whose highest power is greater than 1. You may recall that the equation for a straight line is $y = mx + b$, that is, the x variable is taken to the power of 1. Quadratic equations, on the other hand, are those where x is taken to the power of 2 (i.e. it is squared) such as $y = 3x^2 - 6x + 3$. Quadratic equations may better describe the relation between, for example, production costs and sales than would a straight-line equation.

How, then, do we plot such an equation? All you have to do is follow the same principle that is applied when graphing a straight line—that is, substitute values of x into the equation, solve for y and then graph the results.

Problem: Graph the following function: $y = 3x^2 - 6x + 3$

Solution: First select values for x which you then substitute into the equation.

Let x =   −3   −2   −1   0   1   2   3
then y =   48   27   12   3   0   3   12

Next, plot these coordinates on the graph as seen in Figure 7.15.

Figure 7.15 Graph of $y = 3x^2 - 6x + 3$

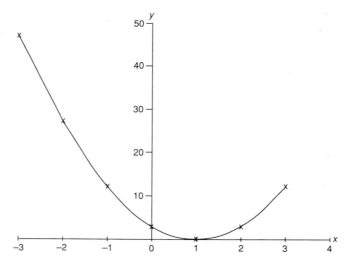

Not only is the graphic representation of the equation not a straight line, but you will also notice that there is a peak and a trough with the former at $y = 48$ and the latter at $y = 0$. This graph illustrates a **turning point** where the y value is at its maximum (peak) or minimum (trough)—in this case, its minimum is at 0 when $x = 1$.

Chapter 7 – Line graphs

Problem: Table 7.2 sets out the number of washing machines manufactured in Australia from 1987/88 to 1995/96. Graph the results and discuss.

Table 7.2 *Number of washing machines manufactured in Australia 1987/88 to 1995/96*

Year	Units manufactured ('000)
1987/88	394
1988/89	397
1989/90	330
1990/91	326
1991/92	296
1992/93	308
1993/94	326
1994/95	305
1995/96	297

Source: Australian Bureau of Statistics, *Australian Economic Indicators* (Cat. No. 1350.0).

Solution: The graphic representation illustrates the cyclical nature of the manufacturing sector. There are upturns and downturns in production, with the turning points corresponding to the peak and trough of activity. That is, the turning points indicate those points where activity is at its high and low points on the graph. For example, the peak production run occurred in 1988/89, while the trough (low) was hit in 1990/91 (296 units). Notice that the peak of 397 was realised at a time of high employment and economic growth (1988/89), while the trough occurred in the midst of Australia's severe recession.

Figure 7.16 *Graphic representation of washing machines produced in Australia, 1987/88 to 1995/96*

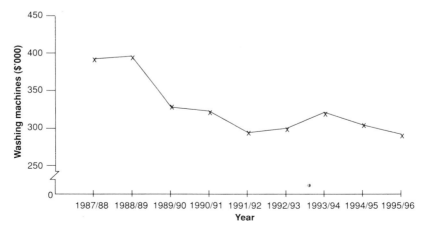

Source: Australian Bureau of Statistics, *Australian Economic Indicators* (Cat. No. 1350.0).

Business mathematics and statistics

If you consider the visual representation of the rate of change in economic growth as defined by GDP (see Section 17.1 in Chapter 17), you will also observe how this cycle is defined by the expansion and contraction of business activity with the turning points the boom and bust in the cycle.

Problem: In an attempt to attract potential investors, a local businessman advertised that investment in his new venture will lead to exponential growth over the next five years. In fact, an investment of $5000 will grow at a rate of 15% compounded annually. Graphically illustrate this function.

Solution: This is referred to as an exponential function which is defined as $y = ab^x$, which is similar to the compound interest formula. It may also be used for determining the trend line of data which exhibits geometric growth such as the population or GDP.

Therefore, with regard to the investment proposition:

$a = \$5000;$ $b = 1.15$ p.a. $x = 5$

Thus solve for $x = 1$ to 5 years for the equation and then plot the points as found in Figure 7.16.

$y = ab^x$
$y = \$5000(1.15)^1 = \$5750$
$y = \$5000(1.15)^2 = \$6612.50$
$y = \$5000(1.15)^3 = \$7604.38$
$y = \$5000(1.15)^4 = \$8745.03$
$y = \$5000(1.15)^5 = \$10\,056.79$

Figure 7.17 *Graphic representation of $y = \$5000(1.15)^5$: exponential growth*

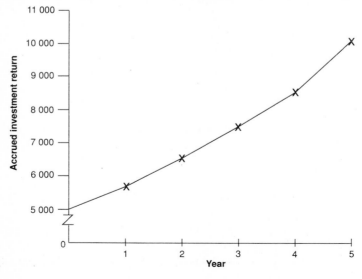

Note that exponential function exhibits constant relative growth (such as the $5000 investment in the above example growing at an annual rate of 15%).

7.4 Competency check

1. What is the difference when graphing a linear function compared to a non-linear function?
2. What is meant by the *turning point*?
3. In graphing a linear function, are there any turning points? If so, how many?
4. Graphically illustrate the sales function (in thousands of dollars) of Hide'N Seek Mystery Tours if its sales performance can be described by the quadratic equation:

    Sales $(S) = 2x^2 + 3x + 1$ (let the x value $= -3, -2, -1, 0, 1, 2, 3$ when solving for S)
5. Estimate the approximate turning points in the form of (x, S).

7.5 Summary

Equations may be represented graphically, and should the equation be presented in the form of a linear equation, that is, $y = mx + b$, its graphic representation is that of a straight line. To plot an equation, construct a table of values and then plot it, drawing a line through a minimum of two points.

In the equation $y = mx + b$, m, the coefficient of x, is the slope or gradient of the line and indicates the change in value of y compared with the change in the value of x. Point b represents the y intercept—that is, the point where the straight line intersects the y axis; hence at the point where the x value is zero (i.e. $(0, b)$). The graph of an equation $y = mx$ is a straight line intersecting the point of origin, as there is no y intercept since b does not exist. The graph of an equation $y = b$ is a horizontal line parallel to the x axis intersecting the y axis $(0, b)$ since b is a constant. Every horizontal line has a slope of zero since the mx is absent in the equation for a straight line—for example, $y = 2$.

Understanding the graphic representation of a linear equation permits further application to simultaneous equations as well as break-even analysis.

7.6 Multiple-choice questions

1. The point of origin is:

    (a) the first point drawn on the graph;
    (b) the point where the simultaneous equations intersect when graphically illustrated;

(c) the point of intersection of the x and y axis;
(d) none of the above.

2. When the equation $y = mx + b$ is graphically presented, it represents:

   (a) a straight line;
   (b) a parabola;
   (c) either (a) or (b), depending on the values of each of the variables;
   (d) a quadratic equation.

3. In the equation, $y = mx + b$, b represents:

   (a) the slope of the graphed function;
   (b) the abscissa;
   (c) the y intercept;
   (d) none of the above.

4. Find the slope and y intercept, respectively, in the equation, $y = 2x - 6$.

   (a) 2, –6   (b) 2, –3   (c) –6, 2
   (d) –3, 2   (e) 1, –6

5. Why is the graphing of simultaneous equations to solve for x and y coordinates not the preferred option to using the substitution or elimination method?

   (a) There is a loss of precision in using the graphic method.
   (b) The graphic method is far more time-consuming.
   (c) There is a greater chance of error when graphing simultaneous equations.
   (d) All of the above.

7.7 Exercises

1. What is meant by the *Cartesian coordinates*?

2. Plot the following points on the same graph:

   A(0, 5)   B(5, 0)   C(–5, 1)   D(1, –5)

3. The equation $y = mx + b$, when graphically presented, depicts a straight line. In the equation, m is defined as the slope or gradient of the line. How is the slope calculated?

4. In the same equation, $y = mx + b$, what does b represent and how is it calculated?

5. Determine the slope and y intercept for the equation $y = 4x + 2$.

6. Determine the slope and y intercept for the equation $2y = 6x - 8$.

7. Determine the slope and y intercept for the equation $-3y = 3x - 9$.

8. Determine the slope and vertical intercept for the equation $\frac{9}{5}C + 32 = F$.

9. Draw the line graph of $y = 4x$.

Chapter 7 – Line graphs

10. Draw the line graph of $3y = 4x + 5$.

11. Draw the line graph of $-2y = 3x - 2$.

12. Draw the graph of $y = 2x - 2$.
    (a) Find the value of x when $y = +8$.
    (b) Find the value of y when $x = +4$.

13. Solve the following simultaneous equation using the graphic method:

    $2x - 3y = 9$
    $3x + 4y = 5$

14. Solve the following simultaneous equation using the graphic method and check your answer by algebraically solving the equations.

    $x - 2y = 2$
    $3x + 1 = 7y$

15. Solve and check the following using the algebraic and graphic methods:

    $y + 7x = -24$
    $2x - 2y = 16$

16. Convert the following equation into the straight-line equation of $y = mx + b$:

    $$\frac{y + 2}{x - 5} + 3 = 1$$

17. Draw the graph for the equation in exercise 16.

18. After analysing a company's cost structure, Con found that the fixed costs were $550 000 p.a. with a constant variable cost per unit of $9.50. If the product is selling for $15 per unit, what is the break-even point? Illustrate graphically.

19. Calculate and graphically illustrate the number of units sold in exercise 18 if a profit of $110 000 was made last year.

20. Two years ago, the same company cited in exercise 18 incurred a loss of $82 500. Graphically illustrate the number of units sold that year. (You may use the same graph drawn to answer exercise 18.)

Part 2
BUSINESS STATISTICS

The profit makers at Sportsgirl

Most companies, when new owners come in, trumpet loudly about new directions, new visions, and new identity. It has been the opposite with the Melbourne-based Sportsgirl Sportscraft ("Sportsgirl"). Since the South African retail giant Wooltru Ltd, and a former Coles Myer Ltd (CML) executive, Frank Whitford, bought the clothing outfit from a grateful banking syndicate 19 months ago, they have engineered a quiet—but effective—revolution.

Blank spaces on the Hawthorn boardroom walls, where former proprietor David Bardas's paintings included fabulous contemporary art, reflect the new management's preoccupation with substance. It would not take long to find something to utilise the empty nails. But Mr Whitford has not given it a thought.

The group's managing director and 10 per cent equity owner, Mr Whitford has overseen its turnaround in his characteristically quiet way—a manner which has inspired his employees to dub him "The Reverend Frank".

The group's expansion has been just as discreet, like its purchase last month of New Zealand's 13 "Warehouse" stores from the Delacca family. Eleven stores are being rebadged as Sportsgirl and two as David Lawrence, the upmarket label named after David Bardas.

In September, Sportsgirl's first genuine profit since the late 1980s will be announced at Wooltru's annual meeting in Capetown, South Africa.

Following a $20 million 1994 loss and $13.7 million 1995 loss, net profit for the year to June 1996 will come in close to $700,000.

Life has not been kind to Australian retailers lately but Sportsgirl Sportscraft is bucking the trend. After some perilous times the company is now back in the black and going for growth, as **Rowena Stretton** reports.

Steady group sales, around $165 million, reflect an ongoing tough retail environment. The profit has been generated by reduced costs of sales and higher margins.

The examination and transformation of every aspect of the group has happened without the usual army of consultants: "We got rid of consultants actually," Mr Whitford says, "and that saved us $1 million a year."

His first task was to refinance the group. Formerly funded, and eventually controlled, by Westpac and the ANZ, the new deal is with National Australia Bank.

The group was crippled by the $100 million debt incurred by its former owner in a visionary but ill-timed Collins Street development (the banks sold the building before they sold the business).

Performance analysis was obscured by a tangled financial web spreading assets, liabilities and cash flow between 12 companies. Even today, accumulated losses amount to about $80 million.

"The first key was to get control of financial information," Mr Whitford said. "That meant getting back to simple accounting systems and building in the key ratios to see how the business was really going."

Exposed against industry norms, all the ratios were ailing: gross retail margins and sales per square metre were between 5 and 10 per cent below. Salary and occupancy costs per sales were far too high. Manufacturing, formerly run to break even, showed potential through better labour utilisation, manufacturing processes, fabric yields and volume.

Rectification called for fresh management. A former CML executive, Judy Coomber was plucked from her managing director role at New Zealand's Hallensteins menswear chain to head marketing and administration. Motorola Australia's chief financial officer, Mark Ashby, was appointed to head finance. Geoff Brown—poached from Myer to run property—now heads international. He oversees the four Singapore stores and is planning a blueprint for further Asian expansion.

The former managing director of women's retailer Jacqueline Eve, Kevin Winterburn was brought in to run retail, including the 14 New Zealand stores.

"They are top class executives with clear objectives, individual accountability and a brief to each to significantly revolutionise their area. Adversity tends to force a team together," Mr Whitford said. A generous incentive program tied to group performance further sharpens resolve.

In the past year head office costs have been reduced by more than 25 per cent; 20 unprofitable stores have been closed and 20 new stores opened; refurbishment costs are down from an average $2000 to $1000 per square metre and the floorspace in new stores averages

170 square metres rather than the former 280 square metres: "We're moving from big is better to small is beautiful," Mr Whitford said.

Margins are up by 5 per cent and like-for-like store growth (distinguishing between old and new stores) is around 6 per cent.

"Our advertising expenditure is down by more than $1 million, but we've improved audience reach and sales yield. The disastrous Girl Power campaign that we inherited has been stopped and our new agency, Samuelson Talbot, has developed the successful "A Girl Like You" campaign."

Outsourcing distribution to Mayne Nickless Ltd—the company now run by Mr. Whitford's former CML colleague Mr Bob Dalziel—have saved the group about $500,000 a year.

Manufacturing operations, shut in 1993-94 when Sportsgirl's former bankers started flexing their muscles, has been revived, including its old Crestknit brand. Manufacturing now boasts $9 million external sales, with clients including David Jones' home brand. Management wants $60 million external sales on top of $50 million internal sales within five years.

"Buying the Glenferrie Road manufacturing building last year was buying back part of our history and that was important to us," Mr Whitford said.

Over the past 18 months $6.3 million capital expenditure has upgraded group systems and technology. This money has come from cash flow and shareholders funds, not borrowings.

Mr Whitford turned to demographic analysis to drive brand repositioning. First under the microscope was the Sportsgirl label which had been aimed at 15–19 year-olds and getting younger: "But when you looked at the statistics you saw this was the smallest demographic, and with girls staying longer at school, it was selling into a shrinking market," Mr Whitford said. It now targets working girls in their early 20s.

The Sportscraft label was also offcourse: "It was getting older in appeal whereas 30-34 year-olds are the biggest demographic. We are taking it back to the early 30s but, don't worry, the baby boomers will love it."

David Lawrence was a successful and "well-understood" label but, like Sportscraft needed development. Now both brands are sold in stand-alone stores as well as at Ahearns in Western Australia, South Australia's John Martin and a larger number of David Jones stores.

The former Myer/Grace executive describes these brands and David Jones as "a good fit": "We have bucked the trends and scored 12–13 per cent sales growth in department stores."

But you will not find Sportscraft or David Lawrence at Myer or Grace Brothers. And, ironically, it was when Mr Whitford was head of retail operations, Myer/Grace (from 1988 to 1994) he was involved in *that* decision—one he would now like reversed.

> ' **Rent costs the group $30 million a year, which is more than its labour costs.** '

"A department store presence always adds to a brand's value. Some people are pessimistic about department stores but customers will continue to shop in them because of their services and product range."

The biggest group headache is rent. Rent costs the group $30 million a year, which is more than its labour costs: "Getting occupancy costs down is an ongoing issue that will take four years," Mr Whitford says.

He has been dealing personally with the retail property czars, trying to sell the message that despite the substantial combined turnovers and spending power of small players like Sportsgirl, Just Jeans, Country Road and Portmans, the department stores tended to get the best deals: "We'd love department store rents," he quips.

Wooltru directors Eddie Parfett and Michael Mark sit on Sportsgirl's board, which meets quarterly, with executive directors Whitford, Mark Ashby and Judy Coomber.

So what influence had the majority shareholders had wielded in all that had occurred? "They have brought a sense of urgency into the organisation," Mr Whitford says. "They are not looking to impose anything except focus.

"They are retailers and we all talk the same language. You have to get your sales per square metre right, your costs as a percentage of sales right, your volumes and your products right.

"If there is 'a South African way' 1 don't know about it. They have been prepared to run with our management decisions. As a small company we are lucky to have people of that calibre (Wooltru turnover is around $A3 billion a year) with us."

Mr Whitford said that although Sportsgirl was back in the black there was a long haul ahead: "The whole consumer psychology has changed in the past decade. In the 1980s people talked about how much they paid for things. Today they boast about how little. That's something the retailer has to understand—the desire for value."

He says the ageing population will undermine the fashion industry at one end: "But fashion will remain extraordinarily significant to the young female shopper through to age 40–45.

"We are in a very tough game but just watch us compete."

Source: Rowena Stretton, 'The profit makers at Sportsgirl', *Australian Financial Review*, 5 August 1996. Reproduced with permission.

Chapter 8
INTRODUCTION TO STATISTICS

LEARNING OUTCOMES

After studying this chapter you should be able to:
- Identify and describe the breadth of statistics
- Identify the key steps in statistical research
- Recognise the many uses of statistics in business
- Identify and apply data collection techniques
- Define the major sources of statistical information
- Recognise and interpret the use and abuse of statistics.

'Experts differ on pace of economic recovery'

On the same day, two respected forecasters provided conflicting signals about where the Australian economy was headed, with their forecasts based on their respective surveys.

'Living standards under Labor—up or down?'

During the 1996 election campaign Australians were provided with conflicting messages, with Labor insisting that living standards had risen by 1% in real terms since they came into office in March 1983 as a result of non-cash benefits such as free or low-cost government services (e.g. education, health, housing and child-care). However, according to the Business Council of Australia, living standards in fact declined by 13% in real terms, as real household income fell 9% and the number of hours worked rose 4%. Which stance is correct or more correct?

Institutions blamed for hotel hiatus'[1]

The lack of institutional investment in hotels could result in infrastructure bottlenecks, as the number of tourists is expected to increase. For example, there were 3.75 million international visitors in 1995, a number expected to grow to 8.8 million by 2005,

Business mathematics and statistics

representing an annual growth rate of 8.9% since 1985. Australian superannuation and insurance companies had $275 billion under management, with 12.5% of these funds invested in property—about $34.4 billion, but only 1% of that ($340 million) directed to the tourism sector.

'Sleep tight, but is it all right?'

Before any new drug is put on the market, there is a great deal of clinical testing to ensure its safety. A particular sleeping pill was examined by the US Food and Drug Administration because the pharmaceutical company distributing it had been accused of covering up evidence of the drug's side effects—in particular, a significant number of serious psychiatric problems. Coincidentally, the files of over 10% of patients tested disappeared, and accusations were levelled that these files were 'lost' deliberately in order to cover up the side effects of the drug.

'Top of the Bill could pay off'

This hot tip for Randwick races was based on the gelding's workout over 1200 metres with a time of 1 minute and 18 seconds recorded, with the last 600 metres covered in 36.75 seconds and the last 200 metres in 12 seconds.

'Rubbery robbery numbers misleading'

Figures which painted Sydney as the mugging capital of Australia (claimed to be 70% above the national average) were found to be highly exaggerated. It was subsequently discovered that the increase in the national crime statistics was actually an increase in reporting rates.

8.1 Statistics

The above headlines referring to commerce, health, sport and crime, all have a common denominator: they all deal with the use of quantitative or numerical information in describing an outcome or prospective result. The word '**statistics**' can also be used to describe this collection of data which is ultimately presented and then analysed (as we noted above). However, the word 'statistics' is used not only in this context; a summary measure or characteristic of a sample taken from a population is also referred to as a statistic. Finally, statistics is also the entire field of study incorporating the collection, classification, summarisation, presentation, analysis and interpretation of numerical data. Access to statistics allows you, the decision-maker, to improve the prospects of a well-considered, educated appraisal of the current state of operations and future growth prospects; in other words, statistics assist in the decision-making process.

As noted above, you can find numerical or quantitative information (i.e. data) in any discipline, but the point is, what is one to do with it? As a businessperson you will at times be confronted with a particular problem (e.g. increased competition, low margins, high labour turnover, whether to relocate or expand operations) and

Chapter 8 – Introduction to statistics

you must determine the severity of the problem and how to resolve, or at least alleviate, it. What you will ultimately have to do is analyse relevant statistical data by using statistical methods.

Descriptive statistics

When you collect data and proceed to classify, summarise and present them, you are basically describing or summarising the collected quantitative information. For example, you have certainly used measures of central tendency (e.g. averages) or come across graphs (e.g. bar graphs, pie charts, line graphs); these are part of **descriptive statistics**, a technique devoted to the accurate representation of data with graphs and summary measures. A major source of descriptive statistics is the Australian Bureau of Statistics, the analysis being left to the reader.

Chapters 9, 10 and 11 all deal with descriptive statistics. Once these have been read and understood, you will find yourself wary of visual presentations (i.e. graphs), and far more aware of how to prepare an accurate presentation, what statistical measures to use and when, how to evaluate the content of articles and in-house reports and, in the process, how to improve your decision-making.

Inferential statistics

Although descriptive statistics is a widely used and integral branch of statistics, much of the published statistical information is derived from samples of a population, which means that its analysis requires conclusions to be drawn which go beyond the data. Subsequently, the information age has generated yet further growth in the statistics discipline, with emphasis increasingly on those methods which assist in drawing conclusions based on a data sample rather than on methods which merely describe (e.g. descriptive statistics). For example, before the next federal election, Morgan Gallup polls will be taken of a small section (known as a 'sample') of the population. Based on the responses from this sample, the popularity of each party will be estimated. In other words, a conclusion will be drawn from the results of collecting, analysing and interpreting data; this constitutes **inferential statistics**. All the chapters from Chapter 11 onwards deal with inferential statistics; however, before studying them, you will need to understand several key terms, including:

(a) *Population*—every observation in the group being analysed; you may have a population of millions or of only a few units.
(b) *Sample*—a subset of the population, which should display the characteristics of the population it is representing.
(c) *Parameter*—a numerical characteristic or measure describing the population.
(d) *Statistic*—a numerical measure describing the characteristic of a sample.
(e) *Continuous data*—variables that can take any value that presumes measurement is possible to the required level of accuracy (e.g. measures of central tendency and dispersion can be measured in any number of digits such as with weights and heights).
(f) *Discrete data*—variables that are expressed only in terms of whole numbers (e.g. the number of pages in this book).

The application of inferential statistics will allow you to (1) improve your decision-making ability in times of uncertainty; (2) select a sample of a population and generalise about how the entire population of observations operates on the basis of the sample results (e.g. quality control); and (3) test your forecast results against the actual results.

Any person operating a business without applying statistical inference is not maximising the firm's ability to anticipate and/or respond to changes.

Statistical methods

The conduct of statistical research with its accompanying analysis is an invaluable tool for any business manager because we all know how business decisions in any firm are inherently affected by the collection and analysis of data. Naturally, the first and most important step is to identify the problem. Next, the proper use of **statistical methods**, conveniently divided into five basic steps, will permit us to deal with the problem.

1. **Collection of data**
   There are two basic sources of information: internal and external. Internal information includes all quantitative information obtained from within the firm (e.g. sales turnover, profit margin). External information is data received from outside the organisation, and may be published data (referred to as *secondary data*) or originally surveyed data (known as *primary data*). For your convenience, the gathering of data, both published and original, is covered in Section 8.2 of this chapter.

2. **Organisation of collected data**
   Not all collected information may be relevant or useful, so you must vet the data and subsequently classify, tabulate and summarise it, often with the use of graphs.

3. **Presentation**
   Once the data have been collected and grouped in an understandable and useful format, they should now be presented. You can choose from a wide variety of visual presentations, each with its own intrinsic advantages and disadvantages. These are all visually displayed and discussed in Chapter 9, and if you remember the guidelines set out in that chapter you will be able to present data competently to management.

4. **Analysis**
   The complexity of analysing data naturally depends on what data are collected. You may find that the table or graph doesn't satisfactorily reveal or describe an important characteristic of the information. Consequently, you may have to do several calculations to describe the data, such as calculating percentages, averages or trends; this branch of statistics is known as *descriptive statistics*. Or you may want to base a forecast on your data—in other words, to draw some conclusion; this is commonly referred to as *statistical inference*. You will find further reference to and examples of both descriptive and inferential statistics throughout the remaining chapters.

Chapter 8 – Introduction to statistics

5. **Interpretation of the results**
    Once you are convinced that the analysis is complete, you must then interpret the findings by evaluating the results and drawing what we hope are logical conclusions. Sometimes the interpretation will be fairly simple. However, in other cases you may have to apply certain statistical measures which require a sound understanding of statistics on your part. A full understanding and correct application of these measures will enhance the likelihood of a valid conclusion to your study—an invaluable tool for decision-making.

If you can master the material contained in this and subsequent chapters, you will find that gaining this knowledge of statistics will be of invaluable assistance in (1) understanding and describing statistical relationships; (2) improving the decision-making process; and (3) improving your ability to cope with the constant changes in our information-oriented world.

Uses of statistics

We are currently in the midst of the Information Age and, as a consumer and decision-maker, you are exposed to an ever-increasing flow of quantitative information from which you must try to extract and digest those data relevant to your concerns. Therefore, you must be able to determine what relationships and accompanying data are necessary for decision-making purposes.

Business applications

The correct application of statistical methods results in better business decisions because of the improved accuracy of the forecasts. The major tools used in forecasting are sampling, probability distribution (e.g. normal distribution), correlation and regression analysis, index numbers and time series (covered in Chapters 12, 14, 15, 16 and 17 respectively). These tools, once learned, can be applied to the major areas of any business, such as production, sales, finance, personnel, accounting, market research and exports, and a number of them are exemplified below.

At any business meeting, discussion about costs, sales and profit performance is inevitable, with results often visually presented in the form of a table and/or graph. If a firm is considering introducing a new product to the market, surveys of consumer preferences are taken. Thus, a sample is taken from the targeted population, and a statistical inference or conclusion about the potential viability of the product is made on this basis. As quality control manager for a parts manufacturer, for instance, you have to ensure that the items being produced are of a certain standard. You cannot physically check each unit that comes off the production line. But what you can do is take a sample of the production run and, on the basis of the results, draw conclusions or inferences about the standard of the production batch.

Annual company budget forecasts rely on past and current company statistics and acknowledge that changes in key variables, such as the current state of the economy, the inflation rate, the Australian dollar, interest rate levels and levels of unemployment can also have a major impact on future performance.

The banking industry is in the throes of further reducing their branch infrastructure, claimed to be excessive by international standards. How will management determine the number and location of those branches to be closed—through a statistical analysis of the catchment area, turnover, competition and the like.

How does a retail chain decide where to locate their shops, a business college its classrooms, an accountant her office, a council new child-care facilities or a government department its office? In each case, the demography of the targeted market would be a key consideration—that is, the concentration of the population within local government areas, the age and income distribution of the inhabitants, and so on. If, however, an economic slowdown is forecast, businesses may well postpone any planned capital expenditure (i.e. new retail outlets or larger premises).

What about possible investment options? How would you assess the alternatives? You would certainly compare available statistics on returns, whether from property, shares, antiques or gold, and assess the degree of risk of each.

Therefore, the business community needs and has access to statistical information more readily available than ever before, with current data on sales turnover, expenditure, inventory and profits of critical importance in establishing effective budgetary strategies.

Other applications

We have all heard about the hazards of passive smoking, high levels of cholesterol, silicon implants, drug-taking, and many more public health concerns, all of which are based on the application of descriptive and inferential statistics. How often were we told in 1991 that the economic recovery had begun, such claims being based on 'statistical evidence'? Compass Airlines apparently had a fair market share of the routes it serviced but still managed to go into receivership. Its management realised, perhaps a bit too late, the need to attract business travellers in order to keep the airline going—an example of the application of statistical inference.

Imagine that you have been one of those lucky enough to be accepted for a course, while others have been excluded because of the lack of TAFE or university places. Your admission was probably based on your high-school academic performance, with educators determining by statistical methods that a minimum grade of, say, 85% for the Higher School Certificate would be necessary for a prospective student to handle the course load successfully.

The railways have always been a drain on Australia's public purse. In an attempt to stimulate greater patronage which would simultaneously reduce traffic congestion, one state government reduced fares by 20% only to see the number of passengers rise by a mere 6%. The lacklustre public response could be attributed to the suburbanisation of jobs precluding many from using public transport, the convenience of driving to work, the train timetable, the safety issue, etc. In other words, the government had miscalculated the public response by its erroneous assumption that it was only the cost factor detracting prospective passengers, rather than a large variety of factors. A lack of statistical sophistication seems to have been present at this time, allowing yet another expensive mistake.

The state of the Australian economy is defined in terms of the GDP, whose value is mostly based on statistical methods because statistical censuses and surveys are

Chapter 8 – Introduction to statistics

used. Sample methods are also used for unemployment statistics, average weekly earnings, the Consumer Price Index (which measures inflation), the balance of payments, and just about all other major economic variables, since the population is far too large for all observations to be measured. That is the obvious reason, along with excessive costs as to why a population census is taken only once every five years.

According to the statistician Ross Dundas, in the 1992 one-day cricket matches Australian cricketers scored 30 fewer runs in their 50 overs than in previous years' matches, averaging 212 in 1992 versus 245 in 1987/88, 226 in 1988/89, 240 in 1989/90 and 241 in 1990/91. Why the decline in runs? Slower outfields, unkind pitches and weather, senior batsmen out of form, top bowling by the opposition, over-cautiousness at the wicket? What about the chances of a trifecta at Randwick? Of Australia regaining the America's Cup? Of Australia improving its medals tally at the next Olympics? Which iron should be used for a 6-metre putt? Now, in every one of these sports examples, the correct use of statistics would help the decision-maker (although it may not necessarily win the game, as we are often dealing with the realm of probabilities).

8.1 Competency check

1. What is meant by the term *statistics*?

2. List the statistics you recall coming across over the past 24 hours.

3. (a) What is meant by *descriptive statistics*?
   (b) Give five examples of descriptive statistics.
   (c) What is meant by *inferential statistics*? How does it differ from descriptive statistics?

4. (a) What are *statistical methods*?
   (b) Using the basic steps outlined above, briefly explain how you, as regional manager for Cooky's Biscuits Pty Ltd, would attempt to analyse your region's sales performance which has apparently been falling behind that of other regions.

5. Give one example (not already mentioned) of the type of statistics collected in each of the following areas:

   (a) medical research   (b) car racing   (c) accounting
   (d) economic analysis   (e) geography   (f) demography

8.2 Data collection

In January 1992 the British astronomers who in 1991 had announced the discovery of the first planet outside our solar system admitted that it was a mathematical error and that no such planet existed. When the discovery was announced in July 1991, it appeared to be the first evidence of a planet associated with a star other than the sun. Using a complicated mathematical calculation, the astronomers had calculated that there existed a planet 10 times the size of Earth. However, it has now been revealed that they forgot to factor in another variable; this resulted in an unusually large error, and in reprocessing the data with the correct figures they found no planet.[2]

If this sort of mistake can happen to eminent scientists, what chance do we mere mortals have when collecting and analysing data? Well, first of all, it is doubtful that many of you will be observing regular interruptions of radio pulsations emitting from a neutron star (pulsar).[3] Second, you are not studying statistics to become a statistician, but rather to develop the knowledge and skills necessary to improve your decision-making prowess.

The example of the embarrassed team of scientists clearly illustrates that the validity of any published statistical information is a function of the actual data collected, their analysis and their visual presentation. The methods of data collection must therefore be studied and understood if you are to collect and present data in an accurate, honest, reliable and informative fashion. Otherwise, irrespective of the expertise of the analyst, erroneous data collection will render any subsequent analysis null and void, as we saw in the above example.

Pre-collection thoughts

The careful selection and recording of data is a prerequisite to ensuring that the statistics drawn from them are valid. The following general principles should be applied when collecting and recording information:

1. Before collecting data, make sure that you know the precise purpose of the exercise. The types of data gathered and the methods of collection will, for the most part, be determined by the purpose. For example, if an educational institution wants to determine the attitude of its students to the cafeteria facilities, the collection of data will be restricted to its students and the information can be gathered through, for example, a questionnaire.
2. Be aware, before data collection begins, of how the data will subsequently be analysed. This determines the questions you hope to discover answers for. If, for example, the educational institution is interested in the attitudes of part-time and full-time students towards available parking and sport facilities, questions must be specially addressed to these two distinct groups of students. Otherwise it will have been a fruitless exercise if, upon completion of the data-gathering, it is discovered that certain key questions could not be statistically examined.

Internal and external data

Sources of statistical information are vast, and the particular sources appropriate for your problem naturally depend on the area investigated. According to the location of information, such statistical data can be categorised as **internal** or **external** data. Internal data are obtained from your own organisation, be it from the balance sheet, the profit and loss statement, or individual departments within the organisation—sales, personnel and production, for example. External data are obtained from outside your organisation, and are readily available from published sources as diverse as the Australian Bureau of Statistics (with approximately 979 publications), the Reserve

Bank (from its annual financial report, monthly bulletin and press releases), the Treasury's Economic Roundup (a quarterly summary of the state of the economy and key economic indicators), the Budget Papers (incorporating a review of the past year's major economic variables, the country's economic performance, and estimates of the forthcoming year), bank, business and/or investment newsletters (if you, your firm or your local library subscribe), international organisations (e.g. the International Monetary Fund, the United Nations, the World Bank and the Organisation for Economic Co-operation and Development (OECD)), trade journals, newspapers, magazines, year books (e.g. *Year Book Australia*), business organisations (e.g. Chambers of Commerce and the Retail Traders' Association), and private research organisations (e.g. BIS-Shrapnel, Access Economics and the Roy Morgan Research Centre).

Secondary data

Generally, the quickest and least costly method of collecting data is to locate data which have already been collected and disseminated by another—this is referred to as **secondary data**. Should such data be found inappropriate to your particular problem, they may nevertheless suggest previously unthought-of ways of attacking your problem or may even suggest altering the strategy or approach originally contemplated.

In Australia, official statistics are predominantly the responsibility of a national statistical agency, the Australian Bureau of Statistics (ABS) whose independence is written into law. The ABS provides both primary data (such as the Population Census, an agricultural census—data on every farm, a retail census—data on every shop, and other censuses) and what is referred to as administrative by-product data.[4] The latter is the method of producing statistics from data that have already been collected for some other, usually administrative, purposes such as the ABS publications on: export and import statistics (port and airport documentation for customs purposes is completed and then used to provide trade data); building approval statistics (extracted from local government authorities' provision of such data to the ABS); and demographic statistics such as births, deaths and marriages (which the ABS receives from the Registrar of Births, Deaths and Marriages).[5] The basis of the ABS relying on these sources is to reduce the need for further surveys and censuses and to avoid duplication of data collection.

If you are considering incorporating secondary data into an analysis, take the following steps:

1. Examine the data thoroughly, noting the definitions, number of observations, representativeness of the data and the criteria used in its presentation.
2. If possible, determine the method of collection.
3. Note any limitations of the data which will detract from the appropriateness, accuracy and, therefore, validity of your summation.
4. Ask yourself whether the data is comparable with other collected evidence.
5. Beware of the purposes of the secondary data as they may conceal bias or prejudice, perhaps because of the way they were collected in the first place. Subjective compilation and interpretation may cloud the published results.
6. Take care in understanding the purpose of the original study, as the choice of market and the sample selected can unduly influence results.

Secondary data provide an immediate and inexpensive method of analysis—often the only possible method. (For example, for certain demographic statistics such as home ownership per local government region, religion, age of population, level of qualifications and the like, the Australian Census is the only source.) However, the limitations of secondary data should be recognised: the data may not have been designed specially for the problem at hand; verification of the original source of information may not be possible (e.g. you extract a table from a publication that reproduced a portion of data from another primary source); the results may be influenced by the purpose of the original study (e.g. over the past few years, several prominent researchers have doctored their results in order to substantiate their hypotheses); and the data may exclude key terms and methods of collection. From the above, it is obvious that no student, firm or businessperson operates in an environment devoid of secondary data.

Primary data

In an attempt to overcome the problems encountered in using secondary data, some organisations initiate their own data collection. Such data, referred to as **primary data**, are considered more useful and reliable, although their collection is far more time-consuming and therefore more costly than it is for secondary data.

Before collecting any information, you must first determine the purpose of the study since, as mentioned earlier, this indicates the areas to be investigated and the appropriate methods to be used. Once this has been established, the method of collection must be considered. Data may be collected by observing or recording business operations, for example. Data on staff numbers, monthly staff wages, the productivity per worker on an hourly basis, the sales turnover per subsidiary, monthly sales over the past five years, monthly inventory levels, monthly profit and loss results, monthly cost of materials, fixed costs and the monthly advertising expenditure per department are usually available from the appropriate personnel. However, when data are unavailable, questionnaires are often compiled and these may be administered in a variety of ways (discussed in Section 12.3 in Chapter 12).

8.2 Competency check

1. What is the difference between internal data and external data? Provide one example of each.
2. Cite one external source of statistical information relevant to your industry and briefly discuss the usefulness of the specific data.
3. As manager of a small retail outlet, what internal data would be useful for you to convince head office of the need for bigger premises?
4. Most of the assignments you will undertake in your tertiary career will contain secondary rather than primary data. Why the dependence on the former?
5. Primary data, if appropriately collected, are often claimed to be more useful and reliable than secondary data. Do you agree with this suggestion? Briefly discuss.

Chapter 8 – Introduction to statistics

8.3 Pitfalls in the use of statistics

'Statistics are no substitute for judgment.'—Henry Clay

'Statistics are like alienists—they will testify for either side.'
— Fiorello H. La Guardia

'Statistics can be made to prove anything—even the truth.'—author unknown

'There are lies, damn lies, and statistics.'—Benjamin Disraeli

'A statistician is a specialist who assembles figures and then leads them astray.'
— author unknown

'Facts are stubborn things, but statistics are more pliable.'—author unknown

All of the above quotations bring into question not so much the reliability of statistics but, rather, the intention of the statistician in presenting the numerical information. For example, there is the story of a two-car race between the Americans and the Russians, with the former declared the winner. However, in *Pravda,* the report read: 'Russian car finishes second, the American car next to last.'[6] This biased interpretation of the result implies that the Russians outperformed the Americans in the race. Unfortunately, this abuse of statistics is not an isolated instance; books have been written solely on the abuse of statistics.

Collecting, classifying, summarising and presenting data does not guarantee success in solving your problem, particularly if the statistical methods were not correctly applied. For example, you may have non-comparable data, incorrect graphing, incorrect assumptions or even bias. Furthermore, the presenter of statistics may deliberately manipulate them to convince the reader of a certain viewpoint or result. Since all of us are bombarded by all sorts of statistics at all hours of the day, you should be aware of the more common abuses of statistics.

This section examines the most common types of statistical abuse, some of which you may find will overlap. Your understanding of these pitfalls, combined with your understanding of statistical techniques/methods, should provide you with the necessary background to ensure that your report or analysis is well presented and factual, unbiased in its data-collection technique, and hence devoid of any mistreatment of the collected numerical information.

The misused and overused average

*'Don't put too much stock in figures;
a man can drown in a stream that averages only two feet in depth.'*

Average weekly earnings for full-time working adult males in Australia were approximately $788 in November 1996. Women's average weekly earnings were 65% of men's average weekly earnings. In 1996, per capita income in Australia was below that of Brunei.

Business mathematics and statistics

The March quarter 1997 Australian Consumer Price Index (CPI—a weighted average) result indicated that inflation on an annual basis was 1.3%. The median price of a house in April 1997 in Sydney, Melbourne and Brisbane was $259 500, $165 000 and $143 500 respectively.[7] All these statistics exemplify the use and misuse of averages. For example, the average weekly earnings for adult full-time working males of $788 is the mean of incomes for males, with the data received from a sample of selected companies. Therefore, since the entire adult working male population is not surveyed, there is the possibility of a difference between the sample result and what would have been the average earnings had all eligible males been surveyed. Furthermore, as the mean is affected by extremes (see Chapter 10 on central tendencies), the unequal distribution of wealth in Australia may lead to an inflated average for weekly earnings.

When observing the average total weekly earnings for women, allowance must be made for the fact that the part-time labour force is dominated by women (in fact, 75% of part-time workers are women), and their average weekly total earnings would therefore be expected to be less. In order to make a comparison of weekly earnings more relevant, only the full-time labour force should be studied. Such a study indicates that women's average weekly total earnings are approximately 80% of men's average weekly total earnings.

There are many Australians who disbelieve the published data that claim an increase in average weekly earnings over the past three or four years. The alleged increase in earnings could be a result of compositional changes in the group surveyed (e.g. a greater number of highly paid people may have been surveyed) or variations in benefit distribution. For example, with the introduction in the 1996/97 Federal Budget of the 15% tax surcharge on superannuation for high income earners, there were those who cashed up their benefits, and this contributes to a rise in the average weekly total earnings. Finally, the ABS's estimates of average weekly earnings are based on information relating to a sample of employers and are therefore subject to sampling errors, as described in Chapter 12.

Finally, when evaluating the average duration of unemployment, the ABS supplies us with two measures, the mean and the median. The former is calculated by taking the sum of the weeks that all the unemployed have been out of work and dividing by the number of unemployed workers, while the median is the time (in weeks) where 50% of the unemployed have been out of work for less than this period and 50% for more. For May 1996, the mean term for an unemployed person was 50.5 weeks while the median was 18 weeks, a marked difference which can be explained by a large number of unemployed workers who have been out of work for far longer than a year, which has resulted in the average (i.e. the mean) for all substantially increasing. (Chapter 10 covers this subject in far greater detail.) This problem of the mean average—being affected by extreme values—also explains why the average prices of residential property are almost always defined in terms of the median.

Percentages

In April 1997 the seasonally adjusted unemployment rate for Australia was 8.7%, compared with 6.1% in January 1990. Thus there was a 2.6 percentage point increase in the unemployment rate, or the unemployment rate rose by approximately 42.6%

Chapter 8 – Introduction to statistics

(i.e. $\frac{2.6}{6.1} \times 100$). The latter observation provides a far more dismal picture of the state of Australia's labour market, a fact that may well be highlighted by the Opposition.

Likewise, assume that you are paying 22% of your income in the form of direct tax (PAYE) versus 20% last year. In this case, there has been an increase of 2 percentage points, or the percentage tax paid has increased by 10% $\left(\frac{22-20}{20} = \frac{2}{20} = 10\%\right)$. A similar 'abuse' of percentages is the case of flat rates of interest (normally cited for personal loans) versus effective rates of interest (see Chapter 3).

The median price of a house in Perth in April 1997 was $136 000 compared to $98 800 in June 1990. Therefore, the median price of houses in Perth rose by 37.7% from June 1990 to April 1997. However, it would be also correct to say that the median-priced house in Perth in June 1990 cost 72.6% of what it cost in April 1997. Therefore, the decision of which year is selected for comparison purposes will determine the percentage change.

Base year selection

Whenever a comparison is made between one period and another, ensure that the base period (i.e. the selected period with which all other periods are being compared) is a normal period and fairly recent; otherwise, results can be highly inflated or deflated. For example, in assessing the annual rate of change in private investment in new capital (equipment, plant, machinery, and buildings and structures), the selection of the base year has a dramatic impact on the evaluation of the performance of this sector. For example, using 1991/92 as the base year, there has been an average growth rate in private capital investment of 14.7% a year to 1995/96; however, nominating 1989/90 as the base period, the average annual rate of growth is a meek 4%.

Similarly, there was a 5.4% reduction in the number of building commencements in Australia in 1994/95 compared to the previous year, yet the third highest number of such commencements in 20 years was recorded in 1994/95. It just so happened that in 1993/94, dwelling commencements were the highest on record; hence, 1993/94 as a base year makes any other year look benign.

Finally, if we turn the clock back to the 'recession we had to have', Australia's sluggish economic performance in 1990/91 was exemplified in the seasonally adjusted Gross Domestic Product's results of −0.5%, −1.0% and −0.3% for the 1991 March, June and September quarters respectively.[8] The September quarter decline was the fifth consecutive quarterly decline. Therefore, the 1991 December quarter's GDP rise of 1.1% is an improvement on the September quarter, but was still well below the result of previous quarters. Why? Because the previous quarter (e.g. the September quarter) was chosen as the base period and the base figure is well below that of, say, a year earlier. Perhaps it would have been more advisable to select the preceding year as the base. In any event, the importance of the nomination of the base period cannot be over-emphasised, as results may differ dramatically depending on the base period selected.

Real terms

When analysing retail sales, salaries, the GDP or interest rates, the effect of changes in the level of prices must be taken into account. For example, depositing your savings in an interest-bearing deposit account attracts, say, 6% p.a. interest. You must ensure that the 6% covers the inflation rate. Otherwise, you have forgone the use of your funds to receive fewer real dollars at the end of a certain period because the rate of interest applying to those funds is less than the rate of inflation, thereby reducing the value of these savings. You will, in this case, have a negative real return on your investment.

Over the period from 1986 to 1996, the average weekly earnings for full-time workers in Australia rose 68% from $425 to $715 a week, but that doesn't suggest that these workers were better off by that amount. One must account for inflation to determine the real growth or decline in wages—in other words, the purchasing power of these earnings. Inflation over the same period was 61.5%, resulting in real growth of 4.2% over the 10 years, or about 0.4% growth in real wages each year rather than the average 5.3% annual growth in money wages.

Sampling

Sampling is the usual method for gathering information to provide an educated estimate of a population which is too diverse in distance and numbers for all its possible respondents to be questioned. The selection of each sample must be random—that is, each member of the nominated population must have an equal chance of selection, because the result of a sample study is only as good as the sample on which it is based.

Problems of sampling may be considered to be of two general types. Since only a portion of the entire population is examined, a sample result may differ from the result that would have been obtained had the entire population been examined (i.e. because of chance errors that may occur in the process of sampling). This is referred to as a *sampling error*. A *non-sampling error*, on the other hand, may be made in the collection and analysis of the data and as such may happen even if the entire population is surveyed; thus, the error was not the result of selecting a sample from the population but may be due to erroneous responses, incorrect data entry, etc.

An often-quoted example of a non-sampling error occurred in the 1936 Presidential election campaign in the United States. Ten million surveys were mailed to those listed in the telephone books, with results suggesting certain victory for Alfred Landon. However, Roosevelt won the election convincingly and the survey was discredited. The sample taken was not random, for in 1936 a telephone was a luxury item owned mostly by the affluent, who normally voted Republican (Landon's party, in this case); hence, the sample was not representative of all voters in the United States. The reliability of a sample normally increases as the size of the sample in relation to the population increases.

What about the change in median-priced houses in Australia's capital cities? How have they fared over the last three years, for example? Well, that depends on whose statistics you examine. Access Economics collected data from the Housing Industry Association (HIA), the Real Estate Institute of Australia (REIA) and the Australian Bureau of Statistics (ABS), all of which depend on survey results to determine the median price of houses in respective capital cities.[9] In examining the percentage change in Sydney house prices for the period June 1993 to June 1996, for instance, the ABS estimated an 8% increase compared to the HIA's 32% and the REIA's 16%. Brisbane's results ranged from 1% to 16%, Perth from 8% to 23%, and all capital cities from

5% to 19%, and this for the same time period. Users of such statistics would have to examine explanatory notes often accompanying such publications in order to determine if the survey method seems appropriate. For example, the REIA relies on the survey returns (often less than 100) from its over 1500 members, and some analysts may find this inadequate.

Have you ever wondered how Australia's best-seller book lists are compiled? The information used by the *Australian Bookseller & Publisher* magazine is collected randomly. Large chain stores are given the same weight as small speciality shops. Approximately 60 bookshops and newsagents are selected randomly throughout the country, and the list is compiled on the basis of replies received. One month 40 responses may be received, and another month only 10 responses may be received. In addition, retailers tend to consider books as best-sellers on the basis of the percentage of books sold from the available stock. For example, if a major retailer sells 400 of 1000 copies (40%), the book may not be included on its best-seller list, while a small shop selling each of its five copies of a book within a short time may include it on the list. In light of the measurement problem, the best-seller list is more of an information guide to available books.

In Australia, surveys are sometimes made by means of phone-ins, especially when initiated by radio commentators about a topic attracting community interest (e.g. tighter gun laws and road traffic laws, early prison release schemes, etc.). However, many statisticians think that responses emanate only from those motivated enough to telephone and thus do not necessarily represent the views of the community as a whole. Besides, do those who listen to the radio truly represent the population, especially during working hours? The theory that phone-ins reflect the views of the community at large is therefore highly questionable, and this invalidates their use for basing policy-making on majority views.

A.C. Nielsen,[10] the research organisation that measures television audiences for the networks, introduced 'people meters' to Sydney and Melbourne on 30 December 1990 and to Brisbane, Perth and Adelaide on 10 February 1991. Each selected household (there are 2200 sample households in the capital city metropolitan areas and 1700 in regional Queensland, northern and southern New South Wales, Victoria and Tasmania; thus 3900 sample homes in total) receives a small eight-button unit, the people meter, which is placed on top of the television set. Each member of the household is assigned a numbered button (and their name is printed above their assigned button number) which he or she activates when viewing, either by pushing the appropriate button on the people meter or by using the remote control. A household member who leaves the room must again press his or her button to indicate no viewing. Even visitors indicate their presence by pushing a visitor button. All people meters and set meters are connected to a microprocessor which constantly records and stores all meter information and automatically downloads this information to the Nielsen mainframe.

This monitoring technique replaced the diary method of measuring viewing audiences, whereby participating households marked off in a two-week diary what programs they watched. This method was obviously open to misuse, as it was easy for people to forget to mark off every show they watched each day, and some may have marked only their favourite programs (the easiest ones to recall). The result in the first year of the people meters indicated that when the diary method had been used, all viewing outside prime time had been heavily underestimated, and that over the entire day viewing levels were underestimated by 15%. This information is obviously invaluable to the networks, all of which have had varying degrees of financial difficulties in the early 1990s.

Bias

The 1996 release of the *World Competitiveness Yearbook* rates Australia as 21st out of 46 nations surveyed which, according to Peter Roberts, doesn't reflect Australia's true economic performance or its prospects.[11] According to Roberts, this survey merely combines hard economic data and subjective views from 'a survey of a few thousand executives spread thinly across 46 countries'. This result contrasts with another report according to which Australia is found to be the world's 12th most competitive nation.[12] This benign appraisal of Australia's economic performance is in spite of the fact that Australia has outshone the seven major Western nations in terms of GDP (i.e. economic growth), industrial production since 1989/90 and lower inflation, which seems to suggest our competitiveness has improved.

Have you ever wondered how the median rent of a house or unit is determined? Well, it depends on which source you decide to use. The Research and Policy Unit of the NSW Department of Housing provides the quarterly *Rent Report* based on new lodgments made with the Rental Bond Board during that quarter. For example, the median rent for the Sydney metropolitan area derived for the different categories of unit and/or house accommodation for the 1996 June quarter was based on 65 418 lodgments made that quarter;[13] therefore, the data are for recently rented accommodation. However, it is known that the total number of bonds held by the Rental Bond Board is not equivalent to the total number of rental properties. There are always some properties vacant and a few landlords do not collect bonds.

The Real Estate Institute of Australia (REIA), on the other hand, based its January to June median rent per housing category for Sydney on an aggregate rent roll of approximately 16 768 properties managed by member firms in Sydney. These rental properties will therefore comprise new as well as old rentals. Thus, the REIA's survey contains accommodation not considered by the Research and Policy Unit—that is, premises which were rented prior to this quarter. Furthermore, the number of properties considered by the latter is almost three times the number enumerated by the former. Therefore, we find that in assessing rental levels, the size of the sample and the method of collection can easily prejudice the median rental result.

Recently, two well-known organisations engaged in economic and business forecasting were at opposite poles on the outlook for the Australian retail property market.[14] One believed that over the 1996 to 2006 decade, real retail rents would decline by 10% and real property values by 12%. This assessment was based on the calculated turnover per square metre which they determined had dropped by 20% over six years and projected future growth, in part, based on these results. The second organisation claimed that the first, while using the ABS data, had failed to recognise a new range of classifications in 1989. This suggests that turnover per square metre was considerably higher because the amount of floor space was considerably less than asserted by the first consultant. If turnover per square metre was, in fact, higher than initially suggested, the outlook looks a bit brighter for the retail sector.

Chapter 8 — Introduction to statistics

You may have realised that the basis of much of the unintended bias may be attributable to problems of definition. Your understanding of relevant terms should correspond to the definitions of the statistician. For example, your definition of an employed person may differ from that of the Australian Bureau of Statistics (ABS), which defines employed people as:

> ... all those aged 15 and over who, during the survey week:
> (a) worked for one hour or more for pay, profit, commission or payment in kind in a job or business, or on a farm (including employees, employers and self-employed persons); or
> (b) worked for one hour or more without pay in a family business or on a farm (i.e. unpaid family helpers). . . .[15]

Further reinforcing the problem of accurately defining and measuring the level of unemployment, the ABS estimated that about 1.1 million Australians were left out of the jobless figures (which stood at 760 000) at the time because they didn't satisfy the official definition of unemployment (as defined above).[16] In fact, the ABS found that there were 1.9 million who are not employed who want to work, rather than the official 760 000 who satisfy the ABS's definition of unemployment. To add insult to injury, there are another 570 000 people who are underemployed—that is, working in jobs beneath their skills or education or employed part-time when they would rather work full-time.[17]

Australia's record of industrial disputes has not been one deemed worthy of emulation overseas. However, when comparing the number of industrial disputes, days lost and the like, it should be understood that in certain countries a strike is not considered to have occurred unless a certain number of hours are lost as a result of an industrial dispute, while in other countries (e.g. Austria) strike time is measured to the minute. Therefore, one must try to ascertain that what is being compared is similarly measured.

Graphic distortion

Perceptual illusions[18]

Let your eyes wander over the following geometrical optical illusions:

1. *Müller-Lyer illusion.* The horizontal lines are the same length.

2. *Ponzo illusion.* The converging lines are seen as parallel lines in depth. The two parallel lines are equal in length.

3. *Jastrow illusion.* The two figures are identical.

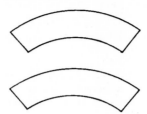

4. *Sander's parallelogram.* The dotted lines are the same length.

5. *Zollner illusion.* The verticals are parallel.

6. *Hering illusion.* The horizontal lines are straight.

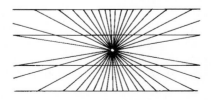

7. *Orbison illusion.* The inner figure is a true circle.

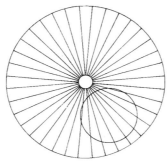

A picture in the wrong hands can paint a thousand misconstrued words!

Graphic distortion
Sales results from Grinn & Bareit are shown in Table 8.1.

Table 8.1 *Grinn & Bareit Company sales, 1991–97 ($ million)*

Year	Sales
1991	20.0
1992	22.0
1993	25.0
1994	26.0
1995	26.5
1996	26.8
1997	27.0

Source: Hypothetical.

A graphic representation for future investors is given in Figure 8.1.

Is there anything wrong with the presentation in Figure 8.1? Why would the company graphically represent its results in this fashion? The company is trying to entice customers to purchase shares in the company, and what better way than to suggest through the graphic presentation of its results that the company is surging ahead by leaps and bounds. What is lacking, in the first instance, is a zero where the vertical (y) axis and horizontal axis (x) axis intersect. Thus the graphic representation could have been drawn as shown in Figure 8.2.

Occasionally you will find a broken line on an axis; this indicates that the area bordered by the broken line is out of scale with the rest of the graphic representation, as shown in Figure 8.3. Note that Figures 8.1 and 8.3 exaggerate the results.

What happens if the data contained in Table 8.1 were presented in a simple bar chart, such as Figure 8.4? Does this chart accurately reflect the true performance of Grin & Bareit? How do you suggest the graph be improved to more adequately describe the actual sales results?

Figure 8.1 *Graphic representation for future investors, 1991–97*

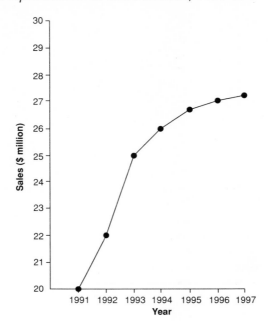

Figure 8.2 *Graphic representation for future investors, 1991–97*

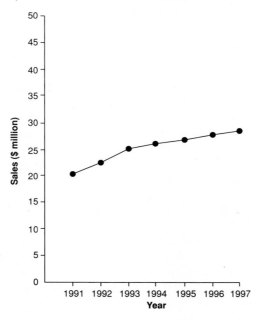

Chapter 8 – Introduction to statistics

Figure 8.3 *Graphic representation for future investors, 1991–97*

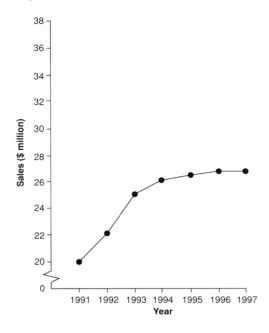

Figure 8.4 *Grin & Bareit Company sales, 1991–97*

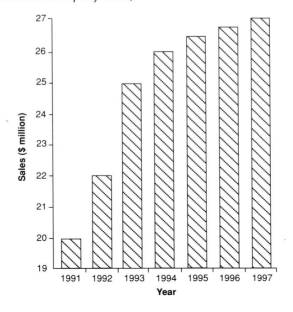

Seasonal fluctuations

You will find that economics statistics are often influenced by seasonal factors which must be accounted for in evaluating, for example, your company's sales or profit performance. For example, your December results for 1997 may indicate that sales are 20% higher than in November; but before you bring out the bubbly, you may find that you budgeted for an even greater turnover for the month. This can, of course, be explained by recognising that December is usually the busiest month of the year, with turnover comprising perhaps 25% of the year's sales; therefore, an increase of 20% on November may be below budget. You may also find that Easter, for instance, falls in March one year and April the next, which could be used to explain the increase in sales turnover in April of the second year.

Consider some of the products that are directly affected by seasonal variations—electrical appliances such as heaters and air conditioners, foodstuffs such as ice cream, umbrellas, swimming apparel, clothing, air travel and greetings cards, to name but a few. In order to allow for such seasonal variations, the ABS uses a statistical tool referred to as seasonal adjustment to standardise the data. For example, a list of seasonally adjusted data presented by the ABS includes the GDP, the balance of payments, retail sales, private investment, industrial production, manufacturers' sales and stocks, building approvals (for new dwellings), construction activity, unemployment rate, job vacancies, household income, and financial aggregates (i.e. components of total credit). Therefore, these data emanating from the Bureau will allow the user to analyse the results had seasonal factors not influenced them. That is, the impact of these seasonal factors upon the firm's performance can be accounted for.

Seasonal variation has a major impact not only on sales turnover but also on inventory or stock control. Improved budgetary prowess should assist in reducing holding costs, warehouse expenses and borrowings, and thus improve the firm's profitability. Chapter 17 examines seasonal adjustment techniques.

Statistical distortion

The presentation of statistics is sometimes (intentionally or unintentionally) distorted. For example, estimates of crowd numbers at the premiere of a film, at a demonstration or greeting the arrival of an overseas rock star can vary quite considerably, especially when reported by those with a vested interest.

Unintentional distortion of statistics can arise from the use of inaccurate or incomplete data, as a result of arithmetic error, or due to a change in the status of a key area. Revisions to quarterly estimates of Australia's national income and expenditure accounts sometimes significantly alter our perception of the state of the economy. Recent surveys of the number of Australians linked to the Internet have ranged from 250 000 to 1.1 million, while in the United States the range is between 10 million and 22 million.[19] In other words, even the Internet experts are unsure of the actual size of the network, and therefore of their market.

Several recent studies claim that the annual cost of traffic congestion in Sydney and Melbourne is between $3 and $5 billion, a hefty difference of 67% between the two figures.[20] This is no small wonder when the difficulty of quantifying such a cost (e.g. lost productivity, pollution, accidents, stress, etc.) is taken into account.

Chapter 8 – Introduction to statistics

When the goods and services tax (GST) was presented in the run-up to the 1993 federal election, Dr John Hewson indicated that the inflationary impact on the economy would be approximately 4.8%; but the Treasury calculated that it would be closer to 9.5%. Why the dramatic contrast in estimates? The basket of goods that Dr Hewson used to measure the inflationary impact of the GST contained products with wholesale tax rates greater than the proposed GST; hence, it allegedly understated the inflationary effect of the GST by deflating it for the pre-existing sales tax. Furthermore, in examining the basket of goods used by the Opposition to assess the inflationary consequences of the GST, the Treasury found that it included 7 kilograms of pet food, disinfectant, dishwashing liquids, stain removers, starch, laundry detergents, toilet cleaner, bleach, tissues, toilet paper, toilet soaps, toothpaste, shampoo and conditioner, and baby powder. In terms of the bare necessities of life, the basket contained only one loaf of bread and a packet of bread rolls. There was baby powder, but not baby food—a clean but hungry child the result. The only meat was a chicken and a small packet of bacon. It is interesting that the various cleaning agents, personal toiletries and pet food, all of which attract 20% wholesale tax which would be abolished with the implementation of a GST, made up some 40% of the total bill. 'So the message of all this', the then Treasurer Mr Willis said, 'is if you want to survive under the goods and services tax, you're going to have to learn how to eat Pal and drink Harpic.'[21]

So what we have here are two estimates of the all-important inflationary impact of the proposed GST, with one estimate more than double the other. Perhaps we can refer to the Opposition's estimate as 'selective distortion', due in large part to the sample of goods selected.

Non-comparable data

How often have you come across figures or statistics which were a bit difficult to swallow? It probably happens fairly frequently, and sometimes the problem can be attributed to the lack of comparable data. For example, the living standards of different countries can be compared by looking at GDP on a per capita basis. However, this is inherently deceptive since per capita income does not indicate a country's distribution of income, take into account its cost of living, or measure quality of life.

It is claimed that Brunei has the highest per capita income in the world. But are its living standards also the highest? Comparing the per capita incomes of the citizens of different countries is fraught with problems, since merely citing per capita incomes and ignoring the cost of living gives an incomplete picture, at best, of the standard of living. Also, if you consider the level of home ownership, the size of homes, access to parks and other recreational facilities, car ownership, travelling time to and from work, the work week, and so on, it would indeed be difficult to convince many Australians that other countries offer the high level of living standards they are accustomed to.

The retail trade normally looks forward to Christmas, as sales over that period often account for as much as 40% of annual sales. Therefore, any goods affected by, for example, festive seasons or climatic conditions must be seasonally adjusted (see Chapter 17 for a discussion on the calculation of seasonal adjustment) to assist in forecasting and evaluating the results. However, being aware of and allowing for seasonal patterns does not prevent errors from occurring. For example, consider a production manager who is called to the supervisor's office because April sales were below the

March results. The manager is able to link the downturn to the Easter holidays in April, which reduced the number of working days for the month below those for March. In fact, on a daily basis, production may have been up in April.

Cause and effect relationships

Statistics prove that you can prove anything by statistics.

Two effects occurring together sometimes result in a pronouncement that one of the two events influenced or caused the other. This may be taken to extremes—as in the United States, for example, where statistics indicated that such a relationship existed between the stockmarket and the winner of the Super Bowl, with the latter said to influence the former. On a more serious note, the December 1987 issue of the *Lancet,* a British medical journal, contained a summary of a 15-year study of Swedish soldiers which suggested that heavy marijuana users are six times more likely than non-users to develop schizophrenia. However, the study's authors claimed that the statistical association between schizophrenia and marijuana did not necessarily mean that the drug caused the mental problem. On the contrary, according to the researchers, cannabis consumption could have been caused by an emerging schizophrenia. In other words, either variable may have caused the other, or both may be a result of a third, unknown factor.

The October 1987 stockmarket crash was to herald a recession, according to a number of prominent commentators, since a fairly marked recession almost always followed a severe sharemarket setback (e.g. in 1929). However, there resulted neither accelerating inflation nor excessive inventories.

Over the 1990–93 period, the Reserve Bank of Australia reduced interest rates no less than 16 times in an attempt to stimulate the economy, particularly the business sector and, to a lesser extent, the household sector. However, an economic recovery was far more elusive than the Federal Government had expected. Although the business and household sectors are interest-rate sensitive, there are other variables that play a major role in business and household spending. These include profit levels, company debt levels, stock on hand, wage levels, the value of the Australian dollar, consumer and business confidence, expectations, and the level of unemployment. In other words, declining interest rates alone may not guarantee an upswing in investment or consumer spending.

An advertisement might claim that daily use of a certain acne cream reduces pimple penetration. However, proper hygiene, nature and time lead to the same result. Similarly, high doses of vitamin C are said to assist in combating a cold, but when allowed to take its normal course a cold will clear up after a certain number of days without the use of vitamin supplements.

The Australian Associated Brewers, in a Budget submission, indicated that beer consumption had fallen by an average of about 2 litres per capita a year over the past decade, and that this decline was due to an increased excise tax on beer. However, was it the excise increase alone that contributed to the fall in consumption? Changing lifestyles (more emphasis on fitness) and/or tastes (increased consumption of wine) and/or the introduction of random breath-testing may also have affected consumption.

Similarly, a tobacco company spokesperson might compare average life expectancy today with that at the time of the first recorded cultivation of the tobacco plant. As longevity

Chapter 8 – Introduction to statistics

has increased since the first cultivation of the tobacco plant, the spokesperson might go on to suggest that, contrary to popular belief, smoking is good for your health!

Meaningless (faked) statistics

How often do we hear of country areas plagued by mice or kangaroos? Various estimates of the mice or kangaroo populations are accordingly forthcoming. How was such a census undertaken? Were all mice or kangaroos available for comment? Was it based on the number of mouse-traps set versus the number of mice successfully trapped, or on the number of kangaroos claimed to have been killed per square kilometre? It has been estimated that Australia's underground economy (i.e. payments in cash without deduction of income tax) is 10% of the GDP (i.e. approximately $49 billion was not included in the 1995/96 GDP). If it is an underground economy, who dug up the statistics?

In business, it may be an advantage to inflate forecasts. For example, in the past, Japanese coal importers have been accused of overestimating their forecast demand. Since Japan is the major importer of Australian coal, the Australian coal industry invested in and budgeted for Japan's forecast demand. However, there was an oversupply because of alleged intentionally excessive forecasts by the Japanese, which forced Australia to reduce its price in order to clear the surplus.

8.3 Competency check

1. What is meant by defining economic data in *real terms*?
2. (a) What is meant by the *base year*?
   (b) Why is the selection of a particular base period considered important?
3. Discuss the claim that research based on a sample is no better than the sample on which it is based.
4. Briefly explain two statistical presentations you have come across which you feel may have been contrived to deceive the reader.
5. Can you identify two other statistical presentations which you feel are inaccurate or poorly presented, although there is no obvious evidence of manipulation by the presenter?

8.4 Summary

'Figures don't lie, but liars figure.'

The term 'statistics' means many things to many people. Most people associate a collection of numerical information or data with statistics. But this discipline also encompasses principles and procedures developed for the collection, classification, presentation, analysis and interpretation of data. Finally, we also have such a thing as a sample statistic, which is a characteristic or measure taken from the sample of a population.

Once you have decided that a particular statistical problem must be confronted, there are certain basic steps to follow:

1. *Collect the data.* There is a range of sources of quantitative information, both internal (within the firm) and external (outside the firm). You may have to use published data (secondary data) or to survey and collect your own (primary) data.
2. *Organise the data.* Edit, classify, tabulate and summarise the data.
3. *Present the data.* Present the results in an easily understood form such as a table, a pie chart or even a word statement.

The collecting, classifying, summarising and presenting of data form a branch of statistics known as descriptive statistics. However, estimating and/or drawing inferences or conclusions based on presented data is another branch, known as inferential statistics, and this incorporates both analysis and interpretation.

4. *Analyse the data.* This may vary from a simple, straightforward analysis (e.g. of average sales turnover) to the more complicated probability analysis (e.g. what term should the warranty period be to ensure that no more than 1% of units sold are returned?) or time-series analysis (recognising the impact of seasonal variations on sales and taking this into account when forecasting).
5. *Interpret the data.* Again, the findings and conclusions may be obvious, or they may suggest that further study is necessary for a valid conclusion.

Statistics are tools which no decision-maker can do without. Irrespective of the size of the business, statistics form an inherent part of the explanation of the success or failure of the venture. Understanding statistics and statistical methods will allow you (1) to understand better the important relationships between such statistical variables as advertising and sales, or seasons and sales; (2) to make better decisions, such as forecasts incorporating seasonal variations; and (3) to cope better with the changing economic environment to which your business is exposed, be it competition, pricing, movements in the Australian dollar or interest rates.

The proliferation of statistics is not confined to any one discipline. Your company probably uses statistical techniques to help it analyse accounting and financial records, quality control, sales management, market research and personnel management, since every section of a business is evaluated on the basis of statistical analysis. In fact, you may be asked to provide current data on your company's or industry's performance and the validity of your presentation and analysis of the data will depend on the method of collecting the data. The improper gathering of information immediately nullifies the usefulness of any evaluation based on the data.

Data can be considered either secondary (i.e. data used by a party other than the original collector) or primary (i.e. data as originally collected and presented). The sources of such numerical information may be internal (i.e. data obtained from within one's own organisation, be it primary or secondary) or external (i.e. material relating to and available from outside the organisation), such as the Australian Bureau of Statistics' publications, investment newsletters, newspapers, magazines, trade journals or original data collected personally.

Secondary data, although far more readily available and less costly than primary data, should be examined carefully for the method of collection, definitions, criteria used in the data presentation and the purpose of the original presentation, as this could contain bias. However, it may also be the only means of obtaining certain information.

Primary data, although more valuable (as it caters to the particular problem at hand), takes far longer to collect and is, consequently, more costly than secondary data. Normally, a statistical survey forms the basis of an analysis, with the survey itself, its content, wording and presentation, a key factor in determining the ultimate usefulness of the evaluation. Questionnaires should be as short and concise as possible; questions should be easy to answer, presented in a logical sequence and be devoid of technical jargon; and an assurance of confidentiality should be given.

In assessing the reliability of collected information, one must be aware of the sampling techniques used. Since the number of observations (i.e. the population) is normally too large and/or too diverse to survey in its entirety, a portion or sample of the population is selected. The sample must represent the elements making up the entire population. Using the data generated by the sample statistic allows us to estimate the population parameter by making statistical inferences (i.e. conclusions about the population based on the sample results).

Look around you and you will see statistics being presented by sports commentators, medical authorities, economists, teachers, demographers, insurance companies, retail companies and stock exchanges, to name but a few. Therefore, your understanding and collection of data and appropriate application of the proper statistical techniques and methods will be valuable to you irrespective of your interest or place of work. However, don't blindly accept statistics, because sometimes statistics are unintentionally or even intentionally misused. Therefore, when evaluating another's statistics (i.e. secondary data) or collecting your own (i.e. primary data) for analysis, watch for the following pitfalls:

1. *Averages.* These ensure that the appropriate measure of central tendency (i.e. mean, median or mode) is used.
2. *Percentages.* Distinguish between absolute percentage changes (e.g. the change from 6% to 9% is a 3% absolute percentage rise) and relative percentage changes (e.g. the change from 6% to 9% is an increase of 50%).
3. *Base year selection.* It must be fairly recent and a normal period.
4. *Real dollar results.* Allow for inflation when you determine your true (real) return.
5. *Sample.* If a sample, is it representative? How was it collected?
6. Bias. Why were the statistics presented? Which organisation collected the figures? Was there any vested interest in the results?
7. *Proper graphic presentation.* Is it a zero-based graph?
8. *Seasonal variations.* Have data which are influenced by seasonal variations been seasonally adjusted?
9. *Definitions.* Are relevant definitions included, and do you understand them?
10. *Causality.* If a relationship between two variables exists, it does not necessarily indicate that one has definitely contributed to the other. Judgment must be used with correlation analysis.

Business mathematics and statistics

8.5 Multiple-choice questions

1. Descriptive statistics is about:

    (a) drawing conclusions from results of a collection, analysis and interpretation of data;
    (b) explaining the different statistical techniques available;
    (c) summarising collected data;
    (d) the link between statistics and respective disciplines such as economics, psychology, etc.

2. Internal data is collected:

    (a) from within the industry in which the company operates;
    (b) from within the company itself;
    (c) by the Australian Bureau of Statistics;
    (d) to assess the industry's performance.

3. Secondary data is often preferred to primary data for all of the following reasons except:

    (a) it is less expensive to collect than primary data;
    (b) it is faster to collect than primary data;
    (c) it may be the only method for obtaining the required data;
    (d) it is far more accurate than primary data.

4. A sampling error may occur because:

    (a) only a portion of the population is examined (i.e. the sample), and the result could vary from what would have been contained if the entire population had been examined;
    (b) there was an error in collecting the data;
    (c) there was an error in analysing the data;
    (d) the respondents filling out the form were dishonest in their answers.

5. Retail businesses should take account of seasonal fluctuations for:

    (a) budgetary purposes;
    (b) inventory control;
    (c) staff recruitment;
    (d) all of the above reasons.

8.6 Exercises

1. Briefly discuss the different ways in which the word 'statistics' has been used.

2. (a) What is meant by a sample? A population?
   (b) Explain the relationship between a sample and a population.

Chapter 8 – Introduction to statistics

3. Define and give an example of each of the following terms:
   (a) sample
   (b) population
   (c) discrete variable
   (d) continuous variable.

4. How can sample data be used in decision-making?

5. Important decisions which don't work out as planned often draw a fair bit of public comment. Can you think of two such examples in Australian business over the past five years?

6. Has it ever happened that nothing worked as you planned, even though you made what seemed to be a sound decision? Was the problem in the decision-making process? Briefly discuss.

7. Can you name two uncertain decision-making situations that you have confronted in the past week? How did you confront them?

8. In what context have you come across descriptive statistics (besides this book, of course)?

9. When you are collecting information on the state of the Australian economy, what are the possible sources?

10. If you wanted to find secondary data about the industry in which you are employed, what would be the major sources of such information?

11. Which forms of presentation do you think are the easiest to understand?

12. If you were anticipating importing toys from Asia, which statistics would you try to collect and analyse?

13. What steps would you take if you were asked by your supervisor to examine absenteeism in your division over the past 12 months?

14. Assume that you are the owner and manager of a very successful shoe shop and are considering opening a second shop. You have just been offered a lease on a shop-front on a main street in a suburb approximately 5 kilometres from the first shop. What statistics would be most useful in helping you make the 'right' decision?

15. Assume that you are considering a move to Cairns where you are contemplating starting a business leasing out your yacht with you as the skipper. What type of statistics would you try to collect prior to undertaking any funding commitments?

16. You have just been informed that you are going to be transferred from Adelaide to Sydney for 12 months, employed in the same capacity as you are now. However, you are informed that the salary will remain the same. What statistics would be useful for you to convince your manager that you need a loading added to your usual salary?

17. There have been a number of major railway accidents over the past decade, which seems to imply that it is, indeed, dangerous to take the train on extended journeys. What kind of statistics would you examine to accept or reject this claim?

18. A 1995 World Bank report found Australia to be the wealthiest nation in the world with its endowments of arable land, minerals, machinery, infrastructure and educated labour adding up to more than US$835 000 for each Australian.[22] Interestingly, in second place was Canada; and the United States was in 12th position. Can we therefore infer that Australia has the highest living standards in the world? Briefly discuss your answer.

19. Some journalists, when evaluating released economic statistics for the last quarter, sometimes multiply the quarterly result by four in order to get the annual equivalent. Do you think this is an appropriate method of determining the year's result? Briefly discuss your reasons for accepting or rejecting this technique.

20. The CPI for June 1996 stood at 118.7 (using 1989/90 as the base period, i.e. 1989/90 = 100). If the CPI by the June quarter 1997 is 123.9, does this mean that inflation was 5.2% from June 1996 to June 1997? Briefly explain your answer.

Notes

1. Fiona Smith, 'Institutions blamed for hotel hiatus', *Australian Financial Review*, 23 October 1996, p. 33.
2. 'Astronomers retract new planet discovery', *Sydney Morning Herald*, 17 January 1992, p. 9.
3. Ibid.
4. Australian Bureau of Statistics, *Surviving Statistics: A User's Guide to the Basics* (Cat. No. 1332.0), pp. 4, 10.
5. Ibid, pp. 10, 11.
6. Robert S. Richard, *The Figure Finaglers* (New York: McGraw-Hill, 1974), p. 6.
7. Real Estate Institute of Australia, *Market Facts*, June 1997, p. 1.
8. Australian Bureau of Statistics, *Australian National Accounts* (Cat. No. 5206.0).
9. Access Economics, *Economics Monitor*, September 1996, p. 12.
10. Many thanks to A.C. Nielsen for their comprehensive guide on the Nielsen People Meter.
11. Peter Roberts, 'The ratings silly season', *Australian Financial Review*, 29 May 1996, p. 31.
12. Ibid.
13. Office of Housing Policy, *Rent Report*, June 1996, No. 36.
14. Robert Harley, 'Analysts claim "hole" in the forecast', *Australian Financial Review*, 4 June 1996, p. 52.
15. Australian Bureau of Statistics, *The Labour Force, Australia* (Cat. No. 6202.0).
16. Mark Riley, 'Figures leave out 1.1 million jobless', *Sydney Morning Herald*, 2 July 1996, p. 6.
17. Ibid.
18. Donald H. McBurney and Virginia B. Collings, *Introduction to Sensation/Perception* (Englewood Cliffs, NJ: Prentice-Hall, 1977), pp. 250–53.
19. David Crowe, 'Web of intrigue: companies leap in, site unseen', *Australian Financial Review*, 29 April 1996, p. 1.
20. Dennis O'Neill, 'Sydney's traffic snarls', *Sydney Morning Herald*, 9 September 1996, p. 13.
21. Mike Secombe, 'Willis pulls the wheels off Hewson's GST shopping trolley', *Sydney Morning Herald*, 20 December 1991, p. 4.
22. Michael Stutchbury, 'Australia: luckiest country on Earth', *Australian Financial Review*, 18 September 1995, p. 1.

Chapter 9
VISUAL PRESENTATION OF BUSINESS DATA

LEARNING OUTCOMES

After studying this chapter you should be able to:
- Identify the benefits of presenting data visually
- Use the four key principles of data presentation:
  Principle 1: truthful
  Principle 2: simple
  Principle 3: interesting
  Principle 4: explained
- Prepare, plot and interpret a table of data
- Prepare, plot and interpret pictures and pictograms
- Prepare, plot and interpret pie charts
- Prepare, plot and interpret the different types of bar charts (simple, compound and multiple)
- Prepare, plot and interpret frequency distributions, histograms, frequency polygons and ogives (cumulative frequency curves)
- Prepare, plot and interpret simple two-axis, cartesian graphing techniques (line graphs)
- Prepare, plot and interpret a range of scales to demonstrate different aspects of your data
- Compare the effectiveness of different methods of visual presentation.

The visual presentation of data has always been an integral part of descriptive statistics. However, with the written media (e.g. newspapers and magazines, as well as scholarly journals) presenting more graphs and tables in their reports, business people, students and the general public are all expected to read and interpret the content of such presentations. With the computer now a mainstay of most businesses and a growing number of homes, computer graphics allow us to print out any of those presentations discussed in this chapter (except for the pictogram).

Situation: project approval

Jane Isaac is a systems development analyst with Transcrete, a major construction and transport company. Jane has just completed a project which involved collecting information from many different departments and individuals in order to establish a company-wide database. To obtain final approval to complete the project, she must make a short presentation to a senior management committee. As she sits in her office considering the situation, Jane realises that although she has the facts and figures to justify support for her project, in its present form the information is too voluminous and uninspiring to attract the attention and interest of the committee. Jane concludes that she will have to prepare charts and graphs to support her presentation.

Situation: progress report

Jack Leon is assistant manager of the Newfield branch of the Arrow Spectacle Company, a nationwide optical dispensing organisation. Jack has been responsible for introducing a coordinated fashion dispensing program for his branch and the sub-branches in his area. This program has resulted in a dramatic increase in business. The area manager has asked Jack to prepare a report on the program for circulation to other branches and interstate. Faced with the task of writing his report, Jack sees that he must present the results visually as well as in writing.

Business people are helped in their task of providing information to clients, colleagues, supervisors, subordinates, and so on by visual presentation techniques. The primary role of statistics—communicate information to other people—is more easily fulfilled if information is presented in a suitable visual form. In this chapter you will learn not only the different kinds of visual presentations available, but also the key principles involved. Finally, there is a checklist to help you assess your own or others' visual presentations. This chapter will also help you to 'see through' many of the so-called statistical presentations in newspapers and reports, and on television.

9.1 Why present information visually?

There are many reasons for presenting information visually. The principal reason, however, is the incredible ability of our eyes and brain to detect and act on visual patterns. For example, when you walk into a business meeting, a busy airport lounge or a crowded sports stadium, you absorb, almost at a glance, a vast amount of information about the characteristics of the people present and organised it into patterns. Usually you express this information as a description of the group as a whole, and you are not particularly aware of any great mental feat in doing so. To you, most of your comments would seem to develop naturally from your observation of what was happening. Typical observations might be 'mostly young people', 'smartly dressed people', or 'about an equal proportion of males and females'. You might qualify these observations with comments such as 'mostly a quiet group, but a few noisy ones over in the far corner' or 'unusually tall people'. This sort of description can be communicated quickly and accurately enough for other people to understand it.

There is nothing very special about this. We are simply doing what centuries of evolution have taught us to do 'naturally'. To a great extent, we have been sorting, classifying, rearranging and grouping data (the incoming information from our senses) from birth. In particular, we have had to learn the correct names for things we observed and which characteristics to select. Furthermore, we have had to learn the different ways in which this information can be presented. These straightforward processes are what you have to use when presenting statistical information visually.

Selectivity

It is important to note that your descriptions are selective in your role both as a human being and as a statistician. We communicate to others only those things that seem relevant to our particular purpose. You will learn shortly that one of the major pitfalls of statistical presentation is trying to cram too much information into a single figure, chart, graph or table. If you do this, your readers are unlikely to get the message you want to convey and you will lose the opportunity to make use of the marvellous pattern-recognition properties of the human visual system.

The point of selectivity highlights a key challenge that the visual presentation of information poses: despite there being a number of tried and tested selection guidelines, you have considerable creative scope when it comes to determining the appropriate categories and selecting whatever means of presentation will ensure that your readers easily perceive the implications you intend your data to convey.

Four principles of visual presentation

Since selectivity is a fundamental process in the presentation of statistical information, every statistician needs some rules, or at least guiding principles, to follow when presenting statistical information visually. In question form, these principles are:

1. Is it *truthful*?
2. Is it *simple*?
3. Is it *interesting*?
4. Is it *explained*?

Let us consider each of these in turn.

Is it truthful?
All visual presentations must be truthful. We have all heard the story that 'you can prove anything with statistics', and it is often very tempting to 'massage' your data to allow you to 'prove' your assertion. It is regrettable when unscrupulous individuals resort to such devices. The major problem is that when their dishonesty is detected, which it mostly is, the rest of their argument(s) will not be trusted either. Thus for the responsible communicator the risk is too great to be worth the short-term candle. However, this necessity to be truthful does not prevent you from selecting the data and method of presentation to help your case; it merely requires that you follow the approved practices and conventions to allow other people to understand and interpret your presentations correctly.

Is it simple?
A visual presentation must be as simple as possible. The main reason for using visual presentations in the first place is to be able to distinguish the 'wood from the trees', making use of the human eye and brain to pick out patterns. If you cram too much information into one chart or diagram, you will prevent this pattern recognition from happening.

The usual reasons given for trying to cram a lot of information into a graph or chart are 'saving space', 'economics', and so on. However, remember that it is your reader, not you, who has to understand the visual presentation. If your reader cannot do so quickly and easily, then you have failed to communicate with him or her. It is up to you to manage the right balance of detail, on the basis of your knowledge of the reader's ability to understand the topic.

Is it interesting?
A visual presentation must be interesting from the reader's point of view. It must convey worthwhile information, and be presented in a manner appropriate to the reader's level of sophistication. In assessing whether something is interesting, consider your purpose in presenting it and your reader's most likely reason for studying it.

Is it explained?
All visual presentations must be explained or interpreted in the text. You have to tell your readers what it is they are supposed to 'see' and why they are supposed to 'see' it. All too often, tables, charts and graphs are included in a piece of writing without the author explaining what they are supposed to show. This omission is common, probably because these points are very obvious to the writer. Remember that your reader is seeing the information for the first time and that your job is to help him or her understand your message.

9.1 *Competency check*

1. What is the basic purpose of presenting data in the form of a visual presentation?

2. When presenting statistical data visually, what should you as the creator of the presentation be mindful of?

3. During our daily routine, we are constantly bombarded with masses of data. Jot down a list of two different sets of data you came across today. In what form was each presented?

4. When visual presentations are included in reports or assignments, it is highly recommended that a running commentary accompany the presentation. Why is this practice suggested when readers can see for themselves what information you are presenting?

5. Find one visual presentation in a newspaper or magazine which you feel does not adhere to the four guiding principles of visual presentation and briefly explain your reasoning behind the choice.

9.2 The range of visual presentations

Let us now suppose that you have some data that you want to show to the rest of the world. How are you going to do it? What range of presentations can you choose from? Which form is best? These are decisions that you will have to make. However, you must remember the following:

- What is your purpose in displaying the data?
- Do you have the data available and in the appropriate form?
- Are there any particular conventions or 'usual ways' of displaying this information which your readers would expect or be familiar with?

The principal visual presentation formats that are frequently used and make good starting-points for most statistical purposes are outlined below. A comparison table (Table 9.15) also summarises this information. We shall consider:

- tables;
- pictures and pictograms;
- pie charts;
- bar charts;
- frequency distributions:
  - histograms
  - frequency polygons
  - ogives;
- line graphs (two-axis graphing);
- different scales.

Tables

A **table** is often the first step in the process of describing data. It locates characteristics, relates them to each other, and identifies specific groups and characteristics. Essentially, a table is just a list of related pairs of items. Sometimes this list is obvious, as in Table 9.1. Often, however, the table contains multiple-column displays (see Table 9.2), and this can obscure the pairing unless readers have been properly trained to interpret such a table.

If you are interested in providing specific and detailed data, construct a table. The guidelines for its presentation include the following:

1. Number the table (use a cumulative number or a decimal number that corresponds to the chapter or section).
2. Give it a suitable descriptive title.
3. Label the rows and columns clearly and unambiguously; indicate units of measurement (e.g. millions of tonnes).

Business mathematics and statistics

Table 9.1 *Where we work: now and next century*

Industry	Percentage share of workforce		
	1984/85	1994/95	2004/05
Agriculture	6.1	5.0	3.9
Mining	1.4	1.1	1.3
Manufacturing	17.1	13.8	12.8
Electricity, gas and water	2.1	1.1	1.0
Construction	7.1	7.3	7.2
Wholesale trade	6.2	6.1	5.2
Retail trade	13.6	14.7	13.9
Accommodation, cafes and restaurants	3.4	4.7	5.5
Transport and storage	5.4	4.7	4.9
Communication services	2.2	1.8	2.1
Finance and insurance	4.1	3.9	3.7
Property and business services	6.7	9.3	10.8
Government administration and defence	4.9	4.4	4.0
Education	6.7	6.9	7.6
Health and community services	7.8	8.9	9.0
Cultural and recreational services	1.8	2.3	2.6
Personal and other services	3.4	3.8	4.5
All industries	100.0	100.0	100.0
TOTAL number employed (million)	6.61	8.06	9.88

Source: Australian Bureau of Statistics, DEET, IBIS (extracted from Phil Ruthven, 'A bountiful future if you know where to look', *Australian Financial Review*, 17 January 1996, p. 13).

4. Make sure that your categories are complete and do not overlap.
5. Use a realistic level of precision (i.e. round off or combine numbers).
6. Use summary statistics such as totals and subtotals at the end of rows and at the bottom of columns as appropriate. (*Note:* Make sure these add up correctly!)
7. Use double lines to highlight certain figures (e.g. totals).
8. Don't clutter the table with too much data.
9. Define terms, in footnotes if necessary.
10. Give the source at the foot of the table if using secondary information. Ensure that the source is explicit.
11. Arrange categories in order of importance or alphabetically.

A table serves to locate characteristics in relation to each other and to identify specific groups and characteristics. A table is essentially just a list of related parts of items. In two-column displays the pairing is obvious, but in multiple-column displays the pairing can be obscured unless readers have been trained to interpret them. Sometimes it is better to order the variables by some aspect of the data and to use space to separate or highlight particular data.

In other words, just because you record the data in a particular order does not mean that that order is the most appropriate. Furthermore, if the data are not placed in their usual position, this in itself can be new and useful information for the reader. This sort of rearrangement can be usefully considered when displaying budget documents, details of annual reports, and so on.

When you are reading tables, particularly the multiple-column variety, it is a good idea to use a ruler to avoid jumping a line, although some tables are conveniently presented with data separated by line spaces. Tables 9.1 to 9.4 contain different units of measurement (e.g. percentages, hours worked, millions of dollars) or a combination of different units of measurement. In other words, you are not restricted to one unit of measurement in tabular presentations.

The large array of data that can be contained within a table should be tempered with the realisation that, if cluttered, this will detract from the presentation and may affect the targeted audience's comprehension, interest or concentration. For example, Tables 9.2 and 9.3 provide a diverse range of data which is easily understood, but if extended further it may undermine the reader's understanding. Note how in Tables 9.2 and 9.3 respectively, one can present the average hours worked in nine industry sectors over a 29-year period and Federal Government expenditure over a three-year period in a simple, easily understood fashion.

The use of percentages to immediately indicate market share is widespread and can be seen in Table 9.1 where manufacturing's reduced share of the workforce and the enhanced job opportunities in the property and business services is readily discernible.

Finally, a wide range of different units of measurement can be incorporated in a single table, as seen in Table 9.4 which includes data on ratios, millions of dollars, millions of people, and hundreds of movie screens presented in the form of a movie screen symbolising the content of the table. Note that the population per screen has declined, while attendance since 1990 has risen over 50% corresponding to a similar average rise in annual admissions.

Pictures and pictograms (pictographs)

Many people find the rows and columns of figures in tables hard to read, as well as uninspiring. However, show them a picture or pictorial presentation (referred to as a **pictogram** or **ideogram**) and then all can become much clearer. Pictures help to give meaning to numerical complexity at a single glance. They enable a person to see the wood before getting involved in the trees. A typical example is the hypothetical information illustrated in Figure 9.1. Newspapers are particularly aware of this difficulty, and one of their most common methods of overcoming the problem is to use a picture. If you decide to use a picture in your presentation, remember to describe what you expect the readers to grasp from it.

Pictograms are also frequently used to present statistical information to a wide range of readers. Essentially, they are pictorial representations of whatever unit or category is displayed. These items must be the same size (see Figure 9.1), otherwise readers may interpret (or misinterpret) them in terms of area or volume.

Another type of pictogram can be seen in Figure 9.2, where a Coca-Cola bottle incorporates a bar chart to indicate the largest consumers of the soft drink. The bottle cap, used to represent the sources of Coca-Cola's worldwide revenue in the form of a pie chart, is another eye-catching presentation that the general public can understand.

Table 9.2 Average weekly hours worked by employed persons by industry (a), 1966–95

August	Agri-culture (b)	Mining	Manu-facturing	Electricity, gas and water	Cons-truction	Wholesale and retail trade	Transport and storage	Commun-ication	Finance (c)	Public admin-istration	Community services	Recreation (d)
1966	47.0	38.8	39.3	38.3	39.7	39.4	39.0	37.6	37.0	35.7	35.5	35.8
1967	44.4	39.6	39.6	37.7	40.2	39.2	39.4	37.6	36.9	35.9	35.0	36.0
1968	47.4	39.2	39.0	37.3	40.1	38.6	39.5	36.6	36.5	35.6	33.6	35.8
1969	47.8	38.9	39.2	37.7	39.5	38.3	40.2	36.8	35.6	35.6	32.9	35.5
1970	48.4	40.6	39.1	37.7	39.3	38.5	39.1	36.7	36.4	35.4	33.9	35.0
1971	47.9	40.4	39.4	37.7	40.1	39.3	39.5	37.2	36.7	35.4	34.7	35.0
1972	49.4	40.4	38.9	37.3	39.3	38.7	39.8	36.4	36.4	35.8	32.1	34.9
1973	47.3	39.2	39.4	37.0	39.6	38.0	39.7	36.1	35.9	35.3	32.7	33.8
1974	46.6	38.1	38.4	36.2	38.9	37.8	38.4	34.8	36.5	35.0	32.6	33.3
1975	47.5	36.9	37.9	36.2	38.7	36.7	38.2	35.0	35.9	35.2	32.2	31.9
1976	47.9	38.0	37.8	36.2	37.4	36.7	38.6	34.0	35.6	34.6	32.7	31.4
1977	46.2	39.2	37.8	36.3	37.5	36.6	38.6	34.1	35.2	34.6	32.5	31.7
1978 (e)	44.5	38.3	38.1	36.8	37.2	36.5	38.0	32.6	35.6	34.4	33.7	31.8
1979	44.8	39.3	38.1	35.9	37.7	36.7	38.3	33.8	36.0	34.6	34.2	31.9
1980	45.0	35.7	38.1	36.8	37.1	36.5	38.7	34.0	34.9	33.6	32.9	32.0
1981	43.1	39.2	37.9	35.8	36.6	35.6	38.1	33.0	35.6	34.3	33.6	32.6
1982	42.7	38.3	37.1	35.2	36.5	35.6	37.8	32.7	34.5	33.6	33.3	33.1
1983	43.7	37.2	37.1	35.0	36.1	36.0	37.9	32.6	35.3	33.9	33.5	31.9
1984	43.2	36.5	37.9	35.2	37.1	35.9	38.6	32.9	36.2	34.5	33.2	33.4
1985	43.0	38.4	37.3	34.2	36.6	36.0	38.8	34.6	35.8	33.0	32.4	33.1
1986 (f)	43.0	38.3	37.6	34.7	35.9	35.9	38.2	33.1	35.4	33.8	32.7	32.9
1987	43.1	38.1	37.7	35.3	36.8	35.1	37.7	32.9	35.8	33.7	32.4	32.5
1988	42.0	41.2	38.5	33.5	38.0	35.1	38.3	33.2	36.7	34.4	32.9	32.4
1989	40.5	41.5	38.8	36.0	38.3	34.9	39.1	35.2	37.1	34.1	32.8	32.9

1990	41.9	41.5	38.8	35.2	36.5	34.8	39.3	34.0	36.1	34.9	32.8	32.5
1991	42.3	40.5	37.4	36.0	36.2	34.3	39.0	33.9	36.1	33.3	32.8	31.4
1992	40.8	40.7	38.6	35.5	36.1	34.1	39.2	34.6	36.1	34.8	32.7	31.4
1993	42.7	42.4	38.5	36.3	37.6	34.4	39.8	35.0	36.5	34.1	32.9	32.1
1994	42.8	44.3	39.3	37.8	39.0	34.9	39.6	37.6	37.4	35.6	32.9	31.7
1995	40.6	43.6	39.1	38.0	38.2	34.4	40.4	36.3	37.3	34.9	n.o.	n.o.

Sources: Australian Bureau of Statistics, *The Labour Force, Australia* (Cat. No. 6204.0); Australian Bureau of Statistics, *The Labour Force, Australia* (Cat. No. 6203.0); and Australian Bureau of Statistics, *The Labour Force, Australia* (Cat. No. 6101.0).

Notes: (a) All estimates refer to the average weekly hours actually worked during the survey reference week by employed persons in the civilian population aged 15 and over. Industry classification used is the Australian Standard Industrial Classification (ASIC).
(b) Agriculture and services to agriculture, forestry, logging, fishing and hunting.
(c) Finance, property and business services.
(d) Recreation, personal and other services.
(e) From 1978 to 1983 inclusive, conforms with the 1981 census. Also from 1978 the survey methodology and questionnaire were changed. From 1984 to December 1988, estimates conform to the 1986 census benchmarks. From January 1989, estimates conform to the 1991 census benchmarks.
(f) From April 1986 inclusive, the definition of employed persons was extended to include unpaid family helpers working between 1 and 14 hours per week. Previously, these workers were counted as either unemployed or not in the labour force.

Table 9.3 Federal budget, expenditure, 1995/96 to 1997/98

	1995/96	1996/97		1997/98
	Actual ($ million)	Budget ($ million)	Change (%)	Estimate ($ million)
Legislative and executive affairs	593.9	524.0	−11.8	481.3
Financial and fiscal affairs	1 698.5	1 811.6	6.7	1 624.7
Foreign economic aid	2 268.0	2 065.6	−8.9	1 979.1
General research	1 081.2	1 174.0	8.6	1 234.8
General services	222.5	128.9	−42.1	−295.1
Government super benefits	989.9	1 492.4	50.8	1 415.7
Total general public services	**6 854.0**	**7 196.5**	**5.0**	**6 440.7**
Defence	10 010.6	10 027.1	0.2	10 389.0
Public order and safety	926.6	1 491.6	61.0	882.4
Education	10 644.2	11 063.7	3.9	11 050.4
Health	18 633.5	19 408.4	4.2	20 659.3
Social security and welfare	46 698.6	48 896.9	4.7	49 994.1
Housing and community amenities	1 204.7	1 121.7	−6.9	1 124.7
Recreation and culture	1 416.0	1 390.4	−1.8	1 286.2
Fuel and energy	37.2	−725.4	n/a	24.5
Agriculture, forestry and fishing	1 875.6	1 878.0	0.1	1 787.4
Mining and mineral resources (other than fuels), manufacturing and construction	1 651.1	1 773.6	7.4	1 699.8
Transport and communication	754.3	1 190.6	57.8	1 321.5
Tourism and area promotion	104.4	97.9	−6.2	93.8
Labour and employment affairs	3 896.8	3 235.4	−17.0	2 815.7
Other economic affairs n.e.c.	308.6	316.1	2.4	303.0
Total other economic affairs	**4 309.8**	**3 649.5**	**−15.3**	**3 212.5**
Public debt interest	9 125.9	9 781.0	7.2	9 815.0
General purpose intergovernment transactions	13 797.9	16 797.0	21.7	17 405.6
Natural disaster relief	−4.3	29.8	n/a	30.8
Contingency reserve	—	−183.9	n/a	1 311.5
Asset sales	−1 230.5	−5 100.0	n/a	−8 010.0
Total other purposes	**21 689.0**	**21 323.9**	**−1.7**	**20 552.9**
Total expenditure	**126 705.2**	**129 686.5**	**2.4**	**130 425.5**

Source: The Treasury, *Budget Statement, 1996–97*, Budget Paper No. 1. Commonwealth of Australia copyright, reproduced by permission.

Chapter 9 – Visual presentation of business data

Table 9.4 *The box office boom*

Year	Number of screens	One screen per head population	Cinema admissions (millions)	Average annual admissions (per person)	Gross box office ($ millions)
1980	829	17 700	38.6	2.6	154.2
1985	742	21 300	29.7	1.9	160.6
1990	851	20 000	43.0	2.5	270.2
1991	885	19 500	46.9	2.7	314.4
1992	906	19 300	47.2	2.7	322.6
1993	940	18 800	55.5	3.1	369.4
1994	1 028	17 400	68.1	3.8	476.0
1995	1 137	16 000	69.9	3.9	501.4

Note: Adjusted by MPDAA to realign its reporting period to a calendar year.
Source: MPDAA/AFC/ABS (adapted from *Sydney Morning Herald*, 30 March 1996, p. 12).

Figure 9.1 *Fredonian petroleum exports, 1996 and 1997*

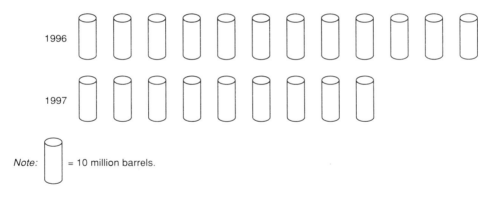

Source: Hypothetical.

The pictogram in Figure 9.3 illustrates a line graph representing the price of crude oil depicted in a 45 gallon drum. The US dollar price per barrel is found on the rim of the drum, beginning at $16.50 a barrel with deviations of 50 cents a barrel up to $21.50 a barrel. The volatility in prices over the period is obvious to the reader, with the pictorial presentation rendering the data more effectively eye-catching.

Figure 9.4 uses poker chips to represent the height of bars in a bar chart, indicating the NSW public's preference for TAB in its weekly expenditure on gambling. Also note the inclusion of tables containing data on the impact of such gambling from the human perspective, with productivity loss over 40% of the total estimated human cost.

Business mathematics and statistics

Figure 9.2 *The world of Coca-Cola*

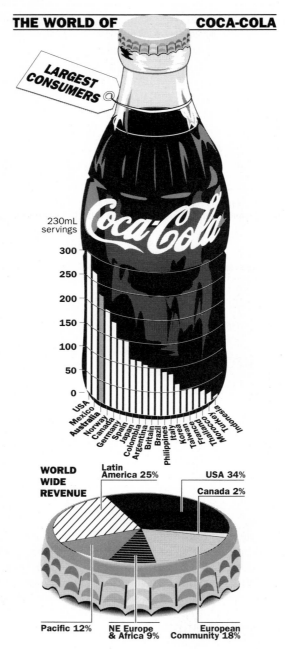

Source: Sun-Herald, 17 November 1991, p. 72. Graphic by Edi Sizgoric.

Chapter 9 – Visual presentation of business data

Figure 9.3 *Crude oil prices*

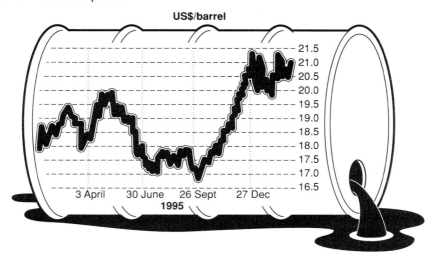

Source: Adapted from an article by Stephen Wyatt, 'Tight stocks of crude turn market precarious', *Australian Financial Review*, 18 March 1996, p. 50.

Figure 9.4 *Where the money goes*

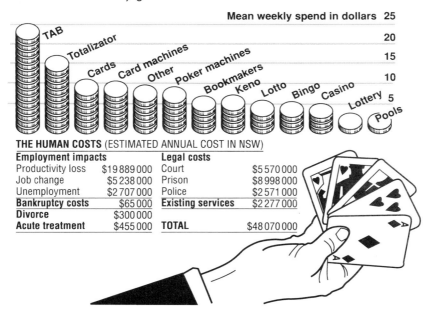

Source: Australian Institute for Gambling Research (adapted from an article by Daniel Lewis, 'A sickness the State can't afford to cure', *Sydney Morning Herald*, 13 April 1996, p. 33.)

Business mathematics and statistics

Finally, Figure 9.5 provides a wide range of data comparing world road fatalities among eight countries, Australian State variations in such fatalities, and the number of deaths on NSW roads over the first five months of the years 1986 to 1995. New Zealand has the poorest road fatalities record of the eight countries, while the Northern Territory has, by far, the worst record of the Australian States. The figure also shows that road deaths have dropped markedly since the 1980s, this in spite of an increasing number of drivers on the road.

Pictograms are less likely to be used in business or by the individual, since some degree of artistic skill is required in their construction and arrangement. Also, the numbers that may be extracted from the presentation lack precision and detail.

Pie charts

Preparing pictograms or ideograms can require a fair degree of artistic talent. A much easier method used by statisticians is to illustrate data about proportions in the form of segments of a **pie** (or **circle**) **chart**. While not suitable for conveying precise numerical data, the pie chart is easy to construct, distinguishes important features and allows

Figure 9.5 *Australians aren't the world's worse drivers—but they aren't the best*

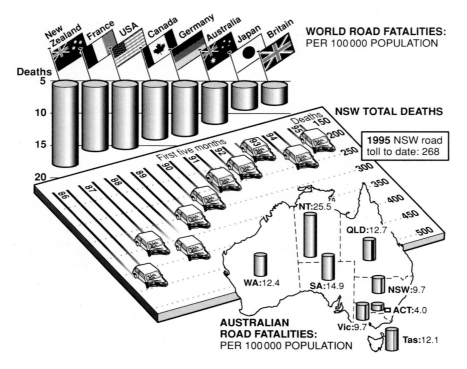

Source: NRMA and Roads and Traffic Authority (extracted from Frank Walker and Martin Warneminke, *Sydney Morning Herald*, 18 June 1995, p. 13).

Chapter 9 – Visual presentation of business data

an easy visual comparison of relative size. A typical pie chart is shown in Figure 9.6. According to this chart, ownership of dwellings is, by far, the single largest asset of the Australian private sector, with business capital a distant second.

Figure 9.6 *Composition of Australian net private sector wealth by asset type, June 1996*

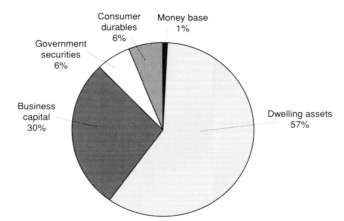

Source: Commonwealth Treasury Australia, *Economic Roundup, Summer 1997*, p. 48. Commonwealth of Australia copyright, reproduced with permission.

Construction of such a pie chart is relatively simple. The steps involved are as follows:

1. Order your data from largest to smallest. (This presentation is best with no more than about six items.)
2. Calculate each item or portion as a percentage of the total.
3. Convert the percentages into degrees of a circle—that is, multiply the percentage of a circle that each section corresponds to by 360 degrees (e.g. Dwelling assets is 57% of 360 degrees, or 205.2 degrees). The slices correspond in size to the various percentages to be compared.
4. If only one pie chart is presented, make the largest slice begin at 12 o'clock on the circle, with the other slices following in a clockwise direction in descending size order. Many pie charts show no rationale for the order of the slices, so a logical sequence is helpful. The use of a protractor is advisable.
5. Label each segment with the percentage of the total it represents.
6. Title the chart appropriately, indicating, for example, how much 100% is equivalent to (i.e. 100% = 87 428 settler arrivals in Figure 9.7). Indicate your source.

Sometimes you may have a number of small elements at the 5 minutes to 12 position. You have several options:

1. You can combine them into one larger category and explain its components in the text.

Business mathematics and statistics

Figure 9.7 Settler arrivals by occupation, 1995/96

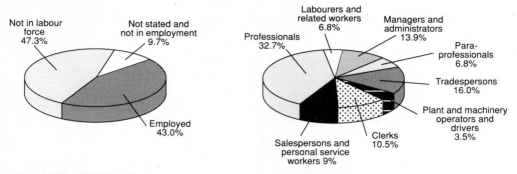

Note: 100% = 87 428 settlers
Source: Department of Immigration and Multicultural Affairs, *Immigration Update*, June Quarter 1996.

Figure 9.8 Transport and storage industry turnover, 1994/95 ($48 billion)

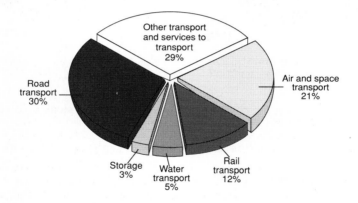

Source: IBIS estimates based on ABS statistics (taken from Jennifer Mead, 'Transport and storage needs more microeconomic reform', *Australian Financial Review*, 1 May 1996, p. 54).

2. You can add to the diagram an enlarged version of the 5-to-12 slice showing the individual items, or add a three-dimensional aspect (such as Figure 9.8 in which the smaller sections are emphasised by extracting them from the rest of the pie for the purposes of clarity).
3. If you have a lot of small components that should not be combined, a pie chart may be inappropriate for your purposes and you should choose another type of visual presentation.

When comparing the results of two pie charts with identical categories in each, if you find that starting at the 12 o'clock position and/or ordering the data from largest to smallest is impractical (e.g. if there are dramatic changes in results over the period, or if the chart is not visually pleasing), then colour or shading may be used to assist

Chapter 9 – Visual presentation of business data

Figure 9.9 *The top four foreign investors in Australia at 30 June 1990 and 1995*

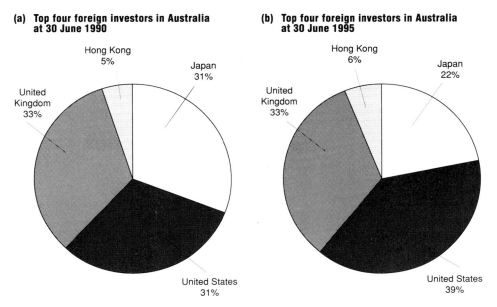

Source: Australian Bureau of Statistics, *Balance of Payments and International Investment Position Australia* (Cat. No. 5363.0).

in fast and easy comparison (see Figure 9.9). Note that Japan's share of total foreign investment in Australia has declined sharply while the USA's has increased markedly, Hong Kong's share has risen and the UK's has remained the same.

At times, you may want to elaborate on a particular section of a pie chart. For example, Figure 9.7 shows that 43% of settler arrivals in 1995/96 entered the workforce, while the occupations of those 43% of arrivals are shown in the pie chart on the right. Note the relatively low share of such arrivals whose occupation in their former home country was unskilled or semi-skilled in nature.

Once again, for effect, a pictorial slant can be introduced to a pie chart, such as that representing greenhouse gas emissions from the rear of a motor vehicle in Figure 9.10. Note how easy it is to determine the major source of such emissions, with cars followed by commercial vehicles guilty of emitting over three-quarters of all greenhouse gas emissions. Such devices are often used by newspapers, television and government departments to show broad indications to unsophisticated audiences or readers.

Bar charts

If you have a number of categories or classes of objects to compare, then the **bar chart** is a very effective diagrammatic representation. Bar charts use rectangular bars or 'blocks' of equal width but varying height or length to represent the magnitude of particular data. Bar charts are generally flat, two-dimensional representations with the bars running either vertically or horizontally. Horizontal bar charts have an advantage

Business mathematics and statistics

Figure 9.10 *Greenhouse gas emissions from Australian domestic transport, 1993 (total CO_2 equivalent emissions: 80.4Mt)*

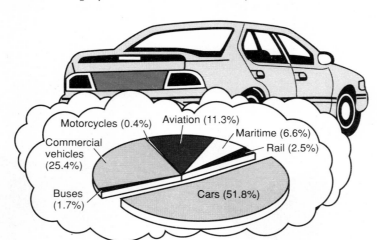

Source: Bureau of Transport and Communications Economics (taken from Michael Dwyer, 'Cabinet decides that voluntary cuts will solve the problem', *Australian Financial Review*, 29 March 1995, p. 10).

in that you can write inside the bar whatever it represents. However, some people prefer to see the variation displayed vertically. It is customary to leave a space equal to half the width of a bar between successive periods. Since it is also one of the most commonly used charts, readers tend to be familiar with it. Difficulties arise only when you burden the bar chart with too much information. The comparison of the variables is by means of the horizontal or vertical bars. For categorical responses that are qualitative, horizontal bars should be used (see Figure 9.11, according to which the largest employment growth was in the Sales assistants positions), while vertical bars should be constructed when such responses are numerical (see Figure 9.12, with tourism's share of service exports reaching about 43% in 1994/95). Both of these charts are referred to as **simple bar charts**, as only one quantity is represented on the chart.

When drawing a bar chart, observe the following:

1. Squared or graph-type paper will be a great help in judging correct lengths.
2. Check the relative quantities or sizes to be represented. The largest category requires the longest bar. This should go to the full width or height of the paper, or as far as you wish. If the largest category is much larger than any of the others, consider splitting it into two or three bars of equal size. This has the effect of magnifying the smaller categories and so makes comparisons between them much easier.
3. Label the axes or scales on your charts.
4. Make sure there is a suitable title and reference to the source of your data.
5. Label each block and indicate the exact amount it represents.
6. Any 'legend' introduced to interpret the chart may be included within the body of the chart (see Figure 9.13).

Chapter 9 – Visual presentation of business data

Figure 9.11 *Occupations with largest employment growth, 1987–95 (persons)*

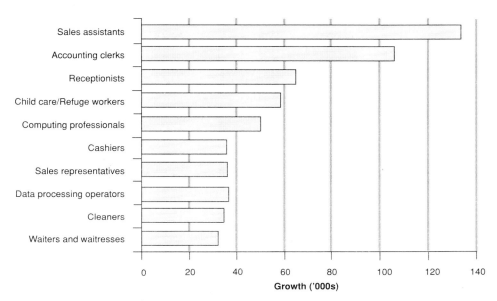

Source: Australian Bureau of Statistics (in Access Economics, *Economics Monitor*, April 1996, p. 10).

Figure 9.12 *Tourism: percentage of service exports, 1974/75 to 1994/95*

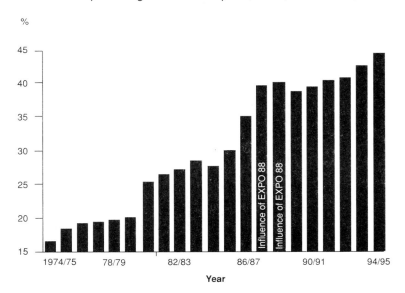

Source: Australian Bureau of Statistics (taken from Nina Field, 'Services make mark on economy', *Australian Financial Review*, 15 January 1996, p. 14).

Business mathematics and statistics

Multiple bar charts

Comparison can sometimes be made by incorporating your data on a single chart instead of placing two bar charts side by side. For a single chart it is usual to place comparable items—succeeding years, for example—side by side as in Figure 9.13 (with tax on gambling comprising an increasing share of all State and Territory Government coffers except Western Australia and the Northern Territory). A glance can then reveal how much items have changed over the period covered. Figure 9.14 illustrates comparable data side by side in order to determine whether there appears to be a relationship between educational attainment and status in the labour force.

Component bar charts

There are two additional types of bar chart, both of which involve dividing each bar into sections to represent the components that make up the total bar (see Figure 9.15); accordingly, they are referred to as component bar charts. The purpose of the first is basically to represent each component figure. The simplest way of doing this is to represent each component figure with a key or legend. The purpose of the second is to convert the components to percentages of the total bar (see Figure 9.16). This can be useful for year-to-year comparisons, as in this example with the general government's share of gross external debt increasing since 1991, while the public trading enterprise

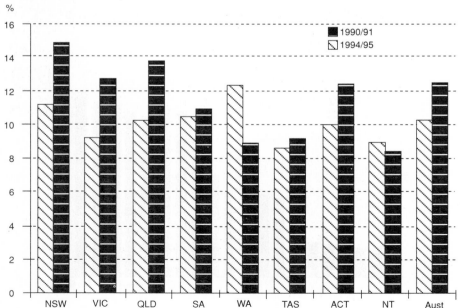

Figure 9.13 *State gambling taxation as a percentage of own-tax revenue, 1990/91 and 1994/95*

Source: Commonwealth Grants Commission (taken from Access Economics, *Economics Monitor*, May 1996, p. 19).

Chapter 9 – Visual presentation of business data

Figure 9.14 *Civilian population aged 15 and over: labour force status and educational attainment, February 1995*

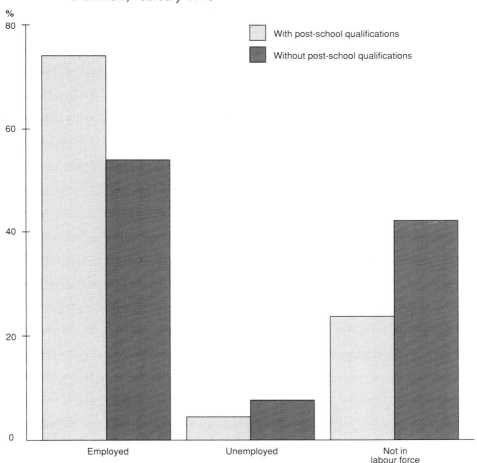

Sources: Australian Bureau of Statistics, *Labour Force Status and Educational Attainment, Australia, February 1996* (Cat. No. 6235.0) and *Australian Economic Indicators* (Cat. No. 1350.0).

share has been sharply reduced. Which one to use depends upon your purpose: are you interested in percentages or the actual figures?

Note that in multiple and component bar charts, you must ensure that the different bars or parts thereof are clearly distinguished. The use of a key or legend will help to explain the categories.

Frequency distribution

Histograms, frequency polygons and ogives are all possible presentations of a frequency distribution. What, then, is a frequency distribution? You have probably come across it in your workplace or in written media presentations without realising what it is

Figure 9.15 *Settlement distribution of country of birth groups, by State, 1991*

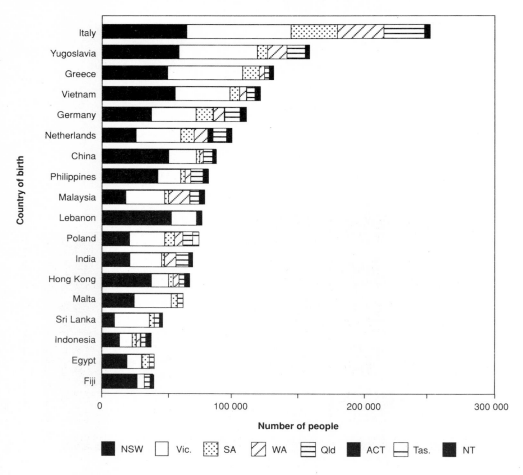

Source: I.H. Minas, T.J.R. Lambert, S. Kostov and G. Boranga, *Mental Health Services for NESB Immigrants* (BIMPR, 1996), p. 5.

called. Essentially, a **frequency distribution** is a tool for grouping data items into classes and then recording how many observations appear in each class. For example, demographic statistics such as household income Australia-wide, age-specific death rates, and age and sex of settlers in Australia (see Tables 9.5 and 9.6) may be presented in the form of a frequency distribution. Why present such information in this form? Well, can you imagine having to list all the ages of the 9 million males and 9 million females (Table 9.5) or the 99 139 settlers into Australia in 1995/96 (Table 9.6)? A frequency distribution allows us to reduce or condense the amount of data we have to list and work with. Therefore, data analysis and interpretation are made more manageable with less chance of human error.

Chapter 9 – Visual presentation of business data

Figure 9.16 *Shares of Australia's gross external debt, 1986–95*

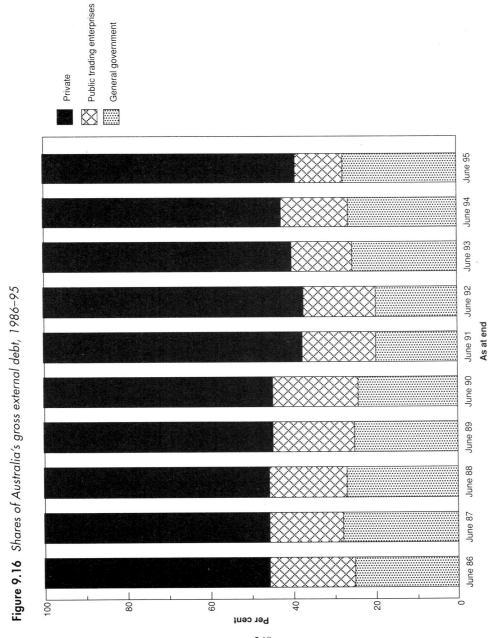

Source: Australian Bureau of Statistics, *1994–95 Balance of Payments and International Investment Position, Australia* (Cat. No. 5363.0).

Table 9.5 *Estimated resident population by age groups, 30 June 1992 to 1995*

Age group (years)	Male				Age group (years)	Female			
	1992	1993	1994	1995		1992	1993	1994	1995
0–4	656 935	659 773	661 464	661 805	0–4	624 086	626 274	627 683	628 608
5–9	655 720	654 439	655 429	659 788	5–9	622 730	622 687	623 797	626 550
10–14	642 650	648 963	654 722	662 205	10–14	608 137	613 688	619 375	628 655
15–19	679 645	665 409	656 284	651 818	15–19	645 481	631 565	622 558	617 324
20–24	726 476	736 291	740 128	736 706	20–24	706 416	713 510	713 739	709 247
25–29	692 546	683 601	680 879	693 177	25–29	688 676	679 403	677 648	687 735
30–34	725 568	729 998	733 993	727 859	30–34	724 750	730 189	733 853	730 584
35–39	673 702	682 000	691 065	706 837	35–39	675 653	685 300	695 109	709 311
40–44	654 565	654 391	656 094	661 701	40–44	642 605	647 937	654 395	664 091
45–49	561 608	595 889	619 079	637 551	45–49	538 595	573 621	599 084	619 294
50–54	447 166	456 880	474 887	496 795	50–54	424 543	433 827	451 745	475 759
55–59	373 830	383 862	395 977	409 242	55–59	365 621	375 251	386 324	396 172
60–64	362 272	356 357	352 123	348 852	60–64	365 165	357 954	353 294	352 039
65–69	325 240	330 957	333 955	336 623	65–69	352 908	356 272	356 636	355 367
70–74	239 249	250 169	262 919	269 755	70–74	292 925	303 430	315 508	322 237
75–79	162 310	164 013	164 977	170 567	75–79	229 500	231 352	231 270	235 702
80–84	88 162	92 644	97 651	102 499	80–84	151 095	157 096	164 858	171 551
85 and over	47 295	50 210	53 111	56 701	85 and over	115 247	121 225	126 788	133 282
All ages	8 714 939	8 795 846	8 884 737	8 990 481	*All ages*	8 774 133	8 860 581	8 953 664	9 063 508

Source: Australian Bureau of Statistics, *Australian Demographic Statistics* (Cat. No. 3101.0). Commonwealth of Australia copyright, reproduced by permission.

Chapter 9 – Visual presentation of business data

Table 9.6 Settler arrivals by age and sex, 1993/94 to 1995/96

Age groups	Males						Females						Persons					
	1993/94		1994/95		1995/96		1993/94		1994/95		1995/96		1993/94		1994/95		1995/96	
	No.	%	No.	%	No.	%	No.	%	No.	%	No.	%	No.	%	No.	%	No.	%
0–4	3 775	11.4	4 480	11.0	4 473	10.0	3 585	9.7	4 506	9.7	4 453	8.2	7 360	10.5	8 986	10.3	8 926	9.0
5–9	2 677	8.1	3 404	8.3	4 148	9.3	2 631	7.2	3 415	7.3	4 289	7.9	5 308	7.6	6 819	7.8	8 437	8.5
10–14	2 315	7.0	2 903	7.1	3 507	7.8	2 134	5.8	2 874	6.2	3 643	6.7	4 449	6.4	5 777	6.6	7 150	7.2
0–14	8 767	26.6	10 787	26.4	12 128	27.1	8 350	22.7	10 795	23.1	12 385	22.8	17 117	24.5	21 582	24.7	24 513	24.7
15–19	1 923	5.8	2 189	5.4	2 835	6.3	2 490	6.8	2 961	6.3	3 540	6.5	4 413	6.3	5 150	5.9	6 375	6.4
20–24	2 974	9.0	3 448	8.5	3 473	7.7	4 759	12.9	5 475	11.7	5 973	11.0	7 733	11.1	8 923	10.2	9 446	9.5
25–29	5 336	16.2	6 350	15.6	5 812	13.0	6 341	17.2	7 662	16.4	7 944	14.6	11 677	16.7	14 012	16.0	13 756	13.9
30–34	4 905	14.9	6 225	15.3	5 876	13.1	4 956	13.5	6 631	14.2	7 197	13.3	9 861	14.1	12 856	14.7	13 073	13.2
35–39	3 039	9.2	4 195	10.3	4 343	9.7	3 031	8.2	4 252	9.1	5 326	9.8	6 070	8.7	8 447	9.7	9 669	9.8
40–44	1 783	5.4	2 506	6.1	3 070	6.8	1 811	4.9	2 607	5.6	3 275	6.0	3 594	5.2	5 113	5.8	6 345	6.4
45–49	1 123	3.4	1 440	3.5	1 869	4.2	1 171	3.2	1 553	3.3	1 958	3.6	2 294	3.3	2 993	3.4	3 827	3.9
50–54	768	2.3	924	2.3	1 178	2.6	934	2.5	1 033	2.2	1 503	2.8	1 702	2.4	1 957	2.2	2 681	2.7
55–59	697	2.1	779	1.9	1 132	2.5	871	2.4	1 096	2.3	1 713	3.2	1 568	2.2	1 875	2.1	2 845	2.9
60–64	632	1.9	751	1.8	1 308	2.9	809	2.2	1 002	2.1	1 445	2.7	1 441	2.1	1 753	2.0	2 753	2.8
15–64	23 180	70.3	28 807	70.6	30 896	68.9	27 173	73.9	34 272	73.5	39 874	73.4	50 353	72.2	63 079	72.1	79 770	71.4
65+	1 045	3.2	1 195	2.9	1 807	4.0	1 253	3.4	1 572	3.4	2 049	3.8	2 298	3.3	2 767	3.2	3 856	3.9
TOTAL	32 992	100.0	40 789	100.0	44 831	100.0	36 776	100.0	46 639	100.0	54 308	100.0	69 768	100.0	87 428	100.0	99 139	100.0

Source: Department of Immigration and Multicultural Affairs, *Immigration Update, June Quarter, 1996.* Commonwealth of Australia copyright, reproduced with permission.

In constructing a frequency distribution table, consider the following:

1. Select the appropriate number of class groupings for the table. Between 5 and 12 classes are usually nominated, the decision being a function of the number and range of observations. If there are not enough or too many class groupings, you will find the information far less meaningful.
2. The size of each class interval must be equal, and each class must be distinctly defined so that any observation falls neatly into one class and one class only. There must be no overlap and/or gaps between successive classes.

    The **class interval** (or size) is equal to the difference between the real lower and real upper limit of a particular class. Therefore, in Table 9.8 on the next page, when examining the class interval $100 to $109, you must realise that the lower limit of that class is $99.50 and the upper limit is $109.50; therefore, the size of the class interval is $10. Table 9.8 also has a class size interval of $10 since the lower limit of, say, $100 to under $110 is $100, and the upper limit is $109.99, close enough to an interval of $10.

    If possible, avoid **open-ended intervals**—that is, where the lower or upper boundaries of the class are undefined (e.g. see Table 9.6 where '65+' is the open-ended interval). These intervals make it difficult to graph the data and impossible to calculate important descriptive measures such as the arithmetic mean or standard deviation (unless the analyst makes certain assumptions, e.g. that the upper limit of 65+ is 69).

    To determine the width of the classes, find the range of the data and divide by the nominated number of classes. For example, if you were interested in constructing a frequency distribution of houses sold in the northern suburbs within a price range of $175 000 to $850 000 and you wanted to have eight classes, the width of the class interval is approximated by:

$$\text{Width of the class} = \frac{\$850\ 000 - \$175\ 000}{8} = \frac{\$675\ 0000}{8} = \$84\ 375$$

   Rather than using a class interval of $84 375, an interval of $85 000, $90 000 or even $100 000 could be appropriate.
3. Tabulate the number of observations in each class, and—voilà!—you have a frequency distribution.

    An example incorporating all of the above should reinforce those points.

Problem: Below are the room tariffs (rates) for 30 4-star hotels in the metropolitan area. Determine the range and then construct a frequency distribution with six class intervals.

Table 9.7 *Hotel room tariffs in metropolitan area, 1997*

95	122	128	104	105	110
111	103	101	98	99	115
88	97	103	127	86	97
100	112	83	101	108	98
103	91	135	96	118	108

Source: Hypothetical.

Chapter 9 – Visual presentation of business data

Solution: First, let's find the range: $135 − $83 = $52. Since we want six classes, $52/6 = 8.67. For convenience, we round the class interval to $9 or even $10. As an interval of $10 is more convenient in constructing the frequency distribution, two different frequency tables are presented, each with an interval of $10.

Table 9.8 *Room tariffs for 30 4-star hotels*

Room tariffs ($)	Tallies	Frequency
80– 89	|||	3
90– 99	㎢ |||	8
100–109	㎢ ㎢	10
110–119	㎢	5
120–129	|||	3
130–139			1

Source: Hypothetical.

Table 9.9 *Room tariffs for 30 4-star hotels*

Room tariffs ($)	Tallies	Frequency
80 and under 90	|||	3
90 and under 100	㎢ |||	8
100 and under 110	㎢ ㎢	10
110 and under 120	㎢	5
120 and under 130	|||	3
130 and under 140			1

Source: Hypothetical.

To make sure you understand the construction of a frequency distribution, try the next problem.

Problem: Table 9.10 shows the number of bicycles sold by Kamakazi Inc. in Sydney and Melbourne during the 1997 calendar year. Construct a frequency distribution containing seven classes.

Table 9.10 *Bicycle sales of Kamakazi Inc. in Sydney and Melbourne, 1997 ('000)*

44	77	47	82	68	44	50	49
56	68	56	48	74	54	78	58
46	89	64	71	62	52	74	63
48	73	64	46	56	69	62	48
66	58	51	70	47	67	52	58

Source: Hypothetical.

Solution: First, determine the range of observations—that is, 89 − 44 = 45. Next determine the class interval: 45/7 = 6.43 interval. Let's use a class interval of seven, as well as the seven classes nominated for this problem (see Table 9.11).

Note that the class interval of each class is 7 units, since the difference between the real upper and real lower limit of each class is 7 (e.g. for 50 to 56, the real lower limit is 49.5 and the real upper limit is 56.5, hence a class interval of 7 units).

Business mathematics and statistics

Table 9.11 *Bicycle sales of Kamakazi Inc. in Sydney and Melbourne, 1997*

Number of bicycles sold ('000)	Tallies	Frequency
43–49	⊞ ⊞	10
50–56	⊞ |||	8
57–63	⊞		6
64–70	⊞ |||	8
71–77	⊞	5
78–84	||	2
85–91			1

Source: Hypothetical.

Histograms

A **histogram** or frequency distribution graph is a way of displaying groups of frequencies and can be thought of as a bar graph of a frequency distribution. It is identical to a bar chart except that the width of each bar is now equal to the class interval, with respective frequencies applied over a continuous range. For example, in Figure 9.17 there are approximately 600 000 females aged between 14.5 and 19.5 years (class interval of 15–19 includes all those over 14.5 to under 19.5 years). This contrasts with a normal bar chart representing discrete quantities such as 150 000 females aged 12 at their last birthday. In histograms the width of the bars must be equal, and the variation in heights of the respective bars must accurately reflect the various frequencies.

Typically, the real class limits are found in ascending order on the horizontal axis of a histogram with the frequency (proportion or percentage per interval) listed along the vertical axis. Since class intervals cannot overlap, this means that the bars touch one another. The advantage of the histogram distribution is that irregularities in the data are smoothed out and the overall shape of the distribution is revealed (see Figures 9.18 to 9.20).

Frequency polygons

A frequency polygon represents the shape of the particular frequency distribution and is generally used for continuous data (i.e. the data forms a continuum of values). It takes the same basic data as in the histogram, but instead of plotting the width of the bars equal to the size of the class intervals, the centres of the class intervals are marked (i.e. the centres of the bars), and these points are joined with straight lines, as in Figure 9.21. Usually an extra class is added at each end of the histogram to permit the polygon to reach the horizontal axis at both ends of the distribution. In other words, the frequency polygon is based directly on the histogram and is drawn by finding the midpoint of each 'bar' and joining these to form a line graph. Bars are not normally shown in a frequency polygon but are illustrated here for your convenience. Figures 9.22 and 9.23 are frequency polygons of Tables 9.9 and 9.11 respectively.

Chapter 9 – Visual presentation of business data

Figure 9.17 *Age and sex distribution, Australia, 1995*

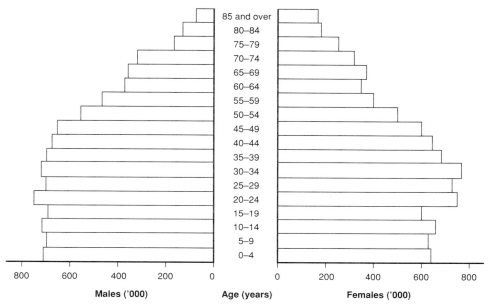

Source: Australian Bureau of Statistics, *Australian Demographic Statistics, December Quarter 1995* (Cat. No. 3101.0).

Figure 9.18 *Histogram of taxable income and respective frequencies of persons earning under $80 000 in Australia, 1992/93*

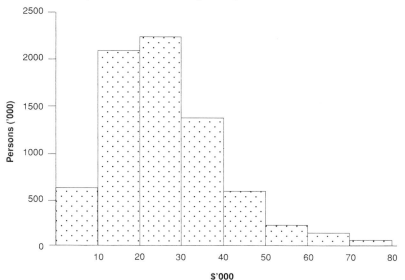

Source: Australian Taxation Office, *Taxation Statistics 1992/93* (1994).

Business mathematics and statistics

Figure 9.19 *Histogram of room tariffs for 30 4-star hotels, 1997*

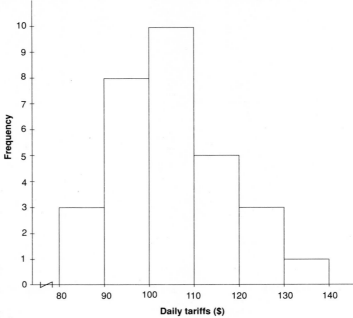

Source: Hypothetical.

Figure 9.20 *Histogram of Kamakazi Inc. sales, 1997*

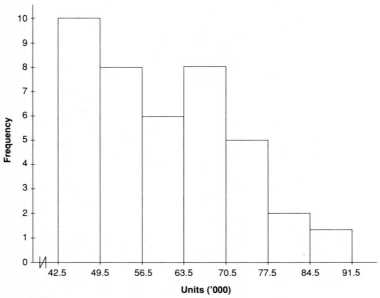

Source: Hypothetical.

Chapter 9 – Visual presentation of business data

Figure 9.21 Frequency polygon of taxable income and respective frequencies of persons earning under $80 000, 1992/93

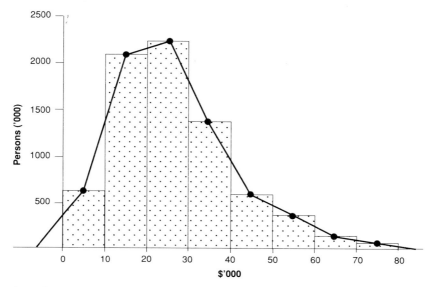

Source: Australian Taxation Office, *Taxation Statistics 1992/93* (1994).

Figure 9.22 Frequency polygon of room tariffs for 30 4-star hotels

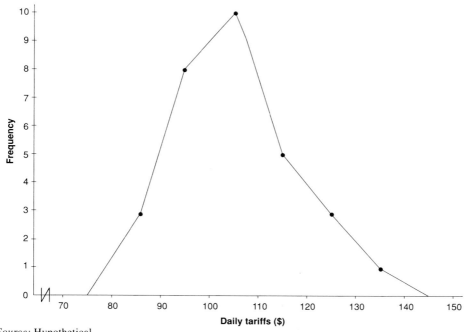

Source: Hypothetical.

Figure 9.23 *Frequency polygon of Kamakazi Inc. sales, 1997*

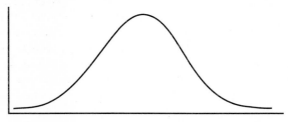

Source: Hypothetical.

Most of the frequency curves, in practice, fall into four main groups:

1. The symmetrical or normal (bell-shaped) curve (such as the taxable income of Australian individuals):

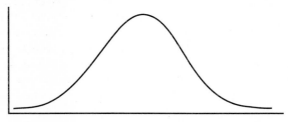

2. The moderately asymmetrical (or skew) curve (for instance, the frequency polygon of room tariffs):

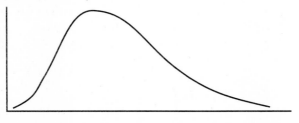

3. The bimodal (or M-shaped) curve:

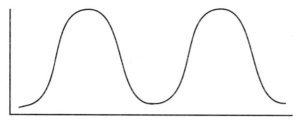

4. The J-shaped (or exponential) curve:

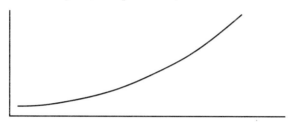

Distributions can be compared visually, but more usually their characteristics are determined mathematically, as seen in Chapter 14.

Ogives or cumulative frequency curves
A further development of the frequency polygon is the cumulative frequency curve—the **ogive** (pronounced 'oh jive'). Sometimes it is more useful to determine the number of data items that fall above or below a certain value rather than within a given interval; to do this, you convert the given or determined frequency distribution to a cumulative frequency distribution—one that adds the number of frequencies. Therefore, as can be seen in Figures 9.24 and 9.25, the curves can be based on either an increasing or decreasing accumulation. Both curves are useful if you wish to know how many items are above or below a certain point. Increasing accumulation shows all the points less than, while decreasing accumulation shows all the points more than. It should be pointed out that the cumulative frequencies are *not* plotted at the class midpoint; instead, depending on whether it is a 'less than' or 'more than' cumulative distribution, the cumulative frequencies are plotted at the upper or lower class limits.

Referring to Table 9.12, a cumulative 'less-than' frequency distribution indicates how many hotels charge under $140 a day for a room. A cumulative 'more-than' frequency indicates the number of hotels that charge more than $80 a day.

According to Table 9.12, no hotel charges under $80 a day, while 11 charge under $100 a day and 21 charge under $110 a day. If we were interested in determining the number of hotels charging more than $80 a day, the final column would provide the information—30 hotels; 19 hotels charge more than $100 a day and only 9 charge more than $110 a day.

Business mathematics and statistics

Table 9.12 *Room tariffs for 30 4-star hotels*

Room tariffs ($)	Frequency	Less than (lower limit of class interval)	More than (lower limit of class interval)
80 and under 90	3	0	30
90 and under 100	8	3	27
100 and under 110	10	11	19
110 and under 120	5	21	9
120 and under 130	3	26	4
130 and under 140	1	29	1
140 and under 150	0	30	0

Source: Hypothetical.

Figure 9.24 *'Less-than' ogive of daily tariffs at 30 4-star hotels*

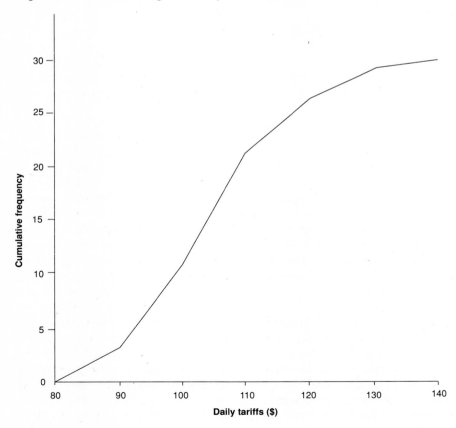

Source: Hypothetical.

260

Chapter 9 – Visual presentation of business data

Figure 9.25 *'More-than' ogive of daily tariffs at 30 4-star hotels*

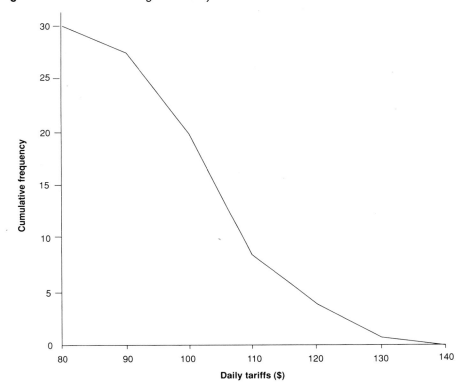

Source: Hypothetical.

The graphic representation of this cumulative frequency distribution is known as an ogive, and each point on the accompanying graphs represents the number of hotels that have tariffs less than (Figure 9.24) or more than (Figure 9.25) the tariffs indicated on the x axis.

Finally, these cumulative frequency columns can also be converted to cumulative percentage columns to determine, for example, the percentage of hotels charging below $120 or $100 a day.

Line graphs (two-axis graphing)

The **line graph** is a particularly useful graph representing the plotting of the relationship between two variables, a concept familiar from elementary mathematics and reviewed in Chapter 7. The connection of data points on a grid indicates immediately the trends in the data, and often implies degrees of change (by the slope of the line). The vertical axis is normally used for plotting quantities or percentages, while the horizontal axis is usually expressed in units of time. Recommended rules to follow for honest graphing include the following:

Business mathematics and statistics

1. Show the scales on both axes.
2. Start scale numbering at zero and ensure that intervals per axis are of equal distance.
3. Indicate the point of origin—that is, draw in the zero line.
4. Place time on the horizontal x axis and the scale for values on the vertical y axis.
5. Leave in the original points when fitting a curve or line to them.
6. Clearly indicate extrapolation of data (usually by a dotted line).

A wide variety of line graphs are used in the business world, the more prominent of which are illustrated in Figures 9.26 to 9.30. The basic type of line graph is shown in Figure 9.26, with this line graph clearly illustrating the trend in profits as a share of GDP. The trend in the movement of the Australian dollar in relation to the US dollar is evident from examining the line graph in Figure 9.27, with the pictorial inclusion of the US dollar in the foreground highlighting its content. Finally, the inclusion of key factors explaining the turning points in the chart assist the reader in understanding the volatility of the Australian dollar.

Figure 9.28 illustrates a multiple graph (more than one quantity being illustrated) and the use of a zero line to distinguish between positive and negative results, in this case the growth of the Gross Domestic Product (GDP) and Gross National Expenditure (GNE). Economic growth is measured by changes in the rate of GDP growth, with two successive quarterly reductions in the real GDP known as a recession. In any event, the recessions of 1982 and to a lesser extent 1990 can be found in the negative portion of the graph (i.e. below the zero line).

The next example of a multiple graph (Figure 9.29) illustrates the trend in the rate of personal income tax over the 1950–95 period. The ratio of tax paid to household income has steadily increased, while the average rate of average weekly earnings

Figure 9.26 *Profits as a share of GDP at factor cost, March 1980 to 1996*

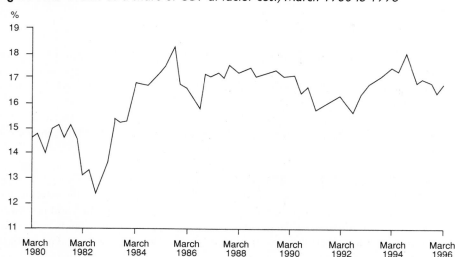

Source: Commonwealth Treasury of Australia, *Economic Roundup*, Winter 1996, p. 15. Commonwealth of Australia copyright. Reproduced by permission.

Figure 9.27 *Commodities, narrowing deficit bolster Australian dollar*

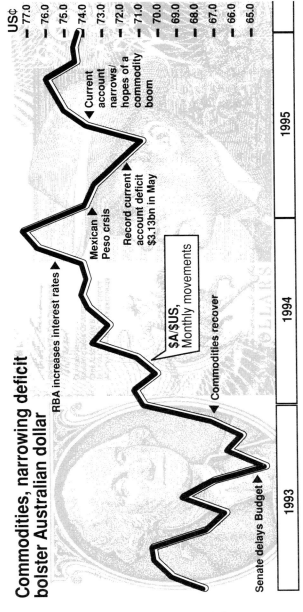

Source: Knight-Ridder Equinet (taken from Andreea Papuc, '$A looks bound to repeat range of '95 in new year', *Australian Financial Review*, 3 January 1995, p. 22).

Business mathematics and statistics

Figure 9.28 *Gross Domestic Product (GDP) and Gross National Expenditure (GNE, September 1978 to 1994)*

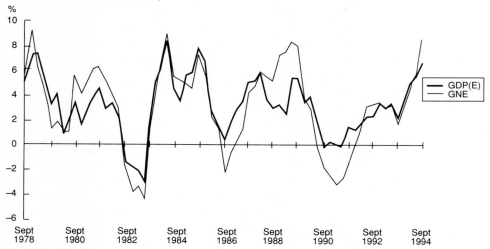

Source: Commonwealth Treasury of Australia, *Economic Roundup*, Summer 1995. Commonwealth of Australia copyright, reproduced by permission.

Figure 9.29 *Rates of personal income tax, 1950–95*

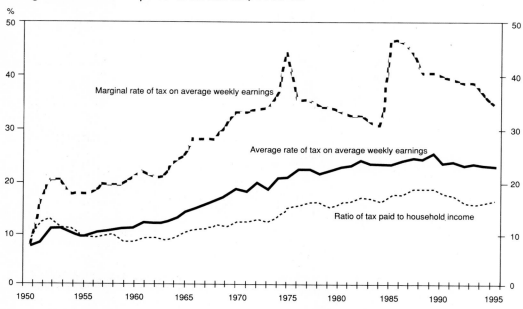

Source: Reserve Bank of Australia, *Australian Economic Statistics 1949–50 to 1994–95* (1996). Commonwealth of Australia copyright, reprinted by permission.

Chapter 9 – Visual presentation of business data

has flattened due to a declining marginal rate of tax. Note that, in this graph, the key is written on the graph itself.

Figure 9.30 shows the components of gambling expenditure—that is, the sum of casinos, other gaming and racing as a share of household disposable income. In the 1990s, 'Other gaming' and 'Casinos' contribute the lion's share of the rise in gambling expenditure as a percentage of household disposable income. Note that although only three lines are drawn, four sets of information are provided: casinos' contribution, the shares of 'Other gaming' and 'Racing', and the total gaming expenditure. This is achieved by using 'Casinos' as the base line or horizontal axis when plotting 'Other gaming'. Therefore, the space between the first line ('Casinos') and the second line is 'Other gaming'. Likewise, the share of household disposable income allocated for 'Racing' is the difference between the third line and the second line. For example, in 1993/94, about 0.06% of household disposable income was spent on racing, 1.75% on other gaming and 0.25% on casinos, although this figure will have increased markedly since 1993/94 due to the opening of casinos in Melbourne and Sydney.

In review, although the line graph does not present detailed data as well as tables do, it visually illustrates relationships more clearly. Often you will find both tables and line graphs used together, the latter being used to clarify or reinforce factors presented in tabular form.

Different scales

Various different scales are available, but they must be used with caution, as you and your readers must understand what they represent. For example, alternatives include

Figure 9.30 *Gambling expenditure as a percentage of household disposable income, 1972/73 to 1993/94*

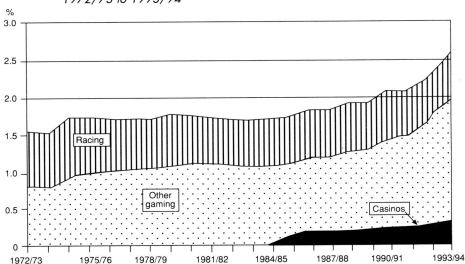

Source: Tasmania Gaming Commission (taken from Access Economics, *Economics Monitor,* May 1996, p. 14).

graph paper available with logarithmic, semi-logarithmic, linear-probability, percentage-probability, hyperbolic and square-root scales. The most frequently used of these is the semi-logarithmic (or ratio) scale, with the horizontal axis arithmetically defined and the vertical axis logarithmically defined. This scale permits comparison of rates of change over time not only between variables that are expressed in the same unit and different magnitudes (e.g. as in Figure 9.31), but also between those expressed in different units (e.g. house prices in $'000 and dwellings commenced in '000s).

To construct a **semi-logarithmic** (or **ratio**) **scale**, the vertical scale is ruled logarithmically, while the horizontal scale is arithmetically constructed. (Ratio scale graph paper can be purchased at newsagents or stationers.) The vertical scale shows graphically the rates of change over the period under consideration, while the horizontal scale shows the periods of time. Observe Figure 9.31 and note the large variations in Sydney house prices (both in nominal and real terms) and that both house prices and number of total dwellings commenced are incorporated into the single chart showing the rate of change of each rather than absolute change. Equal distances on the logarithmic scale represent percentage changes. A straight line indicates a constant percentage change.

In Figure 9.32 the distance between 1 and 2 on the semi-logarithmic scale is the same as the distance between 4 and 8, 10 and 20, 20 and 40, and so on. This is because 2 is twice 1, 4 is twice 2, 8 is twice 4, and so on. Observe how this scale (in Figure 9.31) neatly permits the inclusion of data from $8000 to $300 000.

In conclusion, imagine trying to graph on arithmetic graph paper changes in the size of Australia's population between 1788 and 1988—you would need a scroll! However, with semi-logarithmic paper, the changes could be drawn on one chart. For example, looking once again at Figure 9.31, note not only the varied categories of data presented in the single graph (e.g. house prices and number of dwelling commencements) but also how the median price of Sydney houses was about $12 000 in 1965 and about $240 000 in 1995. Finally, the cyclical nature of the residential property market is exemplified in the upward and downward swings in the number of commencements for New South Wales. Therefore, we are able to portray on the one graph a wide range of data. If, however, a negative or zero value should occur, this could not be depicted on a semi-logarithmic (ratio-scale) graph.

Combining different types of presentation

To illustrate a relationship between two variables which may be measured in different units, or in the same unit but related to a different base (e.g. square metres of construction versus stock vacancy rate), construction of such a presentation may prove useful. For example, Figure 9.33 indicates the change in Sydney CBD office construction (in thousands of square metres on the left-hand side) and total stock vacancy (in percentage terms on the right-hand side). It appears that when the vacancy rate is low, construction begins to climb and when the vacancy rate escalates, construction decelerates. However, you will note that in 1992, construction was high in spite of the high vacancy rate which can be attributable to the long lead time for construction. However, since 1992 there has been a marked lull in such construction as a result of an inordinately high vacancy rate.

Chapter 9 – Visual presentation of business data

Figure 9.31 *Sydney dwellings: prices and activity, 1965–95*

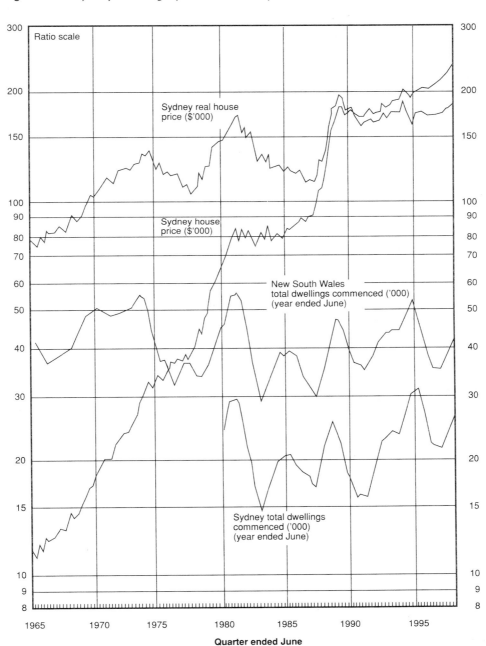

Source: BIS-Shrapnel, *Building in Australia 1996–2011* 16th edition (Sydney, 1996, p. 26).

Business mathematics and statistics

Figure 9.32 *Linear scale and semi-logarithmic scale*

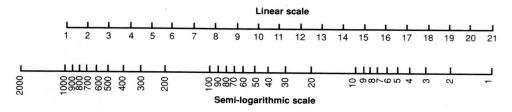

Figure 9.33 *Construction and vacancy in the Sydney CBD, 1980–95*

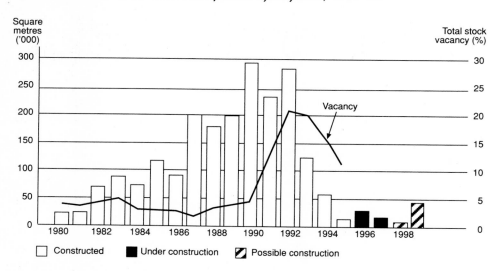

Source: JLW Research (taken from Matthew Kidman, 'Sydney shines in office market', *Sydney Morning Herald*, 27 February 1996).

Finally, the relationship between the rate of change in gross operating surplus of private enterprises and wage growth can also be neatly illustrated on a single graph, as seen in Figure 9.34, using both a bar chart and a line graph. As expected, there seems to be an inverse relationship—that is, when wage, salaries and supplements grow strongly, it is usually at the expense of the gross operating surplus of private enterprises and vice versa.

Chapter 9 – Visual presentation of business data

Figure 9.34 *Wages and profits (% change), 1950–95*

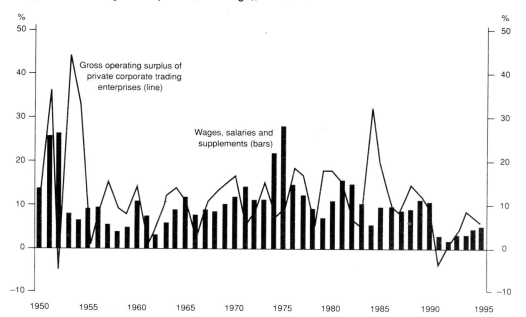

Source: Reserve Bank of Australia, *Australian Economic Statistics 1949–50 to 1994–95*. Commonwealth of Australia copyright, reproduced by permission.

9.2 Competency check

1. Which form of visual presentation would you use to provide precise data on the sales results for each State over each of the last five years for each of the four product divisions? Briefly explain why you chose this mode of presentation.

2. Retail turnover in 1994/95 totalled $99 billion, with food retailing at $37.5 billion; department stores at $10.3 billion; clothing and soft goods retailing at $7.7 billion; household goods retailing at $12.2 billion; recreational goods retailing at $5.2 billion, other retailing at $9.1 billion; and hospitality and services at $17 billion.[1]

   Construct a pie chart illustrating the share of each sector of the retail industry.

3. Construct a bar chart for retail turnover for 1994/95 using the data provided in problem 2.

4. Construct a line graph of Australian investment abroad for the years 1989/90 to 1994/95 (see Table 9.13). Briefly discuss the results.

Business mathematics and statistics

Table 9.13 *Total Australian overseas investment, 1989/90 to 1994/95 ($ billion)*

1989/90	1990/91	1991/92	1992/93	1993/94	1994/95
5.9	4.3	0.8	4.0	15.8	0.4

Source: Australian Bureau of Statistics, *Balance of Payments and International Investment Position* (Cat. No. 5363.0). Commonwealth of Australia copyright, reprinted by permission.

5. Construct a histogram and frequency polygon for the data found in Table 9.14. Briefly comment on the skewness.

Table 9.14 *Salaries of executives in C-U-Later Pty Ltd*

Salary ($)	Number of executives
30 000–39 000	8
40 000–49 000	11
50 000–59 000	22
60 000–69 000	24
70 000–79 000	20
80 000–89 000	9
90 000–99 000	6

Source: Hypothetical.

9.3 Evaluation checklist for visual presentations

Whether you are devising your own chart or graph, or interpreting someone else's, you should run through the following list before acting on the data:

Headings: Is the visual presentation properly and accurately labelled? Is there a title? Does it reflect the content? Are the units being measured clearly defined? Are the axes of the graph labelled?

Source: Does the visual presentation indicate the source of the data? Are any exceptions or irregularities noted? Is the source both reliable and reasonable for such data? Can it be verified?

Scale: Is the scale appropriate to display the type of result claimed? Are the zeros suppressed? Has this been indicated? If the scale is non-linear, is this clearly indicated? Are the class intervals of the grouped data too small or too large? Does the scale exaggerate the meaning of the presentation?

Simplicity: For a person looking at the visual presentation for the first time, is there too much detail? Has the purpose of visual presentation (i.e.

Chapter 9 – Visual presentation of business data

	helping to make it easier to understand) been fulfilled? Would another form of presentation achieve the objective more easily?
Explanation:	Is the visual presentation supported by discussion or interpretation in the text? Is this adequate? Are the key points and any limitations brought out in the discussion?

9.4 Summary

The objective of this chapter has been to show how to efficiently summarise and present collected data in tables and charts. As discussed, there are various possibilities to choose from and for your convenience these are outlined in Table 9.15.

Table 9.15 *Comparison chart of visual presentation methods*

Type	Advantages	Disadvantages
Table	Most comprehensive	Can be cumbersome to construct; hard to read at a glance
Pictogram	Immediate visual recognition of variables and corresponding quantities	Lack of precision
Pie chart	At a glance reveals relative proportions of variables and indicates degree of importance	Too many segments can detract from simplicity of presentation
Bar chart	Direct comparison of the magnitude of variables	Multiple and component bar charts can confuse the reader
Histogram	Displays relative frequencies	The size of class interval chosen must balance precision with simplicity
Frequency polygon	More readily displays trends in the data	Can be a temptation to interpolate values (i.e. forgetting that the line merely joins the midpoints of the bars)
Cumulative frequency polygon (ogive)	Shows number of observations falling above or below a certain value	Interpretation can be confusing for the unsophisticated reader
Line graph (two-axis graphing)	Readily displays trends in the data	Improper choice of scales and labels leads to misinterpreted data
Different scales	Enables specific aspects of the data to be demonstrated	Extremely difficult for uninitiated to understand the graph

Business mathematics and statistics

9.5 Multiple-choice questions

1. Which of the following graphic representations is most suited to illustrate rates of change?

    (a) line graph;
    (b) bar chart;
    (c) semi-logarithmic (ratio) chart;
    (d) pictogram.

2. A frequency polygon is derived from a:

    (a) line graph;   (b) bar chart;
    (c) histogram;   (d) none of the above.

3. According to Figure 9.27, in what year did the Australian dollar reach its low point against the US greenback?

    (a) 1993   (b) 1994   (c) 1995

4. According to Figure 9.16, what percentage of gross external debt in June 1990 was attributable to public trading enterprises?

    (a) 45%   (b) 30%
    (c) 20%   (d) 15%

5. Between 1985/86 and 1994/95, 25% of those living in Australia who decided to move overseas permanently were born in New Zealand and 16% in the United Kingdom. If you were constructing a pie chart, how many degrees would you allocate for New Zealand and the United Kingdom respectively?

    (a) 90 degrees and 58 degrees;   (b) 58 degrees and 90 degrees;
    (c) 45 degrees and 29 degrees;   (d) none of the above.

9.6 Exercises

1. (a) What type of visual presentation would you use if it was important to provide detailed data?
   (b) Which presentation would be most useful for showing an industry's market share?
   (c) How would you recommend illustrating regional sales per quarter to allow the reader to observe any obvious trends?
   (d) What type of graph allows you to present dramatic changes in sales turnover— for example, a 30-fold rise over a 15-year period?

2. When would it be better to use a multiple bar chart than a component bar chart?

3. How does a bar chart differ from a histogram?

4. A cumulative frequency polygon (known as an ogive) is useful for providing what type of information?

Chapter 9 – Visual presentation of business data

5. Define each of the following:

    (a) frequency distribution table;
    (b) class interval;
    (c) upper and lower limits of a class.

6. The annual salaries for accountants listed in today's newspaper range from $27 400 to $65 200. Set up class limits or boundaries for a frequency distribution of:

    (a) five class intervals;
    (b) six class intervals.
    (c) What would be the midpoint of each of the six classes of (b)?

7. Prepare the most appropriate diagram or graph to show the difference in the number of new settlers arriving in Australia from 1991/92 to 1995/96 (Table 9.16).

Table 9.16 *Number of new settlers arriving in Australia, 1991/92 to 1995/96*

Year	Number of settlers
1991/92	107 000
1992/93	76 000
1993/94	70 000
1994/95	87 000
1995/96	99 000

Source: Bureau of Immigration Research, *Australian Immigration Consolidated Statistics, No. 16, 1993–94*, and *Immigration Update, June Quarter 1996*.

8. Study the data supplied in Table 9.17.

    (a) Construct the most appropriate diagram or graph to show differences in mean earnings of males over the years 1991/92 to 1995/96.
    (b) On the same graph, show the differences in mean earnings of females over the same period.
    (c) Construct a percentage graph which shows the relationship between earnings of males and females for the years 1991/92 to 1995/96.

Table 9.17 *Average weekly earnings, June 1991/92 to 1995/96: full-time employees*

Year	Males ($)	Females ($)
1991/92	656	528
1992/93	673	538
1993/94	696	556
1994/95	729	578
1995/96	762	600

Source: Australian Bureau of Statistics, *1996 Labour Statistics* (Cat. No. 6101.0).

Business mathematics and statistics

9. Table 9.18 lists the eligibility categories of settler arrivals in Australia for 1994/95 to 1995/96.

    (a) Construct a component bar chart reflecting the information.
    (b) Construct a multiple bar chart incorporating this same information.
    (c) Construct a pie chart for 1995/96.

Table 9.18 *Eligibility categories of settler arrivals, 1994/95 and 1995/96*

Eligibility category	1994/95	1995/96
Family migration	37 000	46 000
Skill migration	20 000	20 000
Refugees, humanitarian and special assistance	14 000	14 000
Special eligibility	16 000	18 000

Source: Department of Immigration and Multicultural Affairs, *Immigration Update, June Quarter 1996*.

10. Dwelling commencements for Australian States and Territories are listed in Table 9.19 for 1994/95 and 1995/96.

    (a) Construct a chart or charts clearly indicating the proportion of total commencements undertaken in each State or Territory in each year.
    (b) Briefly comment on your observations.

Table 9.19 *Dwelling commencements, 1994/95 and 1995/96: States and Territories ('000)*

Year	NSW	Vic	Qld	WA	SA	Tas	NT	ACT
1994/95	51	29	46	22	10	3	1	3
1995/96	39	23	28	15	6	2	1	2

Source: Australian Bureau of Statistics, *Building Activity Australia* (Cat. No. 8752.0.40.001).

11. The figures in Table 9.20 relate to the income earned by U Beta B-Leave It Inc. for four of its top departments.

Table 9.20 *U-Beta B-Leave It Inc. net profit results, 1993/94 to 1996/97 ($'000)*

	Year			
	1993/94	1994/95	1995/96	1996/97
Department A	45	49	56	71
Department B	30	38	44	52
Department C	24	26	30	33
Department D	23	28	38	50

Source: Hypothetical.

Chapter 9 – Visual presentation of business data

(a) Prepare a graph or diagram to show the total profits arising from these four departments, as well as the contribution made by each of the departments to each year's results.
(b) Prepare another chart that will highlight the relative proportion of income earned by each of the four departments for the financial year 1996/97.

12. The sales achieved by the 62 representatives in the Grimm brothers' business venture for the December quarter 1997 are shown in Table 9.21.

Table 9.21 *Sales for December quarter, 1997*

Sales ($'000)	Number of representatives
1 000 and under 3 000	1
3 000 and under 5 000	5
5 000 and under 7 000	8
7 000 and under 9 000	16
9 000 and under 11 000	12
11 000 and under 13 000	9
13 000 and under 15 000	6
15 000 and under 17 000	3
17 000 and under 19 000	2

Source: Hypothetical.

(a) What type of visual presentation is most appropriate for illustrating the number of representatives with sales per class interval?
(b) Construct that presentation based on the data in the table.

13. The export results per State and Territory are given in Table 9.22.

(a) Construct a multiple bar chart incorporating the data for 1992/93 to 1994/95.
(b) Construct a component bar chart for the three years incorporating both:
   (i) Australia's total exports per year;
   (ii) total annual exports per State/Territory.

Table 9.22 *Exports per State/Territory, 1992/93 to 1994/95 ($ billions)*

State/Territory	1992/93	1993/94	1994/95
New South Wales	13	15	15
Victoria	11	12	13
Queensland	12	12	13
South Australia	4	4	4
Western Australia	15	16	16
Tasmania	2	2	2
Northern Territory	1	1	1
Australian Capital Territory	0	0	0

Source: Australian Bureau of Statistics, *Year Book, 1996* (Cat. No. 1301.0). Commonwealth of Australia copyright, reproduced by permission.

Business mathematics and statistics

14. Table 9.23 shows the sources of Federal Budget taxation receipts for the years cited. The figures are expressed in percentage contributions to total taxation revenue received.

    (a) Prepare a chart which illustrates the distribution between sources of taxation revenue for each year.
    (b) Construct a pie chart for 1996/97 which depicts the relative size of each source of taxation revenue received by the government.

Table 9.23 *Federal Budget taxation receipts percentage, 1995/96 and 1996/97*

Tax	1995/96	1996/97 (estimate)
Income tax		
Individual	49.7	50.7
Companies	15.0	15.1
Other	5.6	5.4
Sales tax	10.6	10.7
Excise duty		
Crude oil and petrol	8.4	8.3
Other	2.1	2.0
Customs duty	2.6	2.3
Other	6.0	5.5
Total (%)	100.0	100.0

Source: The Treasury, *Budget Papers 1996/97, Budget Paper No. 1.*

15. Below are the grades of 40 students of business maths and statistics:

86	89	76	65	51	43	28	37	71	66
57	59	60	58	47	74	62	54	51	79
64	56	44	87	55	65	76	91	69	73
49	77	80	58	64	50	66	77	67	70

    (a) Construct a frequency distribution using seven suitable classes.
    (b) From the frequency distribution, prepare a cumulative frequency distribution.
    (c) Draw a frequency polygon of this distribution.
    (d) Draw a 'less-than' ogive of the cumulative frequency distribution.

16. Below are the waiting times (in minutes) at an airport's baggage claim area for passengers arriving on international flights.

12	20	32	26	17	19	34	38	23	21
46	60	38	27	19	24	32	21	31	11
43	57	20	30	19	37	28	35	16	29
34	26	19	36	42	44	56	32	39	41
27	24	44	40	17	22	31	39	26	21

Chapter 9 – Visual presentation of business data

(a) Construct a frequency distribution with six equal classes. Which interval has the largest frequency?

(b) Construct a histogram based on the frequency distribution of (a).

17. With reference to exercise 16:

   (a) Construct the most appropriate presentation for displaying the number of passengers who waited less than any given waiting time.

   (b) What proportion of passengers waited less than 20 minutes?

   (c) Construct the most appropriate presentation for displaying the number of passengers who waited more than any given waiting time.

18. During the last five years at the Industrial Chemical Company, 121 'lost-time' industrial accidents have been registered. The number of production hours lost per accident is shown in Table 9.24.

Table 9.24 *Number of hours lost because of industrial accidents*

Number of hours lost per accident	Number of accidents
Less than 5	19
5 and less than 10	28
10 and less than 15	36
15 and less than 20	19
20 and less than 25	10
25 and less than 30	6
30 and less than 35	2
35 and less than 40	1

Source: Hypothetical.

(a) Prepare a histogram showing the frequency of 'lost-time' accidents for the given class intervals.

(b) Draw the appropriate frequency polygon.

19. Table 9.25 shows the Australian exports to our major trading partners for each year for the period 1992/93 to 1994/95. Construct a single chart which contains the following information:

   (a) the level of exports at the end of each of the three years;

   (b) the level of exports by each country for each of the three years.

Business mathematics and statistics

Table 9.25 *Amount of foreign investment in Australia by country/markets as of June 1992/93 to 1994/95 (A$ billion)*

Country	1992/93	1993/94	1994/95
Japan	15	16	16
United States	5	5	5
Korea	4	5	5
New Zealand	3	4	5
China	2	3	3
United Kingdom	2	3	2
Hong Kong	3	3	3
Taiwan	3	3	3
All other countries	24	23	25
Total	**61**	**65**	**67**

Source: Australian Bureau of Statistics, *Year Book Australia, 1996* (Cat. No. 1301.0). Commonwealth of Australia copyright, reprinted by permission.

20. Referring to Table 9.25:

    (a) Construct a diagram or graph which indicates the trend in Australia's total exports over the three-year period.
    (b) Construct a presentation which highlights each country's market share of Australian exports for 1994/95.

Note

1. Australian Bureau of Statistics, *Australian Economic Indicators* (Cat. No. 1350.0).

Chapter 10
MEASUREMENT OF CENTRAL TENDENCY

LEARNING OUTCOMES
After studying this chapter you should be able to:
- Calculate the mean for ungrouped and grouped data
- Organise raw data into frequency distribution tables
- Calculate the median for ungrouped and grouped data
- Calculate the mode for ungrouped and grouped data
- Interpret the significance of skewness of a data distribution
- Determine the most appropriate measure(s) for a given situation.

Exposure to averages occurs from the time of our conception to the day of our expiration. For example, over the average nine months of gestation, our mothers visit the gynaecologist to determine if we are growing at the average rate; when we depart, the question is, have we lived the average life expectancy?

Let's look in on an average day in the life of Norm. This morning Norm woke up a bit late (8 a.m.), but otherwise it seemed an average sort of morning, although the temperature was 5 degrees below average. Norm had to rush to the bus stop, as buses come every 15 minutes on average, but he had to catch the 8.15 a.m. bus. He caught it all right, but unfortunately the ride took 10 minutes longer than average because of a traffic snarl.

Norm deals with the public in his job and he noticed that today seemed especially quiet, with customer traffic down for the average Thursday. With some time on his hands, Norm checked with the paymistress why he had received less income than expected for the last week even though he had worked more hours than on average. Norm had failed to allow for the fact that Australia's progressive income tax system results in every dollar earned being taxed at a higher rate, hence increasing Norm's average tax rate. By now, Norm was sick and tired of averages, which had permeated everything from his mother's womb to his pay packet. Dare Norm leave the office for fear of the average befalling him?

Norm decided to go to the pictures to avoid thinking of the unmentionable a——e. On average, the 5 p.m. showings are empty during the week, but this time one of the radio stations was having a special viewing. Foiled again!

Norm walked home in order to avoid confronting the bus again, thereby avoiding the unmentionable. He bought a bag of chips and noticed that not only did it take him longer than average to get served but also there were fewer chips than average. Norm continued to hurry home, upset and still hungry. Upon arriving home he picked up the mail and noted an average number of bills but with above-average amounts. Norm hurriedly changed into his squash gear as he had a match that night in C-grade. (Norm is only an average player.) He was hoping to get home from the squash match by 10 p.m. as there was a replay of highlights of the week's rugby matches. If the squash game took the average 1 hour, Norm would make it back in time. However, a problem arises. If Norm watches the replay, which lasts 2 hours, he won't be able to get the 8 hours' sleep he needs on average. When Norm eventually went to bed and slipped under the covers on this colder-than-average night, he thought to himself that this had certainly been an average sort of day.

It hits us at every angle from morning until night, when we least expect it! How would you describe or define it? An alien force able to leap tall buildings at a single bound? Of course not; this all-pervasive **average** is simply a single value indicating a typical representative figure of the data examined. Being the most representative implies that the value is somewhere in the centre of the group of numbers; therefore, the average is frequently referred to as a **measure of central tendency**. It is used therefore to describe or summarise, in a single figure, the general or typical result of a group of figures. Certainly a degree of precision may be sacrificed in summarising, but the average does provide a most precise summary of the data.

The most commonly used measures of central tendency are the arithmetic mean, the median and the mode. As each measure has certain properties, its appropriate use depends on the distribution of the actual data.

10.1 Arithmetic mean

The **arithmetic mean** is the most commonly used average. In fact, when people use the term 'average' in everyday life, they are referring to the arithmetic mean determination of the average. As the arithmetic mean is often simply referred to as the **mean**, we will do so hereafter.

Ungrouped data

The arithmetic mean for a set of **ungrouped (raw) data** is obtained by adding the values and dividing by the number of items or observations. For example, five houses sold in the same neighbourhood fetched $182 000, $176 000, $187 000, $172 000 and $270 000. To determine the mean of this data, we simply sum the house prices and divide by the number of houses (five).

Chapter 10 – Measurement of central tendency

Thus:

$$\text{Mean} = \frac{\text{Sum of values}}{\text{Number of values}}$$

$$\bar{x} = \frac{\Sigma x}{n}$$

where $\bar{x}$ = the mean for the sample of values
Σ (the Greek letter Sigma) = the symbol representing summation
x = set of values
n = number of values in the sample set.

Therefore:

$$\bar{x} = \frac{\$182\ 000 + \$176\ 000 + \$187\ 000 + \$172\ 000 + \$270\ 000}{5}$$

$$= \frac{\$987\ 400}{5}$$

$$= \$197\ 400$$

Note that the mean average sale price of $197 400 does not appear to be representative of any of the houses sold. In fact, only one house sold for above $187 000. This is due to the fact that the mean is affected by extreme values, and the fifth house, selling for $270 000, skewed the data (i.e. it is not symmetrical), resulting in an excessively high mean average.

You may have observed that the above example and the accompanying formula were for the mean of a sample of values, because those five houses sold did not represent the total number of houses sold in that neighbourhood. If the prices of all houses sold in the neighbourhood were available and included in the calculation of the mean, we would then be looking at the mean of the entire population of houses sold rather than simply the mean of the sample of houses sold. The method used to determine the mean for a population of values is identical to that used to determine a sample mean, with only a couple of the symbols changed, as seen below:

$$\mu = \frac{\Sigma x}{N}$$

where μ (the Greek letter mu) = the symbol representing population and
N = number of values in the population.

Problem: Table 10.1 shows the number of days lost through repairs and maintenance in 1997 for each of the seven taxis belonging to the Miles Behind Taxi Service.

Business mathematics and statistics

Table 10.1 *Days lost during 1997 for Miles Behind Taxi Service*

Taxi	Days lost
1	17
2	21
3	12
4	46
5	26
6	32
7	14

Source: Hypothetical.

Solution: Since we have the total number of taxis belonging to the Miles Behind Taxi Service, we are dealing here with determining the mean of the population of values. Therefore:

$$\mu = \frac{\Sigma x}{N}$$
$$= \frac{17 + 21 + 12 + 46 + 26 + 32 + 14}{7}$$
$$= \frac{168}{7}$$
$$= 24$$

Therefore, 24 days per taxi were lost through maintenance and repairs. Note that the fourth taxi was out of commission for almost twice the average time, which may indicate that the owner should consider replacing it.

Problem: Calculate the mean number of dwellings commenced in Australia for the period 1985–96.

Solution:

Table 10.2 *Number of dwellings commenced in Australia, 1985–96*

Year	Total ('000)
1985	153
1986	136
1987	116
1988	136
1989	175
1990	138
1991	121
1992	140
1993	162
1994	178
1995	166
1996	122

Source: Australian Bureau of Statistics, *Australian Economic Indicators* (Cat. No. 1350.0). Commonwealth of Australia copyright, reproduced by permission.

Chapter 10 – Measurement of central tendency

Since this table represents the total number of dwellings commenced in Australia, we can use the formula for the mean of the population:

$$\mu = \frac{\Sigma x}{N}$$

$$= \frac{153 + 136 + 116 + 136 + 175 + 138 + 121 + 140 + 162 + 178 + 166 + 122}{12}$$

$$= \frac{1743}{12}$$

$$= 145.25$$

The mean number of dwellings commenced per year in Australia over the 1985–96 period was 145 250.

Observe the cyclical nature of commencements, which reflects the sensitivity of dwelling commencements to economic activity, interest rate movements, the level of unemployment and the general level of confidence in the economy. If you recall the severity of the economic slowdown of 1990 and 1991 you will understand the downturn in dwelling commencements at that time and note the marked improvement in dwelling commencements over the following three years and the subsequent marked drop in 1996. If we had provided you with further examples of other business activities over the same period, it is highly likely you would have found similar trends.

Weighted mean

When considering an average, you may find that some items are more important than others. For example, when the Consumer Price Index (CPI) measures inflation it assigns weights to each category of expenditure, relating each category's importance to all the others. For example, as consumers allocate a larger percentage of their expenditure to food than to clothing, a greater weight is applied to food than to clothing when determining the effect of price changes in each category on the spending habits of the consumer. Similarly, if you were enrolled in a course, you would expect the final examination to be worth more than a class test in the determination of your final grade.

Let's have a look at how the **weighted mean** average would be calculated to determine the final mark in your statistics course. Your final mark is a weighted average composite; this means that the scores achieved for a variety of requirements each have a different weight or relative importance in terms of the final grade achieved. Let's assume that five topic tests are worth a total of 15% of your final mark, an essay requirement is worth 30%, the mid-year examination 20% and the final examination 35%. Your grades for each are as follows:

Topic tests	90%
Essay	40%
Mid-year examination	86%
Final examination	50%

Business mathematics and statistics

Is it correct merely to add the four scores and divide by 4? No, it is not correct, since to do this assumes equal importance for each activity. Instead, the appropriate weights per individual category must be allocated as follows:

Topic tests	90% of maximum 15 marks (or %) = 13.5
Essay	40% of maximum 30 marks (or %) = 12.0
Mid-year examination	86% of maximum 20 marks (or %) = 17.2
Final examination	50% of maximum 35 marks (or %) = 17.5
	Final grade = 60.2

If we simply calculated the normal arithmetic mean (i.e. $\mu = \Sigma x/N$), we would expect a grade of 66.5%. However, examining the above results for each category reveals that the higher marks achieved were for lower weighted areas (topic tests and the mid-year examination) and that the lower results were for the more heavily weighted areas; hence, the weighted mean was below the normal mean.

Table 10.3 provides information on five taxis owned by the U-B-Long group of companies. To determine the average distance per fare, a weighted average must be calculated since Taxi 1, for example, would generally be expected to carry fewer fares than the other taxis because its average distance per fare is significantly greater than for the other cabs.

Table 10.3 *Taxis 1 to 5 owned by U-B-Long group of companies: distance per fare*

Taxi	Distance per fare (kilometre) (x)	Number of fares in 8-hour shift (w)	wx
1	11.0	20	220
2	8.5	24	204
3	7.5	25	187.5
4	9.0	21	189
5	5.0	40	200
Total	41.0	130	1000.5

$$\text{Weighted mean} = \frac{\Sigma wx}{\Sigma w} = \frac{1000.5}{130} = 7.7 \text{ kilometres (to nearest 0.1 km)}$$

Source: Hypothetical.

The simple mean distance per fare of the five taxis is 8.2 kilometres ($\frac{41}{5}$). However, a more useful measure is the mean derived from recording the number of fares taken during an 8-hour shift for each taxi (i.e. the weighted mean, 7.7 kilometres). Therefore, the individual weights (in column 3, in the form of the number of fares) are applied to the respective taxi's distance per fare to arrive at the weighted mean.

Grouped data

Data are often assembled into a frequency distribution form (discussed in Chapter 9) and determination of the mean may be required. Study Table 10.4, which contains information on accountants' weekly wages. Notice that precise wages are not recorded; instead, a class interval is provided (e.g. 300 and under 350). To calculate the mean for **grouped data**:

1. Assume that the incomes of each class are evenly distributed around the midpoint of each class. Then calculate the midpoint of each class.
2. Multiply the midpoints by the corresponding frequency of that class.
3. Find the sum of these products.
4. Divide the sum of the products by the sum of the frequencies.

In formula form, we have:

$$\text{(sample)} \quad \bar{x} = \frac{\Sigma fm}{n} \quad \text{and} \quad \mu = \frac{\Sigma fm}{N} \quad \text{(population)}$$

where $\bar{x}$ = the mean of sample
Σ = summation
f = frequency of the individual class
m = midpoint of the individual class
n = number of frequencies.

where μ = population mean
N = number of frequencies in population

Table 10.4 *Weekly wages of accountants*

Weekly wage ($)	Number of accountants (f)	Midpoint (m)	fm
300 and under 350	20	325	6 500
350 and under 400	41	375	15 375
400 and under 450	84	425	35 700
450 and under 500	120	475	57 000
500 and under 550	140	525	73 500
550 and under 600	36	575	20 700
600 and under 650	30	625	18 750
650 and under 700	26	675	17 550
700 and under 750	22	725	15 950
750 and under 800	18	775	13 950
800 and under 850	13	825	10 725
Total	550		285 700

Source: Hypothetical.

Table 10.4 indicates that 20 accountants earned approximately $325 each per week (the midpoint of 300 and under 350) or that $6500 was earned between them, while 41 accountants earned approximately $375 per week each (midpoint of 350 and under

400) or $15 375 between them, and 84 accountants earned $425 per week each or $35 700 between them, and so on. The total income earned during a week by the 550 accountants was $285 700. Therefore, to find the mean:

$$\mu = \frac{\Sigma fm}{N}$$

$$= \frac{\$285\ 700}{550}$$

$$= \$519.45$$

The mean weekly wage of $519.45 for an accountant may be useful information for a prospective employer. On the other hand, it should be realised that no indication is given of work experience, length of service with the firm, qualifications, etc., which may explain why some are earning $300 to $350 a week and others $500 or more a week.

Problem: The following data represent the ages of members of a circuit training class. The instructors want to calculate the mean age to determine whether the target audience of members is attending the classes.

19	24	28	24	32	42	26	30
21	29	37	19	38	31	27	20
17	23	32	20	19	43	18	31
19	27	29	19	49	21	29	36
43	26	28	33	35	54	30	42

Solution: The mean age of this sample of members is:

$$\bar{x} = \frac{\Sigma x}{n}$$

$$= \frac{1170}{40}$$

$$= 29.25 \text{ years}$$

Therefore, the mean age of participants in this class is 29.25 years.

Now suppose that instead of the raw or ungrouped data, a frequency distribution based on this data was requested. How would you go about constructing this? First, we can conveniently construct a frequency distribution of the above data, consisting of eight classes as seen in Table 10.5. Always make sure that no problem arises in classifying each value—that is, that every value fits into one class only. For example, in Table 10.5 each age fits into only one age-group. Had the class intervals and width of the intervals been constructed differently—for example, 15–20, 20–25, etc.—we would have had problems in classifying certain ages such as 20, 30, 35, and so on.

Chapter 10 – Measurement of central tendency

Table 10.5 *Ages of members attending a surveyed circuit training class*

Ages	Number of members
15–19	7
20–24	7
25–29	9
30–34	7
35–39	4
40–44	4
45–49	1
50–54	1

Source: Hypothetical.

Next, we must continue as we did in the previous examples—that is:

1. Find the average age of each member group by making the assumption that the age in each class is equal to the midpoint of that class—for example, the first seven members' ages are 15 to 19 years and the midpoint is 17 years. The next seven members have a midpoint age of 22 years, the third group 27 years, the fourth group 32 years, and so on (see Table 10.6, column 3).

2. Multiply the frequency per class (number of members in that particular class) by the corresponding midpoint of that particular class ($f \times m$) (see Table 10.6, column 4).

3. Total the ages for all classes (Σfm or 1155) and divide this figure by the number of members (40) to arrive at a mean age of 28.875 years per member. That is:

$$\bar{x} = \frac{\Sigma fm}{n}$$
$$= \frac{1155}{40}$$
$$= 28.875 \text{ years}$$

The instructors must now evaluate whether the targeted group has responded and attended classes. If not, a different marketing strategy may have to be introduced.

By the way, did you notice that the mean age for grouped data differs by almost half a year from that for ungrouped data because of the assumption that the midpoint of each class in grouped data represents the result (e.g. age) for each of those members in that class?

Business mathematics and statistics

Table 10.6 *Ages of members attending a surveyed circuit training class*

Ages	Number of members (f)	Midpoint (m)	fm
15–19	7	17	119
20–24	7	22	154
25–29	9	27	243
30–34	7	32	224
35–39	4	37	148
40–44	4	42	168
45–49	1	47	47
50–54	1	52	52
Total	40		1155

Source: Hypothetical.

Problem: Calculate the mean for grouped data on the life of car tyres as per data supplied in Table 10.7.

Table 10.7 *The life of car tyres (kilometres travelled) according to road service survey*

Kilometres travelled	Number of tyres (f)	Midpoint (m)	fm
0 and under 4 000	1	2 000	2 000
4 000 and under 8 000	3	6 000	18 000
8 000 and under 12 000	8	10 000	80 000
12 000 and under 16 000	14	14 000	196 000
16 000 and under 20 000	20	18 000	360 000
20 000 and under 24 000	15	22 000	330 000
24 000 and under 28 000	10	26 000	260 000
28 000 and under 32 000	6	30 000	180 000
32 000 and under 36 000	2	34 000	68 000
36 000 and under 40 000	1	38 000	38 000
Total	80		1 532 000

Source: Hypothetical.

Solution:

$$\bar{x} = \frac{\Sigma fm}{n}$$

$$= \frac{1\,532\,000}{80}$$

$$= 19\,150 \text{ kilometres}$$

Chapter 10 – Measurement of central tendency

Therefore the mean life of a car tyre, according to the data collected in a sample survey, is 19 150 kilometres.

Problem: Calculate the mean examination grade for all the students taking underwater basketweaving.

Table 10.8 Students' examination grades in underwater basketweaving

Grade (%)	Number of students (f)	Midpoint (m)	fm
30–39	6	34.5	207
40–49	20	44.5	890
50–59	64	54.5	3 488
60–69	87	64.5	5 611.5
70–79	51	74.5	3 799.5
80–89	19	84.5	1 605.5
90–99	9	94.5	850.5
Total	256		16 452.0

Solution:

$$\mu = \frac{\Sigma fm}{N}$$
$$= \frac{16\ 452}{256}$$
$$= 64.266$$
$$= 64.3\%$$

Therefore, the mean examination grade for the entire population of students taking this subject was 64.3%.

Review of the arithmetic mean

The advantages of the arithmetic mean include:

1. It is commonly used and easily understood.
2. Calculation is not difficult, although it may be time-consuming.
3. It may be treated algebraically—that is, the sum of the deviations from the mean equals zero, thereby permitting choice of an arbitrary origin to which a correction factor will be ultimately added. This allows the mean for grouped data, in particular, to be determined with far fewer calculations, and hence less chance of arithmetical errors (see Appendix 10A).
4. All values are included in the calculation.

Business mathematics and statistics

The disadvantages of the arithmetic mean include:

1. It is greatly affected by extreme values. For example, the mean of 7, 8, 9, 12, 16, 19 and 94 is 23.6, yet only one figure is above 19. Similarly, in February 1996, average weekly total earnings for full-time working adults were $707.10, but as the mean is affected by extreme values, the $707.10 could well be inflated by extremely high incomes. Interestingly, the average weekly total earnings for all employees (including juniors and part-timers) was $562.60, so the mean has been affected on the downside because of the low wages of juniors and the approximate 25% of the workforce who work part-time.
2. The arithmetic mean cannot be calculated from an open-ended class interval (e.g. '65+' or 'under 20'). Instead, the open-ended class must be omitted, thereby detracting from the precision of the mean as a measure of central tendency.
3. The mean (i.e. the mean of the raw data) will probably not equal any actual value within the distribution; hence, how representative or typical is the mean?

10.1 Competency check

1. Which of the three measures of central tendency do most people associate with the word *average*?
2. What do you see as two major advantages of using the arithmetic mean as a measure of central tendency?
3. What seems to be the greatest disadvantage of using the arithmetic mean as a measure of central tendency? Why do you consider this the greatest pitfall of using the arithmetic mean?
4. Calculate the mean for the ungrouped data provided in:

    (a) Table 10.9 (to the nearest thousand). For whom would such information be useful?
    (b) Table 10.10 (to the nearest thousand). What does this tell you about future job growth prospects with the government?
5. A lecturer has four classes: the first, numbering 45 students, has a mean test score of 70%; the second, numbering 60, has a mean of 65%; the third, numbering 100, has a mean of 56%; and the fourth, numbering 80, has a mean of 82%. What is the weighted average score for the lecturer's classes (to one decimal place)?
6. Calculate the mean for the grouped data provided in:

    (a) Table 10.11 (to the nearest $10);
    (b) Table 10.12 (to the nearest tenth of a cent);
    (c) Table 10.13 (to the nearest dollar).

Chapter 10 – Measurement of central tendency

Table 10.9 Births in Australia, 1987–95

Year	Births ('000)
1987	244
1988	246
1989	250
1990	258
1991	261
1992	256
1993	266
1994	259
1995	258

Source: Australian Bureau of Statistics, *Australian Demographic Statistics* (Cat. No. 3101.0).

Table 10.10 Australian government employees, 1989–94

Year	Government employees ('000)
1989	1732
1990	1755
1991	1726
1992	1683
1993	1654
1994	1588

Source: Australian Bureau of Statistics, *Year Book Australia*, 1990, 1991, 1992, 1993, 1994, 1995 and 1996 (Cat. No. 1301.0).

Table 10.11 Salaries of executives in C-U-Later Pty Ltd

Salary ($)	Number of executives
30 000–39 000	8
40 000–49 000	11
50 000–59 000	22
60 000–69 000	24
70 000–79 000	20
80 000–89 000	9
90 000–99 000	6

Source: Hypothetical.

Business mathematics and statistics

Table 10.12 Price of petrol, Sydney, January 1998

Price per litre (cents)	Number of service stations in surveyed areas
60.0–62.0	22
63.0–65.0	25
66.0–68.0	49
69.0–71.0	64
72.0–74.0	58
75.0–77.0	27
78.0–80.0	11

Source: Hypothetical.

Table 10.13 Weekly rent for all households in Westbridge

Weekly rent ($)	Number of households	
80–99	75	89.5
100–119	82	109.5
120–139	90	129.50
140–159	99	149.50
160–179	79	169.50
180–199	64	189.90
200–219	41	209.50
220–239	20	339.50

Source: Hypothetical.

10.2 Median

The **median** is the positional average, as it is the value of the middle item of a group of numbers. It is found by arranging the numbers in an array, in either ascending or descending order, and then determining which item is the middle or centre value.

Ungrouped data

The median for ungrouped data can be determined merely by arranging the raw data in an array. You then locate the middle item. For example, re-examine the five houses sold in the neighbourhood: this raw, ungrouped data can be arranged in ascending order, as follows:

$172\ 000 \quad \$176\ 000 \quad \$182\ 000 \quad \$187\ 000 \quad \$270\ 000$

As there are five values, the middle value is the third one. A quick method of finding the middle value of ungrouped data is to determine the position of the $(n + 1)/2$ value(s); for example, if there are 17 values, then $n = 17$; the middle is the ninth value (i.e.

(17 + 1)/2 = 9). If there is an even number of values, the median is found by calculating the midpoint of the two middle values; for example, if there are 10 values, the middle position is halfway between the fifth and sixth values (i.e. (10 + 1)/2 = 5.5). In the above example of house sales, the third value ($182 000) is the median.

Let's have another look at Australia's level of dwelling commencements (Table 10.2). We will consider the results only for 1985 up to 1996. Therefore, we have 12 values, so the median is halfway between the sixth and seventh values ((12 + 1)/2). The data can be arranged in ascending order as follows:

116   121   122   136   136   138   140   153   162   166   175   178

$$Md = \frac{138 + 140}{2}$$
$$= 139$$

Note that, unlike the case of the odd number items, the value of the median derived from an even number of items did not, in this case, correspond to any actual value in the original data. Also observe that changing the values of any of the items except the middle values will have no effect on the median value (unlike the mean, which would be affected as its calculation incorporates all values).

Grouped data

When data are presented in a frequency distribution format, the precise median is unknown (as was the arithmetic mean) and the median must therefore be estimated by assuming that the frequencies are evenly distributed throughout the respective classes. Thus the median cannot be determined as a particular value or the midpoint of two values. Instead, the value which divides the total frequencies into halves must be found. For example, Table 10.14 gives the distribution of examination grades for the seven classes of a certain lecturer.

Table 10.14 *Students' examination grades*

Grade (%)	Number of students (f)	Cumulative frequency (C)
30–39	6	6
40–49	20	26
50–59	64	90
60–69	87	177
70–79	51	228
80–89	19	247
90–99	9	256
	256	

Source: Hypothetical.

Business mathematics and statistics

To determine the median, the following steps should be taken:

1. Find the middle position (in this case, the middle student). For grouped data, the middle position is found by $n/2$. Therefore, the 128th student is the median value (256/2 = 128).
2. To discover which of the seven classes the 128th student is in, we must first form a cumulative frequency column. The cumulative frequency of each class is the sum of the frequency of each class plus the sum of the frequencies of all the previous classes. For example, the cumulative frequency of the 50–59 group is 90 students, arrived at by adding the cumulative frequency of the previous classes (26) to the frequency of the 50–59 class (64). This indicates that there are 90 students who achieved grades between 29.5 and 59.5 (the real lower and real upper limits of the first and third class). As 90 students achieved 59.5 or less, to determine how many students achieved 69.5 or less we simply add 87 (the frequency of the class interval 59.5 to 69.5) to 90 to arrive at a cumulative frequency of 177 for the subsequent class of 60–69.
3. As we are searching for the 128th student, we find the student in the 60–69 class, as only 90 student grades were included up to the previous class of 50–59. The 128th student is the 38th student in the median class of 60–69. As there are 87 students in the median class, we want the value of the 38/87th of the class interval.
4. The size of the class interval is 10. (Remember that the size of the class interval is the difference between the real lower and real upper limits of the median class—in this case, 69.5 − 59.5 = 10.)

Therefore:

$$Md = 59.5 + \frac{38}{87}(10)$$
$$= 59.5 + 0.437(10)$$
$$= 59.5 + 4.37$$
$$= 63.87$$

The formula for the above calculation can be written as follows:

$$Md = L + \frac{\left(\frac{n}{2} - C\right)}{f}(i)$$

where Md = the median
 L = real lower limit of the median class
 n = total frequency in the given data
 C = cumulative frequency in the class preceding the median class
 f = frequency of the median class
 i = size of the class interval in the median class.

Examining the students' grades and substituting into the formula:

$$Md = 59.5 + \frac{\left(\frac{256}{2} - 90\right)}{87}(10)$$

$$= 59.5 + \frac{128 - 90}{87}(10)$$

$$= 59.5 + \frac{38}{87}(10)$$

$$= 59.5 + 4.37$$

$$= 63.87$$

$$= 63.9$$

Therefore, the median grade of the students is 63.9. This closely approximates the calculated mean of 64.3% (see p. 289), indicating that the mean was not affected by extreme values in this example.

Problem: The weekly wages of accountants are shown in Table 10.15. Calculate the median average wage.

Table 10.15 *Weekly wages of accountants*

Weekly wage ($)	Number of accountants (f)	Cumulative frequency (C)
300 and under 350	20	20
350 and under 400	41	61
400 and under 450	84	145
450 and under 500	120	265
500 and under 550	140	405
550 and under 600	36	441
600 and under 650	30	471
650 and under 700	26	497
700 and under 750	22	519
750 and under 800	18	537
800 and under 850	13	550
	550	

Source: Hypothetical.

Solution:
1. To determine which accountant is in the middle position, first find how many accountants were surveyed. As 550 accountants were surveyed, we want to calculate the weekly wage of the 275th accountant (550/2 = 275).
2. To discover the median class in which the 275th accountant resides, we must construct a cumulative frequency column (column 3).
3. Find number 275 (in column 3) in the '500 and under 550' class, with the wage closer to the 500 rather than 550 level as there are 140 accountants

receiving between $500 and $550 a week and we want the 10th accountant out of that group of 140 (i.e. the previous groups included 265 accountants and as we want the 275th individual, we need count only 10 more into the next class of 140 accountants).

4. With 140 accountants in the median class, and assuming that the accountants and their wages are spread equally throughout the class, the median accountant must be 10/140th of the way into this class.
5. Since the size of the median class is 50 (the real upper limit is $550 and the real lower limit is $500), then 10/140th × 50 = 3.57.
6. Therefore, the median accountant's wage lies $3.57 into the median class, and since the real lower limit of this class is $500, the median weekly wage must be $503.57.

Alternatively, if you want to use only the formula method:

$$Md = L + \frac{\left(\frac{n}{2} - C\right)}{f} \quad (i)$$

$$= 500 + \frac{\left(\frac{550}{2} - 265\right)}{140}(50)$$

$$= 500 + \frac{10}{140}(50)$$

$$= 500 + 3.57$$

$$= 503.57$$

The median weekly wage for an accountant at $503.57 is below that of the mean wage for an accountant of $519.45 (see Section 10.1 above). Thus the extreme values in the upper class interval have resulted in the mean wage exceeding the median.

Problem: Determine the median life of a car tyre according to Table 10.16.

Table 10.16 *Life of car tyres (kilometres travelled) according to road service survey*

Kilometres travelled	Number of tyres (f)	Cumulative frequency (C)
0 and under 4 000	1	1
4 000 and under 8 000	3	4
8 000 and under 12 000	8	12
12 000 and under 16 000	14	26
16 000 and under 20 000	20	46
20 000 and under 24 000	15	61
24 000 and under 28 000	10	71
28 000 and under 32 000	6	77
32 000 and under 36 000	2	79
36 000 and under 40 000	1	80

Source: Hypothetical.

Solution:

$$Md = L + \frac{\left(\frac{n}{2} - C\right)}{f} \quad (i)$$

$$= 16\,000 + \frac{\left(\frac{80}{2} - 26\right)}{20}(4000)$$

$$= 16\,000 + \frac{40 - 26}{20}(4000)$$

$$= 16\,000 + \frac{14}{20}(4000)$$

$$= 16\,000 + 2800$$

$$= 18\,800$$

The median is therefore 18 800 kilometres; in other words, 50% of tyres lasted 18 800 kilometres or more and 50% lasted 18 800 kilometres or less.

Review of the median

The median is easily calculated and is a positional measure affected by the number of items rather than the value of each item (compared with the mean, which is affected by the value of each item). As such, it is less affected by extreme values. The median value can be calculated from open-ended intervals, as we are more interested in the position than the class value. (For example, in calculating the mean for grouped data, we need the midpoint of each class and this is impossible for an open-ended class such as '20 and under' or '75+'.) However, a drawback in using the median, besides its unfamiliarity for many, is that it is not algebraically defined; hence, there is no shorthand method for its calculation. The question whether the median is representative of the data could certainly be asked, as values outside the middle position are ignored. The median is used when evaluating real estate values (since the mean is affected by extremes, it is not considered appropriate), some demographics statistics, intelligence tests and other types of examinations.

10.2 Competency check

1. If you were presented with raw, ungrouped data and you wanted to determine the median value for this data, what would be your first step?

2. When is the median considered 'more representative' of a set of data than the arithmetic mean?

3. What seems to be the major drawback of using the median as representative of collected data?

Business mathematics and statistics

4. Calculate the median for the ungrouped data in:

   (a) Table 10.17 (to the nearest thousand);
   (b) Table 10.18 (to the nearest thousand).

Table 10.17 *Births in Australia, 1987–95*

Year	Births ('000)
1987	244
1988	246
1989	250
1990	258
1991	261
1992	256
1993	266
1994	259
1995	258

Source: Australian Bureau of Statistics, *Australian Demographic Statistics* (Cat. No. 3101.0).

Table 10.18 *Australian government employees, 1989–94*

Year	Government employees ('000)
1989	1732
1990	1755
1991	1726
1992	1683
1993	1654
1994	1588

Source: Australian Bureau of Statistics, *Year Book Australia*, 1990, 1991, 1992, 1993, 1994, 1995 and 1996 (Cat. No. 1301.0).

5. Calculate the median for grouped data in:

   (a) Table 10.19 (to the nearest $10);
   (b) Table 10.20 (to the nearest tenth of a cent);
   (c) Table 10.21 (to the nearest dollar).

Chapter 10 – Measurement of central tendency

Table 10.19 Salaries of executives in C-U-Later Pty Ltd

Salary ($)	Number of executives
30 000–39 000	8
40 000–49 000	11
50 000–59 000	22
60 000–69 000	24
70 000–79 000	20
80 000–89 000	9
90 000–99 000	6

Source: Hypothetical.

Table 10.20 Price of petrol, Sydney, January 1998

Price per litre ($)	Number of service stations in surveyed areas
60.0–62.0	22
63.0–65.0	25
66.0–68.0	49
69.0–71.0	64
72.0–74.0	58
75.0–77.0	27
78.0–80.0	11

Source: Hypothetical.

Table 10.21 Weekly rent for all households in Westbridge

Weekly rent ($)	Number of households
80–99	75
100–119	82
120–139	90
140–159	99
160–179	79
180–199	64
200–219	41
220–239	20

Source: Hypothetical.

10.3 Mode

A third measure of central tendency is the **mode**, which is the most frequently occurring value in a set of data. The mode can therefore be thought of as the most typical value in a series of data, although it may well be quite different from all the remaining values. Should no value or observation appear more than once, there is no mode in that set of observations.

Ungrouped data

No calculation is necessary to determine the mode for ungrouped data. Instead, you have only to locate the most frequently occurring value. For example, if the grades 80, 60, 40, 80, 100, 80 and 60 were achieved, the modal average would be 80, as 80 is the most frequently occurring score. Data can also be bimodal (two modes) or multimodal (more than two modes). For example, let's assume a survey has been taken of students' cars to determine the year model of each vehicle, hence determining the modal average age of the cars. Assume the following years represent the year model of each vehicle:

1982, 1982, 1983, 1983, 1983, 1984, 1984, 1985, 1986, 1987,
1987, 1987, 1988, 1989, 1990, 1991, 1991, 1992, 1993

This is bimodal as there are two modes, 1983 and 1987 (each of these occurs three times).

Table 10.22 lists the average number of flights arriving per hour each day in November 1997. The modal value is 64, yet except for the three days when an average of 64 flights arrived, there was not a single day when arrivals exceeded 46 per hour. Therefore, the mode as a measure of central tendency in such a case is not representative.

Table 10.22 *Average number of flights arriving per hour each day in November 1997*

36	44	38	31	42	37	39	40	38	30
34	32	46	64	34	42	32	37	29	33
64	31	44	35	36	39	64	43	28	40

Source: Hypothetical.

Had we grouped this data in the form of a frequency distribution, the modal class would certainly have been in the 30s, which seems far more appropriate as a representative figure.

Grouped data

When data are presented in a frequency distribution format, the modal class is the class with the highest frequency. To calculate the value for the mode from the modal class, we use the formula:

$$Mo = L + \frac{d_1}{d_1 + d_2} \quad (i)$$

where Mo = the mode
L = real lower limit of the modal class
d_1 = modal class frequency minus the frequency of the previous class
d_2 = modal class frequency minus the frequency of the class following the modal class
i = class interval of the modal class.

Table 10.23 *Students' examination grades*

Grade (%)	Number of students
30–39	6
40–49	20
50–59	64
60–69	87
70–79	51
80–89	19
90–99	9

Source: Hypothetical.

Therefore, examining the students' grades in Table 10.23, we find the modal class is '60–69' as the highest frequency of grades (87) falls in the range of that class. Substituting into the formula:

$$L = 59.5$$
$$d_1 = 87 - 64 = 23$$
$$d_2 = 87 - 51 = 36$$
$$i = 69.5 - 59.5 = 10$$

Hence:

$$Mo = L + \frac{d_1}{d_1 + d_2}(i)$$

$$= 59.5 + \frac{23}{23 + 36}(10)$$

$$= 59.5 + \frac{230}{59}$$

$$= 59.5 + 3.9$$

$$= 63.4$$

Thus, the modal grade for the students is 63.4%. This compares with a mean of 64.3% and a median of 63.9%; thus, the three measures of central tendency gave very similar values for students' grades.

Business mathematics and statistics

Problem: Apply the formula method to determine the modal weekly wages for the accountants whose wages are listed in Table 10.24.

Table 10.24 *Weekly wages of accountants*

Weekly wage ($)	Number of accountants
300 and under 350	20
350 and under 400	41
400 and under 450	84
450 and under 500	120
500 and under 550	140
550 and under 600	36
600 and under 650	30
650 and under 700	26
700 and under 750	22
750 and under 800	18
800 and under 850	13

Source: Hypothetical.

Solution: The modal class is '500 and under 550', as the largest number of accountants (140) have wages in this particular class. Therefore,

$$L = 500$$
$$d_1 = 140 - 120 = 20$$
$$d_2 = 140 - 36 = 104$$
$$i = 550 - 500 = 50$$

Hence:

$$Mo = L + \frac{d_1}{d_1 + d_2}(i)$$

$$= 500 + \frac{20}{20 + 104}(50)$$

$$= 500 + \frac{1000}{124}$$

$$= 500 + 8.0645$$

$$= 508.0645$$

Thus, the most frequently occurring wage for accountants was $508.06.

Chapter 10 – Measurement of central tendency

Problem: Determine the modal life of car tyres according to Table 10.25.

Table 10.25 *Life of car tyres (kilometres travelled) according to road service survey*

Kilometres travelled ('000)	Number of tyres
0 and under 4 000	1
4 000 and under 8 000	33
8 000 and under 12 000	8
12 000 and under 16 000	14
16 000 and under 20 000	20
20 000 and under 24 000	15
24 000 and under 28 000	10
28 000 and under 32 000	6
32 000 and under 36 000	2
36 000 and under 40 000	1

Source: Hypothetical.

Solution:

$$Mo = L + \frac{d_1}{d_1 + d_2} \quad (i)$$

$$= 16\,000 + \frac{(20 - 14)}{(20 - 14) + (20 - 15)} (4000)$$

$$= 16\,000 + \frac{6}{11} (4000)$$

$$= 16\,000 + 2181.82$$

$$= 18\,181.82$$

Therefore, the modal life of a tyre was approximately 18 182 kilometres versus the mean of 19 150 and a median of 18 800. In this case, the upper extreme values caused the mean to exceed the other two measures of central tendency.

Review of the mode

The mode:

- is the most usual or typical value;
- does not include all values in its calculation;
- can be calculated from open-ended data unless the mode falls in the open-ended class;
- is not affected by extremes unless the modal class happens to be at the extreme;
- is simple to approximate if only a small number of classes exist, but may be of little or no use if no mode exists or if there is a multimodal average;
- is difficult to interpret and compare when it is multimodal.

Business mathematics and statistics

10.3 Competency check

1. How is the mode determined for ungrouped data?
2. When examining ungrouped data, will there always be a modal value? Briefly explain your answer.
3. Determine the modal values in the following sets of data:

    (a) 165 166 167 168 169
    (b) 15 16 16 16 17 17 18 18 19 19 19 20 20
    (c) 25 26 27 28 28

4. How is the modal class for grouped data determined?
5. Calculate the mode of the data provided in:

    (a) Table 10.26 (to the nearest $10);
    (b) Table 10.27 (to the nearest tenth of a cent);
    (c) Table 10.28 (to the nearest dollar).

Table 10.26 *Salaries of executives in C-U-Later Pty Ltd*

Salary ($)	Number of executives
30 000–39 000	8
40 000–49 000	11
50 000–59 000	22
60 000–69 000	24
70 000–79 000	20
80 000–89 000	9
90 000–99 000	6

Source: Hypothetical.

Table 10.27 *Price of petrol, Sydney, January 1998*

Price per litre (cents)	Number of service stations in surveyed areas
60.0–62.0	22
63.0–65.0	25
66.0–68.0	49
69.0–71.0	64
72.0–74.0	58
75.0–77.0	27
78.0–80.0	11

Source: Hypothetical.

Table 10.28 *Weekly rent for all households in Westbridge*

Weekly rent ($)	Number of households
80–99	75
100–119	82
120–139	90
140–159	99
160–179	79
180–199	64
200–219	41
220–239	20

Source: Hypothetical.

10.4 Relationship between the mean, median and mode

When the values of the mean, median and mode of a frequency distribution are the same, it is known as a **normal** (or **symmetrical**) distribution (see Figure 10.1). The data are graphed in the form of a frequency distribution and, when approaching a normal curve or distribution, the peak of the data is in the centre of the frequency polygon with slopes on either side almost equal to each other. When the distribution is not symmetrical—that is, when the peak is either on one side or the other of the centre of the frequency polygon—the frequency polygon is said to be *skewed*. When the distribution is skewed to the right it is referred to as being *positively skewed* (see Figure 10.2), with the mode at the highest point of the distribution, the median to the right of that, and the mean to the right of both the mode and the median. When the distribution is *negatively skewed* (skewed to the left, see Figure 10.3), the extreme values of the mean at the lower end result in its being to the left of the median and the mode, with the median in the middle and the mode to the right of both other measures.

When we recall the major problem of the mean being the impact of extremes on the result, as well as the problem of the mode seldom being truly representative of the sample or population, it is not surprising that the median, being in the middle of the two, is often the preferred measure of central tendency under these circumstances.

Figure 10.1 *Symmetrical frequency polygon*

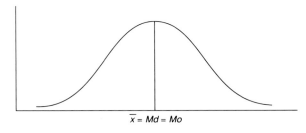

$\bar{x} = Md = Mo$

Figure 10.2 *Positively skewed frequency polygon*

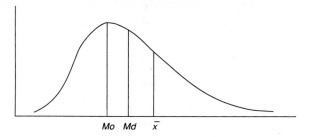

Figure 10.3 *Negatively skewed frequency polygon*

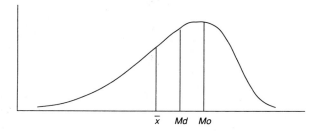

Figure 10.4, the frequency polygon portraying the students' grades listed in Table 10.23, shows a distribution approaching symmetry, but the extreme grades at the upper end make the distribution slightly skewed to the right, with the mean having the highest value, and the mode the lowest; the median is in between the two.

Note: As the median is in the middle and divides the curve into two equal parts, you normally find the median between the mode and the mean when it is not a normal curve (i.e. when it is skewed).

Figure 10.4 *Frequency polygon of students' examination grades*

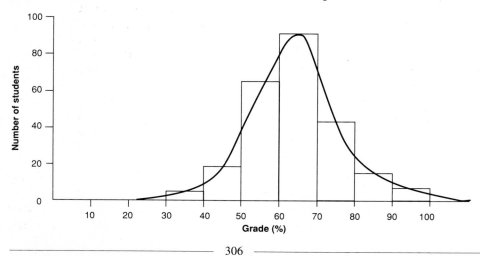

10.5 Summary

A measure of central tendency is often necessary to indicate the representative value of data with a single figure. The most commonly used measures of central tendency are the arithmetic mean ($\bar{x}$), the median (*Md*) and the mode (*Mo*).

Arithmetic mean

The arithmetic mean is the most commonly referred-to 'average'. In fact, when the term 'average' is used in everyday discussions, it is the arithmetic mean that is being referred to. When the mean is calculated, all items of the distribution are included; if any of these items are extreme values, they render the mean less appropriate as a representative value. As all items are included in its calculation, the mean is algebraically defined (i.e. the sum of the deviations of the individual values from the mean equals zero). If any class is open-ended (i.e. no finite lower or upper limit, e.g. under 20 or 65+), the mean cannot be accurately calculated.

The formula for calculating the mean of a sample of ungrouped data is:

$$\bar{x} = \frac{\Sigma x}{n}$$

where $\bar{x}$ = the mean
Σ = summation
x = set of all values in the sample
n = number of values in the set.

The formula for calculating a population mean of ungrouped data is:

$$\mu = \frac{\Sigma x}{N}$$

where μ = the population mean
Σ = summation
x = set of all values
N = number of values in the population.

The formula for calculating the mean for grouped data is:

$$\bar{x} = \frac{\Sigma fm}{n}$$

where $\bar{x}$ = the mean
Σ = summation
f = frequency of the individual class
m = midpoint of the individual class
n = number of frequencies.

We could substitute μ for $\bar{x}$ and N for n in the above formula if we are calculating the mean of the population rather than a sample of the population.

Median

The median is the positional average, as it is the value of the middle item of a group of numbers. For ungrouped data it is found by arranging the numbers in an array, in ascending or descending order, and then determining which item is the centre value. For grouped data, the formula is:

$$Md = L + \frac{\left(\frac{n}{2} - C\right)}{f} (i)$$

where Md = the median
L = real lower limit of the median class
n = total frequency in the given data
C = cumulative frequency in the class preceding the median class
f = frequency of the median class
i = size of the class interval in the median class.

The median is fairly easily calculated; it is less affected by extreme values than the mean, since it is a positional value and can be determined if the data contain open-ended intervals. However, as the median ignores the values of items outside the centre, it is not always representative of the data.

Mode

The mode is the most frequently occurring value in a set of data and can therefore be thought of as the most typical value in a series of data. For ungrouped data, no calculation is necessary, as one needs only to locate the most frequently occurring value. For grouped data, apply the formula:

$$Mo = L + \frac{d_1}{d_1 + d_2} (i)$$

where Mo = the mode
L = real lower limit of the modal class
d_1 = modal class frequency minus the frequency of the previous class
d_2 = modal class frequency minus the frequency of the class subsequent to the modal class
i = class interval of the modal class.

The mode is easily calculated; it can be determined for data containing open-ended intervals and it is not affected by extremes. However, it ignores values outside the modal class. In addition, the mode may be of little use if there is no mode in the data or if the data are multimodal.

On a final note on the omnipotent average, its importance is certainly not lost amongst employers as they count the average cost in lost productivity due to smoking. According to an unofficial estimate of the costs to Australian industry of 'smoko', Australia's 2 million smokers smoke an average of seven cigarettes a day at work, taking an average of 70 minutes, or 10 minutes per cigarette. This adds up to a loss of about 35 working days a year. Based on a national average wage of $33 233, this means a loss of $4486 per year per smoker, or almost $10 billion for the 2 million smokers.[1] According to estimates by Morgan and Banks, the human resources firm, smoking breaks equated to a 14.6% productivity loss among smokers, prompting at least one firm to pay its non-smoking staff a 15% bonus. Apparently, there are companies overseas awarding non-smokers extra annual leave to reward them for spending more time at their desk than their smoking colleagues.[2] That may be reason enough for smokers to quit smoking!

10.6 Multiple-choice questions

1. Which of the measures of central tendency is affected by extreme values?

   (a) arithmetic mean (b) median
   (c) mode   (d) None of the above.

2. The mean number of settlers arriving in Australia from 1986/87 to 1994/95 can be found in Table 10.29.

   Table 10.29 *Settler arrivals, 1986/87 to 1994/95*

Year	Number of settler arrivals
1986/87	113 000
1987/88	143 000
1988/89	145 000
1989/90	121 000
1990/91	122 000
1991/92	107 000
1992/93	76 000
1993/94	70 000
1994/95	87 000

   Source: Bureau of Immigration and Population Research, *Settler Arrivals, Statistical Report*, No. 10, 1992–93 and *Immigration Update*.

   The mean number of settler arrivals over the period was:

   (a) 109 300 (b) 113 000
   (c) 122 000 (d) None of the above.

3. The median number of settler arrivals for the period was:

   (a) 109 300 (b) 113 000
   (c) 122 000 (d) None of the above.

4. Referring to Table 10.30, in which class interval is the median class of fees found?

   (a) 2500 and less than 5000   (b) 5000 and less than 7500
   (c) 7500 and less than 10 000 (d) 10 000 and less than 12 500

Table 10.30 *Fees charged for postgraduate studies in Australian universities*

Fee range for full-time students ($)	Number of courses
0 and less than 2 500	49
2 500 and less than 5 000	364
5 000 and less than 7 500	241
7 500 and less than 10 000	152
10 000 and less than 12 500	56
12 500 and less than 15 000	34
15 000 and less than 17 500	4
17 500 and less than 20 000	5
20 000 and over	3

Source: Council of Australian Postgraduate Associations, *1994 Profiles Data/Report of Fee-paying Arrangements for Postgraduate Courses*.

5. Referring to Table 10.30, in which class interval is the modal class found?

   (a) 2500 and less than 5000
   (b) 5000 and less than 7500
   (c) 7500 and less than 10 000
   (d) 10 000 and less than 12 500

10.7 Exercises

1. Briefly discuss the characteristics, advantages and disadvantages of the arithmetic mean.

2. How does the calculation of the median differ from that of the mean?

3. Under what circumstance(s) would the mean be preferable to the median as representative of the data?

4. Under what circumstances would the median be preferable to the mean as the measure of central tendency?

5. In 1996, it was estimated that conference delegates from overseas spent on average $4202 during their seven-day stay in Sydney, while the average overseas tourist in Sydney spent $1895 for their 24-day stay in Sydney. What is the estimated daily expenditure for delegates compared to that of the average overseas tourist?

6. In Australia during 1994, there were approximately 47 000 road accident victims requiring medical treatment at an estimated cost of $559.3 million. What was the average cost per accident victim? As this estimated average is the mean, what might you ask yourself?

7. Calculate the mean, median and modal number of motor cycles registered in Australia (according to Table 10.31) during the period 30 June 1990 to 30 June 1995.

Table 10.31 *Number of motor cycles registered on 30 June 1990 to 1995*

Year	Number
1990	304 000
1991	284 000
1992	292 000
1993	292 000
1994	292 000
1995	297 000

Source: Australian Bureau of Statistics, *Motor Vehicle Registrations, Australia* (Cat. No. 9304).

8. Table 10.32 indicates the amount of foreign investment arriving in Australia each year during the period 1989/90 to 1994/95. Find the mean and median level of foreign investment in Australia over the period.

Table 10.32 *Foreign investment in Australia, 1989/90 to 1994/95*

Year	($ billion)
1989/90	25
1990/91	24
1991/92	19
1992/93	29
1993/94	32
1994/95	30

Source: Australian Bureau of Statistics, *Balance of Payments and International Investment Position, Australia* (Cat. No. 5363.0).

9. A student recorded a test average of 40%, a mid-term examination grade of 80%, an essay assignment of 84% and a final examination grade of 85%. If the tests were worth 15% of the final grade, the mid-term examination 25%, the essay assignment 25% and the final examination 35%, what was the weighted average of the student's grades (to the nearest per cent)?

10. A property investor must determine which of two units she should buy. Although a rating sheet is a subjective matter, an example of one is given in Table 10.33, showing the investor's evaluation of each category. Which unit should the investor buy?

Table 10.33 Features of unit development

	Location	Size of development	Car parking	Views	Aesthetics and quality	Desirability
Weighting (0–10)	7	6	4	10	8	10
			Unit 1		Unit 2	
Investor's ratings: Location			3		4	
Size of development			4		3	
Car parking			1		5	
Views			5		2	
Aesthetics			5		3	
Desirability			4		5	

Assessment: Excellent (5) to Poor (1).

Source: P. Waxman and D. Lenard, *Investing in Residential Property: Understanding the Market* (Melbourne: Wrightbooks, 1997).

11. The record for absenteeism (in number of days) for a firm over 12 consecutive months was as follows: 43, 41, 40, 41, 42, 88, 45, 51, 46, 42, 40, 38. Find the mean, median and mode (to one decimal place). Which do you think is the most appropriate? Why?

12. A spot check of the number of students attending a lecture over a 21-day period was made, with the following results:

    78 87 94 92 79 78 67
    82 88 62 97 80 82 62
    90 79 82 90 87 86 62

    Determine the mean, median and mode (to the nearest whole number). Which do you think is the most appropriate? Why?

13. An analysis of a company comprising four divisions is shown in Table 10.34.

    (a) Sketch the relative shape of the distribution of wages for each of the four divisions, indicating skewness (where it exists).
    (b) What is the weekly wage bill per division?

Table 10.34 Company analysis

	Division			
	Wine/spirits	Sporting goods	Graphics	Beef
Number of employees	48	52	34	42
Mean wage	405	360	490	402
Median wage	390	370	480	402
Modal wage	370	380	465	402

Source: Hypothetical.

14. Determine the median age (to the nearest tenth of a year) of participants in the circuit training class. How does the median compare to the mean age of 28.9 as calculated in Table 10.6?

Table 10.35 Ages of members attending a surveyed circuit training class

Ages	Number of members
15–19	7
20–24	7
25–29	9
30–34	7
35–39	4
40–44	4
45–49	1
50–54	1

Source: Hypothetical.

15. The dollar amounts of travellers cheques taken out of the country by 120 randomly selected Australians travelling overseas in November 1997 are given in Table 10.36. Calculate to the nearest $10 the mean, median and mode amount of travellers cheques taken out of the country by Australians travelling overseas.

Table 10.36 Amount of travellers' cheques taken out of the country by 120 Australians travelling overseas in November 1997

Amount of travellers' cheques ($)	Number of Australians travelling overseas
0 to less than 1 000	1
1 000 to less than 2 000	5
2 000 to less than 3 000	9
3 000 to less than 4 000	16
4 000 to less than 5 000	22
5 000 to less than 6 000	26
6 000 to less than 7 000	19
7 000 to less than 8 000	13
8 000 to less than 9 000	6
9 000 to less than 10 000	2
10 000 to less than 11 000	1

Source: Hypothetical.

16. By the end of its first year of operation the Up-in-Arms Gymnasium had members aged as listed in Table 10.37. Calculate the mean, median and modal ages of the members to the nearest whole year.

Table 10.37 *Age of Up-in-Arms Gymnasium members*

Age of members (years)	Number of members
15–19	34
20–24	84
25–29	146
30–34	72
35–39	40
40–44	16
45–49	4
50–54	2
55–59	1
60–64	1

Source: Hypothetical.

17. The rents for one-bedroom units in the southern suburbs are shown in Table 10.38. Calculate the mean, median and mode to the nearest dollar. Which of the three measures do you feel is most appropriate in this case? Why?

Table 10.38 *Rents for one-bedroom units in southern suburbs, January 1998*

Rent ($)	Number of one-bedroom units
40 and under 50	12
50 and under 60	14
60 and under 70	28
70 and under 80	46
80 and under 90	50
90 and under 100	25
100 and under 110	15
110 and under 120	10

Source: Hypothetical.

18. A multinational computer company has 150 marketing representatives throughout Australia. Sales results for 1997 are listed in Table 10.39.

   (a) Calculate the arithmetic mean, median and mode to the nearest $100.
   (b) Comment on the nature of the distribution on the basis of your results.

Chapter 10 – Measurement of central tendency

Table 10.39 *Sales for Nouvelle Computer Co., 1997*

Sales ($'000)	Number of representatives
100 to less than 200	2
200 to less than 300	9
300 to less than 400	12
400 to less than 500	25
500 to less than 600	29
600 to less than 700	31
700 to less than 800	24
800 to less than 900	10
900 to less than 1000	7
1000 to less than 1100	1

Source: Hypothetical.

19. The following data indicate the working days lost over a one-year period for 30 workers:

8	2	3	17	15	11	6	20	11	42	13	5	7	14	19
4	12	18	22	1	28	10	7	16	13	16	9	6	4	14

    (a) Determine the mean, median and mode for the raw data (to the nearest tenth of a day).
    (b) Construct a frequency distribution with intervals 0–4, 5–9, 10–14, 15–19, 20–24, 25–29, and 30 and over.
    (c) Determine the mean, median and mode (to the nearest tenth of a day for the the grouped data. (When calculating the mean, assume '30 and over' to be 30–34.)
    (d) Compare the results for ungrouped and grouped data for each of the measures of central tendency.

20. Below are the home loan approval amounts (in thousands of dollars) for the first week in January 1998 from a local branch of the Verbundes Bank:

    (a) Calculate the mean, median and mode for the ungrouped data.
    (b) Construct a frequency distribution with intervals 60–69, 70–79, 80–89, 90–99, 100–109, 110–119, 120–129, 130–139, and 140 and over.
    (c) Calculate the mean (assume that '140 and over' is 140–149), median and mode for the grouped data (to the nearest $100).
    (d) Which of the measures appears the most useful in this case? Why?

92	65	120	85	75	68	75
195	110	78	88	100	130	94
92	104	95	105	103	82	95
98	95	110	118	97	107	97
101	84	80	102	109	99	96

Notes

1. Jane Freeman and Geoff Thompson, 'Smokers at work under increasing pressure to quit' *Sydney Morning Herald*, 18 September 1995, p. 5.
2. Ibid.

Chapter 10 – Measurement of central tendency

APPENDIX 10A
Geometric mean

The **geometric mean** is not widely used, as its application is normally restricted to averaging ratios, rates and geometric progressions. The geometric mean is defined as the nth root of the product of n values. The formula is:

$$\text{Geometric mean } (Gm) = \sqrt[n]{\text{Product of } n \text{ numbers}}$$

For example, the CPI results for the years 1992/93 to 1995/96 were 1.0%, 1.8%, 3.3% and 4.2% respectively.

$$(Gm) = \sqrt[4]{1.0 \times 1.8 \times 3.3 \times 4.2}$$

If a calculator with the roots function key is not available, you may use logarithms:

$$\log G_m = \frac{\log 1.0 + \log 1.8 + \log 3.3 + \log 4.2}{4}$$

$$= \frac{0.0000 + 0.2553 + 0.5185 + 0.6232}{4}$$

$$= \frac{1.397}{4}$$

$$= 0.3492$$

Through interpolation, 0.3492 corresponds to 0.22346. As the characteristic is 0 0.3492 translates to 2.23. Therefore, according to the geometric mean, the inflation rate per annum over the period from 1992/93 to 1995/96 was 2.23%.

What happens if there has been a negative rate of change (e.g. profitability, growth in workforce, real wages, etc.)? How is the geometric mean calculated in such instances? Table 10A.1 lists the percentage change in the size of the employed Australian male labour force between 1991 and 1996.

Table 10A.1 *Percentage change in the size of the employed Australian male labour force, 1991–96*

30 June	Percentage change
1991	−1.6
1992	−2.6
1993	−0.5
1994	1.7
1995	3.5
1996	2.0

Source: Australian Bureau of Statistics, *Australian Economic Indicators* (Cat. No. 1350.0).

Business mathematics and statistics

To determine the geometric mean when confronted with negative rates, each rate of change must be expressed in its relationship to 100%. For example, –1.6% indicates that in 1991 the number of males employed was 98.4% of what it was in 1990; –2.6% signifies that in 1992 the employed male labour force was 97.4% of what it was in 1991 and so on. Unless you convert negative percentage changes to the all-up percentage result (e.g. –1.6% becomes 98.4% and –2.6% becomes 97.4%, etc.) the geometric mean cannot be calculated, as a square root of a negative number cannot be determined. Therefore:

$$Gm = \sqrt[n]{\text{Product of } n \text{ numbers}}$$

$$= \sqrt[6]{-1.6 \times -2.6 \times -0.5 \times 1.7 \times 3.5 \times 2.0}$$

The percentage change for each year must be expressed in terms of 100%. Hence:

$$Gm = \sqrt[6]{98.4 \times 97.4 \times 99.5 \times 101.7 \times 103.5 \times 102.0}$$

$$\log Gm = \frac{\log 98.4 + \log 97.4 + \log 99.5 + \log 101.7 + \log 103.5 + \log 102.0}{6}$$

$$= \frac{1.9930 + 1.9886 + 1.9978 + 2.0073 + 2.0149 + 2.0086}{6}$$

$$= \frac{12.0102}{6}$$

$$= 2.0017$$

The antilog of 2.0017 is 100.3922. By subtracting 100 from 100.3922, we find that there has been an increase in the size of the employed male workforce of only 0.39% per year from 1991 to 1996.

Chapter 11
DISPERSION

LEARNING OUTCOMES

After studying this chapter you should be able to:
- Explain the concept and application of dispersion in the workplace
- Calculate dispersion for both ungrouped and grouped data, using range
- Calculate dispersion for both ungrouped and grouped data, using quartile deviation
- Calculate dispersion for both ungrouped and grouped data, using mean deviation
- Calculate dispersion for both ungrouped and grouped data, using standard deviation
- Calculate and analyse the coefficient of variation.

Have you ever watched an American basketball game on television? You might wonder whether this is really the land of the jolly green giant, with centres averaging over the 7-foot (210-centimetre) mark, forwards over 6 feet 7 inches (200 centimetres) and guards over 6 feet 4 inches (190 centimetres). As you would have noticed, there are players above and below the average, with Tyrone Bogues at 5 feet 3 inches (160 centimetres) and Gheorghe Muresan at 7 feet 7 inches (230 centimetres) at the extremes. Similar observations would be made if you watch AFC and NFL American football. You must have seen a Superbowl in your time. It is called Superbowl because you need one to accommodate the players, who average over 125 kilograms in weight. However, here again the quarterbacks and running backs may weigh in as light as 80 kilograms.

Each of these examples implies a deficiency in merely using an average figure. The deficiency is certainly not in vitamins or minerals (at least in the above examples). Instead, the mean average, being affected by extremes, may not provide a typical or representative value for a series of data. However, the good news is that there are ways of determining the spread of results from the mean, which tells you something about the reliability of the mean as an average. The bad news is that the rest of this chapter is all you ever wanted to know about dispersion but were afraid to ask.

First, what is meant by the term 'dispersion'? From our examples, you can see that we are interested in how much individual players (or observations) vary from the average for all players (observations). In other words, **dispersion** measures the variation between individual units and the centre point or average. Our interest, however, is not restricted to the sports world. What about the occupancy rates of hotels or aircraft? The average occupancy of a hotel may be 65%, which may be considered adequate, but because of the variability of reservations (a response to seasonal factors), special rates must be provided during, for example, the low and shoulder seasons.

There has been much talk over recent years of Telstra considering charging on a time basis for all local calls: the user-pays principle. The rationale behind Telstra's thinking is that many of us are on the phone for fairly short periods, while others seem unable to say the word 'goodbye'. As a result, the current charges per call incorporate both sets of callers. Since the length of each call varies greatly from the mean average, charging on a time basis would reduce the costs for some people and force others to consider Australia Post.

It is also important to know the degree of dispersion when, for example, examining the average weekly earnings of males and females, as a low dispersion indicates that wage earners are earning fairly equal wages (since the variation amongst them is small), while a high dispersion indicates the contrary. If you were to evaluate your employees' wages versus the average weekly earnings for other workers in the same industry, the dispersion would indicate how well off or badly off your employees are compared to all the workers in that industry. The degree of dispersion also indicates the degree of reliability of the measure of central tendency—a high dispersion suggests that the 'average' is not truly typical and, as such, cannot be depended on to repeat itself.

Finally, let's try to determine which of two individuals—Ian and Geoff—should be selected for the 4×100 metre relay team from their times in the 100-metre dash. Let's say that Ian's times are 10.7, 10.7, 10.8, 10.75, 10.65 and 10.6 seconds, and that Geoff's times are 11.0, 10.9, 10.4, 10.45, 10.95 and 10.5 seconds. The mean time of both runners is 10.7 seconds.

If you examine the individual results you will find a higher dispersion (in this case, variation in times) for Geoff than for Ian. In fact, none of Geoff's individual times matched his mean average. Therefore, if we needed a runner who consistently ran 100 metres in 10.7 seconds, we would choose Ian.

In summary, being able to assess the spread of observations from the mean gives us a better idea of how well the average conforms to the individual units; in other words, is the average a reliable indicator? For example, the average return for superannuation funds investment over the 1988–92 period would not be representative of the individual years because of the boom-and-bust cycle that occurred over that period.

Another benefit of measuring dispersion is that one can find out its level (or variability) and then attempt to influence or control it. Cinema attendance, although promising on an average weekly basis, has a high level of dispersion. Attendance at screenings early in the week, particularly Mondays and Tuesdays, is poor. In an attempt to influence the nature of cinema viewing (and, consequently, the dispersion level), Mondays and Tuesdays are special low-price viewing days in many cities. The same principle may be applied to other industries such as telecommunications and the airlines.

Lastly, we can use the measure of dispersion for comparison purposes—for example, in the performance of shares on the stockmarket. If you are interested in steadily performing stock, the blue-chip stocks such as BHP or the National Bank may be recommended. However, if you are out for a killing rather than steady growth, you are looking for what may be considered a high-risk, speculative stock.

There are several measures of dispersion, and this chapter considers some of the more commonly used ones.

11.1 Range

The **range** is the simplest and easiest measure of dispersion. It involves finding the difference between the highest and lowest values in the collected data. For example, from 10 September 1995 to 10 September 1996, the share price of Coca-Cola Amatil ranged from $9.39 to $17.02. Or, the annual returns for Australian shares for the calendar years 1991 to 1995 ranged from 45.36% to −8.67% (a range of 54.03%), while bonds returned from 24.42% to −6.7% (a range of 31.12%) and listed property 30.12% to −5.5% (a range of 35.62%).[1] Therefore, bonds had the lowest level of variation or dispersion amongst the three investment options. Finally, the average payroll tax rate on a $10 million payroll ranges from 5% (Queensland) to 7% (Tasmania)—a 2% range and a difference in payroll tax of $200 000.

Returning to the state of the economy, we find that Australia's annual inflation rate for the period 1987/88 to 1995/96 (as measured by the CPI) registered:

> 7.3, 7.3, 8.0, 5.3, 1.9, 1.0, 1.8, 3.3, 4.2

The highest and lowest CPI figures were 8.0 and 1.0 respectively; this gives us a range of 7.0. The range is obviously easy to calculate and understand, but because it looks only at the extremes and ignores all other values, it may give a completely distorted view of the degree of dispersion.

For grouped data, the range is the difference between the real upper limit of the highest class and the real lower limit of the lowest class. For example, in Table 10.23 in Chapter 10, the upper boundary of the highest class is 99.50, whilst the lowest is 29.50, giving a range of 70.00. Temperature fluctuations, blood pressure, daily sharemarket prices and statistical quality control charts are amongst the varying disciplines where the range is commonly used.

The range has the following characteristics:

1. It is the simplest and most easily understood measure of dispersion.
2. It is easy to calculate.
3. It cannot be determined for distributions with open-ended intervals (i.e. sometimes found in grouped data such as 'below 10' or '65 and above').

4. As its value depends on only two values, the highest and lowest, the range will be of limited use if the data contain extreme values. Furthermore, the range reveals nothing about the variation of values between these two extremes.
5. As the range examines only two values, ignoring all other values, it is a fairly rough estimate of dispersion (unless all values are highly concentrated around the measure of central tendency).

11.1 Competency check

1. The range, although easy to calculate, is not considered a precise measure of dispersion. Why?

2. A travel agency offers package tours at the following prices:

    200, 4000, 1100, 900, 500, 800, 1200, 1300

    (a) What is the range of prices for these tours?
    (b) Is the range appropriate as a measure of dispersion in this case? Why?

3. The following are passenger waiting times (to the nearest minute) at a baggage collection point during a peak period:

    11, 7, 19, 17, 22, 5, 10, 40, 5, 11, 14, 13, 12, 6, 9, 15, 4

    (a) Calculate the mean waiting time.
    (b) Determine the range and briefly analyse its appropriateness here.

4. Table 11.1 lists births in Australia for 1987–95.

    (a) Calculate the range of the data.
    (b) In Competency check 10.1, question 4(a) (see Chapter 10), you were asked to calculate the mean births over the same period. Do you think the range is an appropriate measure of dispersion around this mean? If so, why?

Table 11.1 Births in Australia, 1987–95

Year	Births ('000)
1987	244
1988	246
1989	250
1990	258
1991	261
1992	256
1993	266
1994	259
1995	258

Source: Australian Bureau of Statistics, *Australian Demographic Statistics* (Cat. No. 3101.0).

5. Table 11.2 lists the number of individuals employed by the Australian government from 1989 to 1994.

   (a) Calculate the range of the data.
   (b) Is the range a satisfactory measure of dispersion in this case? If so, why?

Table 11.2 *Australian government employees, 1989–94*

Year	Government employees ('000)
1989	1732
1990	1755
1991	1726
1992	1683
1993	1654
1994	1588

Source: Australian Bureau of Statistics, *Year Book Australia*, 1990, 1991, 1992, 1993, 1994, 1995 and 1996 (Cat. No. 1301.0).

11.2 Quartile deviation

To avoid the problem faced by using the range—that is, extreme values upsetting the reliability of the range—the **quartile deviation** of the values can be used. The quartile deviation indicates the degree of variation among the central 50% of values by examining two values—the first and third quartiles of the data. It is calculated as follows:

1. Arrange the data into an array in ascending order.
2. Divide the data into four equal parts.
3. Determine the first quartile (Q_1) and the third quartile (Q_3)—that is, the value below which 25% of the items occur, and the value above which 25% of the items occur.
4. Subtract the first quartile value from the third quartile value; this is referred to as the *interquartile range*.
5. Divide the interquartile range by 2. The result is the quartile deviation. That is:

$$QD = \frac{Q_3 - Q_1}{2}$$

Ungrouped data

The rentals for two-bedroom units in a particular suburb are as follows:

160, 140, 175, 182, 170, 150, 165, 120, 220, 185,
175, 225, 166, 160, 130, 135, 170, 160, 190, 180

You have just bought a two-bedroom unit as an investment and want to rent it out, but you are uncertain what to charge. By calculating the quartile deviation, you will be able to determine the range of the middle 50% of rents.

Business mathematics and statistics

To find Q_1, Q_3 and QD, first arrange the data into an array. That is:

120, 130, 135, 140, 150, 160, 160, 160, 165, 166,
170, 170, 175, 175, 180, 182, 185, 190, 220, 225

To determine Q_1 and Q_3, divide the data into four quarters of five values each. As Q_1 is the value below which 25% of the values occur, the value of Q_1 must be above 150 and below 160; the value of Q_3 must be above 180 but below 182. That is:

$$Q_1 = \frac{150 + 160}{2} \quad \text{and} \quad Q_3 = \frac{180 + 182}{2}$$
$$= 155 \qquad\qquad\qquad = 181$$

Therefore, the interquartile range is 26 (i.e. 181 − 155) and the quartile deviation is:

$$QD = \frac{Q_3 - Q_1}{2}$$
$$= \frac{181 - 155}{2}$$
$$= \frac{26}{2}$$
$$= 13$$

Thus, two-bedroom units with a median weekly rent of $168 (the median is the middle value and, with 20 units surveyed, the middle value is the average rent of the 10th and 11th two-bedroom unit; i.e. the average of $166 and $170 is $168) have a quartile deviation of $13. This means that the middle 50% of rents are ± $13 from the median of $168—that is, the range is $155 to $181. If you feel your investment is better than average but not in the top bracket, then perhaps rent in the region of $181 a week may be in order. If, on the other hand, you could afford only a mediocre unit, then closer to $155 may be in order.

Problem: Below are the CPI results for 1984/85 to 1995/96. Calculate the interquartile range and quartile deviation, and comment on the outcome.

4.3, 8.4, 9.3, 7.3, 7.3, 8.0, 5.3, 1.9, 1.0, 1.8, 3.3, 4.2

Solution: First arrange the data into an array:

1.0, 1.8, 1.9,	3.3, 4.2, 4.3,	5.3, 7.3, 7.3,	8.0, 8.4, 9.3
1st quartile	2nd quartile	3rd quartile	4th quartile

Next, to determine Q_1 and Q_3, divide the data into four equal groups of three CPIs each (since 12 years are listed). As Q_1 is the value below which one quarter of the values lie, the value of Q_1 must be above 1.9 but below 3.3, while the value of Q_3 must be above 7.3 but below 8.0. Thus:

$$Q_1 = \frac{1.9 + 3.3}{2} \qquad \text{and} \qquad Q_3 = \frac{7.3 + 8.0}{2}$$
$$= \frac{5.2}{2} \qquad\qquad\qquad\qquad = \frac{15.3}{2}$$
$$= 2.6 \qquad\qquad\qquad\qquad\quad = 7.65$$

The interquartile range is 5.05 (i.e. 7.65 − 2.6).
The quartile deviation is:

$$QD = \frac{Q_3 - Q_1}{2}$$
$$= \frac{7.65 - 2.6}{2}$$
$$= \frac{5.05}{2}$$
$$= 2.525$$

The median of the CPI for this period is 4.8%. With a quartile deviation of 2.525% from the median, this indicates that the middle 50% of CPI results were between 2.275% and 7.325%.

Grouped data

The method for calculating quartiles for grouped data is basically the same as that used to determine the median for grouped data. However, instead of finding the middle value, the first and third quartile values are calculated. The formulae are:

$$Q_1 = L + \frac{\left(\frac{n}{4} - C\right)}{f}(i) \qquad \text{and} \qquad Q_3 = L + \frac{\left(\frac{3n}{4} - C\right)}{f}(i)$$

where L = real lower limit of the quartile class (i.e. Q_1 or Q_3)
n = number of frequencies in the data
C = cumulative frequency of the class preceding the quartile class
f = frequency of the appropriate quartile class
i = size of the class interval in the appropriate quartile class.

Problem: Determine the quartile deviation of the data in Table 11.3.

Table 11.3 *Weekly wages of accountants*

Weekly wage ($)	Number of accountants (f)	Cumulative frequency (C)
300 and under 350	20	20
350 and under 400	41	61
400 and under 450	84	145
450 and under 500	120	265
500 and under 550	140	405

Business mathematics and statistics

Table 11.3 *(continued)*

Weekly wage ($)	Number of accountants (f)	Cumulative frequency (C)
550 and under 600	36	441
600 and under 650	30	471
650 and under 700	26	497
700 and under 750	22	519
750 and under 800	18	537
800 and under 850	13	550
	550	

Source: Hypothetical.

Solution: To solve for Q_1 (i.e. the income of the 137.5th accountant ($\frac{1}{4}$ of 550)):

$$Q_1 = L + \frac{\left(\frac{n}{4} - C\right)}{f} \quad (i)$$

$$= 400 + \frac{\left(\frac{550}{4} - 61\right)}{84}(50)$$

$$= 400 + \frac{(137.5 - 61)}{84}(50)$$

$$= 400 + \frac{76.5}{84}(50)$$

$$= 400 + 45.54$$

$$= 445.54$$

To solve for Q_3 (i.e. the income of the 412.5th accountant ($\frac{3}{4}$ of 550)):

$$Q_3 = L + \frac{\left(\frac{3n}{4} - C\right)}{f} \quad (i)$$

$$= 550 + \frac{\left(\frac{3 \times 550}{4} - 405\right)}{36}(50)$$

$$= 550 + \frac{(412.5 - 405)}{36}(50)$$

$$= 550 + \frac{7.5}{36}(50)$$

$$= 550 + 10.42$$

$$= 560.42$$

Therefore, to calculate the quartile deviation:

$$QD = \frac{Q_3 - Q_1}{2}$$
$$= \frac{560.42 - 445.54}{2}$$
$$= \frac{114.88}{2}$$
$$= 57.44$$

Therefore, with a median of $503.57 and a quartile deviation of $57.44, the middle range of accountants earn $503.57 ± $57.44, or between approximately $446 and $561. The problem is, of course, that we are unaware of how the wages in the interquartile range are distributed.

Problem: Table 11.4 lists the life of car tyres in kilometres travelled. Find the quartile deviation of the life of a car tyre—that is, the range (in kilometres) of the middle 50% of car tyres.

Table 11.4 *Life of car tyres (kilometres travelled) according to road service survey*

Kilometres travelled	Number of tyres (f)	Cumulative frequency (CF)
0 and under 4 000	1	1
4 000 and under 8 000	3	4
8 000 and under 12 000	8	12
12 000 and under 16 000	14	26
16 000 and under 20 000	20	46
20 000 and under 24 000	15	61
24 000 and under 28 000	10	71
28 000 and under 32 000	6	77
32 000 and under 36 000	2	79
36 000 and under 40 000	1	80

Source: Hypothetical.

Solution: To solve for Q_1:

$$Q_1 = L + \frac{\left(\frac{n}{4} - C\right)}{f} \quad (i)$$

$$= 12\,000 + \frac{\left(\frac{80}{4} - 12\right)}{14}(4000)$$

$$= 12\,000 + \frac{8}{14}(4000)$$

$$= 12\ 000 + 2285.71$$
$$= 14\ 285.71$$
$$= 14\ 286$$

To solve for Q_3:

$$Q_3 = L + \frac{\left(\frac{3n}{4} - C\right)}{f} \quad (i)$$

$$= 20\ 000 + \frac{\left(\frac{3 \times 80}{4} - 46\right)}{15}(4000)$$

$$= 20\ 000 + \frac{(60 - 46)}{15}(4000)$$

$$= 20\ 000 + \frac{14}{15}(4000)$$

$$= 20\ 000 + 3733$$

$$= 23\ 733$$

To calculate the quartile deviation:

$$QD = \frac{Q_3 - Q_1}{2}$$

$$= \frac{23\ 733 - 14\ 286}{2}$$

$$= \frac{9447}{2}$$

$$= 4723.5$$

$$= 4724 \text{ (rounding off)}$$

Therefore, since the median was calculated to be 18 800 kilometres, a quartile deviation of 4724 kilometres indicates that 50% of tyres have a life of $18\ 800 \pm 4724$, or between 23 524 and 14 076 kilometres.

Finally, the Real Estate Institute of Australia publishes on a quarterly basis quartile house prices for all Australian cities. For example, Melbourne's $Q_3 = \$220\ 000$ and $Q_1 = \$115\ 000$ in June 1996.[2] Therefore, the quartile deviation would be:

$$QD = \frac{Q_3 - Q_1}{2}$$

$$= \frac{\$220\ 000 - \$115\ 000}{2}$$

$$= \frac{\$105\ 000}{2}$$

$$= \$52\ 500$$

Chapter 11 – Dispersion

Characteristics of the quartile deviation

1. The quartile deviation is based on two values, Q_1 and Q_3, with 50% of the values contained between Q_1 and Q_3. The QD indicates the degree of variation among the central 50% of values from the mean or median. Obviously, the smaller the quartile deviation, the greater the concentration of the middle half of the values in the distribution.
2. It may be calculated for a frequency distribution with an open-ended interval.
3. Although it is a better measure of dispersion than the range, it is still a fairly rough estimate as it ignores 50% of the values (the lower and upper 25% of the values).
4. It fails to indicate the degree of clustering around the measure of central tendency.

The major weakness of the range and the quartile deviation (i.e. their values being dependent solely on the position of certain items within the set of data) is overcome by the use of the mean deviation and the standard deviation, both of which examine and include all items in their calculations.

11.2 Competency check

1. What does the quartile deviation measure?

2. Joe Dinero's paypacket for the last eight weeks contained the following wages:

    $485, $592, $634, $620, $546, $540, $520, $550

    Determine both the range and the quartile deviation. Briefly comment on the result.

3. The sales results of Gilbert Inc. for this year are indicated as follows. The company, although recognising the seasonal nature of its business, believes the volatility in results is a matter of concern.

$1 250 000	$ 975 000	$1 500 000	$1 950 000
$ 860 000	$1 750 000	$1 100 000	$ 675 000
$ 990 000	$1 400 000	$1 950 000	$2 790 000

    (a) Calculate:

    (i) the median;
    (ii) Q_1 and Q_3;
    (iii) the interquartile range.

    (b) Briefly evaluate the results. Should it be a matter of concern?

4. Calculate the quartile deviation of the students' grades listed in Table 11.5 (to one decimal place) and relate the result to the median.

Business mathematics and statistics

Table 11.5 *Students' examination grades*

Grade	Number of students
30–39	6
40–49	20
50–59	64
60–69	87
70–79	51
80–89	19
90–99	9

Source: Hypothetical.

5. (a) Calculate (to the nearest dollar), according to Table 11.6:

   (i) the interquartile range;
   (ii) the quartile deviation of the weekly rent.

   (b) Relate your result to the median (calculated for Competency check 10.2, problem 5(c)—see Chapter 10).

Table 11.6 *Weekly rent for all households in Westbridge*

Weekly rent ($)	Number of households	*f cumulative*
80–99	75	75
100–119	82	157
120–139	90	247
140–159	99	346
160–179	79	425
180–199	64	489
200–219	41	530
220–239	20	550

Source: Hypothetical.

11.3 Mean deviation

We have found that both the range and quartile deviations ignore the dispersion among certain sections of the data and that, ideally, we should take account of each and every value in the data, because we are interested in discovering the average spread from the mean. To calculate the deviation from the mean, we simply subtract each data point from the mean. To overcome the problem of positive and negative deviations from the mean equalling each other, hence cancelling out, the positive and negative signs of these deviations are ignored. Only the absolute values of the deviations are used in determining the mean deviation. We then take the average of these absolute deviations and derive the **mean (average) deviation**.

Chapter 11 – Dispersion

Ungrouped data

To calculate the mean deviation for ungrouped data:

1. Calculate the arithmetic mean ($\bar{x}$).
2. Find the deviation of each value from the arithmetic mean ($x - \bar{x}$).
3. Add the absolute values of the deviations ($\Sigma|x - \bar{x}|$).
4. Divide the sum in step 3 by the number of values in the data (n).

Therefore, the formula for calculating the mean deviation is:

$$MD = \frac{\Sigma|x - \bar{x}|}{n}$$

For example, let's find the mean deviation of Ian and Geoff's running times in the 100-metre dash. Ian's times are 10.7, 10.7, 10.8, 10.75, 10.65 and 10.6 seconds. Geoff's times are 11.0, 10.9, 10.4, 10.45, 10.95 and 10.5 seconds. Table 11.7 lists the mean deviation of Ian's running times.

Table 11.7 *Mean deviation of running times for Ian*

Running time (x) (seconds)	Deviation ($x - \bar{x}$, where $\bar{x} = 10.7$)	Absolute value of deviation ($|x - \bar{x}|$)
10.7	0	0
10.7	0	0
10.8	+0.1	0.1
10.75	+0.05	0.05
10.65	−0.05	0.05
10.6	−0.1	0.1
64.2	0	0.3

Source: Hypothetical.

Therefore, to calculate the mean deviation, first determine the mean. That is:

$$\bar{x} = \frac{\Sigma x}{n}$$
$$= \frac{64.2}{6}$$
$$= 10.7$$

Then find the deviation of each value in the data from the arithmetic mean ($x - \bar{x}$). The absolute values of the deviations are then added together ($\Sigma|x - \bar{x}|$) and divided by the number of values in the data. That is:

$$MD = \frac{\Sigma|x - \bar{x}|}{n}$$
$$= \frac{0.3}{6}$$
$$= 0.05$$

Thus, the mean deviation from the arithmetic mean is 0.05 of a second (half of one-tenth of a second). This indicates that Ian would be reliable if a runner able to come in with a time of 10.75 or less was needed (or if someone capable of running consistently between 10.65 and 10.75 was needed).

Table 11.8 lists the mean deviation of Geoff's running times.

Table 11.8 *Mean deviation of running times for Geoff*

Running time (x) (seconds)	Deviation $(x - \bar{x}$, where $\bar{x} = 10.7)$	Absolute value of deviation $(\lvert x - \bar{x} \rvert)$
11.0	+0.3	0.3
10.9	+0.2	0.2
10.4	−0.3	0.3
10.45	−0.25	0.25
10.95	+0.25	0.25
10.5	−0.2	0.2
64.2	0	1.5

Source: Hypothetical.

Therefore, to calculate the mean deviation:

$$MD = \frac{\Sigma |x - \bar{x}|}{n}$$
$$= \frac{1.5}{6}$$
$$= 0.25$$

Geoff's mean deviation of 0.25 indicates that he can run 100 metres between a range of 10.45 and 10.95. Thus, if we needed a runner capable of running times under 10.65, Geoff would be our man, although he is also capable of exceeding Ian's times.

Grouped data

When one is presented with data in the form of a frequency distribution, the mean deviation can be determined as follows:

1. Calculate the arithmetic mean ($\bar{x}$).
2. Find the deviation (absolute value) of the midpoints of each class from the arithmetic mean ($|x - \bar{x}|$).
3. Multiply the absolute deviation in each class by the corresponding frequency of that class ($f|x - \bar{x}|$).
4. Find the sum of the absolute deviations ($\Sigma f|x - \bar{x}|$).
5. Divide this sum by the total frequency of the data (*n*).

Therefore, the formula for calculating the mean deviation for grouped data is:

$$MD = \frac{\Sigma f|x - \bar{x}|}{n}$$

Chapter 11 – Dispersion

Problem: Find the mean deviation of the wages listed in Table 11.9.

Table 11.9 *Weekly wages of accountants*

Weekly wage ($)	Number of accountants
300 and under 350	20
350 and under 400	41
400 and under 450	84
450 and under 500	120
500 and under 550	140
550 and under 600	36
600 and under 650	30
650 and under 700	26
700 and under 750	22
750 and under 800	18
800 and under 850	13

Source: Hypothetical.

Solution: To determine the mean deviation:

1. Calculate the arithmetic mean:

$$\bar{x} = \frac{\Sigma fx}{n}$$

$$= \frac{285\ 700}{550}$$

$$= 519.45$$

$$= \$519 \text{ (rounding off)}$$

2. From the arithmetic mean of $519 (see Table 11.10), find the absolute value of the deviation of each class midpoint (column 5) and multiply that deviation by its corresponding frequency (column 2 × column 5 = column 6). For example, looking at the '300 and under 350' class, the deviation of the mean of $519 from the midpoint of that class ($325) is –$194 (i.e. $325 – $519 = –$194). As mean deviations examine only the absolute deviations, the negative sign is dropped; hence, the deviation for that class is $194. Examining the '650 and under 700' class, the deviation of the mean of $519 from the midpoint of that class ($675) is $156 (i.e. $675 – $519 = $156). After carrying out similar calculations for all classes, multiply the absolute deviation by its respective class frequency and then find the sum of the deviations ($46 170).

Business mathematics and statistics

Table 11.10 *Weekly wages of accountants*

Weekly wage ($)	Number of accountants (f)	Midpoint (x)	fx	Absolute deviation ($\|x - \bar{x}\|$)	($f\|x - \bar{x}\|$)
300 and under 350	20	325	6 500	194	3 880
350 and under 400	41	375	15 375	144	5 904
400 and under 450	84	425	35 700	94	7 896
450 and under 500	120	475	57 000	44	5 280
500 and under 550	140	525	73 500	6	840
550 and under 600	36	575	20 700	56	2 016
600 and under 650	30	625	18 750	106	3 180
650 and under 700	26	675	17 550	156	4 056
700 and under 750	22	725	15 950	206	4 532
750 and under 800	18	775	13 950	256	4 608
800 and under 850	13	825	10 725	306	3 978
	550		285 700		46 170

Source: Hypothetical.

3. Divide $46 170 by the total frequency of 550 to determine the mean deviation. Thus:

$$MD = \frac{\Sigma f|x - \bar{x}|}{n}$$

$$= \frac{\$46\ 170}{550}$$

$$= \$83.95$$

In other words, the mean weekly wage of accountants is $519 but, because of the spread of wages among the 550 accountants surveyed, the mean average variation or spread from the mean is approximately $84.

Problem: Determine the mean deviation of the students' grades listed in Table 11.11.

Table 11.11 *Students' examination grades*

Grade (%)	Number of students
30–39	6
40–49	20
50–59	64
60–69	87
70–79	51
80–89	19
90–99	9

Source: Hypothetical.

Solution:

Table 11.12 *Students' examination grades*

Grade (%)	Number of students (f)	Midpoint (x)	fx	Absolute deviation ($\lvert x - \bar{x}\rvert$)	$f\lvert x - \bar{x}\rvert$
30–39	6	34.5	207	29.5	177
40–49	20	44.5	890	19.5	390
50–59	64	54.5	3 488	9.5	608
60–69	87	64.5	5 611.5	0.5	43.5
70–79	51	74.5	3 799.5	10.5	535.5
80–89	19	84.5	1 605.5	20.5	389.5
90–99	9	94.5	850.5	30.5	274.5
	256		16 452.0		2 418

Source: Hypothetical.

1. Calculate the arithmetic mean:

$$\bar{x} = \frac{\Sigma fx}{n}$$

$$= \frac{16\,452}{256}$$

$$= 64.27, \text{ say } 64\%$$

2. Find the deviation per class by subtracting the arithmetic mean from the midpoint of each class. For example, for the class '60–69', subtract the arithmetic mean of 64 from the midpoint for that particular class (in this case, 64.5), which results in a deviation of +0.5.

3. Multiply the deviations per class by the corresponding frequency to arrive at $f\lvert x - \bar{x}\rvert$.

4. Find the sum of $f\lvert x - \bar{x}\rvert$. That is, $\Sigma f\lvert x - \bar{x}\rvert = 2418.0$ (see Table 11.12).

5. Divide this sum by the total number of students (256). Therefore:

$$MD = \frac{\Sigma f\lvert x - \bar{x}\rvert}{n}$$

$$= \frac{2418}{256}$$

$$= 9.45$$

The mean deviation of the students' grades is 9.45 from the mean average of 64%. This indicates that, on average, students' grades varied approximately ±9.5% from the mean grade of 64.

Business mathematics and statistics

In review, the mean deviation examines the value of every item in the data and is calculated from the arithmetic mean or median. It measures the dispersion around the calculated measure of central tendency rather than the dispersion within the confines of two values, which is measured by the range and quartile deviation. The mean deviation—that is, the average of the absolute deviations of the individual items around the mean—therefore takes all points in the data into account and shows the average spread of the data observations about the mean. As simple and easy as it is to calculate, the mean deviation is not often used because further statistical measures cannot be derived from it since it ignores algebraic signs.

11.3 Competency check

1. How is the mean deviation calculated for ungrouped data?
2. Calculate the mean deviation and briefly explain the outcome's significance for the ungrouped data in Table 11.13 (to the nearest 1000 births).

Table 11.13 *Births in Australia, 1987–95*

Year	Births ('000)
1987	244
1988	246
1989	250
1990	258
1991	261
1992	256
1993	266
1994	259
1995	258

Source: Australian Bureau of Statistics, *Australian Demographic Statistics* (Cat. No. 3101.0).

3. Calculate the mean deviation and briefly explain the significance of these results for the ungrouped data in Table 11.14 (to the nearest 1000 employees).
4. Calculate the mean deviation for the grouped data in Table 11.15 (to the nearest 100).
5. Calculate the mean deviation for the grouped data in Table 11.16 (to the nearest tenth of a cent).

(*Hint:* The mean for each of these was calculated in Chapter 10, Competency check 10.1, Problems 6(a)–(b).

Table 11.14 Australian government employees, 1989–94

Year	Government employees ('000)
1989	1732
1990	1755
1991	1726
1992	1683
1993	1654
1994	1588

Source: Australian Bureau of Statistics, *Year Book Australia*, 1990, 1991, 1992, 1993, 1994, 1995 and 1996 (Cat. No. 1301.0).

Table 11.15 Salaries of executives in C-U-Later Pty Ltd

Salary ($)	Number of executives
30 000–39 000	8
40 000–49 000	11
50 000–59 000	22
60 000–69 000	24
70 000–79 000	20
80 000–89 000	9
90 000–99 000	6

Source: Hypothetical.

Table 11.16 Price of petrol, Sydney, January 1998

Price per litre (cents)	Number of service stations in surveyed areas
60.0–62.0	22
63.0–65.0	25
66.0–68.0	49
69.0–71.0	64
72.0–74.0	58
75.0–77.0	27
78.0–80.0	11

Source: Hypothetical.

11.4 Standard deviation

The primary measure of dispersion is the **standard deviation** which, like the mean deviation, is based on and is representative of the individual data items' deviation from the mean of those values. It is seen basically as a refinement of the mean deviation, in that positive and negative signs are not ignored. As expected, the greater the spread of values about the mean, the greater the standard deviation.

In Chapter 14 you will see that when the frequency distribution for many data is plotted in the form of a histogram, it follows a bell-shaped, symmetrical 'normal' distribution. The normal distribution allows us to use the standard deviation in conjunction with the mean to indicate the proportion of the observations in a distribution that fall within specified distances from the mean.

In the following examples we will deal with the standard deviation for a population, and the appropriate formula will be provided and explained. However, if you want to determine the standard deviation for a *sample* of the population of observations, the following formula should be applied for grouped data:

$$s = \sqrt{\frac{\Sigma f(m - \bar{x})^2}{n - 1}}$$

where s = standard deviation of the sample
  $\bar{x}$ = arithmetic mean for the sample
  $n - 1$ = number of observations in the sample minus 1
  m = midpoint of each class
  f = frequency of each class
  Σ = summation.

Ungrouped data

To determine the *population* standard deviation (σ) for ungrouped data:

1. Find the arithmetic mean of the data (μ).
2. Find the deviation of each item from the mean ($x - \mu$).
3. Square each deviation (($x - \mu)^2$).
4. Find the sum of the squared deviations ($\Sigma(x - \mu)$).
5. Find the mean of these squared deviations (referred to as the *variance*) by dividing the sum (i.e. $\Sigma(x - \mu)^2$) by the number of values (N).
6. Find the square root of the mean.

Therefore, the formula for calculating the population standard deviation for ungrouped data is:

$$\sigma = \sqrt{\frac{\Sigma(x - \mu)^2}{N}}$$

Problem: Calculate the standard deviation of the following house values: 64, 66, 68, 72, 74 and 76 ('000s).

Table 11.17 *Value of houses*

Value of houses ($'000)	Deviation $(x - \mu)$	Deviation squared $(x - \mu)^2$
64	−6	36
66	−4	16
68	−2	4
72	2	4
74	4	16
76	6	36
420	0	112
$\mu = 70$		

Source: Hypothetical.

Solution:

1. Calculate the arithmetic mean (70).
2. Subtract the actual house values from the calculated mean of 70 (e.g. 64 − 70 = −6, 66 − 70 = −4, and so on); see Table 11.17.
3. Square each of the deviations.
4. Find the sum of these deviations (112).
5. Divide the sum of the squared deviations of 112 by the number of houses examined (6) to determine the variance (18.67).
6. Determine the square root of the variance—that is, the standard deviation. That is:

$$\sigma = \sqrt{\frac{\Sigma(x-\mu)^2}{N}}$$

$$= \sqrt{\frac{112}{6}}$$

$$= \sqrt{18.67}$$

$$= 4.32$$

As the mean is 70, a standard deviation of 4.32 means that four of the six houses, or approximately 67% of the houses, fall within ±1 standard deviation (i.e. between 65.68 and 74.32). Examination of two standard deviations (4.32 × 2 = 8.64) indicates that 100% of the houses fall within ±2 standard deviations (see Table 11.17).

Problem: Ian's running times for the 100-metre dash are 10.7, 10.7, 10.8, 10.75, 10.65 and 10.6 seconds. Geoff's running times for the 100-metre dash are 11.0, 10.9, 10.4, 10.45, 10.95 and 10.5 seconds. Determine the standard

Business mathematics and statistics

deviation of the running times for both Ian and Geoff. What do the respective standard deviations indicate?

Solution:

Table 11.18 *Standard deviations for running times of Ian and Geoff*

	Ian			Geoff	
Times	$(x - \mu)$	$(x - \mu)^2$	Times	$(x - \mu)$	$(x - \mu)^2$
10.7	0	0	11.0	+0.3	0.09
10.7	0	0	10.9	+0.2	0.04
10.8	+0.1	0.01	10.4	−0.3	0.09
10.75	+0.05	0.0025	10.45	−0.25	0.0625
10.65	−0.05	0.0025	10.95	+0.25	0.0625
10.6	−0.1	0.01	10.5	−0.2	0.04
64.2	0	0.0250	64.2	0	0.3850

Source: Hypothetical.

The mean running time for both Ian and Geoff is 10.7 seconds. To calculate the standard deviation of Ian's running times:

$$\sigma = \sqrt{\frac{\Sigma(x - \mu)^2}{N}}$$

$$= \sqrt{\frac{0.0250}{6}}$$

$$= \sqrt{0.00417}$$

$$= 0.065$$

To calculate the standard deviation of Geoff's running times:

$$\sigma = \sqrt{\frac{\Sigma(x - \mu)^2}{N}}$$

$$= \sqrt{\frac{0.385}{6}}$$

$$= \sqrt{0.06417}$$

$$= 0.253$$

Ian's standard deviation of 0.065 seconds indicates that 67% of his running times occurred between ±1 standard deviation of the mean of 10.7 (i.e. 10.635 to 10.765) (more about this in Chapter 14). Geoff's standard deviation of 0.253 seconds indicates that 67% of his running times were in the range of 10.446 to 10.954 (±1 standard deviation). Therefore, as confirmed previously, Ian is the

Chapter 11 – Dispersion

more consistent runner (i.e. his individual times for the 100-metre dash vary less from the mean of his times than do Geoff's times).

Shorthand method

Since the mean is algebraically defined with the deviations from the mean equal to zero, we can once again use an assumed mean, which will be particularly helpful when applied to grouped data. First, however, let's look at its application to ungrouped data.

Shorthand method—ungrouped data

The standard deviation of ungrouped data can be calculated by using an assumed mean. The shorthand method formula is:

$$\sigma = \sqrt{\frac{\Sigma d^2}{N} - \left(\frac{\Sigma d}{N}\right)^2}$$

where σ = standard deviation of a population
Σ = summation
d = deviation of each value from the assumed mean
N = number of values.

The correction factor $(\Sigma d/N)^2$ is always negative because the sum of the assumed mean deviations squared will always exceed the sum of the true mean deviations squared, and we must therefore always subtract the correction factor.

Problem: Calculate the standard deviation of the following house values by the shorthand method: 64, 66, 68, 72, 74 and 76 ($'000s).

Solution: To calculate the standard deviation for the house values by the shorthand method, we must first find the values for the variables in the above formula. These are shown in Table 11.19.

Table 11.19 *House values*

Value of house ($000)	Deviation from assumed mean of 68 (d)	Deviation squared (d^2)
64	−4	16
66	−2	4
68	0	0
72	4	16
74	6	36
76	8	64
420	+12	136

Source: Hypothetical.

Substituting into the formula:

$$\sigma = \sqrt{\frac{\Sigma d^2}{N} - \left(\frac{\Sigma d}{N}\right)^2}$$

$$= \sqrt{\frac{136}{6} - \left(\frac{12}{6}\right)^2}$$

$$= \sqrt{\frac{136}{6} - \frac{144}{36}}$$

$$= \sqrt{22.67 - 4}$$

$$= \sqrt{18.67}$$

$$= 4.32$$

This is, of course, the same answer as derived earlier.

Grouped data

The procedure for calculating the standard deviation for a population of grouped data is similar to that used for ungrouped data except that the frequency of the classes is included in the formula.

To determine the standard deviation (σ) for grouped data:

1. Find the arithmetic mean of the data (μ).
2. Find the deviation of the midpoint of each class from the mean ($m - \mu$).
3. Square each deviation (i.e. $(m - \mu)^2$).
4. Multiply the squared deviation for each class by the frequency of each class (i.e. $f(m - \mu)^2$).
5. Find the sum of all the $f(m - \mu)^2$ (i.e. $\Sigma f(m - \mu)^2$).
6. Calculate the mean of the sum of the deviations squared by dividing the sum by the number of values (N)—this is referred to as the variance.
7. Find the square root of the mean of the sum of the standard deviations (i.e. the square root of the variance).

Therefore, the formula for calculating the standard deviation for grouped data of a population is:

$$\sigma = \sqrt{\frac{\Sigma f(m - \mu)^2}{N}}$$

Problem: Calculate the standard deviation for the weekly wages of accountants listed in Table 11.20.

Chapter 11 – Dispersion

Table 11.20 *Weekly wages of accountants*

Weekly wage ($)	Number of accountants
300 and under 350	20
350 and under 400	41
400 and under 450	84
450 and under 500	120
500 and under 550	140
550 and under 600	36
600 and under 650	30
650 and under 700	26
700 and under 750	22
750 and under 800	18
800 and under 850	13

Source: Hypothetical.

Solution: The mean was previously calculated to be approximately $519. To calculate the standard deviation, construct a table as shown in Table 11.21.

Table 11.21 *Weekly wages of accountants*

Weekly wage ($)	Number of accountants (f)	Midpoint (m)	$(m - \mu)$	$(m - \mu)^2$	$f(m - \mu)^2$
300 and under 350	20	325	−194	37 636	752 720
350 and under 400	41	375	−144	20 736	850 176
400 and under 450	84	425	−94	8 836	742 224
450 and under 500	120	475	−44	1 936	232 320
500 and under 550	140	525	6	36	5 040
550 and under 600	36	575	56	3 136	112 896
600 and under 650	30	625	106	11 236	337 080
650 and under 700	26	675	156	24 336	632 736
700 and under 750	22	725	206	42 436	933 592
750 and under 800	18	775	256	65 536	1 179 648
800 and under 850	13	825	306	93 636	1 217 268
	550				6 995 700

Source: Hypothetical.

Therefore:

$$\sigma = \sqrt{\frac{\Sigma f(m-\mu)^2}{N}}$$

$$= \sqrt{\frac{6\,995\,700}{550}}$$

$$= \sqrt{12\,719.45}$$

$$= 112.78$$

The standard deviation is approximately 113.

Shorthand method—grouped data

The shorthand method formula for calculating the standard deviation of a population for grouped data is:

$$\sigma = (i)\sqrt{\frac{\Sigma fd^2}{N} - \left(\frac{\Sigma fd}{N}\right)^2}$$

where σ = standard deviation of the population
 Σ = summation
 d = deviation of the midpoint of the class interval from an arbitrary origin in terms of class intervals
 f = frequency of the class interval
 N = total frequency of the data
 i = class interval size.

Problem: Calculate the standard deviation of the accountants' wages by using the shorthand method.

Solution:

Table 11.22 *Weekly wages of accountants*

Weekly wage ($)	Number of accountants (f)	Deviation (d)	fd	fd²
300 and under 350	20	−4	−80	320
350 and under 400	41	−3	−123	369
400 and under 450	84	−2	−168	336
450 and under 500	120	−1	−120	120
500 and under 550	140	0	0	0
550 and under 600	36	+1	+36	36
600 and under 650	30	+2	+60	120
650 and under 700	26	+3	+78	234

Table 11.22 *(continued)*

Weekly wage ($)	Number of accountants (f)	Deviation (d)	fd	fd²
700 and under 750	22	+4	+88	352
750 and under 800	18	+5	+90	450
800 and under 850	13	+6	+78	468
	550		−61	2 805

Source: Hypothetical.

Substituting into the formula:

$$\sigma = (i)\sqrt{\frac{\Sigma fd^2}{N} - \left(\frac{\Sigma fd}{N}\right)^2}$$

$$= (50)\sqrt{\frac{2805}{550} - \left(\frac{-61}{550}\right)^2}$$

$$= (50)\sqrt{5.1 - \left(\frac{3721}{302\,500}\right)}$$

$$= (50)\sqrt{5.1 - 0.0123}$$

$$= (50)\sqrt{5.0877}$$

$$= (50)(2.2556)$$

$$= 112.78$$

Notice that not only is the number of calculations reduced dramatically, but the degree of difficulty of each has also been substantially reduced.

Problem: Determine the standard deviation of the students' grades listed in Table 11.23. The arithmetic mean is 64.

Table 11.23 *Students' examination grades*

Grade (%)	Number of students
30–39	6
40–49	20
50–59	64
60–69	87
70–79	51
80–89	19
90–99	9

Source: Hypothetical.

Business mathematics and statistics

Solution:

Table 11.24 *Students' examination grades*

Grade (%)	Number of students (f)	Midpoint (m)	$(m - \mu)$	$(m - \mu)^2$	$f(m - \mu)^2$
30–39	6	34.5	−29.5	870.25	5 221.5
40–49	20	44.5	−19.5	380.25	7 605.0
50–59	64	54.5	−9.5	90.25	5 776.0
60–69	87	64.5	0.5	0.25	21.75
70–79	51	74.5	10.5	110.25	5 622.75
80–89	19	84.5	20.5	420.25	7 984.75
90–99	9	94.5	30.5	930.25	8 372.25
	256				40 604.00
	$\mu = 64$				

Source: Hypothetical.

Therefore:

$$\sigma = \sqrt{\frac{\Sigma f(m - \mu)^2}{N}}$$

$$= \sqrt{\frac{40\ 604}{256}}$$

$$= \sqrt{158.61}$$

$$= 12.59$$

To minimise the chance of mathematical error, use the shorthand method as shown in Table 11.25 to determine the standard deviation of students' grades.

Table 11.25 *Students' examination grades*

Grade (%)	Number of students (f)	Deviation (d)	fd	fd²
30–39	6	−3	−18	54
40–49	20	−2	−40	80
50–59	64	−1	−64	64
60–69	87	0	0	0
70–79	51	1	51	51
80–89	19	2	38	76
90–99	9	3	27	81
	256		−6	406

Source: Hypothetical.

Chapter 11 – Dispersion

$$\sigma = (i)\sqrt{\frac{\Sigma fd^2}{N} - \left(\frac{\Sigma fd}{N}\right)^2}$$

$$= (10)\sqrt{\frac{406}{256} - \left(\frac{-6}{256}\right)^2}$$

$$= (10)\sqrt{1.5859 - \left(\frac{36}{65\ 536}\right)}$$

$$= (10)\sqrt{1.5859 - 0.00055}$$

$$= (10)\sqrt{1.58535}$$

$$= (10)1.25911$$

$$= 12.59$$

It was obviously far easier to calculate it this way than by the longhand method.

From the fact that many data, including student examination grades, approximate a normal distribution, we can conclude that about 68% (the middle two-thirds of the distribution) of the grades can be found between ±1 standard deviation (σ) (i.e. 12.59— say, 13) from the mean average of 64%. Therefore, approximately 68% of students received grades of between 51% and 77%. About 94% of students received grades of between mean ± 2σ or 38% and 90%. To confirm this, if you examine Table 11.25 and calculate the middle 68% of results (by calculating the grade for the bottom 16% and top 16% of students), you will find it between approximately 52% and 77%.

You will find in the next chapter, as already implied and briefly explained, that the beauty of a distribution approximating 'normal' is that the standard deviation can be used with the mean to indicate the proportion of observations in a distribution that fall within specified distances from the mean.

Problem: Determine the standard deviation for the life of car tyres (see Table 11.26).

(a) Use the longhand method.
(b) Use the shorthand method.
(c) Interpret the results.

Business mathematics and statistics

Table 11.26 *Life of car tyres (kilometres travelled) according to road service survey*

Kilometres travelled	Number of tyres
0 and under 4 000	1
4 000 and under 8 000	3
8 000 and under 12 000	8
12 000 and under 16 000	14
16 000 and under 20 000	20
20 000 and under 24 000	15
24 000 and under 28 000	10
28 000 and under 32 000	6
32 000 and under 36 000	2
36 000 and under 40 000	1

Source: Hypothetical.

Solution:

(a) See Table 11.27, remembering that the mean was calculated to be 19 150 kilometres. Therefore, the deviations of column 4 were derived by subtracting each respective class midpoint (column 3) from 19 150.
(b) See Table 11.28.

Table 11.27 *Life of car tyres (kilometres travelled) according to road service survey*

Kilometres travelled	Number of tyres (f)	Midpoint (m)	$(m - \mu)$	$(m - \mu)^2$	$f(m - \mu)^2$
0 and under 4 000	1	2 000	−17 150	294 122 500	294 122 500
4 000 and under 8 000	3	6 000	−13 150	172 922 500	518 767 500
8 000 and under 12 000	8	10 000	− 9 150	83 722 500	669 780 000
12 000 and under 16 000	14	14 000	− 5 150	26 522 500	371 315 000
16 000 and under 20 000	20	18 000	− 1 150	1 322 500	26 450 000
20 000 and under 24 000	15	22 000	2 850	8 122 500	121 837 500
24 000 and under 28 000	10	26 000	6 850	46 922 500	469 225 000
28 000 and under 32 000	6	30 000	10 850	117 722 500	706 335 000
32 000 and under 36 000	2	34 000	14 850	220 522 500	441 045 000
36 000 and under 40 000	1	38 000	18 850	355 322 500	355 322 500
	80				3 974 200 000

Source: Hypothetical.

$$\sigma = \sqrt{\frac{\Sigma f(m-\mu)^2}{N}}$$

$$= \sqrt{\frac{3\,974\,200\,000}{80}}$$

$$= \sqrt{49\,677\,500}$$

$$= 7048.23$$

Table 11.28 The life of car tyres (kilometres travelled) according to road service survey

Kilometres travelled	Number of tyres (f)	Deviation (d)	fd	fd²
0 and under 4 000	1	−4	−4	16
4 000 and under 8 000	3	−3	−9	27
8 000 and under 12 000	8	−2	−16	32
12 000 and under 16 000	14	−1	−14	14
16 000 and under 20 000	20	0	0	0
20 000 and under 24 000	15	1	15	15
24 000 and under 28 000	10	2	20	40
28 000 and under 32 000	6	3	18	54
32 000 and under 36 000	2	4	8	32
36 000 and under 40 000	1	5	5	25
	80		23	255

Source: Hypothetical.

(b)
$$\sigma = (i)\sqrt{\frac{\Sigma fd^2}{N} - \left(\frac{\Sigma fd}{N}\right)^2}$$

$$= (4000)\sqrt{\frac{255}{80} - \left(\frac{23}{80}\right)^2}$$

$$= (4000)\sqrt{3.1875 - (0.2875)^2}$$

$$= 4000\sqrt{3.1875 - 0.082656}$$

$$= 4000\sqrt{3.10484375}$$

$$= 4000(1.762056682)$$

$$= 7048.23$$

(c) Tyres had a mean life of 19 150 kilometres and with a standard deviation of approximately 7050 kilometres. This indicates that 68% of tyres would have a life of between 12 100 and 26 200 kilometres (mean ± 1s of 19 150); and 95% of between 5050 and 33 250 kilometres (mean ± 2s of 19 150).

Characteristics of the standard deviation

1. The standard deviation is based on every value in the data. The larger the spread about the mean, the larger the standard deviation.
2. By first squaring the deviations and, subsequently, finding the square root, the standard deviation is mathematically correct and is therefore the most widely used form of dispersion. It is particularly useful when used in conjunction with the normal probability curve.
3. If the frequency distribution contains an open-ended interval, the standard deviation cannot be calculated.
4. The standard deviation of a population for grouped data can be calculated using the formula:

$$\sigma = \sqrt{\frac{\Sigma f(m - \mu)^2}{N}}$$

while the standard deviation for a sample is determined by using the formula:

$$s = \sqrt{\frac{\Sigma f(m - \bar{x})^2}{n - 1}}$$

11.4 Competency check

1. Why is the standard deviation considered superior to the mean deviation as a measure of dispersion?
2. Use the formula for the standard deviation of a population to calculate the standard deviation for the ungrouped data in Table 11.29 (to the nearest 100).

Table 11.29 Births in Australia, 1987–95

Year	Births ('000)
1987	244
1988	246
1989	250
1990	258
1991	261
1992	256
1993	266
1994	259
1995	258

Source: Australian Bureau of Statistics, *Australian Demographic Statistics* (Cat. No. 3101.0).

Chapter 11 – Dispersion

3. Use the formula for the standard deviation of a population to calculate the standard deviation for the ungrouped data in Table 11.30 (to the nearest 1000th employee).

Table 11.30 *Australian government employees, 1989–94*

Year	Government employees ('000)
1989	1732
1990	1755
1991	1726
1992	1683
1993	1654
1994	1588

Source: Australian Bureau of Statistics, *Year Book Australia*, 1990, 1991, 1992, 1993, 1994, 1995 and 1996 (Cat. No. 1301.0).

4. Using whatever method you prefer, calculate the standard deviation for the grouped data in Table 11.31 (to the nearest $100).

Table 11.31 *Salaries of executives in C-U-Later Pty Ltd*

Salary ($)	Number of executives
30 000–39 000	8
40 000–49 000	11
50 000–59 000	22
60 000–69 000	24
70 000–79 000	20
80 000–89 000	9
90 000–99 000	6

Source: Hypothetical.

5. Using whatever method you prefer, calculate the standard deviation (of a population) for the grouped data in Table 11.32 (to the nearest half cent).

Business mathematics and statistics

Table 11.32 *Price of petrol, Sydney, January 1998*

Price per litre (cents)	Number of service stations in surveyed areas
60.0 to 62.0	22
63.0 to 65.0	25
66.0 to 68.0	49
69.0 to 71.0	64
72.0 to 74.0	58
75.0 to 77.0	27
78.0 to 80.0	11

Source: Hypothetical.

11.5 Coefficient of variation

The measures of dispersion discussed so far have all been absolute—that is, the dispersion is expressed in the same unit as the data from which it has been extracted (e.g. standard deviation of the students' grades is in percentages, while the accountants' standard deviation of wages is in dollar amounts). If you wanted to compare results which used different units of measurement or used the same units but vastly different sample sizes, then the measure of dispersion by itself would be inadequate. To overcome this problem, the measure of dispersion (e.g. standard deviation of the data) must be related to its corresponding measure of central tendency (e.g. the mean)—that is, the degree of relative dispersion must be compared. That relationship or measure is then expressed as a percentage. This measure is referred to as **Pearson's coefficient of variation** and, in formula form, is written as:

$$V = \frac{s}{\bar{x}} \times 100 \text{ for a sample and } V = \frac{\sigma}{\mu} \times 100 \text{ for a population}$$

where V = coefficient of variation
  s = standard deviation of the sample
  $\bar{x}$ = sample mean
  σ = standard deviation of the population
  μ = population mean.

For example, if you compare the accountants' mean income of approximately $519 and the standard deviation of approximately $113 with those of clerks, whose mean income is $245 with a standard deviation of $74, you may conclude that the accountants have the higher degree of dispersion (higher standard deviation). This is not correct, however, as you must relate the standard deviation to the mean from which it has been derived. Therefore, the coefficient of variation for the accountants is:

Chapter 11 – Dispersion

$$V = \frac{\sigma}{\mu} \times 100$$

$$= \frac{113}{519} \times 100$$

$$= 21.77\%$$

The coefficient of variation for the clerks is:

$$V = \frac{\sigma}{\mu} \times 100$$

$$= \frac{74}{245} \times 100$$

$$= 30.2\%$$

Therefore, the relative dispersion of the clerks' wages is larger than that of the accountants' wages, in spite of the fact that the accountants' absolute dispersion was much larger (53% larger) than the wage dispersion for clerks.

Problem: Grunnge Jeans, which has outlets Australia-wide, found that during the 1996/97 financial year, monthly sales averaged $3 675 000 with a standard deviation of $547 000 in the Queensland sample survey compared with monthly sales of $1 168 000 and a standard deviation of $195 000 in South Australia. Which of the two States is the more consistent performer (i.e. illustrates less volatility in sales)?

Solution: The absolute variation was, obviously, greatest in Queensland, but then again, its sales exceeded that of South Australia. To determine the relative variability, we must determine the degree of variation for each State and then compare the results. Therefore:

Queensland:

$$V = \frac{s}{\bar{x}} \times 100$$

$$= \frac{547\ 000}{3\ 675\ 000} \times 100$$

$$= 14.88\%$$

South Australia:

$$V = \frac{s}{\bar{x}} \times 100$$

$$= \frac{195\ 000}{1\ 168\ 000} \times 100$$

$$= 16.70\%$$

Queensland's relative variation in monthly sales is less than South Australia's in spite of Queensland's standard deviation being almost three times as much as South Australia's. In other words, the result suggests that when comparing the standard deviation to the mean, South Australia's monthly sales were more variable than Queensland's.

Problem: The results of a hypothetical survey examining IQs and income were published, with participants having a mean IQ of 116 and a standard deviation of 12, while their mean income was $29 500 with a standard deviation of $5400. Which of the two measures illustrates the greater degree of variation?

Business mathematics and statistics

Solution: To determine which of these two measures (which may be considered samples) exhibits the greater degree of variability, we once again turn to our friendly coefficient of variation and discover:

IQ:

$$V = \frac{s}{\bar{x}} \times 100$$

$$= \frac{12}{116} \times 100$$

$$= 0.1034 \times 100$$

$$= 10.34\%$$

Income:

$$V = \frac{s}{\bar{x}} \times 100$$

$$= \frac{5400}{29\,500} \times 100$$

$$= 0.1831 \times 100$$

$$= 18.31\%$$

Therefore, incomes illustrate a greater degree of variation than do IQs.

11.5 Competency check

1. When is it advisable to use the coefficient of variation?
2. In 1992/93 a particular share had a mean price of 60 cents with a standard deviation of 21 cents, while in 1996/97 the mean price was $2.50 with a standard deviation of 75 cents. What was the coefficient of variation for this share during the two years, and in which year was it most volatile?
3. The mean work week in a Collingwood factory is 42 hours with a standard deviation of six hours. The mean work week in a Hawthorne factory is 36 hours with a standard deviation of five hours. What are the respective coefficients of variation (to one decimal place) indicating which factory has the smaller relative dispersion?
4. A firm has a factory in Maitland and another in Ipswich. The former produces an average of 23 500 units per month with a standard deviation of 1500 units, while the latter produces 38 000 units per month with a standard deviation of 2250 units. What are the respective coefficients of variation (to one decimal place)? Which factory has the smaller relative dispersion? Which is therefore the more consistent performer?
5. Using the data provided in problems 4 and 5 in Competency check 11.4, calculate the coefficient of variation for each of the two problems.

11.6 Summary

Examination of the central tendency (average) alone does not indicate how representative the calculated average is when compared with the actual data. The use of dispersion permits evaluation of the degree to which the calculated measure of central tendency approximates the original values. The analysis of dispersion suggests the degree of reliability of the measure of central tendency; a high dispersion implies that the 'average' is not truly typical and cannot therefore be depended on to repeat itself.

There are various methods available to measure dispersion, including the range, quartile deviation, mean deviation, standard deviation and the coefficient of variation. The range and quartile deviation may be considered positional measures, as their determination depends on the position of specific observations within a frequency distribution. The mean deviation and standard deviation include all observations in their calculations, thus measuring dispersion around the average. To measure the degree of relative dispersion, most useful when comparing two or more distributions, the coefficient of variation is used.

Range

The range is the simplest and easiest measure of dispersion. It involves subtracting the lowest value from the highest value of the collected data. Although quick and easy to determine, as its value is based on only two values, any extremes in the data prevent the range from being considered as a proper measure of dispersion. In addition, as all but two values are ignored, the range can be thought of only as a fairly rough estimate of the degree of dispersion.

Quartile deviation

Since the range is affected by extremes, the quartile deviation, which examines the values of the first and third quartiles of the data, the middle 50% of values, is sometimes used. To calculate the quartile deviation for ungrouped data:

1. Arrange the data into an array in ascending order.
2. Divide the array into four equal parts.
3. Determine the first and third quartiles.
4. Subtract the first quartile value from the third quartile value, the result being known as the interquartile range.
5. Divide the interquartile range by 2, resulting in the interquartile deviation *(QD)*. That is:

$$QD = \frac{Q_3 - Q_1}{2}$$

To solve for Q_1 and Q_3 for grouped data, use the same method as that used for determining the median value. That is:

$$Q_1 = L + \frac{\left(\frac{n}{4} - C\right)}{f} \quad (i)$$

$$Q_3 = L + \frac{\left(\frac{3n}{4} - C\right)}{f} \quad (i)$$

Then substitute the calculated Q_1 and Q_3 into:

$$QD = \frac{Q_3 - Q_1}{2}$$

The determination of QD indicates the degree of variation among the central 50% of values. Therefore, although it is an improved measure of dispersion on the range (which observed only two values), the quartile deviation ignores 50% of the values (the lower and upper 25%).

Mean deviation

After the mean of individual values in a series of data has been calculated, the mean deviation can be calculated and used as an indicator of dispersion. To determine the mean deviation for ungrouped data:

1. Calculate the arithmetic mean ($\bar{x}$).
2. Find the deviation by subtracting the arithmetic mean for each value in the data $(x - \bar{x})$.
3. Add the absolute values of the deviations ($\Sigma|x - \bar{x}|$).
4. Divide the sum by the number of values in the data. That is:

$$MD = \frac{\Sigma|x - \bar{x}|}{n}$$

To determine the mean deviation for grouped data:

1. Calculate the arithmetic mean ($\bar{x}$).
2. Find the deviation (absolute value) of the midpoints of each class from the arithmetic mean $(x - \bar{x})$.
3. Multiply the absolute deviation in each class by the corresponding frequency for that class ($f|x - \bar{x}|$).
4. Find the sum of the absolute deviations ($\Sigma f|x - \bar{x}|$).
5. Divide this sum by the total frequency of the data (n). That is:

$$MD = \frac{\Sigma f|x - \bar{x}|}{n}$$

The mean deviation examines the values of every observation in the data and is normally calculated from the arithmetic mean. It is a measure of how the amount of dispersion values typically vary from the mean or median. As it measures the dispersion of values around the calculated measure of central tendency, it is preferred to the range or quartile deviation which measure dispersion within the confines of two values. However, as all deviations are shown to be absolute (i.e. negative signs become positive), the mean deviation lacks mathematical soundness and its further application is therefore restricted.

Standard deviation

The standard deviation is the most widely used measure of dispersion, measuring the variation of values around the mean, but not ignoring signs. The standard deviation for a population is denoted σ (the Greek letter sigma), and the standard deviation for a sample is denoted s.

To determine the population from standard deviation (σ) for ungrouped data:

1. Find the arithmetic mean (μ).
2. Find the deviation of each item from the mean ($x - \mu$).
3. Square each deviation (($x - \mu)^2$).
4. Find the sum of the squared deviations ($\Sigma (x - \mu)^2$).
5. Find the mean of these squared deviations (referred to as the variance) by dividing the sum (i.e. $\Sigma (x - \mu)^2$)) by the number of values (N).
6. Find the square root of the mean. That is:

$$\sigma = \sqrt{\frac{\Sigma(x - \mu)^2}{N}}$$

To determine the *population* standard deviation (σ) for grouped data:

1. Find the arithmetic mean (μ).
2. Find the deviation of the midpoint of each class from the mean ($m - \mu$).
3. Square each deviation (($m - \mu)^2$).
4. Multiply the squared deviation for each class by the frequency of each class ($f(m - \mu)^2$).
5. Find the sum of $f(m - \mu)^2$ (i.e. $\Sigma f(m - \mu)^2$).
6. Calculate the mean of the sum of the deviations squared (i.e. the variance) by dividing the sum by the number of values (N).
7. Find the square root of the mean of the sum of the squared deviations. That is:

$$\sigma = \sqrt{\frac{\Sigma f(m - \mu)^2}{N}}$$

The standard deviation is based on every value in the data and, as signs are not ignored, it is mathematically sound. It is therefore used extensively in further statistical analysis, particularly in conjunction with the normal probability curve.

Note: To calculate the standard deviation of a *sample* for grouped data, use the formula:

$$s = \sqrt{\frac{\Sigma f(m - \bar{x})^2}{n - 1}}$$

There may be times when the dispersion of different units of measurement or markedly different sample sizes of the same measure need to be compared. The measure of dispersion—that is, the standard deviation—should then be related to its corresponding measure of central tendency, thus allowing the degree of relative dispersion of these different units of measurement to be compared. This measure is known as the *Pearson's coefficient of variation* and in formula form is written as:

$$V = \frac{s}{\bar{x}} \times 100 \text{ for a sample and } V = \frac{\sigma}{\mu} \times 100 \text{ for a population}$$

Business mathematics and statistics

Note: It is important to have an idea of how to work out sample statistics or population parameters, but most people let the calculator do the work.

11.7 Multiple-choice questions

1. The following is total spending by the government and private sector on education for the years 1989/90 to 1994/95 in billions of dollars.[3]

   - 1989/90: $17.4
   - 1990/91: $19.1
   - 1991/92: $21.1
   - 1992/93: $22.8
   - 1993/94: $24.1
   - 1994/95: $24.9

   The range of spending on education in Australia over the six years was:

   (a) $7.5 billion   (b) $6.7 billion
   (c) $5.8 billion   (d) $1.7 billion

2. The mean deviation of Australia's spending on education over the period described in question 1 is:

   (a) $2.8 billion   (b) $2.4 billion
   (c) $2.1 billion   (d) $3.2 billion

3. The standard deviation of education expenditure in question 1 was:

   (a) $3.2 billion   (b) $2.7 billion
   (c) $2.1 billion   (d) $3.5 billion

4. The first and third quartile house prices for Sydney in June 1996 were $168 000 and $326 000 respectively.[4] What was the quartile deviation?

   (a) $158 000   (b) $79 000
   (c) $39 500   (d) $494 000

5. Of the methods available to measure dispersion, which one is algebraically defined and can therefore be used in further statistical analysis?

   (a) range   (b) interquartile deviation
   (c) mean deviation   (d) standard deviation

11.8 Exercises

1. Why is the standard deviation, rather than the mean deviation, the preferred measure of dispersion for statisticians?

2. If you had to choose the most consistent performer of three basketball players with the following averages and standard deviations, which would be your pick?

Mike: 16 points with standard deviation of 4 points
Charles: 19 points with standard deviation of 5 points
Patrick: 23 points with standard deviation of 5.3 points

For exercises 3 and 4, determine:

(a) the range;
(b) the interquartile range and quartile deviation;
(c) the mean deviation (to the nearest whole number);
(d) the standard deviation (to the nearest whole number).
(e) Briefly explain which measure is most appropriate and why.

3. The record for absenteeism for a firm over 12 consecutive months was as follows (you can consider this to be a population):

   43   41   40   41   42   88   45   51   46   42   40   38

4. A spot check of the number of students attending a lecture was made with the following results (note that a spot check implies sample):

   78   87   94   92   79   78   67   90   82   87
   82   88   62   97   80   82   62   79   90   86

5. The net migration of people from New South Wales moving to other Australian States during the period 1989/90 to 1994/95 is as follows (rounded off to the nearest 100):[5]

   36 000   17 200   15 200   19 100   13 500   15 000

   Calculate to the nearest 100:

   (a) the range;
   (b) the standard deviation;
   (c) the coefficient of variation.
   (d) Briefly evaluate the coefficient of variation results.

6. The *Australian Financial Review* normally asks a number of leading economists for their annual forecasts on a variety of economic indicators, including the Australian dollar. Assuming that the following 10 forecasts were made (in relation to the US dollar), calculate (to the nearest cent) the mean, standard deviation and coefficient of variation (to one decimal point).

   A$1 = US$0.78   US$0.81   US$0.70   US$0.75   US$0.71
         US$0.68   US$0.71   US$0.75   US$0.76   US$0.74

7. The following data indicate the number of days lost over a one-year period for 30 workers:

   8   2   3   17   15   11
   6   20   11   42   13   5
   7   14   19   4   12   18
   22   1   28   10   7   16
   13   16   9   16   4   14

   Determine the standard deviation.

8. Table 11.33 contains the monthly number of hospital beds used in Westbridge Hospital during 1997.

Table 11.33 *Monthly use of hospital beds in Westbridge Hospital, 1997*

Jan.	Feb.	Mar.	Apr.	May	June	July	Aug.	Sept.	Oct.	Nov.	Dec.
980	650	760	805	700	925	950	840	825	790	810	995

Source: Hypothetical.

Calculate the following measures (to the nearest whole number):

(a) the mean and median;
(b) the range and quartile deviation.

9. From the data provided in exercise 8, calculate:

(a) the mean deviation;
(b) the standard deviation.

10. Two students were comparing their grades to determine which was the more consistent. Albert averaged 84% on exams, with a standard deviation of 18%, while Charles had a 66% average and a 10% standard deviation. Which student was the more consistent?

11. You are considering investing in one of four companies, whose results are listed in Table 11.34. Which company offers the highest degree of risk? The lowest?

Table 11.34 *Average investment returns over the past three years*

Company	Investment return (%)	Standard deviation (%)
Trust N'us	32	16
C-Saw PL	26	12
I-B-4-U PL	18	7
R-U-Nuts PL	9	5

Source: Hypothetical.

12. Calculate the standard deviation of the ages of participants in a recent survey on eating habits.

Chapter 11 – Dispersion

Table 11.35 *Ages of participants in a recent survey on eating habits*

Ages	Number of members
15–19	7
20–24	7
25–29	9
30–34	7
35–39	4
40–44	4
45–49	1
50–54	1

Source: Hypothetical.

13. Table 11.36 lists the weekly rent for all households in Westbridge.

    (a) Calculate the standard deviation (to the nearest dollar) and coefficient of variation (to the nearest per cent).
    (b) Briefly comment on the results.

Table 11.36 *Weekly rent for all households in Westbridge*

Weekly rent ($)	Number of households
80–99	75
100–119	82
120–139	90
140–159	99
160–179	79
180–199	64
200–219	41
220–239	20

Source: Hypothetical.

14. Rents for two-bedroom units in the western suburbs are listed in Table 11.37.

    (a) Determine the standard deviation (to the nearest 10 cents).
    (b) Determine the coefficient of variation (to the nearest per cent).
    (c) Briefly evaluate the results.

Table 11.37 Rents for two-bedroom units in the western suburbs, December 1997

Rent ($)	Number of two-bedroom units
40 and under 50	12
50 and under 60	14
60 and under 70	28
70 and under 80	46
80 and under 90	50
90 and under 100	25
100 and under 110	15
110 and under 120	10

Source: Hypothetical.

15. (a) The manager of Los Locos, a local Mexican restaurant, collected the information shown in Table 11.38 from yesterday's takings. Calculate (to the nearest dollar):

    (i) the mean and median;
    (ii) the mean deviation;
    (iii) the standard deviation;
    (iv) the coefficient of variation (to the nearest per cent).

    (b) Briefly discuss the results.

Table 11.38 Los Locos takings for yesterday

Final bill ($)	Number of orders
10–19	12
20–29	16
30–39	23
40–49	43
50–59	21
60–69	15
70–79	10

Source: Hypothetical.

16. A company dealing in electrical repairs to household appliances listed the time spent per repair over a one-month period. The results are shown in Table 11.39.

    (a) Calculate the quartile deviation (to the nearest minute) and briefly explain the results.
    (b) Calculate the standard deviation.
    (c) Determine the coefficient of variation and briefly discuss the result.

Chapter 11 – Dispersion

Table 11.39 *Time required to repair appliances*

Time (minutes)	Number of customers
5 and under 15	196
15 and under 25	244
25 and under 35	280
35 and under 45	300
45 and under 55	140
55 and under 65	96
65 and under 75	42
75 and under 85	20
85 and under 95	8

Source: Hypothetical.

17. Table 11.40 gives final exam results.

   (a) What are the mean and median grades (to the nearest per cent)?
   (b) What are the mean deviation and standard deviation (to the nearest per cent)?
   (c) What is the coefficient of variation (to the nearest per cent)?

Table 11.40 *Final exam results*

Grade	Number of students
40 and under 50	5
50 and under 60	11
60 and under 70	22
70 and under 80	6
80 and under 90	2

Source: Hypothetical.

18. The dollar amounts of travellers cheques taken out of the country by 120 Australians travelling overseas in June 1997 are given in Table 11.41.

   (a) Calculate the standard deviation (to the nearest dollar).
   (b) Calculate the coefficient of variation (to the nearest per cent).
   (c) As a sales promotion, a special gift voucher worth $100 was given to all those travellers whose cheque amounts were +2 standard deviations or more above the mean.

      (i) How many travellers received a voucher (to the nearest whole number)?
      (ii) How much did the promotion cost the firm in gift vouchers?

Table 11.41 *Amount of travellers cheques taken out of the country by 120 Australian travelling overseas in 1997*

Amount of travellers cheques ($)	Number of Australians travelling overseas
0 to less than 1 000	1
1 000 to less than 2 000	5
2 000 to less than 3 000	9
3 000 to less than 4 000	16
4 000 to less than 5 000	22
5 000 to less than 6 000	26
6 000 to less than 7 000	19
7 000 to less than 8 000	13
8 000 to less than 9 000	6
9 000 to less than 10 000	2
10 000 to less than 11 000	1

Source: Hypothetical.

19. By the end of its first year of operation, the Up-in-Arms Gymnasium had members of the ages listed in Table 11.42. (*Hint:* Treat as a population.)

   (a) Calculate the standard deviation (to the nearest year).
   (b) Calculate the coefficient of variation (to the nearest per cent).
   (c) The owners of the gym are determined to increase their market penetration to the age-group within ±1.5 standard deviations of the mean. What is the range of ages that the gym owners are targeting?
   (d) Comment on the validity of applying the standard deviation to this distribution.

Table 11.42 *Age of Up-in-Arms Gymnasium members*

Age of members (years)	Number of members
15–19	34
20–24	84
25–29	146
30–34	72
35–39	40
40–44	16
45–49	4
50–54	2
55–59	1
60–64	1

Source: Hypothetical.

20. A multinational computer company has 150 marketing representatives throughout Australia with sales results for 1997 as shown in Table 11.43.

Table 11.43 *Sales results for 1997*

Sales ($'000)	Number of representatives
100 and under 200	2
200 and under 300	9
300 and under 400	17
400 and under 500	25
500 and under 600	26
600 and under 700	29
700 and under 800	24
800 and under 900	10
900 and under 1000	7
1000 and under 1100	1

Source: Hypothetical.

(a) Calculate the arithmetic mean and the standard deviation of the sales per representative for the year (to the nearest $1000).
(b) What is the coefficient of variation (to the nearest per cent)?
(c) As an incentive scheme, all representatives with sales exceeding +2 standard deviations above the mean were given an overseas trip. What was the minimum sales amount necessary to earn the overseas trip (to the nearest $1000)?
(d) Management decided to retrench the bottom 16% of the representatives. What was the minimum amount of sales a representative had to make to avoid retrenchment (to the nearest $1000)?

Notes

1. *Rothschild Report*, No. 31, September 1996.
2. Real Estate Institute, *Market Facts*, June 1996.
3. Australian Bureau of Statistics, *Expenditure on Education, Australia* (Cat. No. 5510.0); *Australian National Accounts* (Cat. No. 5204.0).
4. Real Estate Institute of Australia, *Market Facts*, June 1996.
5. Australian Bureau of Statistics, *Australian Demographic Statistics* (Cat. No. 3101.0).

APPENDIX 11A
Coefficient of skewness

In a symmetrical normal distribution, the mean, median and mode are equal in value. However, when these three measures of central tendency are not equal, we have a distribution that is tapered or, more precisely, skewed. That is, the peak of the normal distribution curve (plotted via the midpoints of the histogram of grouped data) no longer falls in the centre of the distribution, as does a normal distribution, but approximates a positively skewed distribution (see Figure 11A.1) or a negatively skewed distribution (see Figure 11A.2). In other words, the values in the frequency distribution are concentrated in either the low or the high end of the measuring scale on the horizontal axis and, as a result, are not equally distributed.

Figure 11A.1 *Positively skewed distribution*

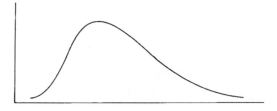

Figure 11A.2 *Negatively skewed distribution*

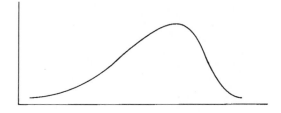

The degree of skewness should be calculated to determine whether a particular frequency distribution can be described by its mean and standard deviation. A large degree of skewness indicates that the distribution does not approach a normal distribution, therefore rendering the mean and standard deviation inappropriate, as these two measures are most useful when describing a symmetrical bell-shaped distribution. When the data are skewed, the mode is approximately three times further away from the mean than is the median. As the mean is affected by extremes and the mode is not, the greater the distance between the mean and the mode, the greater the skewness. The coefficient of skewness can be calculated for a population or sample by applying the following formulae:

$$\text{Skewness} = \frac{3(\text{Mean} - \text{Median})}{\text{Standard deviation}}$$

$$Sk = \frac{3(\bar{x} - Md)}{s} \quad \text{(sample)}$$

$$Sk = \frac{3(\mu - Md)}{\sigma} \quad \text{(population)}$$

For example, the accountants' weekly wages in the XYZ survey revealed central tendencies of $519 for the mean, a median of $504 and a standard deviation of $113. Therefore, the degree of skewness is:

$$\begin{aligned} Sk &= \frac{3(\mu - Md)}{\sigma} \\ &= \frac{3(519 - 504)}{113} \\ &= \frac{45}{113} \\ &= 0.398 \end{aligned}$$

The 0.398 indicates that the distribution is positively skewed, as the skewness is positive and the larger the coefficient, the greater the degree of skewness.

Chapter 12
SAMPLING

LEARNING OUTCOMES
After studying this chapter you should be able to:
- Recognise the proliferation of sample data in the private and public sectors
- Differentiate between a population and a sample
- Understand the basis for using a sample
- Identify and understand the major sampling methods and their respective applications
- Examine common survey techniques and learn how to design questionnaires
- Realise the potential failings of surveys.

Imported cars 'more reliable'[1]

Imported cars are still more reliable than locally made models according to a repeat survey undertaken by *Choice* magazine and confirming the same conclusion reached in 1992, 1994 and 1995. However, the executive director of the Federal Chamber of Automotive Industries claimed the result totally contradicted all other survey results of motoring organisations and the Automotive Industry Authority (the Federal car watchdog). *Choice* magazine's conclusions were based on a reader poll from which useful information on 5765 cars was obtained with varying sample sizes, such as 56 Hondas and as many as 343 Holdens.

As you can appreciate, local manufacturers were not enthused with these results as they vehemently believe that although there has been an improvement in import quality, there have also been major quality gains in local production. Was *Choice*'s appraisal valid? How are surveys and survey respondents selected, and what sampling method(s) can be used? By a strange coincidence, the answers to these questions await your reading.

12.1 Everyday observations of sampling

Sampling is part and parcel of our everyday life. When you least expect it, it's there right before your eyes (or mouth). Have you ever been approached in a supermarket to taste, for example, a new cheese, cracker, spread or bread? What you are asked to do is to take a sample of the product. If you like it, well then, hopefully you will buy it rather than repeatedly returning to the counter for further samples (so that you don't have to buy lunch). The sample therefore represents the quality of the product you would be buying.

In the above example, you have really been exposed to statistical **sampling**. It happens all the time, not just at supermarket counters. Before a new product line is launched nationwide, cities such as Newcastle and Adelaide are frequently nominated as test or sample areas in order to gauge customer response (or lack of). GIO Australia used an innovative method of marketing its range of home loans: in a shopping mall, it linked up data on its loan options with residential real estate listings and access via a touch screen in a kiosk. The experiment lasted six months, after which the results were reviewed by GIO Australia management to determine if it was suitable for wider application.

You cannot possibly test all produced units of televisions, for example, to ensure that the quality of each is up to scratch. Instead, samples are selected which are expected to be representative of the entire batch. For example, the last car you bought was probably bought after you had test-driven it or tested another of the same model.

In Chapter 8 we discussed the 'people meters' used for recording people's television-viewing habits. As you may remember, only 3900 households were selected in Australia out of a population of over 6 350 000 households; in other words, each household represents close to 1630 households. Likewise, radio ratings are extracted from a small sample of the radio-listening population.

In medical science we find many instances of samples being used in the current worldwide attempt to develop a cure for AIDS. Various drugs (such as AZT and ritonavir) were initially tested on a sample of AIDS patients before being approved for use in Australia. Before vaccines such as the polio vaccine (developed in the 1950s), the gamma globulin vaccine against hepatitis and the BCG vaccine against tuberculosis were introduced worldwide, a sample of the population was studied.

In 1996, a comprehensive survey of convention delegates in Sydney found that they spend an average of $4200 during their visit to Australia, more than double the expenditure of ordinary overseas tourists. These findings were then used by the convention industry to back up their requests for infrastructure development and funding.[2]

In mid-1996, a Telstra survey of 1000 households indicated that more than one million Australians had access to the global computer network and were using it far more frequently than experts had assumed. In fact, Telstra's estimate that one million Australians are connected to the Internet is up to three times higher than earlier estimates. The survey also supported the popular view of an overwhelming dominance of male Internet users—80%.[3]

Before any election, we are bombarded with poll results from various polling organisations informing us of the frontrunner, and yet these polls may be based on fewer than 1500 respondents—in a nation of over 13 million eligible voters. AGB

McNair opinion polls give approval ratings of parties and key politicians based on a usual sample size of 2050. The polls are taken nationwide using a telephone survey, with results claimed to have a margin of error of only 2%.[4]

The statistics published by the Australian Bureau of Statistics can be more precisely defined as sample statistics, since most are in fact based on a sample. For example, unemployment statistics are determined on the basis of a telephone survey of one-half of 1% of Australian households, or about 30 000 households each month. It would be impossible, both economically and within the time constraint, to conduct such a survey of all Australian households each month. Other publications, such as those associated with the Consumer Price Index, the average weekly earnings of males and females, the business expectations survey and company profit results, are all based on sample surveys.

From the above, and throughout this chapter, you will see that samples are a matter of everyday life. You have seen that a **sample** is a representative unit of the **population** (as statisticians refer to the larger unit). However, in this instance the term 'population' relates not only to people but also to the collection of all possible observations of any variable, be it cheese, television sets, bank accounts, mice or men. The sample is thus any portion of the population selected for study.

Why then take a sample of a population instead of testing the entire population? Won't the results of a sample differ from the results of the entire population? Certainly they will. However, as long as certain procedures and the correct statistical methods and statistical analyses are followed, a sample's results can closely mirror those of the entire population. The reasons for using a sample rather than testing the entire population therefore include the following:

1. The cost incurred in getting the entire population's reaction to a new toothpaste, for example, would be prohibitive and the reaction impossible to gather. We have seen how the population census, undertaken every five years (because of the cost factor), still falls short of tabulating the entire Australian population.
2. It makes no sense to engage in destructive testing beyond the necessary point. If you tested each motor vehicle to see if it could withstand a crash at 60 kilometres an hour, you would have no undamaged vehicles left to sell. Or closer to home, in the kitchen in fact, if you sampled all the soup, what would be left for the guests? When you think of a chef, often a rather large person comes to mind, one who because of the hazards of the job must ensure that the meals are just right—through sampling.
3. Time is money. Since decisions must often be made fairly quickly, it is not possible to contact an entire population of, for example, customers. Or you may be coming up with a new product line but are fearful that your competitors are doing likewise. A feasibility study on the product's likely level of acceptance would not only be too expensive if the entire population of potential buyers were canvassed, but would also take too long.
4. Sample results can give a very accurate measure of the population because there are methods that can produce highly representative samples of the population.

Therefore, the information generated by the collection of the sample data will be used for statistical analysis. Conclusions will be drawn about the population on the basis of the sample results, this process being referred to as **statistical inference**. We can even measure the degree of accuracy of our estimates by statistical methods (e.g. sampling distribution of the mean).

The distinction between sample and population is further reinforced by the statistical symbols used in describing each. The characteristics of a population are referred to as *parameters*, while measures describing sample items are referred to as **sample statistics**. Therefore, statistics are used to estimate parameters. Greek or capital letters are usually used to denote population parameters and lower-case roman or italic letters to denote sample statistics. Table 12.1 illustrates the difference in symbols used for population and sample.

Table 12.1 *Different symbols used for population and samples*

Population	Sample
Greek letters or capitals	Lower-case and/or italic letters
μ = population mean	$\bar{x}$ = sample mean
N = population size	n = sample size
σ = population standard deviation	s = sample standard deviation

A selected sample of a population must exhibit the characteristics of the chosen population so that there is a high degree of reliability, and hence confidence, when inferences about the population are based on the sample estimates. The larger the sample size in relation to the targeted population, the more precise, and therefore the more reliable, the information will be. However, research has shown that increases in the sample size do not provide a proportional increase in data reliability. For example, the Australian Bureau of Statistics found that by increasing the household expenditure survey sample from 7500 to 10 000 households there would be a reduction in the standard error (i.e. the dispersion of a sample statistic about the population value) of up to 13%. Any item with a relative standard error of 10% from a sample of 7500 would have a relative standard error of about 8.7% if the sample had been increased to 10 000. Thus, the costs of increasing the sample size by 33.3% would have been, in this case, far out of proportion to the improved data reliability.

To determine the size of the sample, the acceptable degree of sampling error must be determined, since the former directly influences the latter. The smaller the sampling error, the greater the reliability or precision of the population estimate based on the sample study. The smaller the desired sampling error, hence standard error, the larger the sample size must be.* However, it should be recognised that a representative sample must be large enough to exhibit the same, or a similar, amount of variation as the population. If the population exhibits no variation, a sample size of one is representative. The greater the variation in the population, the larger the sample must be to be representative. For example, if you are interested in examining a few variables which may vary (such as what share of Melbourne residents own a television and a video machine), a representative sample of 75 to 100 households would be adequate.[5] However,

---

* Chapter 11 covers dispersion, including standard deviation and standard error of estimate.

Chapter 12 – Sampling

more complicated areas such as the economic state of the Australian manufacturing industry would require larger samples in order to make the results reliable.[6]

Finally, the major disadvantages of sample surveys are that they cannot provide as much detailed data as does a census (which polls the entire population), and the quality of the sample survey's results is a function of the quality of the sample. If it is inadequate in numbers, not representative, or each member of the population is not given an equal opportunity of selection, then the results will be highly questionable.

12.1 Competency check

1. Distinguish between a sample and a population, using an example not mentioned in the text.
2. A survey of the population of observations would be preferable in most cases to a sample. Yet the majority of data collected are of the sample stream. Why?
3. You may recall reading about *descriptive statistics* and *inferential statistics* in Chapter 8. Can you apply either or both of these statistical techniques to the concept of sampling?
4. In selecting a sample of a population, what criteria should you use?
5. What factors influence the size of the sample?

12.2 Types of sampling

Sampling methods are of two basic types: judgment samples and probability samples. **Judgment samples** (also known as **non-random samples**) are just that—the sample selected is based on the opinions of one or more people who believe they can correctly identify a representative sample of the population. This method may be adequate if the judging is done by a reliable expert. For example, a grocer going to the markets may order a couple of crates of vegetables but may sample only a couple of vegetables and these may be from the top of the crate. On the basis of his or her judgment about the taste and condition of the commodity, the grocer will accept or reject it. But even then, bias may enter the selection and decision process. Therefore, although this sampling method is perhaps more convenient, less costly and at times the only practical alternative, it can be seen as subjective and it is difficult to assess how closely it measures reality. (An example of a judgment sample is **quota sampling**, which is described later in this chapter.)

The most common form of sampling is **probability** (or **random**) **sampling**, according to which every element of the population has an equal chance of being selected. Since sample results are used to infer conclusions about the population, samples must be selected in such a way as to ensure that the rules of probability can be applied to them. Therefore, we will look at the commonly used probability samples:

Business mathematics and statistics

1. simple random number samples;
2. systematic samples;
3. stratified samples;
4. cluster samples;
5. multi-stage samples.

Simple random sample

A simple **random number sample** is one in which each member of the sample has an equal chance of being selected and each unit of the population has an equal probability of being included in the sample. Lotto, lotteries and a spinning wheel exemplify numbers selected from a population, with each number having an equal opportunity, or probability, for selection. For convenience, there are random number tables, as in Table 12.2.

Table 12.2 *Random numbers*

67575	21672	60614	99692	53643	15237	75497	04891	95940	36404
57818	50250	64143	21454	43778	29215	16767	78664	61052	20792
89370	29936	73564	**89039**	87520	42617	69540	16116	77089	63482
60620	00577	36139	68962	20100	96971	07889	41046	88774	31898
01437	57324	04732	16695	25734	90554	04152	62329	25129	87459
20928	92014	33849	56275	87661	89532	62634	05451	83992	28510
10983	02117	53472	85916	28897	23339	55006	41144	53299	19156
23325	57478	09349	70354	51570	89669	51861	94388	38384	55181
12363	93425	01716	62032	65662	41733	75346	23219	31664	88299
16521	15881	35674	83865	33163	21158	47653	32290	48778	15915

Let us assume that in your current capacity as a manager, you want to determine whether invoices have been properly completed by a recently employed clerk. Rather than examine every invoice, you decide to take a sample of invoices processed over the past week. These are numbered 1 to 400, yet you are interested in looking at only 20 of these. To select the 20 invoices, you can examine Table 12.2 and randomly select a starting-point. We are interested only in the first three digits of any number (as we want to select 20 invoices from a batch numbered from 1 to 400), and there are several different ways of selecting the three digits. We may pick the last three digits, the first three or the middle three of each group of five.

Where to start? Well, we can start anywhere. For example, assume we start at the randomly selected number of 89039 (located in the fourth column, third row). Let's choose our 20 invoices on the basis of the first three digits of each group. The first three digits of 89039 exceed our range of 1 to 400; hence, this number is not selected as one of the 20. From this randomly selected number of 89039 we can move in any direction to determine the sample of 20. Let's look down from 89039, then down the columns to the right and then down the three columns until such time as the sample reaches 20. Therefore, the selected invoices are numbers 166, 201, 257, 288, 331, 152, 292, 233, 211, 167, 78, 41, 48, 161, 54, 232, 322, 251, 383 and 316.

We could just as easily have put the invoices in a barrel and selected 20. However, even here there could be a potential bias unless the numbers were put in the barrel randomly. For example, in the United States in 1970 the first peacetime draft lottery was conducted in order to determine which men were eligible to be drafted into the armed forces. It was supposed to be a simple random sample: wooden balls with January birth dates were put into the barrel first, followed by February birth dates, and so on. After each month's balls had been added to the barrel, the balls were mixed. Therefore, by the time the November balls were added, January had been mixed 10 times. In the end, December balls were mixed only once, November's twice, and lo and behold, December and November birth dates were over-represented during the selection process, because these months had not been mixed nearly as often as the preceding months.

Finally, if the population is large—for example, the number of households in Australia (over six million)—it would be too costly and time-consuming (not to mention nearly impossible) to assign each a number and randomly select a sample. To overcome such a problem, modifications of probability sampling are made and we will now explain these in detail.

Systematic sample

A **systematic sample** is chosen from an orderly list of the population, with the starting-point chosen at random and each successive number selected systematically. For example, suppose a general insurance company contemplates introducing a new insurance package and wants to test its members' attitudes. Suppose it decides to survey 10% of its 50 000 members (i.e. a sample of 5000). If the customers are arranged in alphabetical order, every 10th customer will be selected. To determine the first customer, a number from 1 to 10 must be randomly selected and from that number thereafter every 10th customer is selected. Similarly, election surveys may be administered by selecting an electoral roll, nominating the sample size, and randomly selecting the first and then every nth name. We could have used a systematic sample for selecting the 20 invoices explained earlier. All we had to do was randomly select the first invoice from numbers 1 to 20 and then take every 20th invoice thereafter.

In review, a systematic sample is one in which the first number is chosen randomly, with the remaining numbers selected systematically. The advantages of using systematic selection are its simplicity and relatively low administration costs. It may also, depending on the population's characteristics, be an improvement over random number sampling, as the sample will be more evenly distributed over the entire population and, therefore, more representative.

Using a systematic sample may, however, bias the result if there is an apparent periodic tendency in the population. For example, if you were interested in sampling the daily turnover of a newsagent whose shop is open seven days a week, you would not use every seventh day's receipts because sample receipts would always fall on the same day. Furthermore, with sales of lotto tickets being heaviest on certain days of the week, a systematic survey may understate or overstate the results.

Systematic sampling is frequently used in business applications. It should be used as an alternative to simple random sampling only when the population is randomly ordered (as suggested in instances such as attempting to define customer or electoral attitudes), but should be avoided when periodicities are present (such as the newsagent's receipts).

Stratified sample

To improve precision and reliability, it is possible to divide a population into groups or strata exhibiting common characteristics, rather than take a random or systematic sample of the entire population. For example, if it is known that 80% of food purchases are made by females, then 80% of those questioned in a survey of foodstuff shoppers should be women, and these women should be chosen randomly. In other words, the same proportion as exists in the population should be applied to the sample (in the above case, 80%). This eliminates the bias that would occur if only a random sample was taken, as there would be little likelihood of 80% of respondents being women. Therefore, when a population exhibits discernible or identifiable characteristics, we can improve on random sampling by stratification. After nominating each stratum (e.g. 80% of the sample would be females), the mean of each individual stratum is estimated. Finally, these estimates are combined in the proportion in which the stratum occurred in the population, to provide an overall estimate. Predesigning a sample structure will therefore normally produce greater accuracy and hence reliability of results, assuming that the data on which the stratification itself is based are correct and that appropriate random sample procedures were applied to the selected strata.

Suppose we were interested in collecting the average annual turnover for restaurants in Australia. Not all restaurants would be surveyed; hence, a sample would be required. However, it would need to be recognised that restaurants differ in size, suggesting that they could be stratified—according to turnover, for example. Then a simple random sample of restaurants for each strata would be selected and an estimate of the average turnover calculated from this combined sample. Therefore, as long as sections of a particular population share common characteristics, the sample can be stratified. For instance, company employees can be stratified according to age, gender, education, income, years of service, absenteeism and the like.

However, stratification is not reserved for the private sector. The Australian Bureau of Statistics, for example, in its publication *Overseas Arrivals* and *Departures, Australia* (Cat. No. 3401.0), uses stratified sampling for statistics relating to all movements by air with duration of stay equal to or less than one year. This stratification is shown in Table 12.3.

Table 12.3 *Overseas arrivals and departures by air, Australia*

Country of citizenship	Sample
Australia	1 in 55
USA, UK, NZ	1 in 40
Japan	1 in 50
Germany, Malaysia, Singapore, Taiwan	1 in 20
Canada, France, Indonesia, Italy, Thailand, Republic of Korea, Netherlands	1 in 15

Chapter 12 – Sampling

Similarly, data in *Average Weekly Earnings, States and Australia* (Cat. No. 6302.0), based on a sample of approximately 5000 employers selected from the ABS register of businesses, are collected by having the statistical unit (in this case, all activities of an enterprise in a particular State or Territory) stratified by State, public or private industry, and size of employment, with a simple random sample selected from each stratum.

Therefore, to improve the precision of sample results in cases where population groups show characteristics that are not common to the whole population, stratify the population before random sampling. You will find stratified sampling especially useful when some strata are relatively small and may be overlooked when using only the conventional simple random sample. Thus, the **stratified random sample** ensures that data will be collected from all strata, irrespective of size.

Cluster sample

In Australia, with the population scattered over vast distances, the cost of gathering information may be a serious problem. Furthermore, a list of the population members may be unavailable, thus preventing the use of simple random sampling or stratified sampling. **Cluster sampling**, however, allows us to canvass a sample which may be scattered throughout the country or world, at a far more manageable cost.

With cluster sampling, a specified geographical area is apportioned into smaller geographical areas in which groups or clusters of individual items serve as the sampling unit, with each group representative of the entire population. In other words, the population is divided into groups or clusters and then a random sample of these clusters is selected. The assumption is that the individual units being sampled within the cluster are representative of the population as a whole.

Let's assume that McDonald's, with its 6000-plus outlets worldwide, intends to develop a new worldwide corporate strategy and requires the input of middle and upper management. Working under the supposition that personal interviews with these individuals is the preferred option, would McDonald's engage in a simple random sample? Consider the time, effort and funding. Then why not look at stratified or systematic sampling? All three of these methods would, seemingly, require visits to a number of outlets in each region within every country. To overcome this problem, the population of McDonald's outlets could be divided into groups or clusters (e.g. postcodes per country) and a random selection made. These randomly selected clusters can be divided into further clusters (e.g. streets or city blocks) and a random selection is made of these. Finally, if necessary, yet another sample of these randomly selected clusters could be taken (by building addresses).

Have you ever wondered about published opinion polls which are claimed to be representative of the Australian population? One such well-known poll, the Morgan Gallup Poll, undertaken by the Roy Morgan Research Centre Pty Ltd, is conducted continuously every weekend all year round. Each weekend, 300 interviewers are sent to different randomly selected 'clusters' of eight dwellings in 148 electorates, spread in proportion to the population (hence stratified) over the cities and country areas of all six States. In other words, this organisation has created these clusters by breaking down geographical areas into smaller geographical units (e.g. a block), with each unit comprising a cluster of eight dwellings or households rather than a single household (as in multi-stage sampling) and, from city to household clusters, random sampling is used at every stage of selection.

The fact that individual items (e.g. households) within each cluster are assumed to represent the population explains why consumer surveys of large cities employ cluster sampling. Furthermore, interviewing several households within a short distance of each other reduces travel time and costs. However, this may entail the risk of an increased sampling error since respondents from the same block or neighbourhood, for example, tend to be alike and this raises questions about whether the results represent the total population. Therefore, cluster sampling may require a larger sample size than a simple random sample to provide the same level of information.

Multi-stage sample

An extended form of the cluster sample method can be found in the **multi-stage sample method**. Like the cluster method, in test-marketing the entire population, market research surveys and advertising agencies often divide the geographical area under consideration into a more manageable sample. For example, testing the Melbourne metropolitan market may entail a random selection or stratified sample of 10 local government areas. Next, neighbourhoods within each suburb may be randomly chosen and, finally, a small sample of streets from each nominated neighbourhood may be randomly selected. In other words, multi-stage sampling is an extension of cluster sampling.

The Australian Bureau of Statistics uses this multi-stage (or area-probability) sampling method when requiring statistics of a population spread over a large geographical area. For example, the monthly labour force statistics (Cat. No. 6202.0) are the result of a population survey based on a multi-stage area sample of private dwellings (about 30 000 houses, flats, etc.) and on a list sample of non-private dwellings (hotels, motels, etc.), and covers about one-half of 1% of the population of Australia. Similarly, the *Australian Household Expenditure Survey* (Cat. No. 6527.0), a survey defining the spending habits of Australians, is also based on a multi-stage sample with approximately 8400 dwellings randomly selected from private dwellings and caravan parks throughout Australia. The survey covers both rural and urban areas across all States and Territories, and dwellings include caravans, garages, tents, and any other structure used as a place of residence.

Australia's huge land mass, and the consequent dispersion of the population, make the cost of undertaking a survey by means other than multi-stage sampling prohibitive. Multi-stage sampling structures the population as a hierarchy and makes selections at each level. It can be applied whenever such structuring is possible, and substantially reduces sampling costs.

A multi-stage sample may therefore be introduced as a cost-cutting exercise, dividing the geographical region into areas or towns, then dividing each such area into smaller zones. Finally, a random or systematic sample is selected from these smaller zones. It should also be noted that as certain populations may not be listed or available, there is thus little alternative but to conduct either a multi-stage or a cluster sample survey.

Quota sample

Stratified sampling can get fairly expensive if a random sample must be taken from a large number of different strata. To tackle this problem, interviewers may be given quotas to be filled from each strata with the selection of individual units often left to the discretion of the interviewer. As a result, interviewers tend to select those individuals or units that are readily available but who/which may not be representative of the strata. Indeed, as the interviewer is exercising a judgment in selecting who/which is included in the sample, a quota sample is essentially a non-probability or judgment sample; this implies that, in the final analysis, one cannot assess the probability of reaching a correct conclusion based on the sample information.

For example, on your last visit to the shopping mall did a market researcher or interviewer try to get you to answer a survey? Did you think to yourself, 'Why me? Do I dress funny? Am I what is considered 'normal'? Do I want to be considered 'normal'? These and other questions might be going through your mind. What is going through the mind of the market researcher? He or she has been given a quota (or number) that must be interviewed, the quota often being divided into sub-quotas (males and females, age, education, marital status, ethnic background). The people interviewed may share a characteristic (e.g. people in sporting gear, people munching on sweets, pensioners, well-dressed individuals, women with children).

This type of sample technique, although seemingly straightforward, leaves perhaps a bit too much to the researcher's discretion in nominating location and people. The researcher may well be more interested in meeting his or her quota than in following strict statistical data-collection techniques. Although those interviewed for the quota sample should share certain characteristics, they must still be selected randomly from everyone sharing those same characteristics. It is doubtful, however, whether many such quota samples are in fact randomly selected.

In conclusion, most probability sampling methods are variations and/or combinations of the methods discussed above. Factors determining the nominated technique(s) used will be a function of the type and number of variables being examined, the population's geographical distribution, and general homogeneity within the population, time and funding requirements.

12.2 Competency check

1. What is meant by *probability sampling*? *Non-probability* or *judgment sampling*?

2. A simple random sample, if properly conducted, would be the preferred option for most pollsters as it would be the most accurate. Why, then, are many of the other methods used instead?

3. Assume that a university student body of 20 000 is to be sampled according to faculty and year of study. Which method(s) would you recommend and why?

4. There are populations where strata sampling may be preferable to random sampling. Can you think of any such circumstances?

5. From the information provided at the beginning of this chapter on *Choice* magazine's survey of automobiles, what questions would you now pose about the sample survey?

. . . .

12.3 Questionnaires

As the basis of your analysis will be the statistical survey of your data, which is a function of the responses to your questions, the latter must be thoroughly considered and discussed within your organisation before they are administered to your target audience. You must also decide how you will analyse the data once they are collected, since this too will influence the form, content and number of questions.

The following is a minimum checklist for preparing a questionnaire:

1. Provide precise instructions and examples if necessary.
2. Make sure the survey is as concise as possible.
3. Make sure each question is short, relevant, lucid and easy to answer.
4. Try to start the survey with an easy question that attracts the interest of the respondent.
5. Ensure that the questions follow a logical sequence.
6. Avoid technical jargon (unless, of course, it is assumed that the targeted market understands such terms).
7. Avoid leading questions (e.g. 'Have you stopped beating your wife?').
8. Avoid infringing on the privacy of the respondent.
9. Ensure confidentiality when appropriate.
10. Word the questions so that the results are easy to tabulate.
11. Ensure that each question can be classified and is significant in the final analysis.
12. Try to provide an attractive presentation.
13. Pilot-test the questionnaire to ensure that all of the above steps have been followed.

Assuming that a questionnaire has now been constructed, the problem of how to administer it remains. There are several possibilities (detailed as follows); which one to use will depend on the time and funds available, as well as on what sample size is suitable and representative.

Personal interviews

The benefits of administering questionnaires by means of personal interviews include the following:

1. A personal interview provides immediate feedback.
2. It allows the interviewer to modify his or her approach.
3. It tends to be fairly accurate. Any ambiguity can be cleared up at the interview, and the information is collected directly from the source.
4. It is likely to provide answers to all questions.
5. It is sometimes the only means of obtaining certain information.

Chapter 12 – Sampling

These benefits are not without their costs, however. For example:

1. The financial costs involved in directing personal interviews (e.g. training, transport, salaries) may be prohibitive.
2. The high financial cost may restrict the size and geographical dispersion of the sample.
3. Anonymity is lost, and this may inhibit responses to sensitive questions.
4. Will a respondent answer personal or sensitive questions from a complete stranger?
5. The field worker (the interviewer) may not follow instructions, and may add his or her own interpretation of the question.
6. As only a limited number of people can be personally interviewed, can the result be considered representative? For example, if a questionnaire is administered during working hours to people found at home (unless these individuals are the targeted audience), do the survey results represent the population of that area?

Mailed questionnaires

How many times have you received a mailed questionnaire and tossed it out, only to see it reappear in a different form or packaging the next day? Why do companies persevere in addressing surveys to households if such surveys are often classified by the recipient as junk mail and thrown in the appropriate receptacle?

The advantages of a mailed questionnaire include the following:

1. A larger targeted geographical area and number of respondents can be reached, compared with the personal interview method.
2. The questionnaire achieves a more reliable and less costly result than the personal interview method, since a greater audience can be targeted.
3. Being relatively inexpensive to administer, the questionnaire may be sent out periodically for updating, cross-checking and follow-up purposes.
4. It permits the respondent to complete the form in his or her own good time (unlike the personal interview method), and sensitive questions can be more easily answered.
5. Intimidation by an interviewer is avoided.

However, the problems associated with a mailed questionnaire should also be recognised, namely:

1. There is a relatively low proportion of responses in relation to the number of questionnaires mailed.
2. Answers to questionnaires may be incomplete.
3. A question or questions may be misinterpreted.
4. The individual filling out the questionnaire may not be part of the targeted audience (e.g. a child rather than a parent might complete a questionnaire on shopping habits).

A mailed questionnaire should always include a self-addressed prepaid envelope and perhaps even an incentive to return the completed questionnaire, since the major problem of a mailed questionnaire is non-response, often running at over 90%. In fact, a 15–20% response rate is considered acceptable, but does this 15–20% response group represent the entire market? Analysis is based on received responses rather than perceived responses.

Telephone surveys

An increasingly popular means of surveying a market is the telephone survey. What are the attractions of a telephone survey?

1. Over 90% of Australian households have a telephone.
2. Establishing and administering a telephone survey is relatively easy.
3. Selecting a random sample from the phone book is simpler than drawing a random sample for personal interviews.
4. Telephone surveys permit faster collection of data than do mailed questionnaires.
5. They cost less to administer than the personal interview method.
6. The geographical dispersion of the targeted audience may be similar to the dispersion of an audience receiving a mailed questionnaire.

As we have seen, no survey technique appears infallible, with the telephone survey no exception. The disadvantages of the telephone survey are as follows:

1. If there are too many questions, the respondent may decide not to continue to answer the questions.
2. Since the respondent cannot see the caller, he or she may be reluctant to answer any questions asked over the phone.
3. Since the caller has little means of proving the existence of the company survey group he or she represents—nor of proving that he or she works for it—the respondent may choose not to divulge any information.
4. More complicated or sensitive questions may be not only difficult to ask over the phone but also difficult to answer when asked by an unseen stranger.
5. Almost 10% of households do not have telephones.
6. A further 10% have unlisted phone numbers.
7. Other households may be difficult to reach during normal working hours.

In summary, we find that in spite of the obvious benefits of each of the survey techniques considered, they all have similar drawbacks: potential respondents' lack of interest in answering questions; the possibility that the responses are less than honest; and people's unwillingness to divulge certain personal or sensitive information to a stranger. Consequently, doubt is cast on the validity and reliability of the collected data as a representative sample of the targeted population.

Reliability of sample estimates

Two types of error, sampling errors and non-sampling errors, are possible when estimating a population's characteristics from sample survey results. A **sampling error** occurs because only a portion of the entire population is surveyed (a sample is taken), and there is a chance or probability that the estimates based on the sample may deviate from figures that would have been collected had the entire population been tested. (Standard error of estimate is used to determine sampling error.)

One measure of the likely difference is given by the standard error, which indicates the extent to which an estimate might have varied by chance because only a sample was included. There are about two chances in three that a sample estimate will differ by less than one standard error from the estimate that would have been obtained had the entire population been examined, and about 19 chances in 20 that the difference will be less than two standard errors.[7]

A **non-sampling error** occurs because of human error (e.g. responses incorrectly reported, or incorrect recording, coding or processing of data). However, human error can occur in any collection of information, be it of the entire population or of a sample. To minimise such non-sampling error, careful attention must be paid to questionnaire construction, survey technique and data processing.

12.3 Competency check

1. Without a properly constructed questionnaire, the resulting analysis will be worth less than the paper on which the questionnaire is printed. What are a number of key considerations you should take into account in preparing a questionnaire?

2. Your supervisor has asked you to discover the attitudes of your firm's Australia-wide clientele to a proposed change in corporate image. How would you administer the questionnaire—personal interview or mailed questionnaire? Briefly explain the reasoning behind your choice.

3. Mailed questionnaires are renowned for their poor response rate. Have you ever received a mail questionnaire that was inviting enough for you to complete and return it? What motivated you to do so?

4. Surveying an entire population (such as all computers being shipped to Australia in the past month) is impractical; as a result, a sample of the population is selected. However, one must be aware of potential sampling errors. What is meant by a sampling error? Give an example of this occurring in the case of a survey sample of computers.

5. Every five years a population census is taken in Australia—that is, all inhabitants of Australia are expected to fill out the census form. The census provides a wealth of information for both the public and private sectors. However, it is not without its failings. What are the potential problems even when surveying the entire population?

12.4 Summary

Samples are part and parcel of everyday life for businesses and households. Hotel proprietors order carpeting, fittings, etc., based on samples examined; restaurateurs do likewise with their foodstuffs, as do greengrocers when buying at the markets; and quality control managers examine but a fraction (a sample) of manufactured output. When consumers get a project home built, or select a fridge, or buy a new car, or get their home painted, they are constantly exposed to and use samples.

There are two basic types of sampling techniques: judgment sampling (or non-probability sampling) and random (probability) sampling. The former would be undertaken by most consumers, as it is doubtful that many would take the time and effort to engage in a random number applied sample technique to determine the likelihood of the cheese being well received by their guests. However, for more pressing matters, ideally a sample is randomly selected; this implies that each member of the population has an equal chance of selection. Five methods of random sampling and one of judgment sampling were discussed.

The simple random sample is one where each member of the population has an equal likelihood of being selected, with well-known examples including lotto and the spinning wheel.

A systematic sample is one in which, following the orderly arrangement of the population, a starting-point is randomly determined, with each successive number selected systematically (e.g. at equal intervals). For example, as the administrator at a local university, you are considering introducing extended cafeteria hours but are uncertain whether this would be economically viable. Instead of surveying all 40 000 students, you could survey, say, 10% (or 4000) by examining student names (e.g. in alphabetical order), selecting the first name at random from the first 10 listed, and then taking every 10th name thereafter.

Stratified sampling involves dividing the population into groups that exhibit common characteristics and selecting the sample on the basis of these distinct features. For example, if 80% of the part-time labour force is female and a survey sets out to determine the likelihood of those in part-time work switching to full-time work if it becomes available, 80% of those questioned will be female.

Cluster sampling involves dividing geographical areas into smaller areas, with each such area comprising a cluster (e.g. shops in a suburban shopping centre); clusters are then randomly selected. Cluster sampling is less costly than simple random or stratified sampling, since it lets interviewers question more respondents, with little travelling time and fewer interviewers needed. However, since the individual items in the cluster (be they shops, individuals, or some other item) are in proximity to one another, the results may be biased.

A multi-stage sample, an extension of the cluster sampling technique, such as the monthly unemployment statistics published by the Australian Bureau of Statistics, involves the division of geographical areas into a more manageable sample. For example, a geographical region may be divided into areas or towns, chosen randomly or through a stratified sample, with each area subsequently subdivided into smaller zones. A randomly selected or systematically selected sample is then chosen from these smaller zones.

For a quota sample, each interviewer must collect information from a certain number of individuals (the quota), with the individuals selected by the interviewer from within certain parameters (e.g. certain age-groups, marital status, tenant or owner-occupier). There is no guarantee that those being interviewed are representative of the population, since the interviewees are often selected at the discretion of the interviewer.

Normally, a statistical survey forms the basis of an analysis, with the survey itself—its content, wording and presentation—a key factor in determining the ultimate usefulness of the evaluation. Questionnaires should be as short and concise as possible. Questions should be easy to answer, presented in a logical sequence and be devoid of technical jargon; an assurance of confidentiality should also be given.

Questionnaires can be administered in a number of ways, including:

1. *Personal interview.* This method provides immediate feedback and allows any ambiguity in a question or answer to be cleared up. All questions are likely to be answered. The personal interview may sometimes be the only means of obtaining certain information. However, the financial costs involved often inhibit the selection of an appropriate sample size.
2. *Mailed questionnaire.* This method permits increased geographical penetration and an increased number of respondents to the survey, but it relies on the target audience returning the questionnaire. This method is relatively inexpensive, but its reliability is questionable as a non-response rate of over 85% is not uncommon.
3. *Telephone interview.* This method allows the interviewer to collect data quickly and to question people living in different areas, at no expense in terms of travelling time and transport costs. However, some people do not have telephones, some have unlisted numbers, and some are averse to discussing any matter with strangers over the phone.

In assessing the reliability of collected information, one must be aware of the sampling techniques used. Since the number of observations (i.e. the population) is normally too large and/or too diverse to survey in its entirety, a portion or sample of the population is selected. The sample must represent the elements making up the entire population. Using the data generated by the sample statistic allows us to estimate the population parameter by making statistical inferences (i.e. conclusions about the population based on the sample results).

The reliability of estimates based on sampling can be undermined by sampling and/or non-sampling errors. As a sample is chosen from the available population, the selection of the sample might not mirror the results that would have been obtained if the entire population had been tested; hence, the sampling error. Human errors, such as inappropriate responses and incorrect data processing, contribute to non-sampling errors.

Data must be collected methodically, with particular attention directed to the sources of information and its reliability. When initiating primary data collection, avoid the pitfalls of improper sampling and/or non-sampling errors.

12.5 Multiple-choice questions

1. A sample:

   (a) is a portion of the population being surveyed;
   (b) should be representative of the population being considered;
   (c) if properly selected, allows conclusions to be drawn about the population;
   (d) all of the above;
   (e) (a) and (b).

2. Which of the following is not considered a probability sampling technique?

   (a) simple random sample;
   (b) stratified sample;
   (c) systematic sample;
   (d) multi-stage sample;
   (e) quota sample.

3. Which sampling technique would be most appropriate if you wanted to survey Internet users and knew that 80% of users are male?

   (a) simple random number;
   (b) systematic sample;
   (c) stratified sample;
   (d) cluster sample.

4. If you conducted a sample survey and discovered that, irrespective of how careful you were, the result varied from what it would have been had you tested the entire population, this would be an example of:

   (a) bad luck;
   (b) sampling error;
   (c) non-sampling error;
   (d) poor sample selection.

5. The appropriate size of the sample of a population will be a function of all but which one of the following?

   (a) The larger the variation between units of the population, the larger the sample size.
   (b) The smaller the desired sampling error, the larger the sample size.
   (c) The smaller the variation between units of the population, the smaller the sample size.
   (d) The degree of non-sampling error forecast.

12.6 Exercises

1. When is a probability sample more desirable than a judgment sample?

2. Under what circumstances would a judgment sample be appropriate?

3. Explain the distinction between:

   (a) random sampling and systematic sampling;
   (b) stratified sampling and cluster sampling.

4. Questionnaires may be administered in a variety of ways, such as by personal interview, telephone or mail. Briefly examine one benefit and problem of each method.

Chapter 12 – Sampling

5. Mailed questionnaires are renowned for their low rate of return and yet continue to find their way into your mail boxes. Why do businesses continue to use this method of interviewing when the results suggest that it is less than useful?

6. A quota sample can be seen as a stratified sample gone judgmental. Do you agree with this claim? Briefly discuss.

7. If the entire population is canvassed, rather than just a sample, there is still the possibility of a non-sampling error detracting from the results and conclusions. How is this possible, since the entire population is being surveyed?

8. In 1991 a NSW state election was called. All opinion polls pointed to an easy win for the Greiner government, with talk of a possible replacement for the Opposition leader, Bob Carr, once the election was over. However, the Greiner government scraped home by the narrowest of margins. What are the potential problems of such opinion polls?

9. The owner of a catering company wants to sample customers to see if they are satisfied with the service and food they receive. From the list of 400 customers on its books (who are accordingly numbered), a random sample of 15 needs to be taken from Table 12.2. Starting with the three-digit number in the extreme upper right-hand corner of the table, read down to the end of the column, then start at the top of the next column and read down, and so on until 15 randomly selected numbers have been selected. What are the 15 customers' numbers who will be contacted?

10. Give an example where stratified sampling would be more appropriate than random sampling and explain why this is the case.

11. Invoices received by the Grinn & Bareit Company are filed by date of receipt. If 6000 invoices have been received over the past year and you want to ensure that they have been duly entered and paid (as entries in the ledger would confirm), how would you select a 10% sample?

12. A domestic airline is contemplating carrying out a survey to determine passengers' attitudes towards the introduction of no-frills flights. If 65% of its passengers are travelling on business, devise a sample method for obtaining passengers' attitudes without having to survey the entire population of domestic plane travellers.

13. The results of a recent telephone survey indicated that 75% of respondents were sick and tired of cricket taking up too much time on television. Should the appropriate television network therefore seriously consider cutting down on its cricket coverage? Can you see any potential problems in the telephone survey (even though you may well agree with the results)?

14. The Consumer Price Index survey is used as a measure of inflation in capital cities only. Why might its results be misleading?

15. A Leagues Club surveys 0.5% of its members' attitudes to the refurbishment recently completed. The club has 50 000 members, with entry restricted to those aged

over 18. Briefly outline two types of samples you could use for this particular survey.

16. A sample of mangoes was examined to ensure that the quality was as promised. The shipment comprised 200 crates, with 20 boxes per crate and 16 mangoes per box. To get the sample, five crates were randomly selected. From each crate, four boxes were selected and from each box three mangoes were examined. What type of sampling method was used?

17. A wholesaler with customers nationwide wants to determine the likelihood of acceptance of a new product range by its customer base. With over 10 000 clients, it was decided that a sample would be more cost-efficient than attempting to survey the entire population. What type of sampling method would you recommend if a sample size of 200 was decided upon?

18. Ask five people in your class if they engage in any type of physical exercise (e.g. team sports, gym class, weight training, etc.) during a normal work week.
    (a) Explain the basis for your selection of those five students.
    (b) From their responses, what proportion of your class would you suggest engages in sport on a weekly basis?
    (c) Do you believe your conclusions are accurate? Briefly explain.

19. In 1936 a poll was conducted in the United States to predict who would win the presidential election—Alfred Landon, the Republican candidate, or Franklin Roosevelt, the Democrat. The polls indicated a clear victory for Landon. The rest, as they say, is history, as Roosevelt won not only his second presidential election but two more as well. Ten million surveys had been sent out, a figure equivalent to approximately 25% of the voting population. Most names had been taken from telephone books from all over the United States; others came from car registrations or from telephone subscriber lists. The sample size was certainly large enough—so what was the problem?

20. The following is taken from a request by the Australian Bureau of Statistics to returned Australian travellers to complete a survey the results of which will be used to help 'the ABS prepare Australia's balance of payments statistics'.

    A form is enclosed for you to complete, together with a calendar and an exchange rate conversion table for your reference. The form asks for details of the overseas trip from which you returned in the month of August 1997. The form should only take a short time to complete. If exact information is not readily available, careful estimates can be provided.

    These results are used to estimate the amount of expenditure undertaken by Australians while travelling overseas. Can you see any potential problems in relying on these results to calculate that figure?

Notes

1. This is an abridged version of an article by Phil Scott in the *Sydney Morning Herald*, on 9 September 1996, p. 5.
2. Mark Lawson, 'Conventions rake in dollars', *Sydney Morning Herald*, 13 February 1996, p. 4.
3. Charles Wright, 'Internet pulls in 1 million Australians', *Australian Financial Review*, 14 June 1996, p. 3.
4. Geoff Kitney, 'Polls give Howard dream start in top job', *Sydney Morning Herald*, 8 May 1996, p. 3.
5. Australian Bureau of Statistics, *Surviving Statistics* (Cat. No. 1332.0), p. 9.
6. Ibid.
7. Australian Bureau of Statistics, *The Labour Force, Australia* (Cat. No. 6203.0).

Chapter 13
PROBABILITY*

LEARNING OUTCOMES
After studying this chapter you should be able to:
- Calculate the probability of an event
- Discriminate between the definitions of probability used to structure an experiment
- Distinguish between mutually exclusive, dependent and independent events
- Calculate conditional probabilities.

All of our lives are touched by the desire to know the chance, possibility or likelihood of some event occurring. Questions such as the following are common:

- What is the possibility of rain on Saturday which will affect plans for sport or entertaining? (Meteorology reports can actually quote a percentage chance of rain occurring.)
- If Australia's inflation rate falls further, what is the likelihood of increased overseas investment and a rise in the value of the dollar?
- What is the chance of winning some game in either a social or gaming context? (This is the historical origin of probability studies.)
- What is the chance that sales will increase if the price of a product is decreased?
- How likely is it that a new production method will incur worker dissatisfaction?

Probability is the study of assigning numerical values to questions of chance. *It is also a study which allows informed business decision-making on the basis of numerical data.*

*This chapter was written by Ken Stevenson, BSc, BMath, Dip Ed, Head Teacher (Mathematics), Sydney Institute of Technology.

13.1 Basic definitions and concepts

Experiment

An **experiment** is a process which produces a well-defined outcome. The descriptor 'well-defined' means that any repetition of the process will produce one and only one of the possible experimental outcomes. Consider the examples set out in Table 13.1.

Table 13.1 *Possible experimental outcomes*

Experiment	Experimental outcomes
• Toss a coin	{heads, tails}
• Roll a die	{1, 2, 3, 4, 5, 6}
• Football game	{win, draw, lose}
• Inspection of a part	{satisfactory, unsatisfactory}
• Examination grade	{Withdrawn, Fail, Pass Terminating, Pass, Credit, Distinction, High Distinction}
• Sales promotion	{increase, no change, decrease}

Sample space

The **sample space** is the complete statement of all possible experimental outcomes and is usually denoted by the symbol S. Any one of them is a **sample point** or **element**. The examples above list the sample spaces of the experiments and use the braces {. . .} notation, which is very common. It invokes the mathematical concept of a *set*, which is a collection of objects. (The sample space is sometimes called the *universal set*.) Each element in a sample space is unique—there is no 'doubling up' of elements.

It is exceptionally important that we are certain of the sample space; we must know every possible outcome or our calculations will be in error. The **tree diagram** is a useful visual approach to determine systematically the sample space.

Problem: A coin is tossed three times. Use a tree diagram to list the sample space.

Solution: (See the figure on the opposite page.)

So the sample space is:

{(H,H,H) (H,H,T) (H,T,H) (H,T,T) (T,H,H) (T,H,T) (T,T,H) (T,T,T)}

if the order of occurrence, i.e. heads before tails, is important. If it is not, then the sample space can be written, say, as:

{(H,H,H) (H,H,T) (H,T,T) (T,T,T)},

which just reflects the mixture of three heads, two heads/one tail, one head/two tails, three tails, rather than the order in which they appear.

Chapter 13 – Probability

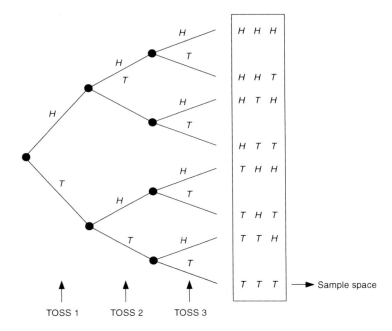

The probability of an experimental outcome
This is written $P[E_i]$, where the subscript i refers to outcome number i out of the complete list of outcomes and the symbol P stands for 'probability'.

Numerical value
The numerical value assigned to a probability ranges from zero to unity (one):

$$0 \leq P[E_i] \leq 1 \text{ for each and every outcome, } i.$$

- An impossible event has a probability of zero.
- A certain event has a probability of unity (one).

Total probability
The sum of the probability for all the outcomes in an experiment is unity: if a sample space has t elements, then:

$$P[E_1] + P[E_2] + P[E_3] + \ldots + P[E_t] = \sum_{i=1}^{t} P[E_i] = 1$$

where $\sum_{i=1}^{t} P[E_i]$ is read as 'the sum of all $P(E_i)$ ranging over all possible values of i, from 1 to t in steps of one'. Thus the sum of the probabilities of all individual events in a sample space totals unity. (The summation notation, Σ, has been met in descriptive statistics.)

It follows that the probability of the sample space is unity. By this we mean that when an experiment is being considered, some outcome in it must happen, and so:

$$P[S] = 1$$

Event

An **event** is a collection of outcomes from a specified sample space. For example, if we read the values when a die is rolled, then the sample space is composed of the outcomes {1, 2, 3, 4, 5, 6}. We could define an event A as 'even values' of the die, and so we write $A = \{2, 4, 6\}$. We also can say, of course, that A is a subset of S, i.e. it is a smaller set drawn from the larger set S. This can be written $A \subset S$, where the symbol $\subset$ stands for the phrase 'is a subset of'. If we read this statement from the right-hand side, it reads 'S contains A'.

Random selection

Random selection is assumed in this study and it means that each and every event has an equal chance of happening—there is no bias or structuring of an experiment to force selected events to happen.

The definition of probability

Classical (theoretical) definition
The probability of an event E is defined as:

$$P[E] = \frac{n(E)}{n(S)}$$

where $n(E)$ = number of elements or outcomes in event E
$n(S)$ = number of elements or outcomes in the sample space S.

Given the example above for the roll of a die, we can say that:

- the sample space has six elements or outcomes: $n(S) = 6$;
- the probability of each individual outcome occurring randomly (or freely, i.e. without any bias or cheating) is hence 1/6 and so we can write:

$P[1] = 1/6$ (the probability of rolling a '1' is 1/6)
$P[2] = 1/6$ (the probability of rolling a '2' is 1/6), and so on.

In general, $P[E_i] = 1/6$ and so we can see that the requirements for probabilities are satisfied:

$\sum P[E_i] = P[\text{rolling a '1'}] + P[\text{rolling a '2'}] + P[\text{rolling a '3'}]$
$\qquad\qquad + P[\text{rolling a '4'}] + P[\text{rolling a '5'}] + P[\text{rolling a '6'}]$
$\sum P[E_i] = 1/6 + 1/6 + 1/6 + 1/6 + 1/6 + 1/6$, and so
$\sum P[E_i] = 1$

and $0 \leq P[E_i] \leq 1$ for each and every one of the $i = 6$ outcomes—that is, each outcome has a probability that is positive, and its value is a fraction between 0 and 1.

Relative frequency (empirical) definition
Suppose that a marketing research analyst surveys the market for a new product. There are only two possible outcomes—a customer will buy or not buy the product—and there is no known reason why the probability of each option should be equal (at 0.50).

It is found that out of 500 customers contacted, 300 bought the new item. The analyst might thus decide to use the relative frequency 300/500 as the probability for success of the new product (i.e. 0.6).

Subjective definition
The subjective method is based on an individual's experience and belief that an event will occur. For example, the result of a sports match is not known before the event. However, if we examine previous performance, weather conditions and players' condition, and so forth, we can give a numerical value of the probability (likelihood) of a team winning. Electoral outcomes, the success of a business venture, and investment outcomes may also be predicted by subjective probabilities.

Subjective probabilities relate to experiments that cannot be repeated. Because the probability value is a personal judgment, the subjective approach has also been referred to as the *personalistic* approach.

The complement

The **complement of an event** is the event which does not happen. The complement of success is failure. The complement of failure is success. If we have an event symbolised by M, then the complement of M contains all the outcomes that are not in M. The complement of M is written M^c or $\tilde{M}$ or $\overline{M}$, and is often referred to as 'not M'.

This is demonstrated visually by the **Venn diagram** approach. The (outer) rectangle shows the sample space S. Event M is shown by the shaded region. The complement of M is everything outside M and inside S. Venn diagrams are used to clarify the relationships between events.

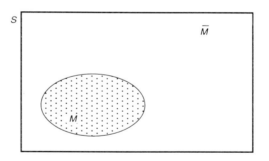

The subtraction rule

It follows that probabilities can be calculated for complementary events:

$$P[\overline{M}] + P[M] = 1$$

or

$$P[\overline{M}] = 1 - P[M]$$

Problem: A die is rolled and the face value is read. Find the probability of:

(a) rolling a '6';
(b) not rolling a '6'.

Solution: $n(S) = 6$

There are six possible outcomes in this experiment: $\{1, 2, 3, 4, 5, 6\}$

(a) The event E we are considering is a '6', and this has one element: $n(E) = 1$.

$$P[E] = \frac{n(E)}{n(S)} = 1/6$$

(b) Not rolling a '6' is the complement of rolling a '6':

$$P[\text{not rolling a '6'}] = 1 - P[\text{rolling a '6'}]$$
$$= 1 - 1/6$$
$$= 5/6$$

Note that this could have been found by a longer (but valid) method—the probability of not rolling a '6' is the sum of the individual probabilities $P[\text{rolling '1'}] + P[\text{rolling '2'}] + P[\text{rolling '3'}] + P[\text{rolling '4'}] + P[\text{rolling '5'}]$.

Hence, $P[\text{not rolling a '6'}] = 5 \times 1/6 = 5/6$.

Problem: The lunchtime card club is in progress at W.E. Playem and Co. What is the probability of the following events occurring as the result of random selection?

(a) $A = \{\text{a heart}\}$
(b) $B = \{\text{a picture ('face') card}\}$
(c) $C = \{\text{a picture ('face') card which is also a club}\}$
(d) $D = \{\text{a red card}\}$
(e) $E = \{\text{a card with an even numeral}\}$
(f) $F = \{\text{no hearts}\}$
(g) $G = \{\text{a numeral card}\}$

Solution: Each of the events must be enumerated carefully, along with the sample space size: $n(S) = 52$.

(a) $A = \{A♥\ 2♥\ 3♥\ 4♥\ 5♥\ 6♥\ 7♥\ 8♥\ 9♥\ 10♥\ J♥\ Q♥\ K♥\}$
$P[A] = 13/52 = 1/4 = 0.250$

(b) $B = \{J♥\ Q♥\ K♥\ A♥\ J♣\ Q♣\ K♣\ A♣\ J♠\ Q♠\ K♠\ A♠\ J♦\ Q♦\ K♦\ A♦\}$
$P[B] = 16/52 = 4/13 = 0.308$

(c) $C = \{J♣\ Q♣\ K♣\ A♣\}$
$P[C] = 4/52 = 1/13 = 0.077$

(d) $D = \{A♥\ 2♥\ 3♥\ 4♥\ 5♥\ 6♥\ 7♥\ 8♥\ 9♥\ 10♥\ J♥\ Q♥\ K♥$
$\quad\quad A♦\ 2♦\ 3♦\ 4♦\ 5♦\ 6♦\ 7♦\ 8♦\ 9♦\ 10♦\ J♦\ Q♦\ K♦\}$
$P[D] = 26/52 = 1/2 = 0.500$

(e) $E = \{2♥\ 4♥\ 6♥\ 8♥\ 10♥\ 2♦\ 4♦\ 6♦\ 8♦\ 10♦$
$\quad\quad 2♣\ 4♣\ 6♣\ 8♣\ 10♣\ 2♠\ 4♠\ 6♠\ 8♠\ 10♠\}$
$P[E] = 20/52 = 5/13 = 0.385$

(f) $P[F] = 1 - P[A] = 0.750$, since A and F are complementary events. We could also enumerate the outcomes for F and use the definition.

(g) By reasoning similar to that in (f) above, G and B are complements; hence: $P[G] = 1 - P[B] = 1 - 0.308 = 0.692$

Joint and marginal probabilities

Problem: A group of executives consists of 10 people, four men and six women. Three of the men smoke, as do two of the women. Find the probability of randomly selecting:

(a) a male smoker;
(b) a female non-smoker;
(c) a male non-smoker;
(d) a female smoker;
(e) a smoker;
(f) a non-smoker;
(g) a female;
(h) a male.

Solution: This problem lends itself to a **contingency table** approach. The sample space has 10 elements and can be broken into four subsets—men, women, smokers and non-smokers. Table 13.2 sets out the raw numbers.

Table 13.2 *Study of executives and smoking*

	Male M	Female F	Totals
Smokers S	3	2	5
Non-smokers NS	1	4	5
Totals	4	6	10

The question is most easily answered by dividing the value in each cell by the sample space value, and hence producing probabilities, as shown in Table 13.3.

Table 13.3 *Study of executives and smoking*

	Male M	Female F	Marginal probabilities
Smokers S	3/10 = 0.3	2/10 = 0.2	5/10 = 0.5
Non-smokers NS	1/10 = 0.1	4/10 = 0.4	5/10 = 0.5
Marginal probabilities	4/10 = 0.4	6/10 = 0.6	10/10 = 1

The answers, reading from the cells, are:

(a) P[a male smoker] = 0.3
(b) P[a female non-smoker] = 0.4
(c) P[a male non-smoker] = 0.1
(d) P[a female smoker] = 0.2
(e) P[a smoker] = 0.5
(f) P[a non-smoker] = 0.5
(g) P[a female] = 0.6
(h) P[a male] = 0.4

The term **joint probability** refers to the fact that the probability is determined from the intersection of two (or more) of the criteria in the question. A joint probability is one that relates to more than one category of classification, as the name suggests, and is the intersection of two events. Thus, the probability of a female non-smoker being chosen is the value of the cell which is at the intersection of the 'female' column and the 'non-smoking' row. And so each cell has a joint probability.

Marginal probabilities refer to single events and are the values in the margins of the table. Each marginal value represents the value of the probability of the event listed at the head of the row or column. It can be obtained by summing the joint probabilities in the row or column. Thus, the answer to parts (e) to (h) are marginal probabilities.

13.1 Competency check

1. State the complement for each of the following events:

   (a) No errors in an account ledger.
   (b) There are six cars in a garage and at least three have flat batteries.
   (c) Jim fails an exam.
   (d) A committee of five people is chosen, of which no fewer than two are women.

2. A survey of 55 households shows that they have the following number of cars:

No. of cars	0	1	2	3	4	5
Frequency	8	22	15	5	3	2

   Find the probability of the following random events:

   (a) A = {exactly two cars};
   (b) B = {at least two cars};
   (c) C = {fewer than two cars};
   (d) D = {no more than two cars};
   (e) E = {more than two cars};
   (f) $\bar{A}$.

3. The data set out in Table 13.4 apply to 10 employees in a company.

Table 13.4 *Study of company employees*

Name	Age	HSC graduate?	Marital status	Previous experience?
Ms Rouen	35	Yes	M	No
Ms Bulic	25	No	S	Yes
Ms Foe	30	Yes	S	Yes
Mr James	42	No	M	Yes
Mr Honeysit	37	No	S	No
Ms Dragin	44	No	S	Yes
Mr Griffine	22	Yes	S	No
Ms Smithers	58	No	M	Yes
Mr Treadbear	37	Yes	M	No
Ms Wales	37	No	S	No

List the following events and calculate their probability of random selection:

(a) HSC graduate;
(b) older than 30;
(c) same age;
(d) no previous experience;
(e) no previous experience and male;
(f) single and female.

4. A die is tossed once. List the sample space and determine the probability of obtaining the following events:

(a) at least a 3;
(b) at most a 4;
(c) more than 4;
(d) 5 or less;
(e) 2 or more;
(f) more than 2 and 5 or less.

13.2 Probability rules and compound events

Rule of addition

The rule of addition is used when we wish to determine the probability of one event or another (or both) occurring in a single observation. There are two variations of the rule of addition, depending on whether the two events are mutually exclusive or not.

Symbolically, we can represent the probability of event A or event B occurring by $P[A$ or $B]$. In the language of set theory, this is the union of A and B and the probability is designated as $P[A \cup B]$.

Business mathematics and statistics

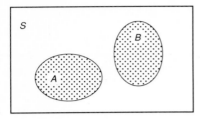

A and B are mutually exclusive or disjoint sets.

The shaded region shows $A \cup B$ in both diagrams.

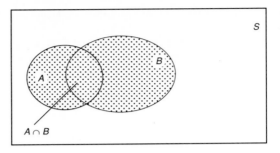

$A \cap B$

A and B overlap—there are some common elements in each set. The common region is called the intersection and written $A \cap B$.

Mutually exclusive or disjoint events

Mutually exclusive or disjoint events cannot occur together—that is, the occurrence of one automatically precludes the occurrence of the other event (or events). For example, if we consider two possible events in baseball—a single or a home run—we see that the two events are mutually exclusive, because a hit cannot simultaneously be both a single and a home run.

For mutually exclusive events, the rule is:

$$P[A \text{ or } B] = P[A \cup B]$$
$$= P[A] + P[B]$$

To determine the probability of the compound event A or B occurring, we simply add the probability of A occurring to the probability of B occurring. For example, when drawing a card from a deck of cards, the events 'ace' (A) and 'king' (K) are mutually exclusive. The probability of drawing either an ace or a king in a single draw is:

$$P[A \text{ or } K] = P\{A \cup K\}$$
$$= P[A] + P[K]$$
$$= \frac{4}{52} + \frac{4}{52} = \frac{8}{52} = \frac{2}{13}$$

Non-exclusive or joint events

Non-exclusive or joint events have an overlap or common region. The probability of joint occurrence is represented by $P[A \text{ and } B]$ or $P[A \cap B]$, which is read as 'A intersects B'. In the language of set theory, this is called the *intersection* of A and B. The probability of the joint occurrence of the two events is subtracted from the sum, otherwise we will have 'double counted' the common region—once from set A and once from set B:

$$P[A \text{ or } B] = P[A \cup B]$$
$$= P[A] + P[B] - P[A \cap B]$$

This formula is often called the *general rule of addition*, because for events that are *mutually exclusive* the last term would always be equal to zero. Thus, the formula is the general case upon which the previous formula for mutually exclusive events is based. In terms of the number of elements in the overlapping sets situation, we can write:

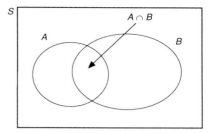

$$n(A \text{ or } B) = n(A) + n(B) - n(A \cap B)$$

as seen in the Venn diagram to the right. When each item is divided by $n(S)$, we have the probability statement.

Problem: A quantity surveyor is assessing the performance of contractors. His database of 50 glass contractors shows:

- 6 regularly default by both being late (B) and giving defective workmanship (D);
- 14 are regularly late, but give good quality work;
- 15 regularly give defective workmanship, but are not late.

If a new surveyor takes over and does not consult the database records thoroughly, what is the probability that he will choose a contractor who:

(a) will be behind schedule?
(b) will have to be called back to fix up his work?
(c) will be unsatisfactory (owing to lateness or poor quality)?
(d) can produce acceptable work?

Solution: Consult the Venn diagram at right, where the numbers in each category have been entered.

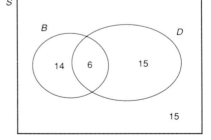

(a) $n(B) = 14 + 6 = 20$

Note how the value of 6 from the intersection is added to give the total value for $n(B)$.

$$P[B] = \frac{20}{50} = 0.4$$

(b) The criterion 'called back to fix up his work' indicates defective workmanship.

$$n(D) = 15 + 6 = 21$$

Hence: $P[D] = \dfrac{21}{50} = 0.42$

(c) The criterion 'unsatisfactory' implies from the surveyor's point of view either B or D—that is, the union of the two sets.

Hence: $n(B \cup D) = n(B) + n(D) - n(B \cap D)$
$= 20 + 21 - 6$
$= 35$

So: $P[B \cup D] = \dfrac{35}{50} = 0.70$ or 70%

Alternatively, we calculate:

$P[B \cap D] = \dfrac{6}{50} = 0.12$
$P[B \cup D] = P[B] + P[D] - P[B \cap D]$
$= 0.4 + 0.42 - 0.12$
$= 0.70$

(d) $P[\text{satisfactory}] = 1 - P[\text{unsatisfactory}]$
$= 1 - 0.70$
$= 0.30$

or $P[\text{satisfactory}] = P[\text{not unsatisfactory}]$
$= \dfrac{15}{50}$
$= 0.30$

Multiplication of probabilities for independent events

It is often necessary to consider probabilities associated with the sequential occurrence of two or more specified outcomes. For instance, in game-of-chance experiments we may be concerned with calculating the probability of obtaining a sequence of one head in a coin toss followed by an ace from a well-shuffled deck of cards.

The drawing of an ace and the toss of a head are completely independent events—neither outcome has any effect on the other. **Independent events** are events which have no influence on each other.

The rule of multiplication for two independent events is:

$P[A \text{ and } B] = P[A]\, P[B]$

Problem: What is the probability of the sequence of outcomes defined by one head in a coin toss followed by an ace from a well-shuffled deck of cards?

Solution:
$P[\text{head in one toss}] = 1/2$
$P[\text{draw an ace}] = 4/52 = 1/13$
$P[\text{head followed by ace}] = P[\text{head}] \times P[\text{ace}]$
$= \dfrac{1}{2} \times \dfrac{1}{13}$
$= \dfrac{1}{26}$

Problem: What is the probability of throwing two heads in a row when a coin is tossed twice?

Solution:
For one toss of one coin, $P[\text{head}] = 1/2$.

For two separate tosses, one following the other, $P[2 \text{ heads}] = \dfrac{1}{2} \times \dfrac{1}{2} = \dfrac{1}{4}$.

You should note that the same result is obtained for the toss of two separate coins at the one time.

Problem: A large college campus has a computer-controlled telephone exchange. It is known from previous studies that at any given moment during the day there is a 15% chance that all extensions are busy and incoming messages will be placed on hold, and that all extensions can be given independent status. What is the probability that two successive incoming calls will be put on hold?

Solution: Let B_1 symbolise a busy signal for call 1, and B_2 a busy signal for call 2. Since the data tell us that there is independent status:

$$P[B_1] = P[B_2] = 0.15$$
$$P[B_1 \text{ and } B_2] = P[B_1] \times P[B_2]$$
$$= 0.15 \times 0.15$$
$$= 0.0225$$

Extended rule: more than two events
When there are more than two independent events involved, the rules for addition and multiplication are extended.

$$P[A \text{ or } B \text{ or } C \text{ or } \ldots] = P[A] + P[B] + P[C] + \ldots$$
$$P[A \text{ and } B \text{ and } C \text{ and } \ldots] = P[A] \times P[B] \times P[C] \times \ldots$$

Sampling with and without replacement and probability tree diagrams

The following problems and notes will demonstrate the above concepts.

Problem: A builder regularly orders consignments of 200 deadlock units for villa developments. Experience shows that supplier quality runs at 5% defective and 95% acceptable. Given a random selection of three from the 200 units supplied, what is the probability that:

(a) all three are defective?
(b) all three are acceptable?
(c) two are defective?
(d) only one is acceptable?

For this problem assume that the builder *replaces* each unit into the consignment after it has been tested. (This is *sampling with replacement*.)

Business mathematics and statistics

Solution: This problem is well suited to a probability tree diagram. The symbol D_1 stands for outcome 1 giving a defective result; G_2 stands for a good unit on the second outcome or test.

(a) $P[3 \text{ defective}] = P[D_1 D_2 D_3] = \dfrac{10}{200} \times \dfrac{10}{200} \times \dfrac{10}{200} = \left(\dfrac{1}{20}\right)^3$

$= \dfrac{1}{8000} \approx 0.00013$

(b) $P[3 \text{ good}] = P[G_1 G_2 G_3] = \dfrac{190}{200} \times \dfrac{190}{200} \times \dfrac{190}{200} = \left(\dfrac{19}{20}\right)^3 = \dfrac{6859}{8000} \approx 0.857$

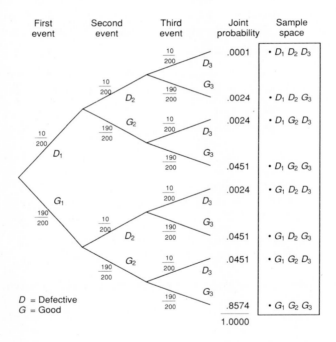

(c) $P[2 \text{ defective}] = P[D_1 D_2 G_3] + P[D_1 G_2 D_3] + P[G_1 D_2 D_3]$

$= \dfrac{10}{200} \times \dfrac{10}{200} \times \dfrac{190}{200} + \dfrac{10}{200} \times \dfrac{190}{200} \times \dfrac{10}{200} + \dfrac{190}{200} \times \dfrac{10}{200} \times \dfrac{10}{200}$

$= 3 \times \dfrac{19}{8000}$

$= \dfrac{57}{8000}$

≈ 0.007

(d) $P[1 \text{ defective}] = P[D_1G_2G_3] + P[G_1D_2G_3] + P[G_1G_2D_3]$

$= \dfrac{10}{200} \times \dfrac{190}{200} \times \dfrac{190}{200} + \dfrac{190}{200} \times \dfrac{10}{200} \times \dfrac{190}{200} + \dfrac{190}{200} \times \dfrac{190}{200} \times \dfrac{10}{200}$

$= 3 \times \dfrac{19 \times 19}{8000}$

$= 3 \times \dfrac{361}{8000}$

$= \dfrac{1083}{8000}$

≈ 0.135

Notes:
- Each of the branches is composed of independent events, since the tested items are replaced into the consignment.
- Each pair of branches at any outcome has a total probability of unity; this is because each time an outcome is reached, there are only two possibilities— *D* or *G*.
- The box down the right lists out the sample space.
- The numbers to the left of the box use the multiplication law to determine the probability of each complete branch.
- The four probabilities in (a), (b), (c) and (d) *completely exhaust* the sample space and hence total to a value of unity.

Problem: The same builder orders from the same supplier a consignment of 200 deadlock units for a villa development, with the supplier's quality still running at 5% defective and 95% acceptable. Given a random selection of three from the 200 units supplied, what is the probability that:

(a) all three are defective?
(b) all three are acceptable?
(c) two are acceptable?
(d) only one is acceptable?

For this problem the builder *does not replace* the tested unit into the consignment. (This is *sampling without replacement*.)

Solution: This problem has a different tree diagram. *Since the inspection is carried out without replacement, the sample space reduces by one every time an item is removed.* Hence, the subsequent inspection is from a sample space of reduced size. But also, if one defective unit has been chosen, and removed, there is also one fewer defective unit available to be chosen in the subsequent choice.

Business mathematics and statistics

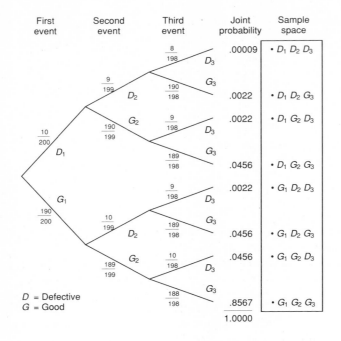

(a) $P[\text{3 defective}] = P[D_1D_2D_3] = \dfrac{10}{200} \times \dfrac{9}{199} \times \dfrac{8}{198} = \dfrac{720}{7\,880\,400} \approx 0.00009$

(b) $P[\text{3 good}] = P[G_1G_2G_3] = \dfrac{190}{200} \times \dfrac{189}{199} \times \dfrac{188}{198} = \dfrac{6\,751\,080}{7\,880\,400} \approx 0.857$

(c) $P[\text{2 defective}] = P[D_1D_2G_3] + P[D_1G_2D_3] + P[G_1D_2D_3]$

$= \dfrac{10}{200} \times \dfrac{9}{199} \times \dfrac{190}{198} + \dfrac{10}{200} \times \dfrac{190}{199} \times \dfrac{9}{198} + \dfrac{190}{200} \times \dfrac{10}{199}$

$\times \dfrac{9}{198}$

$= 3 \times \dfrac{17\,100}{7\,880\,400}$

$= \dfrac{51\,300}{7\,880\,400}$

≈ 0.007

(d) $P[\text{1 defective}] = P[D_1G_2G_3] + P[G_1D_2G_3] + P[G_1G_2D_3]$

$= \dfrac{10}{200} \times \dfrac{190}{199} \times \dfrac{189}{198} + \dfrac{190}{200} \times \dfrac{10}{199} \times \dfrac{189}{198} + \dfrac{190}{200} \times \dfrac{189}{199}$

$\times \dfrac{10}{198}$

Chapter 13 — Probability

$$= 3 \times \frac{10 \times 190 \times 189}{7\,880\,400}$$

$$= 3 \times \frac{359\,100}{7\,880\,400}$$

$$= \frac{1\,077\,300}{7\,880\,400}$$

$$\approx 0.137$$

Notes:
- The four probabilities above *completely exhaust* the sample space and hence total to a value of unity, allowing for the rounding-off process, which gives a total of 1.00109.
- Strictly speaking, the outcomes are dependent, not independent. However, the dependence is removed by specifying the new sample space and outcome size.
- Comparison of these values with those in the preceding problem leads to the conclusion that where large quantities are involved, and the sampling is proportionally small (e.g. three items out of 200), sampling without or with replacement yields comparable answers.

Problem: A survey of 750 residents in a local government area classifies the population by three categories:

I: whether income is over $35 000
P: whether purchasing their home
M: whether they own a motor vehicle.

To make the information easier to present, it is given below in a Venn diagram.
  Find the probability of randomly selecting a person who:

(a) owns a motor vehicle;
(b) has income greater than $35 000 and is purchasing their own home;
(c) has income greater than $35 000 or owns a motor vehicle;
(d) has income greater than $35 000 or is purchasing their home and who also owns a motor vehicle;
(e) has income greater than $35 000 and is purchasing their home but does not own a motor vehicle.

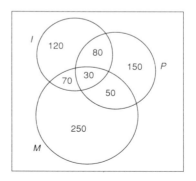

Solution:
Note that this example is exhausted by the contents of the three sets—there are no outcomes outside the sets.

(a) The number of people who own a motor vehicle is shaded. It is the totality of set M.

$$P(M) = \frac{400}{750} = 0.533$$

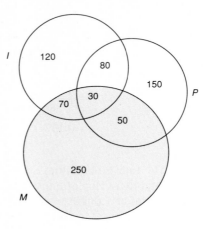

(b) The number of people who have income greater than $35 000 and are purchasing their home is shaded. It is the intersection of sets I and P.

$$P(P \cap I) = \frac{110}{750} = 0.147$$

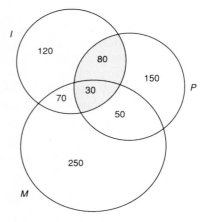

(c) The number of people who have income greater than $35 000 or own a motor vehicle is shaded. It is the union of sets I and M.

$$P(I \cup M) = \frac{600}{750} = 0.8$$

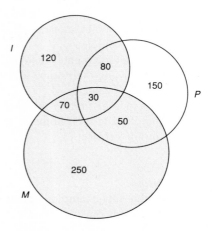

(d) The number of people who have income greater than $35 000 or are purchasing their home and own a motor vehicle is shaded. It is the union of sets I and P which is then intersected with M.

$$P(I \cup P \cap M) = \frac{150}{750} = 0.200$$

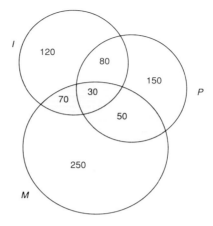

(e) The number of people who have income greater than $35 000 and are purchasing their home, but do not own a motor vehicle is shaded. It is the intersection of sets I and P, which is then intersected with everything outside M.

$$P(I \cup P \cap \tilde{M}) = \frac{80}{750} = 0.107$$

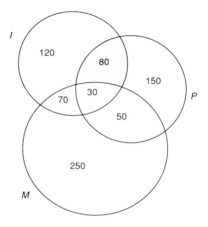

13.2 Competency check

1. State, giving a reason each time, whether the following pairs of events are independent:

   (a) winning first and second prize in a raffle;
   (b) earning a large salary and being subject to a high tax rate;
   (c) any two mutually exclusive events;
   (d) being a police officer and having large feet;
   (e) driving under the influence of alcohol and having an accident.

2. A bag contains 15 red counters and 12 white counters; two counters are taken out in succession, their colours are recorded and neither is replaced. What different results could be obtained, and what is the probability of each result?

Business mathematics and statistics

3. Repeat the question above, but assuming this time that the counters are replaced in the bag after their colour is recorded.

4. Jones Manufacturing Co. has two machines—*A* and *B*. It tenders for 10 contracts. Three contracts will require the use of *A* but not *B*; four contracts will require the use of *B* but not *A*; two contracts require the use of both *A* and *B*. Draw a Venn diagram and find the probability that, if all tenders are won:

   (a) *A* and *B* are used;
   (b) *A* or *B* are used;
   (c) *B* is used;
   (d) no machine is used;
   (e) one machine is used.

5. If $P[A] = 0.6$ and $P[B] = 0.7$, can *A* and *B* be mutually exclusive?

6. In a survey of 120 students at a local high school it was found that 38 played cricket, 35 played football (of some code), 47 played tennis, 11 played cricket and football, 12 played cricket and tennis, 8 played football and tennis and 6 played all three sports. Find the probability of randomly selecting a student who:

   (a) plays no sport;
   (b) plays tennis only;
   (c) plays cricket or tennis;
   (d) plays only cricket or only tennis (and not both);
   (e) plays both cricket and tennis but not football.

13.3 Conditional probability and dependent events

Have you ever found that you displayed symptoms of an illness and thought that perhaps you had that illness? You rushed to the doctor to determine what affliction you really had. Did displaying the symptom mean that you had the illness or not? This type of question, and others we will investigate, can be assessed by using conditional probability. For example, if you know the probability of individuals having illness *X* if they display a certain condition, what is the probability of having this condition if illness *X* has occurred?

In many experiments it is necessary to be able to calculate probabilities if a prior event is known to have happened. The prior event can often affect the probability of the current event. Suppose that there is an event *A* with probability $P[A]$. If another event, *B*, occurs which has an influence on *A*, then we need to take account of the effects of *B*.

The **conditional probability** of *A* is the new probability of *A*, based on the occurrence of *B*, and is written $P[A \mid B]$, which is read as 'the probability of *A* given that *B* has occurred'. The symbol | indicates that we are considering the probability of event *A* given the condition that event *B* has occurred. $P[A \mid B]$ is called a *conditional probability*.

Chapter 13 – Probability

Multiplication rule for dependent events

For dependent events the probability of the joint occurrence of A and B is the probability of A multiplied by the conditional probability of B given A. An equivalent value is obtained if the two events are reversed in position. Thus, the rule for dependent events is:

$$P[A \text{ and } B] = P[A \cap B] = P[A]P[B \mid A]$$
$$P[A \text{ and } B] = P[A \cap B] = P[B]P[A \mid B]$$

and can be restated with the conditional probability as the subject of the formula:

$$P[B \mid A] = P\frac{[A \cap B]}{P[A]}$$

These formulae are often called the *general rule of multiplication* because, for events that are independent, the conditional probability value $P[B \mid A]$ would be the same as the respective unconditional probability value, $P[B]$. Go back to Section 13.2, find the multiplication rule for independent events, and you will see how that rule was a specific case of this newer and general rule.

In fact, the mathematical definition of independent events is:

- $P[B \mid A] = P[B]$—the occurrence or non-occurrence of A does not influence the probability of event B.
- or $P[A \mid B] = P[A]$—the occurrence or non-occurrence of B does not influence the probability of event A.

Problem: In a group of 50 students, 20 take an English subject, 25 take a technology subject and 15 take both. Determine the probability of randomly selecting a student who is taking:

(a) the technology subject, given that the student is already taking English;
(b) the English subject, given that the student is already taking technology.

Solution: These are conditional probabilities—they are based on a prior condition. The Venn diagram is useful here. We have:

$P[E] = 20/50$
$P[T] = 25/50$
$P[E \cap T] = 15/50$

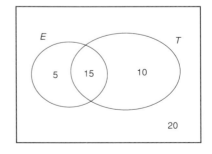

(a) Applying the multiplication rule:

$$P[T \mid E] = \frac{P[T \cap E]}{P[E]}$$
$$= \frac{15/50}{20/50} = \frac{15}{20} = 0.75$$

Note that this could also be found from the diagram by taking the intersection region size of 15 and dividing by the English set size of 20.

(b) Applying the multiplication rule:

$$P[E \mid T] = \frac{P[E \cap T]}{P[T]}$$

$$= \frac{15/50}{25/50} = \frac{15}{25} = 0.60$$

Note that this could also be found from the diagram by taking the intersection region size of 15 and dividing by the English set size of 25. However, it will not always be so easy to use a simple diagram!

The contingency table approach enables us to determine joint, conditional and marginal probabilities.

Problem: Suppose that a survey gives the breakdown of income levels and political affiliation shown in Table 13.5:

Table 13.5 *Income levels and politics*

Income	Political affiliation			Total
	Labor	Liberal	Other	
High	44	92	4	140
Low	90	34	36	160
Total	134	126	40	300

Find the probability that a person randomly chosen from this survey will be:

(a) a Labor voter;
(b) a high-income earner;
(c) a Labor voter and a high-income earner;
(d) a Labor voter, given that he or she has a high income.

Solution: The relevant probabilities for each cell are found by dividing each cell by 300.

(a) P [Labor voter] $= \dfrac{134}{300} = 0.447$. This is a marginal probability.

(b) P [high income] $= \dfrac{140}{300} = 0.467$. This is also a marginal probability, but on the other axis.

(c) P [Labor *and* high income]. This is a joint probability—at the intersection of the Labor column and the high-income row $= \dfrac{44}{300} = 0.147$.

(d) This is a conditional probability—the probability of voting Labor depends on the person already being a high-income earner.

$$P[\text{Labor} \mid \text{high income}] = \frac{P[\text{Labor and high income}]}{P[\text{high income}]} = \frac{44/300}{140/300} = \frac{44}{140} = 0.314$$

Note that this probability is much higher than P[Labor *and* high income] since the denominator is smaller—we are drawing from a smaller set—44/140.

Problem: Assume that there are equal numbers of male and female students in a university. Of all male students, 10% major in economics; and of all female students, 5% major in economics. What is the probability that:

(a) a student selected at random will be a male economics student?
(b) a student selected at random will be a female economics student?
(c) a student selected at random will be an economics student?

Solution:
- Let M = male student; F = female student; E = economics student.
- From the data provided, the probability of a student specialising in economics, if the student is male, is 0.10; hence, $P[E \mid M] = 0.10$.
- Similarly, from the data, $P[E \mid F] = 0.05$.
- There are equal numbers of male and female students; hence, $P[M] = P[F] = 0.5$.

(a) The event 'male economics student' is the event M and E, or $M \cap E$. Applying the multiplication formula, we have:

$$P[M \cap E] = P[M]\,P[E \mid M] = 0.5 \times 0.1 = 0.05$$

(b) The event 'female economics student' is the event F and E, or $F \cap E$. Applying the multiplication formula, we have:

$$P[F \cap E] = P[F]\,P[E \mid F] = 0.5 \times 0.05 = 0.025$$

(c) The event 'economics student' (E) consists of the events 'male economics student' ($M \cap E$) and 'female economics student' ($F \cap E$), which are mutually exclusive:

$$\begin{aligned} P[E] &= P[M \cap E] + P[F \cap E] \\ &= 0.05 + 0.025 \\ &= 0.075 \end{aligned}$$

In many types of situations we are concerned with calculating probabilities associated with the results of several repetitions of some basic experiment—for example, tossing a coin or drawing an object from a well-defined group. We have already encountered sampling without replacement in Section 13.2—taking an object from a group and not replacing it. The sample space of the experiment is altered after each draw, and the events representing the outcomes of the successive drawings are dependent events. The problems in Section 13.2 did not use the notation of conditional probability, but the concept was there.

Problem: A box contains 10 numbered tender documents, of which three are from unknown contractors (U) and seven are from known contractors (K). A tender is drawn at random from the box and then a second tender is drawn at random, without the first tender being replaced in the box. What is the probability that:

(a) the first tender is unknown and the second is known?
(b) exactly one of the tenders is unknown?

Solution:
(a) Let $U1K2$ be the event that the first tender is unknown and the second is known. The probability of drawing an unknown tender first is 3/10. If the outcome of the first draw is indeed 'unknown', then there are nine tenders left in the box, of which seven are known. Thus, $P[K2 \mid U1] = 7/9$, and by the multiplication rule:

$$P(U1K2) = P(U1)\, P(K2 \mid U1)$$
$$= \frac{3}{10} \times \frac{7}{9}$$
$$= \frac{7}{30}$$

(b) The event 'exactly one unknown' can occur in one of two possible ways: 'unknown' first, then 'known' ($U1K2$) or 'known' first, then 'unknown' ($K1U2$). These are mutually exclusive; hence:

$$P(\text{exactly 1 unknown}) = P(U1K2) + P(K1U2)$$
$$= \frac{3}{10} \times \frac{7}{9} + \frac{7}{10} \times \frac{3}{9}$$
$$= \frac{7}{15}$$

This situation is well modelled by a probability tree diagram:

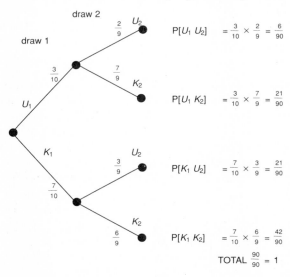

Chapter 13 – Probability

Problem: If three tenders are now drawn from the box without replacement, what is the probability that:

(a) all three are unknown?
(b) the first two are unknown and the third is known?
(c) exactly two of the tenders are unknown?

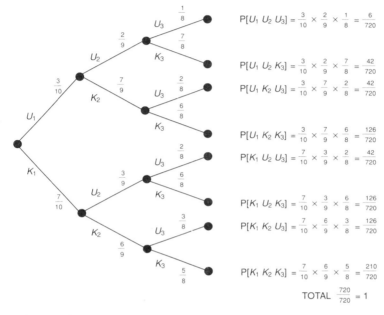

Solution:

(a) Let the symbols $U1$, $U2$ and $U3$ indicate the occurrence of unknown tenders on the first, second and third drawn respectively, and we have:

$$P[U1] = 3/10$$
$$P[U2 \mid U1] = 2/9$$
$$P[U3 \mid U2U1] = 1/8$$

Multiplying these three quantities, we obtain:

$$P[U1U2U3] = 1/120$$

(b) Similarly, $P[U1U2K3] = \dfrac{3}{10} \times \dfrac{2}{9} \times \dfrac{7}{8} = \dfrac{7}{120}$

(c) Recognise that 7/120 is the probability for the first two being unknown and the third known. We can have exactly two unknowns by having the first being known and the second and third unknown, or the first and third unknown and the second known. Hence, there are three ways of attaining the condition, and therefore the probability is: $3 \times \dfrac{7}{120} = \dfrac{7}{40}$.

415

Business mathematics and statistics

Once we reach the complexity of three outcomes or more, it is best to use the subtraction rule which is repeated from Section 13.2:

$$P[A] = 1 - P[\bar{A}]$$

Problem: Of 12 accounts held in a file, four contain a procedural error in posting the account balances. If an auditor samples:

(a) one account randomly, what is the probability that it will contain an error?
(b) two accounts randomly, what is the probability that at least one will contain an error?
(c) three accounts randomly, what is the probability that at least one will contain an error?

Solution:

(a) $P[\text{error}] = \dfrac{\text{number of accounts with error}}{\text{total number of accounts}} = \dfrac{4}{12} = \dfrac{1}{3}$

(b) $P[\text{at least one } E] = P[2 \text{ errors}] + P[\text{one error}]$

$= P[2 \text{ errors}] + P[E \text{ and } \bar{E}] + P[\bar{E} \text{ and } E]$

(allowing for order of error occurrence and the fact that each 'one error' is accompanied by a non-error due to the other choice)

$= P[E1] \times P[E2 \mid E1] + P[E1] \times P[\bar{E} \mid E1] + P[\bar{E}1] \times P[E2 \mid \bar{E}1]$

(all these being conditional probabilities, since there is no replacement of sampled items)

$= \dfrac{4}{12} \times \dfrac{3}{11} + \dfrac{4}{12} \times \dfrac{8}{11} + \dfrac{8}{12} \times \dfrac{4}{11}$

$= \dfrac{12}{132} + \dfrac{32}{132} + \dfrac{32}{132}$

$= \dfrac{76}{132}$

$= 0.576$

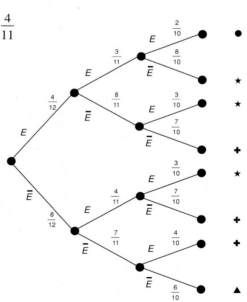

A tree diagram clarifies the choices:

- The branch symbolised by ● gives three errors.
- Each of the branches symbolised by ∗ have two errors.
- Each of the branches symbolised by ✚ have one error.
- The branch symbolised by ▲ has no errors.

Or better still, using the subtraction principle:

$P[\text{at least one error}] = 1 - P[\text{no errors}]$
$= 1 - P[\text{not 2 errors and not 1 error}]$
$= 1 - P[1\,\overline{E} \text{ and } 2\,\overline{E}\,]$
$= 1 - P[1\,\overline{E}\,] \times P[2\,\overline{E}\mid 1\,\overline{E}\,]$
$= 1 - \dfrac{8}{12} \times \dfrac{7}{11}$
$= 1 - \dfrac{56}{132}$
$= \dfrac{76}{132}$
$= 0.576$

(c) We can use the subtraction rule:

$P[\text{at least one error}] = 1 - P[\text{no errors}]$
$= 1 - P[1\,\overline{E} \text{ and } 2\,\overline{E} \text{ and } 3\,\overline{E}\,]$
$= 1 - P[1\,\overline{E}\,] \times P[2\,\overline{E}\mid 1\,\overline{E}\,] \times P[3\,\overline{E}\mid 1\,\overline{E} \text{ and } 2\,\overline{E}\,]$
$= 1 - \dfrac{8}{12} \times \dfrac{7}{11} \times \dfrac{6}{10}$
$= 1 - \dfrac{336}{1320}$
$= \dfrac{984}{1320}$
$= 0.745$

The tree diagram on the previous page also illustrates this clearly.

13.3 Competency check

1. Thirty-five per cent of the sales at W.E. Hookum Real Estate are two-level homes. Records also show that 15% of the clients buy a two-level home *and* pay the full cash value. When a client is selected at random, what is the probability that he has made a full cash settlement, given that he is buying a two-level home?

2. Consider families with two children, and assume that all four possible 'outcomes' are equally likely. What is the probability that both children are boys, if at least one is a boy?

3. It has been estimated by an insurance underwriter that the probability of a liquefied gas tanker having an accident is $\dfrac{1}{5000}$, and that the probability of the tanker spilling its load in an accident is $\dfrac{3}{10\,000}$. Calculate the probability of a tanker having an accident and spilling its load.

Business mathematics and statistics

4. The probability that a real-estate agent makes a sale is 1/5 on a wet day and 7/10 on a fine day. Given that the probability of a wet day is 3/5, find the probability that the day was wet given that the agent does not make a sale. (*Hint:* Use a tree diagram.)

5. A survey of householders in a town showed that 10% were unhappy with the plumbing jobs done in their homes. If 50% of the complaints named the plumber, Mr Rortum, and Mr Rortum does 40% of plumbing jobs in the town, find the probability that:

    (a) a householder will obtain an unsatisfactory plumbing job if the plumber is Mr Rortum;
    (b) a householder will obtain a satisfactory job if the plumber is Mr Rortum.

. . . .

13.4 Bayes' Theorem for conditional probability

Bayes' Theorem is a useful extension of conditional probability when a number of sequential events are occurring. It provides a way of revising our *prior* probabilities in the light of new information in order to provide a *posterior* probability. The probability of event A occurring can be modified when we know that events B and C have occurred—they change our attitude to A.

Often we hear an evening weather forecast that indicates a 50–50 or 50% chance of rain (event A) on the following day; this is a prior probability. Upon wakening, we see a heavy black sky (event B) which causes us to revise our estimate upwards to a 90% chance of rain; this is a posterior probability.

The process of revision follows the line:

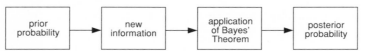

Bayes' Theorem in its simplest form for two events is:

$$P[B \mid A] = \frac{P[A \cap B]}{P[A]}$$

This theorem is a restatement of the general multiplication rule:

$$P[A \cap B] = P[A]P[B \mid A]$$

with $P[A \mid B]$ as the subject. Its aim is to evaluate the conditional probability of two events when we already know the probability of the joint event. We can also use the conditional statement $P[A \cap B] = P[B]P[A \mid B]$ to substitute for the term $P[A \cap B]$ and get:

$$P[B \mid A] = \frac{P[B]P[A \mid B]}{P[A]}$$

as an alternative form of Bayes' Theorem for two events.

Chapter 13 – Probability

The *general statement of Bayes' Theorem* involves n events from the sample space, namely $E_1, E_2, E_3, E_4, \ldots E_n$, and event A, which is also an event in the sample space. Then the posterior probability of event E_k occurring, given that A has occurred, is:

$$P[E_k \mid A] = \frac{P[E_k]P[A \mid E_k]}{P[E_1]P[A \mid E_1] + P[E_2]P[A \mid E_2] + \ldots P[E_n]P[A \mid E_n]}$$

$$P[E_k \mid A] = \frac{P[E_k]P[A \mid E_k]}{\sum P[E_i]P[A \mid E_i]} \quad \text{for } i = 1 \ldots n \text{ and } 1 \leq k \leq n$$

As an illustration of how this complex formula arises, consider the next problem.

Problem: A manufacturer receives supplies from two sources—70% from supplier 1 and 30% from supplier 2. Supplier 1 has a track record of 95% good parts. Supplier 2, with 90% good parts, is not quite so reliable. If a bad part is used in the manufacturing process, it will cause serious malfunctions and loss of revenue. Calculate the probability that a bad part is used and that it came from supplier 2—that is, $P[S_2 \mid B]$.

Solution: Examine the tree diagram below.

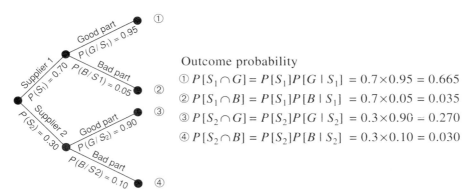

Outcome probability
① $P[S_1 \cap G] = P[S_1]P[G \mid S_1] = 0.7 \times 0.95 = 0.665$
② $P[S_1 \cap B] = P[S_1]P[B \mid S_1] = 0.7 \times 0.05 = 0.035$
③ $P[S_2 \cap G] = P[S_2]P[G \mid S_2] = 0.3 \times 0.90 = 0.270$
④ $P[S_2 \cap B] = P[S_2]P[B \mid S_2] = 0.3 \times 0.10 = 0.030$

② and ④ provide the bad outcomes, so $P[B] = P[S_1 \cap B] + P[S_2 \cap B] = 0.065$ ★

In the next few lines we will start the development of the general Bayes' formula. If we return to the Bayes' formula for two events:

$$P[X \mid Y] = \frac{P[X \cap Y]}{P[Y]}$$

here we have:

$$P[S_2 \mid B] = \frac{P[S_2 \cap B]}{P[B]}$$

We can now substitute from a similar statement: $P[B \mid S_2] = \dfrac{P[S_2 \cap B]}{P[S_2]}$

$$P[S_2]P[B \mid S_2] = P[S_2 \cap B] \; \maltese$$

and also from line ★, to get:

$$P[S_2 \mid B] = \frac{P[S_2]P[B \mid S_2]}{P[S_1 \cap B] + P[S_2 \cap B]}$$

Now, using line $\maltese$ we can substitute for the terms in the denominator:

$$P[S_2 \mid B] = \frac{P[S_2]P[B \mid S_2]}{P[S_1]P[B \mid S_1] + P[S_2]P[B \mid S_2]}$$

which gives an indication of how the general Bayes' formula is derived. It is worthwhile noting that the denominator is the complete exhaustion of all the events which give a 'bad' path.

To return to the numerical calculations, using the tree diagram:

$$P[S_2 \mid B] = \frac{P[S_2]P[B \mid S_2]}{P[S_1]P[B \mid S_1] + P[S_2]P[B \mid S_2]}$$

$$= \frac{0.3 \times 0.1}{0.7 \times 0.05 + 0.3 \times 0.1}$$

$$= 0.462$$

Problem: A builder employs three subcontractors—Gerston, Juncal and Bertrand. Because of their different rates of work, Gerston is responsible for completing 30% of the houses, Juncal 25% and Bertrand 45%. On average 5%, 10% and 12% of their respective work is substandard and has to be redone. If a house is selected at random and is deemed to be substandard, what is the probability that the work was done by Gerston?

Solution:
Let S be the event the work is substandard.
Let G be the event the work is done by Gerston.
Let J be the event the work is done by Juncal.
Let B be the event the work is done by Bertrand.
We wish to find $P[G \mid S]$.

From the data:

$P[G] = 0.30;\ P[J] = 0.25;\ P[B] = 0.45$
$P[S \mid G] = 0.05;\ P[S \mid J] = 0.10;\ P[S \mid B] = 0.12$

$$P[G \mid S] = \frac{P[G]P[S \mid G]}{P[G]P[S \mid G] + P[J]P[S \mid J] + P[B]P[S \mid B]}$$

$$P[G \mid S] = \frac{0.30 \times 0.05}{0.30 \times 0.05 + 0.25 \times 0.10 + 0.45 \times 0.12}$$

$$= 0.160$$

13.4 Competency check

1. A room contains 100 people who possess the following characteristics:

   - 50% are male, of whom 80% are overseas-born;
   - 50% are female, of whom 60% are overseas-born.
   - Find the probability of randomly selecting a male who is overseas-born.

2. Two per cent of the inhabitants of a country suffer from a certain disease. A new diagnostic test is developed that gives a positive indication 96% of the time when an individual has the disease, and a negative indication 94% of the time when the disease is absent. An individual is selected at random, is given the test and reacts positively. What is the probability that the person has the disease?

13.5 Summary

The basic facts for probability calculations are as follows:

1. For one event, X:

   - $0 \leq P[X] \leq 1$
   - $P[X] + P[\overline{X}] = 1$

   where $\overline{X}$ is the event that X does not happen.

2. For two events, X and Y:

   - $P[X \text{ or } Y] = P[X \cup Y]$ is the probability that X happens, Y happens or both X and Y happen—it is the joining of the sets X and Y.
   - $P[X \text{ and } Y] = P[X \cap Y]$ is the probability that both X and Y happen simultaneously.
   - $P[X \cup Y] = P[X] + P[Y] - P[X \cap Y]$.

3. Conditional probability:

   - The probability of X given that Y has already occurred is $P[X \mid Y]$.

   - $P[X \mid Y] = \dfrac{P[X \cap Y]}{P[Y]}$

Business mathematics and statistics

4. Independent events:
   - If $P[X \mid Y] = P[X]$, then the event Y has no influence on the probability of event X.
   - The events are said to be independent.
   - $P[X]\, P[Y] = P[X \cap Y]$ is the multiplication law for independent events.

5. Mutually exclusive events:
   - If $P[X \cap Y] = 0$, then X and Y cannot both occur at the same time—they are said to be mutually exclusive.
   - $P[X \cap Y] = P[X] + P[Y]$ is the addition law for mutually exclusive events.

6. Tree diagrams:

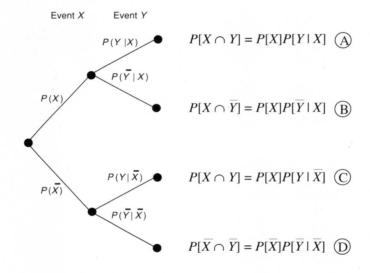

   - The total probability for any one set of branches is unity.
   - The sum of final probabilities (intersections) sums to unity—that is, the sample space.
   - Conditional probabilities are found—for example, for $P(X \mid Y)$—by:

$$P[Y] = \text{\textcircled{A}} + \text{\textcircled{C}}$$

$$\therefore P[X \mid Y] = \frac{P[X \cap Y]}{P[Y]} = \frac{\text{\textcircled{A}}}{\text{\textcircled{A}} + \text{\textcircled{C}}}$$

7. Bayes' Theorem:
   - For a set of conditional events:

$$P[E_k | A] = \frac{P[E_k]P[A|E_k]}{P[E_1]P[A|E_1] + P[E_2]P[A|E_2] + \ldots + P[E_n]P[A|E_n]}$$

where E_k is event number k out of a total of n events.

13.6 Multiple-choice questions

1. Jim has played 3283 games of solitaire on his computer. His percentage of games won has fluctuated around 7% since game number 1220. This situation illustrates:

    (a) the classical definition of probability;
    (b) independent probability;
    (c) the relative frequency definition of probability;
    (d) subjective probability.

2. A pack of 10 cards has a number printed on each card, namely 2, 3, 4, 5, 6, 7, 8, 9, 10 and 12. The cards are well shuffled and a card is drawn at random. The probability that the card will show an even number is:

    (a) $\frac{1}{10}$ (b) $\frac{1}{6}$ (c) $\frac{2}{5}$ (d) $\frac{3}{5}$

3. A standard pack of playing cards is used. If the cards are chosen randomly, which one of the following represents a set of mutually exclusive events?

    (a) jack, red;
    (b) 10, 8, queen;
    (c) diamond, club, red, black;
    (d) jack, queen, picture card.

4. Why are the events of a coin toss independent?

    (a) Both a head and tail cannot turn up on the one toss.
    (b) The probability of obtaining a head and the probability of obtaining a tail are the same.
    (c) The outcome of a toss is not affected by the outcome of any other toss.
    (d) Each of the outcomes is conditional.

5. X and Y are two subsets from the sample space S. It is known that $P[X \cup Y] = P[X] + P[Y]$. Which of the following diagrams best illustrates this situation?

    (a)   (b)   (c)   (d)

6. A card is drawn at random from a standard pack of 52 playing cards. The probability of drawing a red card or an ace is:

   (a) $\dfrac{2}{52}$ (b) $\dfrac{26}{52}$ (c) $\dfrac{28}{52}$ (d) $\dfrac{30}{52}$

7. Three coins are tossed. The probability of obtaining two heads is:

   (a) $\dfrac{1}{8}$ (b) $\dfrac{1}{2}$ (c) $\dfrac{3}{8}$ (d) $\dfrac{2}{3}$

8. A bag contains three red and two green balls. Two balls are chosen at random, without replacement. The probability of choosing one red and one green ball is:

   (a) $\dfrac{3}{5} \times \dfrac{2}{5} = \dfrac{6}{25}$ (b) $\dfrac{3}{5} \times \dfrac{2}{4} \times \dfrac{2}{5} \times \dfrac{3}{4} = \dfrac{12}{20}$

   (c) $\dfrac{3}{5} \times \dfrac{2}{5} \times \dfrac{2}{5} \times \dfrac{3}{5} = \dfrac{12}{25}$ (d) $\dfrac{3}{5} \times \dfrac{2}{5} = 1$

9. A company has 50 employees. Seventeen have tertiary qualifications (10 males, 7 females); 23 have their highest qualifications at Year 12 level (11 males, 12 females); and 10 have their highest qualifications at Year 10 level (3 males, 7 females). The conditional probability of randomly choosing an employee with tertiary qualifications given that it is a female is:

   (a) $\dfrac{17}{50}$ (b) $\dfrac{7}{26}$ (c) $\dfrac{26}{50}$ (d) $\dfrac{7}{50}$

10. Using the data in question 9, the marginal probability of randomly choosing a male is:

    (a) $\dfrac{10}{24}$ (b) $\dfrac{17}{50}$ (c) $\dfrac{24}{50}$ (d) $\dfrac{10}{50}$

13.7 Exercises

1. In a series of tests on the thermal conductivity of glass samples, the following values were obtained, measured in units of W/mK (watts per millikelvin).

1.35	0.83	0.85	1.45	0.69	0.97	1.33	0.84	0.76	0.96	0.95
1.02	1.05	1.15	0.71	1.45	1.37	1.84	1.36	1.27	1.27	1.16
1.08	1.18	1.09	1.12	1.03	0.94	1.04	0.73	1.12	1.79	1.26
1.17	1.15	1.03	0.92	0.99	0.89	0.89	1.05	1.07	1.16	1.11
1.21	1.18	1.34	1.25	1.17	1.37	1.26	1.15	0.93	0.87	0.99
1.45	1.29	1.85	0.87	0.56						

   What is the probability of selecting a glass with thermal conductivity of < 1.05?

Chapter 13 – Probability

2. During 40 working days, five days were lost because of rain, two days because of industrial dispute and three because of supply problems with materials. (The assumption is that these are independent events.) Find the probability, based on this experience, that the next working day will be lost because of:

   (a) rain;
   (b) rain or lack of materials;
   (c) any of these factors.

   Find the probability that in the next three days:

   (d) the first day will be lost because of a dispute, the next because of lack of materials, and the third because of rain.

3. In a survey of 87 students, it was found that:

   - 44 studied valuation;
   - 28 studied surveying;
   - 5 studied valuation and surveying;
   - 8 studied neither economics, surveying nor valuation;
   - 12 studied economics but neither surveying nor valuation;
   - 2 studied economics and surveying but not valuation;
   - 3 studied valuation and economics but not surveying;
   - 1 studied all three subjects.

   Place these facts onto a Venn diagram and use it to find the probability of randomly selecting a student who studied:

   (a) economics;
   (b) valuation only;
   (c) valuation and economics but not surveying;
   (d) valuation or surveying or economics;
   (e) two subjects.

4. In a survey of 100 executives, it was found that:

   - 65 read *Business Review Weekly*;
   - 45 read the *Australian Financial Review*;
   - 55 read the *Asia Times*;
   - 5 read all three magazines;
   - 25 read *Business Review Weekly* and the *Australian Financial Review*;
   - 35 read *Business Review Weekly* and the *Asia Times*;
   - 15 read the *Australian Financial Review* and the *Asia Times*.

   Place these facts onto a Venn diagram and use it to find the probability of randomly selecting an executive who:

   (a) reads none of these magazines;
   (b) reads *Business Review Weekly* only;
   (c) reads *Business Review Weekly* and the *Asia Times*, but not the *Australian Financial Review*;

(d) reads *Business Review Weekly* or the *Australian Financial Review* or the *Asia Times*;
(e) reads two magazines.

5. There is a 65% probability that each toss of an asymmetrical coin will produce a head. Find the probability that three consecutive tosses will produce:

   (a) three heads
   (b) three tails
   (c) two heads followed by a tail
   (d) two heads.

6. Batches of microchips with 60 in each batch are known to contain on average three defectives in each batch. A dealer decides to test the present quality he is receiving and samples three microchips, *with replacement*. What is the probability that:

   (a) all three are defective?
   (b) one is defective?
   (c) two are defective?
   (d) no more than two are defective?

7. A market research company is commissioned to survey where people buy their home improvement supplies. Out of 120 people interviewed, 40 stated that they use only local hardware stores, 50 use only specialist trade stores and 10 use neither. Use a Venn diagram to represent this data. What is the probability of randomly selecting one person who:

   (a) uses both stores?
   (b) uses neither type of store?
   (c) uses a specialist shop but not a local store?
   (d) uses a specialist shop?
   (e) uses one type of shop only?

8. It is known from long-term sales records that the probability of a salesman and a saleswoman failing to make the first sale on any given day are 2/3 and 3/5 respectively. Assuming these events are independent, find the probability that on a given day:

   (a) both will successfully make their first sale;
   (b) neither will make a successful first sale;
   (c) at least one will successfully make a first sale.

9. To determine the ballot results for a release of land, three boxes *A*, *M* and *S* contain coloured tokens.

   - *A* contains five blue tokens and four white tokens.
   - *M* contains seven blue tokens and five white tokens.
   - *S* contains three blue tokens and five white tokens.

   (a) If tokens are drawn from *S* with replacement, find the probability that the third token drawn is the second white token.
   (b) If one of the boxes is selected at random and a token is withdrawn, find the probability that:

(i) box *A* is chosen and the token is blue;
(ii) a white token is chosen.

10. On the basis of past market trends, an investor forecasts the probability that a certain stock will rise to be 0.25; that it will fall to be 0.55; and that it will remain stationary to be 0.30. Does this seem reasonable?

11. A land and stock agent has two vehicles. In winter the probability of the Yotato 4WD starting is 0.8, and the probability of the Sinans 4WD starting is 0.4. There is a probability of 0.3 that both will start. Find the probability that:

    (a) at least one vehicle will start;
    (b) neither will start.

12. The probability that a company representative will have an automobile accident in a given year is 0.08. Find the probability that the representative:

    (a) goes for four years without an accident;
    (b) has at least one accident in four years.

13. A survey of a carpark shows that the probability of a car:

    - having ACT numberplates is 0.68;
    - being an imported vehicle is 0.24;
    - being an imported vehicle with ACT numberplates is 0.18.

    Find the probability that:

    (a) a car chosen at random is either an import or has an ACT numberplate;
    (b) a car with an ACT numberplate is an import;
    (c) an imported car has an ACT numberplate.

14. The Waterfall Wanderers have won the district hurling competition. An unbiased commentator estimates that their probability of winning the state title is 0.65. If they win the state title, their probability of winning the Australian title is 0.75. Find the probability that the Wanderers:

    (a) will win both the state and Australian titles;
    (b) will win the state title but lose the Australian title.

15. A high-rise residential and office complex will have a security system installed. There are two main electronic components—*J* and *G*.

    - The probability of *J* failing is 0.55.
    - The probability of *G* failing, given that *J* has failed, is 0.75.
    - The probability of *G* failing if *J* has not failed is 0.15.

    Find the probability that:

    (a) both *J* and *G* need replacing;
    (b) *G* needs replacement when *J* does not;
    (c) one of the components needs replacing.

16. A doctor has a patient with a complaint that is probably one of two related conditions—A or B. She gives the probability that it is A as 0.38. She estimates a probability of 0.52 that it is B. There is a 10% chance that it is neither of these and hence would be treatable by a simple, old-fashioned remedy. Ninety per cent of patients with condition A require surgery, whereas only 30% of those with B require surgery.

    (a) What is the probability that the patient requires surgery?
    (b) If the patient does not require surgery, what is the probability that he has condition A?
    (c) If the patient does not require surgery, what is the probability that he has neither condition?

17. It is known that 7.7% of drivers on the road are under 20 years of age. It is also known that the probability of a driver under 20 having a fatal accident is 0.00082, and the probability of a driver over 20 having a fatal accident is 0.00039. What is the probability that a driver having a fatal accident will be:

    (a) under 20?
    (b) over 20?

18. A company has decided to discontinue credit to any customer who has been at least one week late in payment at least twice. Records show that 80% of all those defaulting on payments had fulfilled the above conditions. Further examination shows that:

    - 3% of all credit customers default;
    - 40% of those who have not defaulted have been late with at least two payments.

    Find the probability that a customer with two or more late payments will default. Will this be a good policy for the company?

19. In a factory, four machines produce the same product. The statistics are:

    - machine W produces 10% with 0.1% defective output;
    - machine X produces 20% with 0.05% defective output;
    - machine Y produces 30% with 0.5% defective output;
    - machine Z produces 40% with 0.2% defective output.

    An item, selected at random, is found to be defective. What is the probability that it was produced by W?

20. A housing estate is to be built next to a nuclear research plant. A warning device to indicate leakage is accurate 94% of the time. Also, the device gives a false alarm 15% of the time. Environmentalists claim the chance of leakage is 8% at any given time. What is the probability that a leakage actually exists if the warning device indicates there is a leakage?

Chapter 14
NORMAL PROBABILITY DISTRIBUTION CURVE

LEARNING OUTCOMES
After studying this chapter you should be able to:
- Describe the special features of a normal distribution curve
- Identify the characteristics of the standard normal curve
- Read z score tables to find the area under the curve and the area between given standard z scores
- Find the z score given the area under the curve
- Solve business problems that can be represented by the normal distribution curve.

Each year, hordes of people attend the post-Christmas clothing sales in hope of buying that bargain of a lifetime. Many of them are disappointed, however, when they can't find the right size skirt or dress, shoes or suit. The problem is that retailers generally order garments only in sizes corresponding to the 'average' shopper's measurements. If your measurements differ too much from the norm, you may have to shop at stores that cater for non-standard sizes.

You can't blame the retailer for not stocking up on non-standard sizes, because the company buyer has to order stock on the basis of expected sales. Buyers operate under conditions of uncertainty, keeping their fingers crossed that the garments will sell and that they ordered the correct assortment of sizes. It is interesting that collected biological data (such as height, weight, shoe size, waist measurement, life expectancy and the like) approach what is referred to as the **normal probability distribution**. This prevents the buyer from ordering too many women's shoes in size 4B or insufficient women's dresses in size 12. In other words, this distribution provides an insight into the chance or likelihood of an event occurring (e.g. a male with a size 56 chest wanting to buy a shirt) and therefore assists in improving decision-making.

As we are all aware, many business decisions are made under conditions of uncertainty; if there are ways of reducing the degree of uncertainty, we should seize them as they will help our purchasing methods, production, inventory control and, ultimately, profits. The use of the **normal curve** (as the normal probability distribution curve is more conveniently referred to) gives businesspeople the opportunity to improve their chances of success by accounting for the probability of occurrence.

The normal probability distribution is also called the Gaussian distribution after the 18th century mathematician–astronomer Karl Gauss who observed, along with other mathematicians, that discrepancies among repeated measurements of the same physical quantity displayed a surprising degree of regularity; their pattern (distribution), it was found, could be closely approximated by a certain kind of continuous distribution curve, referred to as the 'normal curve of errors' and attributed to the laws of chance.[1] It was found, for instance, that the normal distribution comes close to fitting actual observed frequency distributions of many phenomena, including human characteristics (weight, height, IQ), outputs from physical processes and other measures of interest to management in both the social and natural sciences. The properties of the normal distribution also allow inferences to be made from samples. In fact, the normal probability distribution is considered the most important.

Therefore, the phenomenon of certain types of data approaching the normal probability distribution can be found in most business and economic statistics, such as those relating to the life of plant and equipment, faulty merchandise, and wages of similar occupations, as well as in the social sciences, medical sciences and physical sciences. You will readily find that the characteristics of the normal probability distribution apply in each of these disciplines; for our purposes, we will concentrate on the business sector. However, first you need a more precise explanation of just what this normal curve is.

Figure 14.1 *Normal curves with the same mean but different standard deviations*

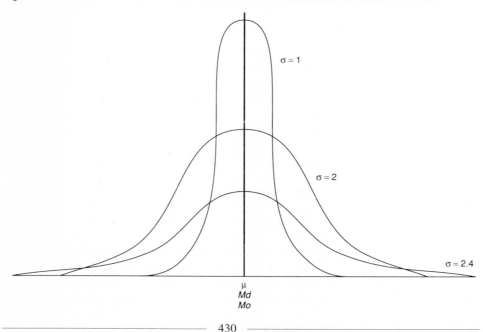

Chapter 14 – Normal probability distribution curve

When plotting most business statistics in a histogram and subsequent frequency polygon, the frequencies of observations tend to approximate a bell-shaped curve. If you look at Figure 14.1, all the curves are considered normal curves—the three measures of central tendency equal one another at the midpoint under the curve, but each normal curve exhibits a different standard deviation (i.e. variation or dispersion from the mean value). Figure 14.2 illustrates the other case, where standard deviations are identical but the means are different.

Figure 14.2 *Normal curves with the same standard deviations but different means*

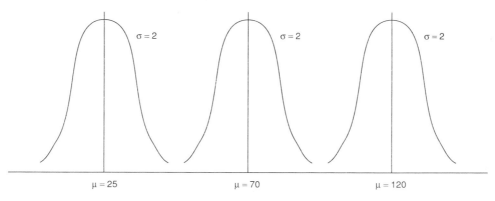

In other words, the family of normal curves is extended, as the normal curves can portray many populations which differ only in terms of their mean and/or standard deviation. Therefore, we can say that a particular normal curve is defined by its mean (μ) and its standard deviation (σ).

The graphic representation illustrates several important properties of the normal distribution curve:

- It is bell-shaped.
- The mean of a normally distributed population lies at the centre of its normal curve.
- The median and mode of the distribution are also at the centre, due to the symmetry of the distribution.
- The two tails of the normal probability distribution extend indefinitely in each direction without ever touching the horizontal axis.
- The area under the curve and above the horizontal axis is equal to 1.00 (or 100%).
- To define a specific normal probability distribution, the mean and the standard deviation are all that is needed.

14.1 Area under the normal curve

Probabilities for the normal curve are represented by **areas under the curve**. The total area under the curve and above the *x* axis is 1.00 (or 100%), regardless of the values of the mean and standard deviation. Figure 14.3 should help to explain

Business mathematics and statistics

this concept. It has been mathematically proved that in a normally populated distribution:

- approximately 68% of all values lie between ±1 standard deviation from the mean and thus consume 68% of the area under the curve;
- approximately 95% of all values lie between ±2 standard deviations from the mean and thus consume 95% of the area under the curve;
- approximately 100% of all values lie between ±3 standard deviations from the mean and thus consume almost 100% of the area under the curve.

Figure 14.3 *Areas under the normal curve here*

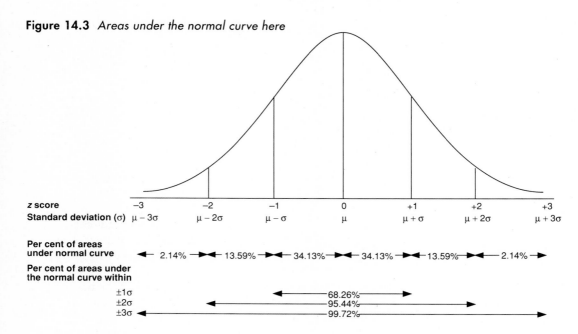

Since the mean is the midpoint of the normal curve, cutting it into halves, obviously makes each half equal 50%. Currently, with the tourism boom Australia-wide, congestion at international and domestic terminals is being examined. Suppose the mean time in clearing customs at the international terminal is 75 minutes with a standard deviation of 15 minutes. There is a 50% probability that a randomly selected traveller would have waited 75 minutes or more to leave the terminal and a 50% probability that the traveller would have waited 75 minutes or less.

If we wanted to discover what percentage of travellers had waited between 75 and 90 minutes to get through the terminal, realising that the mean is 75 minutes, 90 minutes is one standard deviation away from the mean (90 minutes is 15 minutes greater than the mean, hence one standard deviation). Since 90 is greater than 75, it must be one standard deviation to the right of the mean. According to Figure 14.3, 34.13% of observations occur within +1 standard deviation from the mean. It is advisable to sketch the normal curve, since it makes it easier to visualise the location of the

Chapter 14 – Normal probability distribution curve

values or standard deviations you are to determine. Figure 14.4 illustrates the proportion of travellers waiting between 75 and 90 minutes to clear customs (the hatched area).

Figure 14.4 *Areas under the normal curve: customs clearance time for travellers (in minutes)*

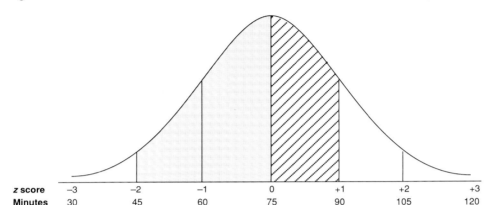

| z score | −3 | −2 | −1 | 0 | +1 | +2 | +3 |
| Minutes | 30 | 45 | 60 | 75 | 90 | 105 | 120 |

Problem: Determine the proportion of travellers who waited between 45 and 90 minutes.

Solution: In Figure 14.4, 45 is −2 standard deviations (to the left) from the mean (since one standard deviation is 15, then the value of 45 which is two deviations or 30 away from the mean of 75), equivalent to 34.13% plus 13.59% (as per Figure 14.3) or 47.72% (the shaded area of Figure 14.4). This indicates that 47.72% of travellers waited between 45 and 75 minutes, and 2.28% of travellers waited less than 45 minutes (since to the left of the mean is 50% of the curve and 50 − 47.72 = 2.28).

Ninety minutes is +1 standard deviation from the mean, or 34.13% of travellers (note the hatched area under the curve between 0 and +1). Therefore, 81.85% (i.e. 47.72% + 34.13%) of travellers waited between 45 and 90 minutes (the combined shaded and hatched areas). Note that we always measure the deviation from the mean, not between 45 and 90 minutes, but between 45 and 75 and between 75 and 90.

Problem: Determine the proportion of travellers who will wait:

(a) longer than 105 minutes;
(b) less than 90 minutes;
(c) less than 60 minutes.

Solution:
(a) In Figure 14.4, 105 minutes is +2 standard deviations to the right of the mean, equivalent to 34.13% + 13.59% (see Figure 14.3) or 47.72%.

Therefore, 47.72% of travellers waited for between 75 and 105 minutes, and another 50% waited for under 75 minutes, leaving 2.28% who waited for over 105 minutes.

(b) Ninety minutes is +1 standard deviation to the right of the mean. Once again, referring to Figure 14.3, you will note that this is equivalent to 34.13% of the area under the curve. Since the question asked for the proportion of travellers who took less than 90 minutes to get their luggage, we must add the 50% of travellers who took less than 75 minutes (where the standard deviations are to the left of the mean). Therefore, 84.13% of travellers took less than 90 minutes.

(c) To discover the proportion of travellers who took less than 60 minutes, first we must find the proportion which took between 75 (the mean) and 60 minutes. Sixty minutes is −1 standard deviation from the mean, giving a proportion of 34.13%. This suggests that of the 50% taking less than 75 minutes to get their luggage, 34.13% took between 60 and 75 minutes, and implies that the rest (i.e. 15.87%) took less than 60 minutes.

Table of the area under the normal curve

As you can appreciate, applications of the normal curve result only infrequently in precise intervals of 1, 2 or 3 standard deviations from the mean. A statistical table is available (Appendix 14A at the conclusion of this chapter) just for this purpose. However, since it would be impossible to create a table measured in the actual units of every normal curve (e.g. in terms of time, weight, age, costs), the table depicts what is known as the *standardised normal curve*, which is then used to determine probabilities for any normal distribution.

To understand why this table of areas can be applied in all instances of normal curves, you should try to understand the relationship of the standard deviation to the normal curve. In the previous example, for instance, there is a 34% probability that travellers had to wait between 75 and 90 minutes for customs clearance, with a mean time of 75 minutes and a standard deviation of 15. In another example—a car taking approximately 36 hours to be manufactured, with a standard deviation of 7 hours—there is also a 34% probability that it will take between 36 and 43 hours to manufacture the vehicle. Therefore, the area under the curve is the same in both examples. The area from the mean to +1 standard deviation is the same (see Figure 14.5 opposite).

Turning to Appendix 14A, we note that the table shows the area between the mean and different standard deviations (referred to as the **z score**). As you can see, the z score, or standard deviation, is defined in increments of one-tenth reading down and increments of one-hundredth reading across. For example, a z score of 1.75 can be located by reading 1.7 down and 0.05 across, with the intersection at 0.45994. This means that 45.994% of observations would occur between the mean and +1.75 standard deviations, or that there is a 45.994% probability that an observation will be between the mean (μ) and 1.75 standard deviations (see Figure 14.6). Note that only half of the area under the curve is covered in the table, with a z score of 2.99 having a value of .49861 or 49.861% of the area under the curve covered by 2.99

Figure 14.5 *Areas under the curve for customs clearance time for travellers (a) and motor vehicle manufacturing (b)*

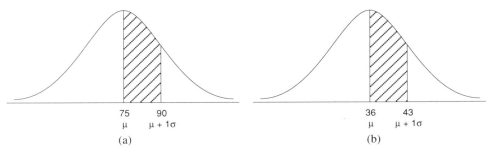

standard deviations from the mean. Since the normal curve is symmetrical, values applying to one half of the curve also apply to the other half (i.e. these z scores may be for positive or negative standard deviations from the mean).

A z score defines the distance from the nominated value to the mean (in terms of standard deviations). It can be calculated by:

$$z \text{ score} = \frac{\text{Selected value } (x) - \text{Mean of all } x \text{ values in the population } (\mu)}{\text{Standard deviation of } x \text{ in the population } (\sigma)}$$

$$z \text{ score} = \frac{x - \mu}{\sigma}$$

Figure 14.6 *Area under the curve: locating z score*

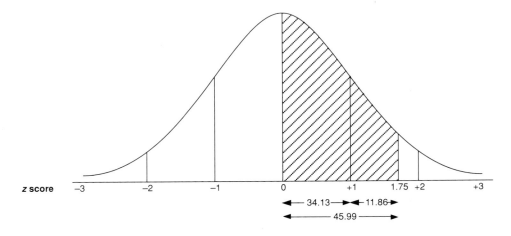

Problem: What proportion of travellers can expect to clear customs in 50 minutes or less?

Solution: First sketch a diagram (Figure 14.7).

Figure 14.7 *Area under the curve for travellers' customs clearance time of 50 minutes or less*

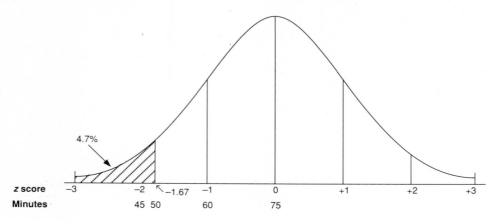

Now solve for z:

$$z = \frac{x - \mu}{\sigma}$$

$$= \frac{50 - 75}{15}$$

$$= \frac{-25}{15}$$

$$= -1.67 \text{ standard deviations}$$

A negative z score indicates that the result is less than the mean, and is hence to the left of the mean; continue to read the table as usual with a z score of −1.67 approximately equivalent to 0.453. Therefore, 45.3% of travellers would take between 50 and 75 minutes to clear customs. To discover what proportion would take less than 50 minutes, we must find what area remains under the curve to the left of the mean; hence, 50% − 45.3% = 4.7%, or approximately 5% of travellers would clear customs in less than 50 minutes.

Problem: Let's revisit our wandering international minstrels and this time determine the probability of their having to wait for between 85 and 95 minutes to clear customs.

Solution:

1. Sketch a diagram (see Figure 14.8). Since only areas immediately adjacent to the mean can be found in the table, we must solve for two z scores, 75 to 85 and 75 to 95.

Figure 14.8 *Areas under the curve for travellers' customs clearance time of between 85 and 95 minutes*

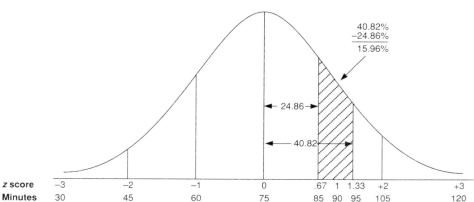

2. To find the z score for 95, apply the formula:

$$z = \frac{x - \mu}{\sigma}$$
$$= \frac{95 - 75}{15}$$
$$= \frac{20}{15}$$
$$= 1.33$$

3. Looking up 1.33 in Appendix 14A, we arrive at 0.40824, which indicates that 40.82% of travellers have to wait for between 75 and 95 minutes.

4. Find the probability of waiting for between 75 and 85 minutes:

$$z = \frac{x - \mu}{\sigma}$$
$$= \frac{85 - 75}{15}$$
$$= \frac{10}{15}$$
$$= 0.666$$
$$= 0.67$$

5. Looking up 0.67, we find that there is a 24.86% probability of waiting for between 75 and 85 minutes.

6. We determined that there was a 40.82% probability of waiting for between 75 and 95 minutes and a 24.86% probability of waiting for between 75 and 85 minutes. Therefore, there is an approximate 16% probability (i.e. 40.82%

– 24.86%) of travellers having to wait for between 85 and 95 minutes for customs clearance.

Problem: Assume that the accountants' weekly wages followed a normal distribution. An accountant not surveyed wants to earn $750 per week. How can we calculate the likelihood of his or her achieving that, given that the mean salary was calculated to be $519?

Solution: $750 is $231 more than the approximate mean of $519. The standard deviation was found to be $112.78, say $113. Divide the deviation from the mean (231) by the standard deviation (113). That is:

$$z \text{ score} = \frac{x - \mu}{\sigma}$$

$$= \frac{750 - 519}{113}$$

$$= \frac{231}{113}$$

$$= 2.04$$

To determine the likelihood of earning a salary of $750, look once again at Appendix 14A. The 2.04 deviation corresponds to the area under the curve of 0.47932; this means that to the left of this point, 97.932% (0.47932 + 0.50), approximately 98% of accountants receive less than $750, while only 2% receive more (see Figure 14.9). In other words, only about 2 out of every 100 accountants earn $750 or more per week.

Figure 14.9 *Areas under the normal curve for accountants' wages*

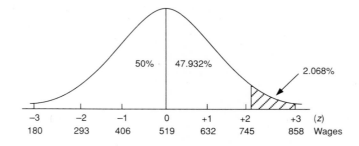

Problem: What is the likelihood of an accountant not surveyed earning less than $450 per week?

Chapter 14 – Normal probability distribution curve

Solution: To determine the z score, substitute into the equation:

$$z = \frac{x - \mu}{\sigma}$$

$$= \frac{450 - 519}{113}$$

$$= \frac{-69}{113}$$

$$= -0.611$$

$$= -0.61$$

This indicates that $450 is –0.61 standard deviations less than the mean. Looking up 0.61 in Appendix 14A, we get 0.22907, or approximately 23%. This figure indicates that 23% plus 50%, or 73%, receive weekly incomes exceeding $450, while approximately 27% receive less than $450 (see Figure 14.10).

Figure 14.10 *Areas under the normal curve for accountants' wages*

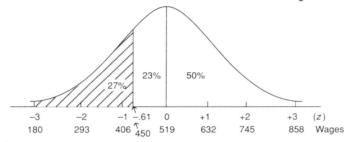

Problem: What is the likelihood of an accountant receiving a weekly wage of between $600 and $700?

Solution: First determine the z score of each of the selected values. To calculate the z score of the selected value of $600:

$$z = \frac{x - \mu}{\sigma}$$

$$= \frac{600 - 519}{113}$$

$$= \frac{81}{113}$$

$$= 0.717$$

$$= +0.72$$

This indicates that $600 is +0.72 deviations to the right of the mean (as 0.72 is positive). From Appendix 14A we can see that a z score of 0.72 corresponds to

0.26424; therefore, approximately 26.4% of accountants would receive between $519 and $600.

To calculate the z score of the selected value of $700:

$$z = \frac{x - \mu}{\sigma}$$

$$= \frac{700 - 519}{113}$$

$$= \frac{181}{113}$$

$$= +1.60$$

A z score of 1.6 is equivalent to 0.44520 or 44.5% (Appendix 14A), which indicates that approximately 45% of accountants receive weekly wages of between the mean ($519) and +1.6 standard deviations ($700).

To determine the percentage of accountants receiving incomes of between $600 and $700, we have to realise that 26.4% receive wages of between $519 and $600 (+0.72 standard deviation) and that 44.5% receive between the mean of $519 and $700 (+1.6 standard deviations). Therefore, approximately 18% (44.5% − 26.4%) would receive incomes of between $600 and $700 (see Figure 14.11).

Figure 14.11 *Areas under the normal curve for accountants' wages of between $600 and $700*

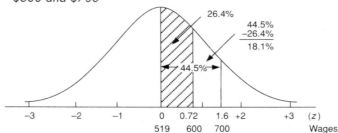

Problem: What is the likelihood of a student receiving a grade below 50% (see Table 11.23 in Chapter 11)?

Solution: As the mean is 64 and the standard deviation is 12.59, substitute into the equation:

$$z = \frac{x - \mu}{\sigma}$$

$$= \frac{50 - 64}{12.59}$$

$$= \frac{-14}{12.59}$$

$$= -1.11$$

Looking up Appendix 14A, we see that −1.11 corresponds to 0.36650, or 36.65%. This indicates that to the left of 0.36650 lies 0.1335 and to the right lies 0.50000. Therefore, about 87% of students received a grade of 50% or more, while approximately 13% received less than 50% (see Figure 14.12).

Figure 14.12 *Areas under the normal curve for students' grades*

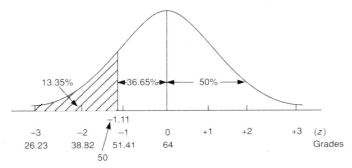

Problem: What is the likelihood of a student receiving a grade between 60% and 70%?

Solution: First find the z score for each selected value. To calculate the z score of the selected value of 60%:

$$z = \frac{x - \mu}{\sigma}$$

$$= \frac{60 - 64}{12.59}$$

$$= \frac{-4}{12.59}$$

$$= -0.318$$

$$= -0.32$$

A z score of −0.32 is equivalent to 0.12552, or approximately 12.6%. This indicates that 12.6% received grades between 60% and 64%, 37.4% received grades below 60%, and 62.6% received grades above 60%.

To calculate the z score of the selected value of 70%:

$$z = \frac{x - \mu}{\sigma}$$

$$= \frac{70 - 64}{12.59}$$

$$= \frac{6}{12.59}$$

$$= 0.48$$

A z score of 0.48 corresponds to 0.18439, or 18.4%. Thus, 18.4% of students received grades between the mean of 64% and 70% (0.48 standard deviations from the mean).

Therefore, 31% of students (12.6% + 18.4%) received grades between 60% and 70% (see Figure 14.13).

Figure 14.13 *Areas under the normal curve for students' grades between 60% and 70%*

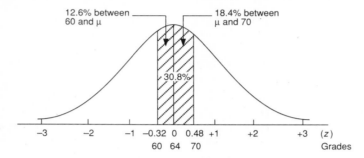

Problem: Management has decided to give wage rises to the bottom 15% of salaried accountants. If the distribution follows a normal curve and the mean is $519 with a standard deviation of $113, what is the cut-off point for eligibility for the raise?

Figure 14.14 *Areas under the curve for the lowest 15% of paid accountants*

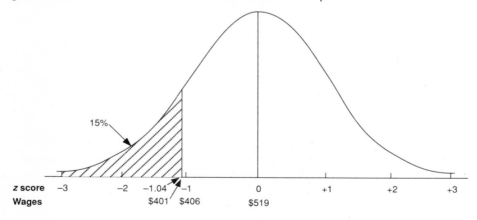

Solution: First sketch a normal curve (Figure 14.14) and observe that since we want the lowest 15%, we are looking at the left-hand side of the mean (hence, a negative standard deviation), and since the area under the curve relates to standard deviations (hence, percentages) from the mean, we locate 35% in the area under the curve (i.e. 50% − 15%) at approximately −1.04 standard deviations. We now substitute into our formula and solve for x:

Chapter 14 – Normal probability distribution curve

$$z = \frac{x - \mu}{\sigma}$$

$$-1.04 = \frac{x - \$519}{\$113}$$

$$-1.04(\$113) + \$519 = x$$

$$-\$117.52 + \$519 = x$$

$$\$401.48 = x$$

$$\$401 = x$$

Therefore, any accountant receiving less than approximately $401 will receive the wage rise.

Problem: A manufacturer must decide what warranty to give on her new product (see Figure 14.15). The mean life is expected to be 10 years with a standard deviation of 2.5 years. Assuming the product's life approaches a normal curve, for how long should the warranty apply if the manufacturer targets a return of faulty products under a warranty of only 2.25%?

Solution: First look up Appendix 14A. The z score for 47.75% (since 50% – 2.25% = 47.75%) is approximately –2.0. Sketch this and note that we are looking for the left-hand tail end of the normal curve.

Figure 14.15 *Areas under the curve for a 2.25% return under warranty for product*

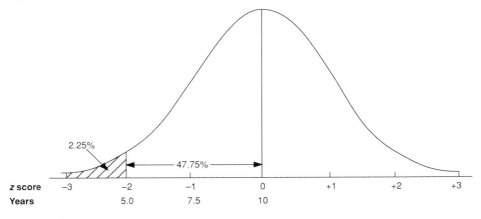

Next, solve for x in the formula:

$$z = \frac{x - \mu}{\sigma}$$

$$-2.0 = \frac{x - 10}{2.5}$$

$$(-2.0)(2.5) + 10 = x$$

$$5 = x$$

Business mathematics and statistics

For the manufacturer to ensure a return of only 2.25%, hence incorporating this expense into the costing at the time of production, a warranty of five years is necessary.

Problem: Let's have a look at another situation where the z score is given but we must find the selected value. For example, assume that admission to a particular university is restricted to the top 20% of applicants. If the mean score is 250, with a standard deviation of 50, what is the minimum aggregate needed for admission to this university?

Solution: We must find the grade that the top 20% of applicants achieved—that is, the grade below which 80% of the applicants fell. First, find the 30% area to the right of the mean or zero standard deviation (as to the left of that are 80% of students and to the right the 20% of students we are aiming for) on the normal distribution curve. According to Appendix 14A, 30% corresponds to 0.84. As we now have the value for z, the standard deviation and the mean, we can substitute into the formula:

$$z = \frac{x - \mu}{\sigma}$$
$$z\sigma = x - \mu$$
$$\begin{aligned} x &= z\sigma + \mu \\ &= (0.84)50 + 250 \\ &= 42 + 250 \\ &= 292 \end{aligned}$$

Thus, applicants must score above 292 to be in the top 20% (see Figure 14.16).

Figure 14.16 *Areas under the normal curve for university admission applicants*

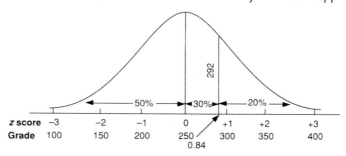

Problem: Scores on IQ tests approximate normal distribution for people of the same age. The standard scaling gives a mean IQ of 100 with a standard deviation of 15. What IQ would be needed to be:

(a) in the top 2% of the population?
(b) in the bottom 10% of the population?

Chapter 14 – Normal probability distribution curve

Solution:

(a) To determine the IQ necessary to be in the top 2%, find the *z* score for the 48% which lie to the right of the centre of the distribution. Looking up Appendix 14A, we find that 48% corresponds to a *z* score of approximately 2.05. Therefore:

$$z = \frac{x - \mu}{\sigma}$$

$$2.05 = \frac{x - 100}{15}$$

$$x = 15(2.05) + 100$$
$$= 30.75 + 100$$
$$= 130.75$$

Therefore, an IQ of approximately 131 is necessary to be in the top 2% of the population (see Figure 14.17).

Figure 14.17 *Areas under the normal curve for IQ results*

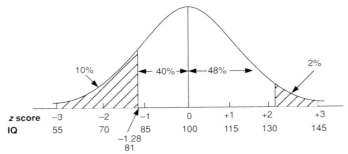

(b) To ascertain the IQ of the lowest 10%, locate the 40% to the left of the distribution centre. The 40% corresponds to a *z* score of approximately −1.28. Therefore:

$$z = \frac{x - \mu}{\sigma}$$

$$-1.28 = \frac{x - 100}{15}$$

$$x = 15(-1.28) + 100$$

$$= -19.2 + 100$$

$$= 80.8$$

Therefore, an IQ of about 81 designates the bottom 10% of the population.

14.1 Competency check

(Note: Assume that all of the following are normal distributions.)

1. Using Appendix 14A, determine the area under the standard normal curve for:

    (a) between $z = 0$ and $z = 1.75$;
    (b) between $z = 1$ and $z = 1.25$;
    (c) between $z = -0.76$ and $z = 2.05$.

2. Customers at Eat and Run Deli spend an average of $5.50 per lunch with a standard deviation of $1.10. What share of customers spend more than $8.20 on average?

3. If you have completed the questions in the two previous chapters, you will have found that executives had an approximate mean salary of $63 300 with a standard deviation of $15 700. What percentage of executives earned (to the nearest per cent):

    (a) between $75 000 and $85 000?
    (b) less than $35 000?
    (c) between $55 000 and $65 000?

4. Referring to question 3, the top 5% of executive salary earners expect a further promotion, and hence a salary increment. What minimum salary must an executive earn to be eligible for this salary increment?

5. A process worker takes a mean time of one hour to complete a job, with a standard deviation of 10 minutes. Find (to one decimal place):

    (a) the percentage of times he or she can complete the job in 45 minutes or less;
    (b) the probability of the worker completing the job within 82 minutes;
    (c) the probability of the worker receiving a commendation for completing the task within 35 minutes.

14.2 Sampling distribution of the mean*

As you are no doubt aware, few marketers attempt to survey an entire population—in many cases, because it would be impossible, or because of time and monetary constraints. Instead, a sample of the population is taken because we intuitively assume that the value of the sample mean tends towards the value of the population mean. But is there a problem with selecting a sample? For instance, if you have a population

---

* Optional

Chapter 14 – Normal probability distribution curve

of one million workers and you took different samples of size 100 to determine their average income, do you think the mean of each sample chosen would vary from sample to sample? Certainly, they probably would. But who has the time or energy to construct a probability distribution of all the possible means of all the possible samples—in other words, to construct the sampling distribution of the mean?

Well, the good news is that we don't have to waste time coming up with different sample sets of 100 workers. Kind statisticians have proved that if the sample size is large enough (a minimum of 30), the sampling distribution approximates the normal distribution and has the same mean as the population, irrespective of whether or not the population is normally distributed. And not only that: the sampling distribution is normally distributed regardless of sample size if the population is normally distributed.

Standard error of the mean

When we looked at the normal curve, we discussed the likelihood of an event being determined by the number of standard deviations away from the mean. Well, we can also describe the standard deviation of the sample means (defined by the symbol $\sigma_{\bar{x}}$) from the population mean (μ), and this is referred to as the **standard error of the mean**. For example, there is roughly a 68% chance that a randomly selected sample will have a mean lying within one standard deviation of the population mean. As you may have suspected, there is a 95.4% likelihood that the sample mean will lie within two standard deviations of the population mean.

It may help to think of the standard error of the mean as the average absolute distance between the sample mean and the population mean. The average distance should decline as the sample size increases; in other words, the larger the sample size, the closer the sample mean approaches the population mean.

There is a simple relationship between the population's standard deviation and the standard deviation of the sample means defined as:

$$\sigma_{\bar{x}} = \frac{\sigma}{\sqrt{n}}$$

where $\sigma_{\bar{x}}$ = standard error of the mean
$\sqrt{n}$ = square root of the sample size
σ = standard deviation of the population

Central Limit Theorem

The **Central Limit Theorem**, the most widely used theorem in probability statistics, incorporates all we have discussed above:

1. The mean of the sampling distribution of means is equal to the population mean irrespective of whether or not the population is normally distributed.
2. The standard error of the mean ($\sigma_{\bar{x}}$) is a measure of dispersion of sample means about the population mean. Even if the population isn't normally distributed, the distribution of the sample mean will be normal as long as the sample size is large enough (at least 30). In other words, the sampling distribution of the mean approaches

Business mathematics and statistics

normality with increasing sample size irrespective of the distribution of the population from which the sample was drawn.
3. The above properties allow surveys such as the Nielsen Television Ratings (with 3900 households selected) and unemployment statistics (with 33 000 households selected) to be considered adequate in sample size. Therefore, the Central Limit Theorem is an invaluable tool (besides being cost-efficient in severely reducing necessary sample-size numbers) because we can make statistical inferences drawn from a sample without knowing whether or not, in fact, the population from which the sample was drawn was normally distributed.

Problem: Let's look back at the workers' average earnings and assume that the mean for the population is $33 000 p.a. with a standard deviation of $4000. If we draw a random sample of 30 workers, what is the probability that their earnings will average above $33 450?

Solution:

1. Find the standard error of the mean from the population standard deviation. This, you may recall, is defined as:

$$\sigma_{\bar{x}} = \frac{\sigma}{\sqrt{n}}$$

where $\sigma_{\bar{x}}$ = standard error of the mean
σ = standard deviation of the population
n = sample size.

Therefore:

$$\sigma_{\bar{x}} = \frac{\sigma}{\sqrt{n}}$$

$$= \frac{\$4000}{\sqrt{30}}$$

$$= \frac{\$4000}{5.477}$$

$$= \$730.33$$

Round off the standard error to $730; this indicates the standard error of the sampling mean from the population mean. Therefore, $\pm 1\,\sigma_{\bar{x}}$ from the population mean of $33 000 indicates that there is a 68.3% likelihood of the sampling mean falling between $\pm\$730$ from $33 000—that is, between $32 270 and $33 730; $\pm 2\,\sigma_{\bar{x}}$ for a 95.4% probability of the sampling mean falling between $31 540 and $34 460. This is not a wide variation around the population mean of $33 000.

2. Remember that we are calculating probability on the basis of a sample, which means that the formula we have been using throughout this chapter

Chapter 14 – Normal probability distribution curve

has to be modified slightly because we are measuring the variation of the sample mean from the population mean ($\bar{x} - \mu$) and dividing by the standard error of the mean ($\sigma_{\bar{x}}$) to get the z score. We continue to use Appendix 14A to look up probabilities and z scores.

$$z = \frac{\bar{x} - \mu}{\sigma_{\bar{x}}}$$

Therefore, to find the probability of the sample of 30 workers having an average income above $33 450, we must find the z score.

$$z = \frac{\bar{x} - \mu}{\sigma_{\bar{x}}}$$
$$= \frac{33\,450 - 33\,000}{730}$$
$$= \frac{450}{730}$$
$$= 0.6164$$
$$= 0.62$$

The 0.62 z score represents a standard deviation from the population mean of a normal distribution. 0.62 corresponds to approximately 23% of the area under the curve; in other words, there is a 23% probability of the average income of the sample of the 30 workers being between $33 000 and $33 450. Therefore, there is a 27% (i.e. 50% – 23%) probability of the average income of the 30 sampled workers being above $33 450.

Problem: The Eatery, a state-wide restaurant chain, wants to estimate the average number of customers visiting its restaurants each week. A random sample of 35 weeks was taken. If we assume that the actual average number of weekly customers is 11 250, with a standard deviation of 200, what are the chances that the sample mean will have a value within 50 customers of the actual mean?

Solution: In this example, since the sample size is greater than 30, we can use the Central Limit Theorem.

This example is asking you about the likelihood of the sample mean being located in the interval $\mu \pm 50$. Therefore:

1. Find the standard error with the formula:

$$\sigma_{\bar{x}} = \frac{\sigma}{\sqrt{n}}$$

and $\sigma_{\bar{x}} = \frac{200}{\sqrt{35}}$
$$= \frac{200}{5.916}$$
$$= 33.81$$

This 33.81 (or shall we say, 34 customers) is 1 standard error of the sampling mean ($\sigma_{\bar{x}}$) from a sample size of 35. It therefore indicates that there is a 68.3% likelihood that the sample mean will fall between 11 250 ± 34, or between 11 216 and 11 284. There is a 95.4% chance that the sampling mean will fall within ±2 $\sigma_{\bar{x}}$ of the population mean or 11 250 ± 2 $\sigma_{\bar{x}}$—that is, between 11 182 and 11 318.

2. Now, substitute into the formula to determine the probability of the sample mean falling within ±50 customers of the population mean by looking at $\bar{x}$ = 11 300.

$$z = \frac{\bar{x} - \mu}{\sigma_{\bar{x}}}$$

$$= \frac{11\,300 - 11\,250}{34}$$

$$= \frac{50}{34}$$

$$= 1.47 \text{ standard errors}$$

A z score of 1.47 corresponds to a probability of 42.92% or 43%. Since this is the probability of the sample mean being within 1.47 $\sigma_{\bar{x}}$ of the population mean, there is an 85.84% or 86% chance of the sampling mean being within ±50 customers of the population mean(i.e. ±50 of actual mean of 11 250 = 42.92% × 2 = 86%).

Let's try a few more examples to illustrate the diverse uses of the Central Limit Theorem.

Problem: A certain brand of truck tyres has a mean life of 80 000 kilometres and a standard deviation of 25 000 kilometres. What is the probability that a random sample of:

(a) 50 tyres will have a mean life of less than 75 000 kilometres?
(b) 100 tyres will have a mean life of less than 75 000 kilometres?
(c) 200 tyres will have a mean life of less than 75 000 kilometres?
(d) 100 tyres will have a mean life of greater than 90 000 kilometres?
(e) 100 tyres will have a mean life of between 78 000 and 86 000 kilometres?

Solution:

(a) We must solve for the z score in the above formula by substituting:

$$z = \frac{\bar{x} - \mu}{\sigma_{\bar{x}}}$$

where $\bar{x}$ = 75 000
μ = 80 000

$$\sigma_{\bar{x}} = \frac{\sigma}{\sqrt{n}} = \frac{25\,000}{\sqrt{50}} = \frac{25\,000}{7.07} = 3536$$

Chapter 14 – Normal probability distribution curve

Hence:

$$z = \frac{75\,000 - 80\,000}{3536}$$
$$= -1.4140$$
$$= 1.41 \text{ standard errors}$$

If you look up Appendix 14A, you will see that 1.41 corresponds to 0.42073. Therefore, a sample of 50 tyres will have a 7.9% (0.50 – 0.42073) probability of having a mean life of less than 75 000 kilometres.

(b) This is the same as the problem above except that the sample size has been doubled, with the result that $\sigma_{\bar{x}}$ is 2500 (25 000/$\sqrt{100}$ = 25 000/10 = 2500). Hence:

$$z = \frac{\bar{x} - \mu}{\sigma_{\bar{x}}}$$
$$= \frac{75\,000 - 80\,000}{2500}$$
$$= \frac{-5000}{2500}$$
$$= -2 \text{ standard errors}$$

A z score of –2 corresponds to 0.47725; therefore, there is a 47.7% probability of the 100 tyres having a life of between 75 000 and 80 000 kilometres and a 2.3% probability that a sample of 100 tyres would have a mean life of less than 75 000 kilometres.

(c) Once again, the sample size has been increased, so it is necessary to recalculate the standard error of the mean $\sigma_{\bar{x}}$ (i.e. 25 000/$\sqrt{200}$ = 25 000/14.142 = 1767.78). Hence:

$$z = \frac{\bar{x} - \mu}{\sigma_{\bar{x}}}$$
$$= \frac{75\,000 - 80\,000}{1768}$$
$$= -2.828$$
$$= 2.83 \text{ standard errors}$$

A z score of –2.83 corresponds to 0.49767; thus, the probability of a sample size of 200 having a mean life of less than 75 000 kilometres is 0.233% (0.50 – 0.49767).

Note: Parts (a)–(c) of this problem varied only in sample size, with each increase in size bringing the sample mean closer and closer to the population mean. Therefore, as one would expect, the larger the sample size, the greater the probability of its mean being closer to the mean of the population from which it is being sampled.

(d) To determine the probability of a mean of a random sample size of 100 having a mean greater than 90 000, first solve for $\sigma_{\bar{x}}$ with $25\,000/\sqrt{100} = 2500$:

$$z = \frac{\bar{x} - \mu}{\sigma_{\bar{x}}}$$

$$= \frac{90\,000 - 80\,000}{2500}$$

$$= \frac{10\,000}{2500}$$

$$= 4 \text{ standard errors}$$

A z score of 4 indicates that there is less than a 0.139% chance that a sample size of 100 will have a sample mean of greater than 90 000 when the population's mean is 80 000.

As Appendix 14A includes a z score up to 2.99, giving a 0.139% probability, then a z score of 4 would give a smaller probability.

(e) To determine the probability of a mean between 78 000 and 86 000 with a sample size of 100, we must calculate the z scores separately for 78 000 and 86 000. The standard error of the mean is 2500, as the sample size is still 100 and the standard deviation is still 25 000. Thus for 78 000:

$$z = \frac{\bar{x} - \mu}{\sigma_{\bar{x}}}$$

$$= \frac{78\,000 - 80\,000}{2500}$$

$$= \frac{-2000}{2500}$$

$$= -0.8 \text{ standard errors}$$

A z score of 0.8 corresponds to an area of 0.28814; thus, there is a 28.814% probability of the mean of the random sample size of 100 lying between 78 000 and 80 000 kilometres.

$$z = \frac{\bar{x} - \mu}{\sigma_{\bar{x}}}$$

$$= \frac{86\,000 - 80\,000}{2500}$$

$$= \frac{6000}{2500}$$

$$= 2.4 \text{ standard errors}$$

A z score of 2.4 gives 0.49180; hence, there is a 49.18% chance that the mean of the random sample of size 100 lies between 80 000 and 86 000 kilometres.

In aggregate, there is an approximate 78% probability (49.18% + 28.814%) that the mean life of the random sample of size 100 tyres will lie between 78 000 and 86 000 kilometres.

Let's have another look at its application in the workplace.

Problem: The *Overnight Herald* has approximately 4000 unpaid advertising accounts. The mean amount owing is $52.50 with a standard deviation of $7.00. What is the probability that for a sample size of 40, the sample mean will be less than $50.00 with a sample size of 40?

Solution:

$$\bar{x} = 50.00$$
$$\mu = 52.50$$
$$\sigma = 7.00$$
$$\sigma_{\bar{x}} = \frac{7.00}{\sqrt{40}} = \frac{7.00}{6.3246} = 1.107$$
$$z = \frac{\bar{x} - \mu}{\sigma_{\bar{x}}}$$
$$= \frac{50.00 - 52.50}{1.107}$$
$$= -2.258$$
$$= -2.26 \text{ standard errors}$$

A z score of 2.26 indicates that 48.809% of sample means from a sample size of 40 would be between $50 and $52.50; hence, there is a 1.191% probability, or about one chance out of 100, that the sample mean will be less than $50.

Problem: Since starting your own business two years ago, you have calculated that the average invoice is $86, with a standard deviation of $18. How large a sample of invoices must there be for a 95% probability that the sample mean will fall between $84 and $88?

Solution: This is, of course, a different problem and much easier than it may at first appear. We are asked to find n or the sample size in the formula:

$$\sigma_{\bar{x}} = \frac{\sigma}{\sqrt{n}}$$

However, we are given all other values:

1. According to Appendix 14A, 95% corresponds to ±1.96 standard deviations of the sampling mean (or the standard error) from the population mean.

2. The population mean is given as $86, and the range of values for the sampling mean as between $84 and $88 ($4 in total), or ±$2 from the population mean.

3. Therefore, $1.96\,\sigma_{\bar{x}} = \pm\$2$; therefore $1\,\sigma_{\bar{x}}$ equals about 1.02.
4. Now we can substitute in the formula and solve for n:

$$\sigma_{\bar{x}} = \frac{\sigma}{\sqrt{n}}$$

$$1.02 = \frac{18}{\sqrt{n}}$$

$$\sqrt{n} = \frac{18.00}{1.02}$$

$$\sqrt{n} = 17.65$$

$$n = 312$$

In other words, you need a sample of 312 invoices to have a 95% certainty that the sample mean will fall between $84 and $88 when the population mean is actually $86. Knowing this would be invaluable, because you would save yourself all the trouble of counting what would be thousands of invoices to get an answer so close to the true result.

Problem: Trying to save for that quarter-acre block, you are doing a second job as a waiter in an upmarket restaurant. You depend on the tips, which average $15 per reservation with a standard deviation of $6. You normally have 32 reservations which you personally handle each week. What is the probability of your total tips for the week exceeding $550?

Solution: First, we must calculate the standard error.

$$\sigma_{\bar{x}} = \frac{\sigma}{\sqrt{n}}$$

$$= \frac{6.00}{\sqrt{32}}$$

$$= \frac{6.00}{5.6569}$$

$$= 1.06$$

Now, to calculate the sample mean, the same sample size of 32 reservations results in tips of $550, or $17.19 a reservation (i.e. $550 ÷ 32). Now we can substitute into the formula for the z score.

$$z = \frac{\bar{x} - \mu}{\sigma_{\bar{x}}}$$

$$= \frac{17.19 - 15.00}{1.06}$$

$$= \frac{2.19}{1.06}$$

$$= 2.066$$

$$= 2.07 \text{ standard errors}$$

A z score of 2.07 corresponds to 48%, which is the area between the mean of $15.00 and $17.19. We are interested in the area to the right of $17.19, or 2%. Therefore, there is a 2% probability that tips of $550 or more would be received for the sample week.

In summary, the distribution of sample means will be normal, irrespective of the departure from normality of its population provided that the sample size is greater than or equal to 30. The larger the size of the samples drawn from the population, the smaller the dispersion of the sample's mean from the population's true mean. This dispersion or standard deviation of the means of all possible samples is referred to as the standard error of the mean. And, finally, the mean of all sample means is always equal to the population mean. The application of these principles allows the selection of a relatively small sample size which would nevertheless permit statistical inferences to be made with a high degree of confidence of being correct.

14.3 Summary

When the graphic representation of a frequency distribution approaches the shape of a smooth, symmetrical, bell-shaped curve, the curve is referred to as a normal distribution probability curve, as the properties of such a curve permit evaluation of the likelihood of an event. In a normal curve, the values of the three measures of central tendency (i.e. the mean, the median and the mode) are equal to one another, but the standard deviation varies.

When a frequency distribution does not approach symmetry when graphed, it is said to be skewed, either to the right or to the left. The more skewed the distribution, the greater the difference between the measures of central tendency. The usefulness of the mean and its corresponding standard deviation are determined by the frequency distribution. If the frequency distribution approaches a normal distribution, valid statistical inferences can be made on the basis of the proportion of observations lying within a certain interval. If the graphic representation of the data approaches a normal curve, approximately 68% of frequencies will fall within ±1 standard deviation of the mean, 95% between ±2 standard deviations of the mean, and 99.7% within ±3 standard deviations of the mean.

The z score—that is, the number of standard deviations away from the mean—must be calculated and Appendix 14A examined to find the corresponding percentage or area under the curve. The formula is given as:

$$z \text{ score} = \frac{x - \mu}{\sigma}$$

where x = selected value
 μ = mean value of x in the population
 σ = standard deviation of x in the population.

The importance and application of the normal curve to the business sector is overwhelming, with examples dealing with wage-rise determination, quality control, assessment of applicants, a product's useful life, warranties, production targets and the like.

In review, the main characteristics of the normal curve include the following:

1. It is symmetrical about the mean and is bell-shaped.
2. As the number of standard deviations increase, the curve approaches the x axis but never intersects it. (It is referred to as being asymptotic to the x axis.)
3. It is defined by the parameters of the mean and standard deviation.
4. The mean divides the curve into halves, with a zero standard deviation at the mean.
5. The area within ±1 standard deviation from the mean encompasses approximately 68%, ±2 standard deviations 95% and ±3 standard deviations 99.7% of observations.
6. The total area under the normal curve must equal one (or 100% probability), with the probability being that total area under the normal curve and above the x-axis.

These properties, combined with the many varied disciplines whose results follow a normal distribution pattern (biology, physics, business and production statistics), and the fact that as the sample size increases a sampling distribution approaches a normal curve irrespective of the original population's distribution, make the normal probability distribution invaluable in statistical inference—that is, we can draw conclusions about the population from the results of a small sample of the population.

14.4 Multiple-choice questions

1. The area under the standard normal curve to the left of $z = +0.85$ is:
   (a) 0.69766   (b) 0.30234
   (c) 0.19766   (d) 0.80234

2. The area under the standard normal curve between $z = -0.45$ and $z = 0.75$ is:
   (a) 0.44701   (b) 0.55299
   (c) 0.09973   (d) 0.59983

3. The average number of customers frequenting the Petticoat Junction Mall on a Saturday is 14 500, with a standard deviation of 3200. What is the probability that the coming Saturday will have customer traffic in excess of 20 100 customers?
   (a) 0.04006   (b) 0.45994
   (c) 0.39065   (d) 0.10935

4. What is the average height of an adult male if the standard deviation is 14 centimetres and the probability of a randomly selected male being over 200 centimetres is 0.02275?
   (a) 172 centimetres   (b) 193 centimetres
   (c) 168 centimetres   (d) None of the above.

5. The average grade on the statistics exam was 54% with a standard deviation of 12%. For a student to be in the top 10% of the class, what minimum grade should he or she achieve (to the nearest whole grade)?
   (a) 66%   (b) 69%   (c) 57%   (d) 84%

14.5 Exercises

(Assume that all of the following approach a normal curve. Decimal answers may vary according to the degree of interpolation.)

1. To define a normal probability distribution, only two parameters are needed. What are these?

2. If a random variable has a mean of 20 and a standard deviation of 6, what is the probability that its value is less than 8?

3. Forest Chump is aiming to be in the next Olympics as a marksman, but to be eligible he must be in the top 5% of those undergoing the Olympic trials. The average number of bulls-eyes for these Olympic candidates is 83 out of 100 attempts, with a standard deviation of 6. To the nearest whole number, how many bulls-eyes does Forest need to fell to get into the team?

4. Your lecturer has decided to give the top 5% of students an A, and the next 15% a B. If the aggregate marks for the students are normally distributed around a mean of 65% and a standard deviation of 6%, find the mark (to the nearest per cent) required for an A or a B.

5. Let's assume that the return on investment is normally distributed, with a mean of 8% and a standard deviation of 6%. What is the probability (to the nearest per cent) that the return will be:
   (a) greater than 20%?
   (b) nil?

6. An analysis of a meat-packing company revealed that the standard deviation of the mean weekly wage ($402) was $16.
   (a) Management has decided to award a bonus to the top 5% of wage earners in the beef division. What is the minimum wage that a beef worker must earn to be eligible for the bonus (to the nearest dollar)?
   (b) Management has also decided to give a $5 raise to the lowest 10% of wage earners. Calculate the maximum wage that a beef worker can earn and still be eligible (to the nearest dollar).

7. A multinational computer company's mean sales result per sales representative in 1997 was $581 000.
   (a) As an incentive scheme, all representatives with sales exceeding +2 standard deviations above the mean were given an overseas trip. If the minimum sales amount necessary to earn the overseas trip is $961 000, what is the standard deviation?
   (b) Management has also decided to retrench the bottom 16% of the representatives. What was the minimum amount of sales a representative had to invoice to avoid retrenchment (to the nearest $1000)?

Business mathematics and statistics

8. Locally made watches have a 12-year life with a standard deviation of four years. The manufacturer is unsure of the warranty term (to the nearest year) that she should provide, but she anticipates that no more than:

   (a) 2% will be returned under warranty;
   (b) 5% will be returned under warranty.

   Since this person wants to do the right thing by you, what should the warranty period be for (a) and for (b)? Would you buy a watch from her?

9. The lifetime of dishwashers has a mean of 9.5 years and a standard deviation of three years. What is the probability (as a percentage to one decimal place) that:

   (a) your dishwasher will last for over 15 years?
   (b) you will receive a lemon lasting less than 3 years?
   (c) the dishwasher will last for 8–10 years?
   (d) the dishwasher will last for 10–14 years?

10. A particular salesman's mean sales per month are $80 000, of which he receives a flat 5% commission with no retainer. If the standard deviation of mean sales per month is $21 500, what is the probability (to one decimal place) that his income will be:

    (a) greater than $7000 this month?
    (b) less than $5000 this month?
    (c) between $4500 and $5500 this month?
    (d) between $3250 and $4250 this month?

11. Re-examining the Sydney petrol stations (as in Chapters 10 and 11), you would have found the mean price of a litre of petrol to be 69.8 cents with a 4.5 cents standard deviation. If you are a motorist looking for petrol, what is the probability (to one decimal place) that a randomly selected petrol station will:

    (a) offer petrol at less than 60 cents a litre?
    (b) offer petrol at more than 75 cents a litre?
    (c) offer petrol at between 68 and 72 cents a litre?

12. A restaurant chain advertised a free meal for any customer who did not receive their order within 10 minutes. Before this offer was advertised it was calculated that approximately 2.5% of customers would probably be eligible for the free meal. What was the mean time (to one decimal place) for serving the meal if the standard deviation was:

    (a) 2 minutes?   (b) 80 seconds?   (c) 150 seconds?

13. Assume the mean life expectancy of an Australian female is 76 years, with a standard deviation of 12.5 years. What percentage of females (to one decimal place) will live to:

    (a) over 90 years?   (b) fewer than 55 years?
    (c) between 70 and 80 years?   (d) between 60 and 65 years?

Chapter 14 – Normal probability distribution curve

14. A particular microchip has a life of 4800 hours, with a standard deviation of 1300 hours. What is the likelihood (to one decimal place) of a randomly selected microchip having a life of:

    (a) more than 7500 hours?
    (b) between 4500 and 5000 hours?
    (c) less than 4000 hours?
    (d) between 6500 and 8000 hours?

15. The mean age of an airline pilot for an international carrier is 43 years, with a standard deviation of six years. What is the probability (to one decimal place) that your life is in the hands of a pilot who is:

    (a) under 25 years?
    (b) over 60 years?
    (c) between 30 and 35 years?
    (d) between 44 and 54 years?
    (e) between 40 and 45 years?

16. Assume that the average life of a colour television picture tube is seven years. If a warranty is given for four years and the manufacturer expects to have to replace 2% of tubes under warranty, what is the standard deviation (to two decimal places)?

17. The standard deviation now having been determined in question 16, what is the probability (to one decimal place) of a picture tube:

    (a) lasting for more than 11 years?
    (b) lasting for less than 5 years?
    (c) lasting for between 6 and 9 years?

18. A particular brand of car battery has a life of 4.5 years, with a standard deviation of 0.75 years. What is the likelihood (to one decimal place) of you purchasing one of these batteries and it lasting:

    (a) for less than 6 years?
    (b) for no more than 2.5 years?
    (c) for between 3 and 5 years?

19. Assume that Higher School Certificate students' scores are normally distributed, with a mean of 52% and a standard deviation of 15%. What is the probability that a student will score:

    (a) more than 52%?
    (b) more than 97%?
    (c) between 52% and 75% (to the nearest per cent)?
    (d) between 40% and 49% (to the nearest per cent)?

20. In Chapters 10 and 11, the useful life of car tyres was found to have a mean of 19 150 kilometres and a standard deviation of 7048. Assuming that the useful

Business mathematics and statistics

life of car tyres approaches a normal distribution (to one decimal point), what is the probability of randomly selecting a tyre which will have a useful life of:

(a) more than 30 000 kilometres?
(b) less than 15 000 kilometres?
(c) between 22 000 and 27 500 kilometres?
(d) between 18 000 and 20 000 kilometres?

21. Protractor Airlines is interested in estimating the average number of passengers carried each month. A random sample of 40 months was taken. If we assume that the average number of passengers per month is 45 000, with a standard deviation of 5000, what is the likelihood that the number of passengers according to the sample mean will be within plus or minus 1500 of the actual number? (Give answer to the nearest percentage.)

Note

1. John E. Freund and Gary A. Simon, *Modern Elementary Statistics* (Englewood Cliffs, NJ: Prentice Hall, 8th ed, 1992), p. 223.

APPENDIX 14A
Areas under the normal curve

Example:

$z = 0.34$ (or -0.34)

Area under the curve $(z) = 0.13307$ or 13.307%

$$z = \frac{x - \mu}{\sigma}$$

$z\left(\dfrac{x}{a}\right)$	.00	.01	.02	.03	.04	.05	.06	.07	.08	.09
0.0	.00000	.00399	.00798	.01197	.01595	.01994	.02392	.02790	.03188	.03586
0.1	.03983	.04380	.04776	.05172	.05567	.05962	.06356	.06749	.07142	.07535
0.2	.07926	.08317	.08706	.09095	.09483	.09871	.10257	.10642	.11026	.11409
0.3	.11791	.12172	.12552	.12930	.13307	.13683	.14058	.14431	.14803	.15173
0.4	.15542	.15910	.16276	.16640	.17003	.17364	.17724	.18082	.18439	.18793
0.5	.19146	.19497	.19847	.20194	.20540	.20884	.21226	.21566	.21904	.22240
0.6	.22575	.22907	.23237	.23565	.23891	.24215	.24537	.24857	.25175	.25490
0.7	.25804	.26115	.26424	.26730	.27035	.27337	.27637	.27935	.28230	.28524
0.8	.28814	.29103	.29389	.29673	.29955	.30234	.30511	.30785	.31057	.31327
0.9	.31594	.31859	.32121	.32381	.32639	.32894	.33147	.33398	.33646	.33891
1.0	.34134	.34375	.34614	.34850	.35083	.35314	.35543	.35769	.35993	.36214
1.1	.36433	.36650	.36864	.37076	.37286	.37493	.37698	.37900	.38100	.38298
1.2	.38493	.38686	.38877	.39065	.39251	.39435	.39617	.39796	.39973	.40147
1.3	.40320	.40490	.40658	.40824	.40988	.41149	.41309	.41466	.41621	.41774
1.4	.41924	.42073	.42220	.42364	.42507	.42647	.42786	.42922	.43056	.43189
1.5	.43319	.43448	.43574	.43699	.43822	.43943	.44062	.44179	.44295	.44408
1.6	.44520	.44630	.44738	.44845	.44950	.45053	.45154	.45254	.45352	.45449
1.7	.45543	.45637	.45728	.45818	.45907	.45994	.46080	.46164	.46246	.46327
1.8	.46407	.46485	.46562	.46638	.46712	.46784	.46856	.46926	.46995	.47062
1.9	.47128	.47193	.47257	.47320	.47381	.47441	.47500	.47558	.47615	.47670
2.0	.47725	.47778	.47831	.47882	.47932	.47982	.48030	.48077	.48124	.48169
2.1	.48214	.48257	.48300	.48341	.48382	.48422	.48461	.48500	.48537	.48574
2.2	.48610	.48645	.48679	.48713	.48745	.48778	.48809	.48840	.48870	.48899
2.3	.48928	.48956	.48983	.49010	.49036	.49061	.49086	.49111	.49134	.49158
2.4	.49180	.49202	.49224	.49245	.49266	.49286	.49305	.49324	.49343	.49361
2.5	.49379	.49396	.49413	.49430	.49446	.49461	.49477	.49492	.49506	.49520
2.6	.49534	.49547	.49560	.49573	.49585	.49598	.49609	.49621	.49632	.49643
2.7	.49653	.49664	.49674	.49683	.49693	.49702	.49711	.49720	.49728	.49736
2.8	.49744	.49752	.49760	.49767	.49774	.49781	.49788	.49795	.49801	.49807
2.9	.49813	.49819	.49825	.49831	.49836	.49841	.49846	.49851	.49856	.49861

Chapter 15
CORRELATION AND REGRESSION ANALYSIS

LEARNING OUTCOMES
After studying this chapter you should be able to:
- Describe the nature of correlation analysis
- Construct and interpret a scatter diagram
- Calculate and interpret the coefficient of correlation using the product moment method
- Identify the limitations of correlation calculations
- Calculate and interpret the coefficient of correlation using the rank order method
- Calculate and interpret the line of best fit using the least square method
- Apply regression analysis for prediction
- Calculate and interpret the standard error of estimate.

'Run for your life, say men from Harvard.'[1]

Medical researchers at Harvard and Stanford universities in the United States, who began studying the habits and health of 16 936 middle-aged and older men in 1960, reported the first scientific evidence that even modest exercise helps to prolong life. From the study it was concluded that sedentary lifestyles, even among former athletes, lead to heart and lung disease that shortens lives. The results were not surprising, but this is the first time scientific proof has been offered that there is a 'direct relationship between the level of physical activity and the length of life in the university men studies'. The editor of the *Journal of the American Medical Association,* Bruce B. Dan, said: 'The real discovery of this research is not that people who exercise have strong cardiovascular systems, rather it is that sedentary people have more cardiovascular disease. Sedentary people have shrivelled hearts and most of us who do not exercise have an atrophied body.' This is believed to be the first study linking physical fitness to longer life. Another research report, published in the same journal, was a study undertaken in Dallas, Texas, of the lives and habits of 6000 men and women, investigating the premise that the physically fit were less likely to develop hypertension. It was found that people with low levels of physical fitness were half again as likely to develop hypertension as those who were highly fit. Although the information imparted by these two studies has been common knowledge for years, convincing the public of the need for exercise was difficult without scientific proof. However, as Bruce B. Dan said, 'since we now can show a direct *cause and effect* relationship between fitness and life-span, the message will become much clearer in the minds of most people'.

The message for you as students is twofold:

1. Take an occasional break from studying business mathematics and statistics to engage in physical activity.
2. You have now entered the world of correlation and regression analysis.

Up until now, we have concerned ourselves only with series of data dealing with one variable. However, an examination of two variables (such as physical activity and prolonged life) is often undertaken to determine the nature and degree of relationship between one variable (the dependent) and another (the independent)—that is, the correlation between the two variables. Interest in correlation analysis is widespread, as is evident by its application in most fields—for example, economics (savings ratio and economic growth), medicine (e.g. smoking and lung cancer), government (tax rate levels and taxation revenue), real estate (interest rate levels and residential property prices), agriculture (rainfall and yield per hectare), energy prices (seasons and OPEC prices), energy consumption (urban density and petrol consumption), business (sales experience and sales), stockmarket (share prices and dividends; output growth and share prices), psychology (IQ test and performance in workplace) and education (hours studied and grades received). After the October 1987 stockmarket crash, it was suggested that a direct correlation could exist between sharemarket losses and forced residential property sales.

Finally, the news Australians wanted to hear: two Australian professors who studied the drinking habits of 2805 men and women in Dubbo since 1988 (with all participants over 60 years of age) concluded that beer and longevity were directly linked. For example, the study revealed that men who drank more than three glasses of beer every day had a 60% lower risk of death than abstainers, while women who drank between 8 and 14 glasses of beer each week lived longer than heavy drinkers or teetotallers.[2] In other words, you don't have to walk a straight line to gain a bit of yardage. The list of applications is inexhaustible, and the prevalence of correlation analysis and its relevance to our daily well-being should be obvious.

As a manager and/or decision-maker you have inevitably found that forecasts are often used as a prelude and prerequisite to your ultimate decision, since if the forecasts do not reveal promising prospects, you may decide to abort the project. These forecasts and decisions are frequently based on the relationship between two or more variables. If a relationship between two or more variables can be ascertained, it helps us to forecast sales, for example. For instance, if we found a close positive

Chapter 15 – Correlation and regression analysis

relationship between immigration levels and furniture sales, and we were aware of the government's immigration policy over the next x number of years, our forecast sales are much more likely to be accurate than they would be without such statistics. In addition, determining the relationship between two variables (commonly referred to as *bivariate*) and forecasting results on the mathematical basis of this relationship should be used in conjunction with other forecasting techniques (e.g. time-series analysis), as any forecasting technique is imperfect and the use of two or more different techniques should help to minimise error.

Regression analysis describes, in the form of an equation, the nature of the relationship between variables. The regression (or estimating) equation allows us to predict the values of a dependent variable (e.g. sales) on the basis of the value of an independent known variable (e.g. advertising expenditure).

In order to assess the degree of relationship between two or more variables, we use **correlation analysis**, basing the calculation of the degree of closeness on the regression equation. Regression analysis indicates how the variables are related (i.e. Y in terms of X), and the closer the relationship between the two variables, the greater the confidence in the estimates.

In summary, regression analysis describes the pattern of the existing relationship, and correlation analysis then reveals how strong that relationship is. We will study both simple linear regression and correlation—that is, if and to what extent two variables (a dependent and an independent variable) are related.

Naturally, regression and correlation analysis can extend beyond two variables. For instance, when we desire a measure that incorporates the effect of more than one series of data upon another particular series—for example, sales results affected by population growth, age distribution, consumer confidence and interest rate levels—it is called *multiple correlation*, the scope of which is beyond this book.

15.1 Scatter diagram

In regression analysis, the first, simplest and fastest method of determining whether there is an apparent relationship between two variables is to plot the available data on a graph, with the **dependent variable** *(Y)* always plotted on the y axis and the **independent variable** *(X)* on the horizontal x axis. Each point on the **scatter diagram**, as it is called, represents a pair of values based on the x and y scales.

Table 15.1 indicates the annual maintenance cost for nine vehicles, randomly selected from a fleet of cars used by the Mandrake Corporation. Since it appears logical that the age of the vehicle may influence the annual maintenance cost, we should now plot the data (see Figure 15.1) to confirm that a relationship does indeed exist. The nine points, when viewed together, illustrate a positive (or direct) relationship between the variables—that is, as the age rises, so does the cost of maintaining the vehicle.

Now that we have established that a positive near-linear relationship does appear to exist, we can fit or draw a straight line which then describes more precisely the relationship between the age and cost of maintaining a motor vehicle. This straight line is known as a *regression* (or *estimating*) line and is illustrated in Figure 15.2. The line has been drawn in an attempt to minimise the deviation of each plotted point from the straight line—a bit difficult when using only the naked eye.

Business mathematics and statistics

Table 15.1 Motor vehicles by age and annual maintenance expenditure

Age (years)	Maintenance expenditure p.a. ($ '000)
6	3.1
7	3.6
4	2.5
3	1.9
5	2.5
6	3.3
2	1.8
2	1.6
1	1.0

Source: Hypothetical.

Figure 15.1 Scatter diagram of motor vehicles' age and their corresponding maintenance cost

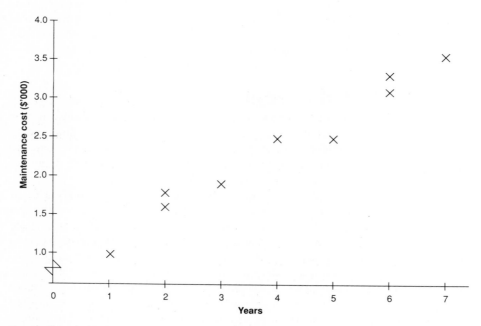

Source: Hypothetical.

Chapter 15 – Correlation and regression analysis

Figure 15.2 *Scatter diagram of motor vehicles' age and their corresponding maintenance cost*

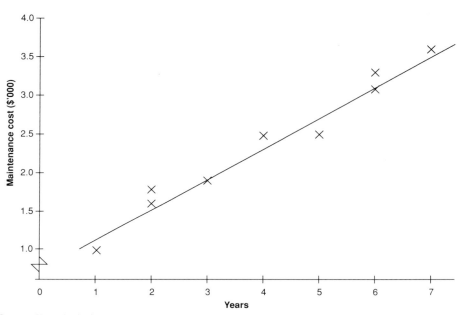

Source: Hypothetical.

If you now wanted to forecast the annual maintenance cost of a four-year-old vehicle, for example, you would find the point of intersection of the straight line and the four-year mark on the x axis. Then you would read the corresponding 'Y' value to get an estimated annual maintenance cost. Note that as there is a mild scatter of data around the line, the estimate is obviously not as reliable as it would be if the points lay directly on the line. This may be due to factors other than the age of the vehicle affecting the maintenance cost (such as poor driving skills and accidents).

Let's now look in on the average weekly total earnings (AWE) of adults and see how its rate of change may affect Australia's inflation. Plotted in Figure 15.3 are the percentage changes in the AWE and the CPI percentage changes for the years 1983/84 to 1995/96 derived from Table 15.2. As many economists believe that change in weekly wages is an inherent factor contributing to inflation (hence, AWE percentage changes constitute the independent variable and the CPI percentage change the dependent variable), there should be a positive correlation between the two variables—that is, with a rise in AWE, there follows an increase in inflation and with a fall in the rate of growth in AWE, an accompanying fall in inflation occurs.

Note that the years themselves are not included in the graph, as we are interested not in time but in the relationship between the growth rate in AWE and the CPI. It is obvious from the graph that a positive relationship exists, as a pattern emerges of

Figure 15.3 *Scatter diagram of Average Weekly Total Earnings (AWE) for adults and Consumer Price Index (CPI) percentage changes, 1983/84 to 1995/96*

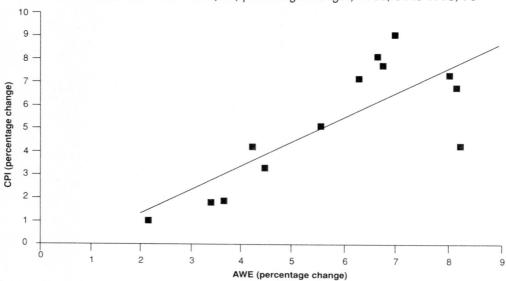

Source: Reserve Bank of Australia, *Australian Economic Statistics 1949–50 to 1994–95*, Australian Bureau of Statistics (Cat. No. 1350.0).

increased percentage changes in AWE resulting in corresponding increases in inflation. It is not a perfect correlation, since the plotted points, if connected, would not result in a near-perfect straight line. The wide scattering of the points on the scatter diagram indicates that there is a lower degree of association between the independent and dependent variables in this example than in the example of motor vehicle age and maintenance expense.

If we were unaware of or unable to use mathematical measures of regression for forecasting, we could draw a line through the pattern of points, as we have done in Figure 15.3, the objective of the line being to minimise the deviation of the plotted points from the line. There are many lines that could be drawn to represent the relationship between changes in AWE and changes in the CPI. The most common is the straight line determined by the application of the least-squares method, chosen because the squared distance between the Y values on the nominated line and the observed Y values is the least. This technique will be explained in Section 15.2.

To forecast the CPI result if, for instance, the AWE grew by 5%, one would locate 5% on the x axis, find its point of intersection with the line of best fit, and determine the corresponding Y value (i.e. the CPI) at this point (found by drawing a perpendicular from the 5% point on the x axis to the point of intersection with the line of best fit and drawing a perpendicular from this point of intersection to the y axis). In this case, the CPI of approximately 4% corresponds to a 5% increase in government outlay.

Figure 15.1 exhibits a positive (direct) linear relationship between the age and maintenance expense of a motor vehicle—that is, as the age of the vehicle increases, so do the maintenance expenses. The changes in the growth rate of AWE and the CPI (see Figure 15.3) also illustrate a positive relationship, but in this case there is a wider scattering of points, which implies that the relationship is not as strong as in the first example. Other types of possible relationships are shown in Figure 15.4.

Table 15.2 *Average Weekly Total Earnings (AWE) for adults and Consumer Price Index (CPI) percentage changes, 1983/84 to 1995/96*

Year	AWE (% annual change)	CPI (% annual change)
1983/84	8.1	6.9
1984/85	8.2	4.3
1985/86	6.6	8.4
1986/87	6.9	9.3
1987/88	6.3	7.3
1988/89	7.9	7.3
1989/90	6.7	8.0
1990/91	5.9	5.3
1991/92	3.7	1.9
1992/93	2.3	1.0
1993/94	3.4	1.8
1994/95	4.5	3.2
1995/96	4.3	4.2

Source: Reserve Bank of Australia, *Australian Economic Statistics, 1949–50 to 1994–95*; Australian Bureau of Statistics, *Australian Economic Indicators* (Cat. No. 1350.0).

A negative or inverse relationship indicates that as the independent variable increases (decreases), the dependent variable declines (rises)—for example, the number of divorces is the independent variable and the number of births is the dependent variable. If there is no relationship between the two variables, then the plotted data will be scattered widely and haphazardly (e.g. the number of theatre tickets sold and the sale of fountain pens).

In review, a scatter diagram is a graph of observations of two variables (a dependent variable and an independent variable), showing the relationship between them. If a pattern does exist and it approaches a linear relationship, we can draw or fit a straight line through the points as a means of subsequently forecasting the value for the dependent variable, given the value of the independent variable. However, as the line of best fit in this case is determined by the naked eye of whoever draws it, the true accuracy of the subjectively drawn line is questionable. In order to forecast the dependent variable with greater accuracy, other, more reliable measures are used. These are discussed in the following sections.

Figure 15.4 *Scatter diagrams: various types of relationships*

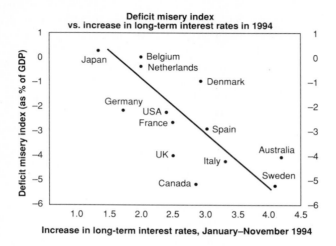

Source: Max Walsh, 'Misery index meets the deficits daleks', *Sydney Morning Herald*, 15 February 1995, p. 39.

(a) Negative linear relationship

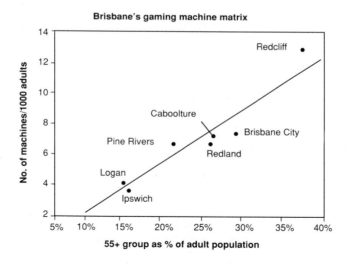

Source: Kathy McDermott, 'Queensland continues to gamble with gaming', *Australian Financial Review*, 5 October 1995, p. 36.

(b) Positive linear relationship

Chapter 15 – Correlation and regression analysis

Figure 15.4 *(continued)*

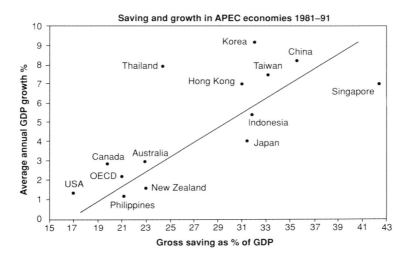

Source: Max Walsh, 'Darwin confused with Confucius', *Sydney Morning Herald*, 25 April 1996, p. 21.

(c) Positive linear relationship with more scatter

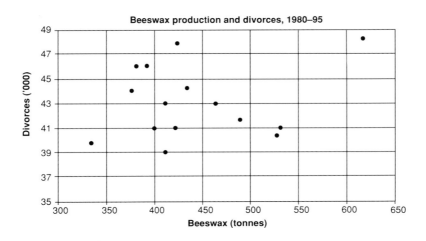

Source: Australian Bureau of Statistics, *Year Book Australia*, 1990, 1993, 1996 (Cat. No. 1301.0).

(d) No relationship

Business mathematics and statistics

15.1 Competency check

1. What is the purpose of constructing a scatter diagram?
2. Differentiate between a dependent variable and an independent variable, giving an example.
3. What type of correlation would you expect when examining:

    (a) video-cassette recorder sales and cinema attendance?

    (b) education level and income?

    (c) pollution and orange juice consumption?

    (d) income and imports?

    (e) economic growth and pollution?

    (f) number of cars on the freeway and their average speed?

    (g) urban density and petrol consumption?

4. The manager of the local race track is interested in determining the effect of the outside temperature on race-day attendances. A random sample of five race-day meetings was selected and the results are presented in Table 15.3.

    (a) Plot this data in the form of a scatter diagram.
    (b) Does a relationship appear to exist between these two variables? If so, describe the relationship.

Table 15.3 *Temperature and attendance at race meetings*

Temperature (°C)	Attendance ('000)
15	64
13	64
17	73
19	79
16	70

Source: Hypothetical.

5. Table 15.4 lists the IQ results and examination results for 10 students.

    (a) Plot these data in the form of a scatter diagram.
    (b) Determine whether a relationship exists and, if so, what type it is.

Table 15.4 *Intelligence quotients (IQ) and examination results*

IQ	Examination result (%)
100	60
110	70
115	75
125	70
105	55
120	70
90	50
95	55
105	60
110	75

Source: Hypothetical.

15.2 Correlation analysis

When examining and/or explaining the relationship between variables, we can determine the degree of association or correlation between the variables being observed (as well as the regression line and corresponding dispersion, i.e. standard error of estimate—both of which are covered in later sections of this chapter). We will be examining two of these correlation measures: the coefficient of correlation and the coefficient of determination.

There are several methods of calculating the coefficient of determination and/or the coefficient of correlation, but for our purposes only one method will be explained—the **product moment formula** for *r*. That is:

$$r = \frac{\Sigma(X - \overline{X})(Y - \overline{Y})}{\sqrt{\Sigma(X - \overline{X})^2 \Sigma(Y - \overline{Y})^2}}$$

where   *r* = coefficient of correlation
   $X - \overline{X}$ = deviation of the *X* value from its mean
   $Y - \overline{Y}$ = deviation of the *Y* value from its mean.

For example, to examine the ages and incomes of 10 individuals, first prepare a table (see Table 15.5) incorporating the above elements from the formula.

To solve for *r* in the formula

$$r = \frac{\Sigma(X - \overline{X})(Y - \overline{Y})}{\sqrt{\Sigma(X - \overline{X})^2 \Sigma(Y - \overline{Y})^2}}$$

take the following steps:

1. Determine $(X - \overline{X})$ by calculating the mean of the *X* variable and subtracting the mean value of *X* from the actual given value of *X*—for example, as the mean age is 26.3, subtract 26.3 from the ages of each of the 10 individuals to

Table 15.5 Age and income of 10 individuals

Age (years) (X)	Income p.a. ($'000) (Y)	$X - \bar{X}$	$Y - \bar{Y}$	$(X - \bar{X})(Y - \bar{Y})$	$(X - \bar{X})^2$	$(Y - \bar{Y})^2$
17	13	−9.3	−7	65.1	86.49	49
19	15	−7.3	−5	36.5	53.29	25
21	16	−5.3	−4	21.2	28.09	16
23	18	−3.3	−2	6.6	10.89	4
23	20	−3.3	0	0.0	10.89	0
25	18	−1.3	−2	2.6	1.69	4
26	21	−0.3	1	−0.3	0.09	1
32	24	5.7	4	22.8	32.49	16
37	27	10.7	7	74.9	114.49	49
40	28	13.7	8	109.6	187.69	64
263	200	0	0	339.0	526.10	228
$\bar{X} = 26.3$	$\bar{Y} = 20$					

Source: Hypothetical.

arrive at $(X - \bar{X})$. For example, age 17 − 26.3 = −9.3, age 23 − 26.3 = −3.3, age 32 − 26.3 = + 5.7.

2. Determine $(Y - \bar{Y})$ by calculating the mean of the Y variable and subtracting the calculated mean (in this case, incomes) from the actual Y value. For example, the mean is 20; hence, at an income level of 13, we have:

   13 − 20 = −7, at 18 − 20 = −2 and 24 − 20 = +4.

3. Multiply the result for $(X - \bar{X})$ by the corresponding $(Y - \bar{Y})$ result to arrive at $(X - \bar{X})(Y - \bar{Y})$ and find the sum of the individual products.
4. Square each $(X - \bar{X})$ value and find the total of the individually squared values (i.e. $\Sigma(X - \bar{X})^2$).
5. Square each $(Y - \bar{Y})$ value and find the total of the individually squared values (i.e. $\Sigma(Y - \bar{Y})^2$).
6. Substitute the calculated values into the formula. That is:

$$r = \frac{\Sigma(X - \bar{X})(Y - \bar{Y})}{\sqrt{\Sigma(X - \bar{X})^2 \Sigma(Y - \bar{Y})^2}}$$

$$= \frac{339}{\sqrt{(526.1)(228)}}$$

$$= \frac{339}{\sqrt{119\,950.8}}$$

$$= \frac{339}{346.34}$$

$$= .979$$

Chapter 15 – Correlation and regression analysis

There is a short-cut method to this product moment formula by which you can avoid all these tiresome calculations.

$$r = \frac{\Sigma XY - \frac{1}{n}\Sigma X \Sigma Y}{\sqrt{(\Sigma X^2 - \frac{1}{n}(\Sigma X)^2)(\Sigma Y^2 - \frac{1}{n}(\Sigma Y)^2)}}$$

The closer the **coefficient of correlation** (i.e. r) is to +1 or –1, the closer the relationship between the two variables; a coefficient of +1 indicates a perfect **positive correlation** of the two variables (represented by the observed points lying on the straight line as determined by the least-squares method) and –1 means a perfectly **negative correlation**, again represented by a straight line containing the actual values. The closer r is to zero, the less close the relationship. In the above example, .979 is certainly close to +1 and, as such, indicates a close direct linear relationship between age and income. However, a problem with the coefficient of correlation is the interpretation of the result. For example, how much closer is .9 than .8? Of course, this means the relationship is more direct or closer to linear, but how much closer? That is the problem. However, the use of the coefficient of determination is more helpful in understanding the association between two variables.

The **coefficient of determination** measures the relative amount of variation in the dependent variable (Y) statistically explained by variations in the independent variable (X) and is the square of the coefficient of correlation. Therefore, the formula for the coefficient of determination is r^2, where r is the coefficient of correlation.

In the above example, we have:

$r = .979$

Hence: $r^2 = .958$

It is most important to interpret this result which, in this example, indicates that 95.8% of the variation in the incomes of these individuals is linearly related to, and can therefore be explained by, the variations in the ages of these individuals.

What about comparing the strengths of relationships between two sets of paired data—for example, $r = 0.35$ and $r = 0.51$ for two sets of paired data?

- First, you need to calculate the coefficient of determination—that is, square the coefficient of correlation (r^2). Therefore, the coefficient of determination of $r = 0.35$ is 0.1225 (i.e. 0.35^2) and for $r = 0.51$, it is 0.2601 (0.51^2).
- Next, divide the calculated coefficients of determination of one set by the other. Thus, $0.2601/0.1225 = 2.1233$. In other words, the second set's relationship is more than twice as strong as the first's.

Problem: Referring to Table 15.6:

(a) determine the coefficient of correlation;
(b) determine the coefficient of determination;
(c) interpret the results.

Business mathematics and statistics

Table 15.6 *Motor vehicles by age and annual maintenance expenditure*

Age (years) (X)	Maintenance expenditure p.a. ($'000) (Y)	$X - \bar{X}$	$Y - \bar{Y}$	$(X-\bar{X})(Y-\bar{Y})$	$(X-\bar{X})^2$	$(Y-\bar{Y})^2$	XY	X^2	Y^2
6	3.1	2	0.7	1.4	4	0.49	18.6	36	9.61
7	3.6	3	1.2	3.6	9	1.44	25.2	49	12.96
4	2.5	0	0.1	0.0	0	0.01	10.0	16	6.25
3	1.9	−1	−0.5	0.5	1	0.25	5.7	9	3.61
5	2.5	1	0.1	0.1	1	0.01	12.5	25	6.25
6	3.3	2	0.9	1.8	4	0.81	19.8	36	10.89
2	1.8	−2	−0.6	1.2	4	0.36	3.6	4	3.24
2	1.6	−2	−0.8	1.6	4	0.64	3.2	4	2.56
1	1.0	−3	−1.4	4.2	9	1.96	1.0	1	1.00
36	21.3	0	−0.3	14.4	36	5.97	99.6	180	56.37

$\bar{X} = 4 \quad \bar{Y} = 2.4$

Source: Hypothetical.

Solution:

(a) $$r = \frac{\Sigma(X - \bar{X})(Y - \bar{Y})}{\sqrt{\Sigma(X - \bar{X})^2 \Sigma(Y - \bar{Y})^2}}$$

$$= \frac{14.4}{\sqrt{(36)(5.97)}}$$

$$= \frac{14.4}{\sqrt{214.92}}$$

$$= .98225$$

Using the shorthand method, we have:

$$r = \frac{\Sigma XY - \frac{1}{n}\Sigma X \Sigma Y}{\sqrt{(\Sigma X^2 - \frac{1}{n}(\Sigma X)^2)(\Sigma Y^2 - \frac{1}{n}(\Sigma Y)^2)}}$$

Looking up the corresponding values in Table 15.6, and substituting into the formula:

$$r = \frac{99.6 - \frac{1}{9}(36)(21.3)}{\sqrt{(180 - \frac{1}{9}(36)^2)(56.37 - \frac{1}{9}(21.3)^2)}}$$

$$= \frac{99.6 - \frac{1}{9}(766.8)}{\sqrt{(180 - 144)(56.37 - 50.41)}}$$

$$= \frac{99.6 - 85.2}{\sqrt{(36)(5.96)}}$$

$$= \frac{14.4}{\sqrt{214.56}}$$

$$= \frac{14.4}{14.647408}$$

$$= 0.98308$$

Note that the coefficient of correlation (r) differed only slightly (.982 and .983) when using the two methods, and this can be attributed to rounding. In any event, it is obvious that there appears to be a strong positive correlation between the age of the motor vehicle and the maintenance expense.

(b) To determine the coefficient of determination (r^2), simply square the coefficient of correlation. Hence:

$$r^2 = (.98225)^2 = .9648$$

Using the shorthand method, $r^2 = .9664$. In other words, approximately 96% of the variability in maintenance expense can be explained or accounted for by the age of the motor vehicle, with only 4% of the variability in expenses explained by other factors.

(c) There is a strong positive correlation (.98225) between the age of a motor vehicle and its maintenance expense—that is, as the age of the motor vehicle increases, so do the maintenance costs. In fact, 96% of the maintenance costs can be explained or attributed to the age of the motor vehicle.

Problem: Solve the coefficient of correlation and coefficient of determination of AWE and the CPI as per Table 15.7. Interpret your results.

Table 15.7 *Average Weekly Earnings (AWE) and Consumer Price Index (CPI) percentage changes, 1983/84 to 1995/96*

Year	AWE (%) change (X)	CPI (%) change (Y)	$X - \bar{X}$	$Y - \bar{Y}$	$(X - \bar{X})(Y - \bar{Y})$	$(X - \bar{X})^2$	$(Y - \bar{Y})^2$
1983/84	8.1	6.9	2.3	1.6	3.68	5.29	2.56
1984/85	8.2	4.3	2.4	−1.0	−2.40	5.76	1.00
1985/86	6.6	8.4	0.8	3.1	2.48	0.64	9.61
1986/87	6.9	9.3	1.1	4.0	4.40	1.21	16.00
1987/88	6.3	7.3	0.5	2.0	1.00	0.25	4.00
1988/89	7.9	7.3	2.1	2.0	4.20	4.41	4.00
1989/90	6.7	8.0	0.9	2.7	2.43	0.81	7.29
1990/91	5.9	5.3	0.1	0	0	0.01	0
1991/92	3.7	1.9	−2.1	−3.4	7.14	4.41	11.56
1992/93	2.3	1.0	−3.5	−4.3	15.05	12.25	18.49
1993/94	3.4	1.8	−2.4	−3.5	8.40	5.76	12.25

Table 15.7 (continued)

Year	AWE (%) change (X)	CPI (%) change (Y)	$X - \bar{X}$	$Y - \bar{Y}$	$(X - \bar{X})(Y - \bar{Y})$	$(X - \bar{X})^2$	$(Y - \bar{Y})^2$
1994/95	4.5	3.2	−1.3	−2.1	2.73	1.69	4.41
1995/96	4.3	4.2	−1.5	−1.1	1.65	2.25	1.21
	74.8	68.9	−0.6	0	50.76	44.74	92.38
	$\bar{X} = 5.8$	$\bar{Y} = 5.3$					

Source: Reserve Bank of Australia, *Occasional Paper No. 8, Australian Economic Statistics 1949–50 to 1989–90*; Australian Bureau of Statistics, *Australian Economic Indicators* (Cat. No. 1350.0).

Solution: To solve for r, substitute the appropriate calculations into the formula:

$$r = \frac{\Sigma(X - \bar{X})(Y - \bar{Y})}{\sqrt{\Sigma(X - \bar{X})^2 \Sigma(Y - \bar{Y})^2}}$$

$$= \frac{50.76}{\sqrt{(44.74)(92.38)}}$$

$$= \frac{50.76}{\sqrt{4133.0812}}$$

$$= \frac{50.76}{64.29}$$

$$= .79$$

The coefficient of correlation is 0.79, suggesting that a positive (direct) relationship exists between AWE and inflation. The coefficient of determination (e.g. r^2 or 62%) implies that 62% of the variance in inflation can be explained by variations in AWE, and this assumes that all other factors affecting inflation (e.g. the Australian dollar and government spending) remain constant.

In review, in the case of age versus income of the individual and age of the motor vehicle versus maintenance expense, there appears to be a high positive correlation (evidenced by relatively linear scatter diagrams and high coefficients of correlation and determination). It is a high coefficient correlation because in both these examples it was very close to 1. And finally, the high coefficient of determination in each of these two examples indicates that over 95% of the variation in the dependent variable could be explained by variations in the independent variable (in both cases, the age). On the other hand, a lower correlation exists—that is, the degree of closeness between the variables is marginal—between AWE and inflation (as shown by a greater scattered positive correlation, and lower coefficients of correlation and determination). This is not surprising since inflation is often the result of interrelated factors, with the Australian dollar, government charges and excess demand at the forefront of concern.

Spurious correlation

In Sweden in the mid-1960s, a high positive correlation was observed between the birth rate and nesting storks. Similarly, in England, a high positive correlation was discovered between cases of dysentery and the money supply. Can we therefore declare that in Sweden, should the storks decide to fly south for the winter, the birth rate will accordingly decline? What about the unfortunate consequences in England if the money supply increases? What these relationships and their corresponding high correlations illustrate is that although a high correlation may exist (e.g. a near-linear scatter diagram, and high coefficients of correlation and determination), there may be no direct relationship between the variables or there may be a common link to another factor or event. In the above two cases, there is a coincidental appearance of a relationship. This is referred to as a **spurious correlation**.

In other words, just because two variables may be associated in a statistical sense does not guarantee the existence of a causal relationship. There may be some other factors which cause changes in both dependent and independent variables. For example, if there is a strong inverse correlation between the number of umbrellas and swimsuits sold, this will be a result of a third factor, such as climatic conditions. Since statistical correlation does not in any way prove causality, correlation analysis requires you to think more in terms of the *association* between the two variables rather than a cause-and-effect *relationship*. Also, always consider the possibility of other factors not explicitly taken into account which may have contributed to the 'apparent' strong correlation.

15.2 Competency check

1. What is meant by *correlation analysis*?

2. Differentiate between *coefficient of correlation* and *coefficient of determination*.

3. Find the coefficient of correlation and the coefficient of determination of the IQ results and exam grades listed in Table 15.8 (to two decimal places). Discuss your results, incorporating the scatter diagram in your final summation of the relationship between these two variables.

Table 15.8 *Intelligence quotients (IQ) and examination results*

IQ	Examination result (%)
100	60
110	70
115	75
125	70
105	55
120	70
90	50
95	55
105	55

Source: Hypothetical.

4. (a) Find the coefficient of correlation and determination of the temperature and corresponding attendance patterns at the local race track, as per Table 15.9.
   (b) Briefly interpret the results.

Table 15.9 *Temperature and attendance at race meetings*

Temperature (°C)	Attendance ('000)
15	64
13	64
17	73
19	79
16	70

Source: Hypothetical.

5. (a) Find the coefficient of correlation and determination of the kilometres travelled, as well as the duration of the trip, by the coaches of Burke and Wills Coach Line, as per Table 15.10.
   (b) Briefly interpret the results.

Table 15.10 *Distance travelled by 10 coaches and time taken*

Coach number	Distance (kilometres)	Time (hours)
1	1000	12
2	750	8
3	2000	26
4	250	4
5	500	6
6	1500	20
7	1250	18
8	1000	10
9	800	8
10	1800	25

Source: Hypothetical.

15.3 Rank correlation

There may be occasions where you want to discover the relationship between two variables, but information in the form of numerical values may be unavailable. However, all is not lost as long as rankings can be allocated to each of the two variables under review. Once this is done, a **rank correlation coefficient** (r_s) can be calculated measuring the degree of association between the two variables. Often this measure is referred

Chapter 15 – Correlation and regression analysis

to as the Spearman rank correlation in honour of the statistician who is credited with developing it in 1904.

It should be noted that another reason for calculating the rank correlation may be that if you are working with a large array of data, it can be extremely cumbersome and time-consuming to calculate the correlation coefficient. For example, each year the International Institute for Management Development in Lausanne, Switzerland, publishes a competitive index based on surveys of 46 countries with 225 individual categories of competitiveness measured for each country. Imagine the trials and tribulations of working with such data. Instead, however, we are provided with a ranking and can determine the level of association based on these ranks rather than on the vast range of data.

The formula for determining the value of the rank correlation coefficient (r_s) is:

$$r_s = 1 - \frac{6\Sigma D^2}{n(n^2 - 1)}$$

where r_s = rank order correlation coefficient
D = difference between the rankings for the two variables for each individual item in the series
n = number of individuals or pairs of ranks.

In other words, if D^2 (which measures the difference between the rankings) is zero, this indicates that the rankings were identical in each case, and this is known as a perfectly positive correlation (+1). A perfect negative (or inverse) correlation (–1) illustrates the situation where, for example, the top five National Basketball League (NBL) teams in 1997 come in the complete reverse order with regard to each other in 1998.

Let's calculate the rank correlation for the most recent Swiss study on international competitiveness. Table 15.11 contains the rankings for 1995 and 1996. Is there a significant correlation between the two results?

To calculate the rank correlation coefficient:

1. Calculate the difference between ranks of the two variables for each country (D).
2. Square these differences (D^2).
3. Find the sum of the squared differences and then apply the following formula to calculate the rank order correlation coefficient:

$$r_s = 1 - \frac{6\Sigma D^2}{n(n^2 - 1)}$$

Table 15.11 *International competitive sweepstakes*

Country	1995 Rank	1996 Rank	D	D²
United States	1	1	0	0
Singapore	2	2	0	0
Hong Kong	3	3	0	0
Japan	4	4	0	0
Denmark	7	5	+2	4

Business mathematics and statistics

Table 15.11 *(continued)*

Country	1995 Rank	1996 Rank	D	D²
Norway	10	6	+4	16
The Netherlands	8	7	+1	1
Luxembourg	12	8	+4	16
Switzerland	5	9	−4	16
Germany	6	10	−4	16
New Zealand	9	11	−2	4
Canada	11	12	−1	1
				74

Source: AP Dow Jones, AAP, 'US leads world while we settle for 21st place', *Sydney Morning Herald*, 12 May 1996, p. 43.

To calculate the rank order correlation coefficient, substitute into the formula:

$$r_s = 1 - \frac{6\Sigma D^2}{n(n^2 - 1)}$$

$$= 1 - \frac{6(74)}{12(12^2 - 1)}$$

$$= 1 - \frac{444}{12(144 - 1)}$$

$$= 1 - \frac{444}{12(143)}$$

$$= 1 - \frac{444}{1716}$$

$$= 1 - 0.26$$

$$= +.74$$

A rank order correlation coefficient of 0.74 suggests a substantial positive correlation between the rankings in 1995 and the competitive rankings in 1996, at least among the 12 countries listed in Table 15.11. In other words, the positive result indicates there is a direct relationship between the two years' performances.

Problem: Determine the rank correlation coefficient of IQs versus examination results for the 10 individuals listed in Table 15.12. Briefly interpret the result.

Solution: In calculating the rank correlation, the following steps would have to be taken:

1. Rank the scores for each of the 10 individuals.
2. Calculate the difference between ranks of the two variables for each student (*D*).

Chapter 15 – Correlation and regression analysis

3. Square these differences (D^2).
4. Find the sum of the squared differences and then apply the following formula to calculate the rank order correlation coefficient:

$$r_s = 1 - \frac{6\Sigma D^2}{n(n^2 - 1)}$$

Table 15.12 *Intelligence quotients (IQ)*

Student	IQ	Rank	Examination result (%)	Rank	D	D^2
Bob	100	8	60	6	2	4
Ted	110	4	70	3	1	1
Carol	115	3	75	1	2	4
Alice	125	1	70	3	−2	4
Edwina	105	6	55	8	−2	4
Frederick	120	2	70	3	−1	1
Gary	90	10	50	10	0	0
Hollie	95	9	55	8	1	1
Kristina	105	6	60	6	0	0
Lawrence	110	4	75	1	3	9
						28

Note: In case of identical scores, assign the same ranks to those involved in the tie and assign to the next individual the rank that would have been assigned if there had been no tie.

Source: Hypothetical.

To calculate the rank order correlation coefficient, substitute into the formula:

$$r_s = 1 - \frac{6\Sigma D^2}{n(n^2 - 1)}$$
$$= 1 - \frac{6(28)}{10(10^2 - 1)}$$
$$= 1 - \frac{168}{990}$$
$$= 1 - 0.1697$$
$$= +0.83$$

The high positive result suggests that there is a direct relationship between the two variables—that is, an increase in one variable (IQ) is associated with an increase in the other (exam results). Those students who performed well in the IQ test normally performed well in the examination, as no single student with a high IQ test did poorly in the examination.

Problem: Table 15.13 lists the rankings of five companies in Australia in terms of market capitalisation. Calculate the rank correlation coefficient and comment on the results.

Business mathematics and statistics

Table 15.13 *Ten companies in Australia in terms of market capitalisation*

Company	Rank (12/87)	Rank (10/96)	D	D²
BHP	1	1	0	0
National Australia Bank	2	2	0	0
ANZ	3	3	0	0
CRA	4	4	0	0
WMC	5	5	0	0
				0

Source: Sydney Morning Herald, 23 December 1987 and 1 October 1996.

Solution: Substitute into the formula:

$$r_s = 1 - \frac{6\Sigma D^2}{n(n^2 - 1)}$$

$$= 1 - \frac{6(0)}{5(25-1)}$$

$$= 1 - \frac{0}{120}$$

$$= 1$$

The correlation coefficient of 1 suggests that there is a perfect correlation (or relationship association) between the two time periods.

In summary, the calculation of the rank correlation coefficient is appropriate if you are presented with ordinal data (e.g. scales measuring popularity and political orientation or international competitiveness), or when you are concerned only with the position of scores of two variables. Note also, if the population is not normally distributed, then the observations should be ranked and the rank correlation coefficient calculated.

The obvious advantage of using the rank correlation coefficient is the simplicity of the calculation, as well as providing a means of measuring the correlation between characteristics which cannot be quantitatively expressed but can, instead, be ranked. However, as the scores are placed in ranking order, the accuracy of the actual values is sacrificed since only positional variation is accounted for. For instance, if the highest score was 99 and the second highest was 56, the rankings in order of performance would be 1 and 2. This is less accurate than when it is being calculated from the actual data (e.g. company sales instead of rankings), because the magnitude of the difference between scores is lost in the rankings. On the other hand, a correlation coefficient may be ineffectual as a measure of the degree of association between two variables if faced with extreme numerical values. In such a case, the more appropriate measure may be the rank correlation which is not affected by such extreme observations because results are merely ranked in order of magnitude.

Chapter 15 – Correlation and regression analysis

15.3 Competency check

1. What is the difference between a *correlation coefficient* and a *rank correlation coefficient*?
2. What are the main advantages of using rank correlation coefficient as compared to the correlation coefficient?
3. A firm made it a practice only to hire applicants who graduated with honours from university. In evaluating the performance of last year's intake of five applicants, a rank correlation coefficient of –0.8 was calculated between the university grades and the graduates' performance in the workplace (in relation to their five peers). Briefly interpret this result.
4. The top 10 constraints in investment are found in Table 15.14. What is the rank correlation coefficient of the two years of rankings? Briefly discuss the result.

Table 15.14 *Top 10 constraints on investment, December 1994*

	1994 Rank	1993 Rank
Business taxes and government charges	1	1
Federal government regulations	2	2
Non-wage labour costs	3	3
State government regulations	4	4
Interest rates	5	10
Union resistance to change	6	8
Wage costs	7	7
Local competition	8	6
Insufficient demand	9	5
Charges by lending institutions	10	9

Source: ACCI Survey of Investor Confidence (taken from Rowan Callick, 'Business fears taxes, rate rises', *Australian Financial Review*, 6 February 1995, p. 3.

5. Table 15.5 lists the heights and average points scored in basketball for each of six players on a neighbourhood team. Calculate the rank correlation coefficient and interpret the result.

Table 15.15 *Basketball players from the neighbourhood*

Name	Height (centimetres)	Average points per game
Charles	204	22
Michael	206	29
Patrick	212	20
Dennis	205	16
Scott	207	19

Source: Hypothetical.

15.4 Regression analysis

In the construction of the scatter diagram, we were able to get an idea whether an obvious relationship between the two variables existed. The relationship can take many forms, but we will examine only the simplest—a straight-line or linear relationship. The straight line drawn in Figures 15.2 and 15.3, drawn (you may recall) by fitting a line as close to the points as possible, is called a **regression** (or **estimating**) **line**.

To improve the accuracy and reliability of the regression line, we can apply mathematical formulae which relate the two variables studied. The proper mathematical model selected will be influenced by the distribution of the X and Y values on the scatter diagram. However, as we are interested only in linear relationships, we will apply the equation for a straight line—that is:

$$Y = a + bX$$

where $Y =$ value of the dependent variable
 $b =$ slope of the line—that is, the increase or decrease in Y with one unit change in X
 $X =$ a given value of the independent variable
 $a =$ Y intercept (value of Y when $X = 0$).

This equation requires the determination of two coefficients, a (the Y intercept) and b (the slope), in order to predict values of the dependent variable (i.e. Y). Once a and b have been obtained, the straight line is known and can be plotted on the scatter diagram. We can make a visual comparison of how well our particular statistical model (a straight line) fits the original data. That is, we can see whether the original data are close to the fitted line or deviate greatly from the fitted line, as this obviously has a direct bearing on the forecast's degree of accuracy and reliability.

Simple linear regression analysis is concerned with finding the straight line that 'fits' the data best. What we mean by this is that we try to find a straight line which minimises the deviations between the estimated data points on the regression line and the actual data points. A mathematical technique used to find the values of a and b that best fits observed data is known as the **least-squares method**. Any values for a and b, other than those determined by the least-squares method, would result in a greater sum of squared deviations between the actual value of Y and the predicted value of Y.

In regression analysis, the straight-line equation reveals how a change in a series of data tends to compare with a given change in another series of data—in other words, it describes the average relationship between the variables of X and Y.

The formulae for determining the respective unknowns (a and b) in the straight-line equation are:

$$a = \frac{\Sigma X^2 \Sigma Y - \Sigma X \Sigma XY}{n\Sigma X^2 - (\Sigma X)^2} \quad \text{or} \quad \frac{\Sigma Y - b\Sigma X}{n} \quad \text{or} \quad \overline{Y} - b\overline{X}$$

$$b = \frac{n\Sigma XY - \Sigma X \Sigma Y}{n\Sigma X^2 - (\Sigma X)^2}$$

Chapter 15 – Correlation and regression analysis

where $\overline{Y}$ = mean of the *Y* variable
$\overline{X}$ = mean of the *X* variable
n = number of items.

For example, Table 15.16 lists the income and age of 10 individuals. Let the age of the individuals be the *X* variable (the independent variable) and let income be the *Y* variable (the dependent variable).

Table 15.16 *Age and income of 10 individuals*

Age (years) (X)	Income p.a. ($'000) (Y)	XY	X²
17	13	221	289
19	15	285	361
21	16	336	441
23	18	414	529
23	20	460	529
25	18	450	625
26	21	546	676
32	24	768	1024
37	27	999	1369
40	28	1120	1600
263	200	5599	7443
$\overline{X}$ = 26.3	$\overline{Y}$ = 20		

Source: Hypothetical.

Therefore, to solve for *b*:

$$b = \frac{n\Sigma XY - \Sigma X \Sigma Y}{n\Sigma X^2 - (\Sigma X)^2}$$

$$= \frac{10(5599) - (263)(200)}{10(7443) - (263)^2}$$

$$= \frac{55\,990 - 52\,600}{74\,430 - 69\,169}$$

$$= \frac{3390}{5261}$$

$$= 0.644$$

To solve for *a*, substitute 0.644 for *b* into the formula:

Business mathematics and statistics

$$a = \frac{\Sigma Y - b\Sigma X}{n}$$

$$= \frac{200 - .644(263)}{10}$$

$$= \frac{200 - 169.37}{10}$$

$$= \frac{30.628}{10}$$

$$= 3.0628$$

Note: If using a scientific calculator, the answer may differ slightly from the above due to rounding.

You could also solve for a by substituting 0.644 for b and the arithmetic mean for each of the two variables (i.e. $\overline{X} = 263/10 = 26.3$, and $\overline{Y} = 200/10 = 20$) into the formula:

$$a = \overline{Y} - b\overline{X}$$

$$= 20 - .644(26.3)$$

$$= 20 - 16.9372$$

$$= 3.0628$$

Hence, the straight-line regression equation is:

$$Y = a + bX$$
$$Y = 3.0628 + .644X$$

where Y = income
X = age

3.0628 is the value of Y when X equals zero, and 0.644 indicates the annual average increase in income (in thousands) for each additional year of age.

Forecasting

Now that we've got this regression line equation, what do we do with it? Its primary use is to estimate or predict values of the dependent variable, given values of the independent variable. Therefore, we can first draw the appropriate straight line. To do this, we need only two points. For example, if 19 and 26 years of age are nominated as the X values (independent variable), the respective incomes that can be expected, according to the regression line $3.0628 + .644X$, are 15.3 (i.e. $3.0628 + 0.644(19) = 15.3$) and 19.8 ($3.0628 + 0.644(26)$). These points are plotted with a straight line drawn through them. Thus, the line of best fit, free of any subjective assessment, provides the means for an objective description of the nature of the relationship between the two variables.

Figure 15.5 illustrates the presentation of the scatter diagram and regression line of the age and income of the 10 individuals. Note how the scatter diagram portrays a very high positive relationship between age and income, and observe how the deviations of the individual values above and below the regression line are minimal. If we now examine the estimated income level versus the actual income levels for the ages of 19 and 26, for example, we find that the estimates are reasonably close, confirming this line of best fit and also allowing us to use the equation for forecasting purposes with reasonable confidence.

Figure 15.5 *Scatter diagram and regression line relating age and income as determined by the least-squares method*

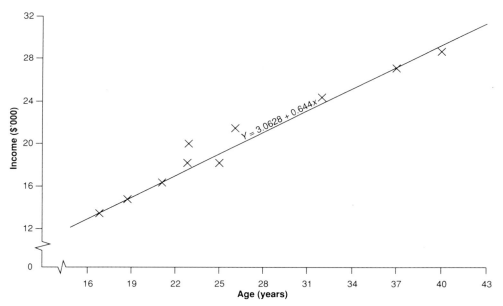

Source: Hypothetical.

Problem:
(a) Assuming a linear relationship exists, use the least-squares method to predict the maintenance expense from the age of the vehicle (refer to Table 15.17).
(b) Predict the annual expense the owner can expect to incur on a six-year-old motor vehicle.
(c) Draw the regression line as determined in (a).

Table 15.17 Motor vehicles by age and annual maintenance expenditure

Age (years) (X)	Maintenance expenditure p.a. ($'000) (Y)	XY	X^2
6	3.1	18.6	36
7	3.6	25.2	49
4	2.5	10.0	16
3	1.9	5.7	9
5	2.5	12.5	25
6	3.3	19.8	36
2	1.8	3.6	4
2	1.6	3.2	4
1	1.0	1.0	1
36	21.3	99.6	180
$\bar{X} = 4$	$\bar{Y} = 2.37$		

Source: Hypothetical.

Solution: First solve for b in the equation $Y = a + bX$:

$$b = \frac{n\Sigma XY - \Sigma X \Sigma Y}{n\Sigma X^2 - (\Sigma X)^2}$$

$$= \frac{9(99.6) - (36)(21.3)}{9(180) - (36)^2}$$

$$= \frac{896.4 - 766.8}{1620 - 1296}$$

$$= \frac{129.6}{324}$$

$$= .4$$

$$a = \bar{Y} - b\bar{X}$$
$$= 2.37 - .4(4)$$
$$= 2.37 - 1.6$$
$$= 0.77$$

The regression line is therefore $Y = 0.77 + 0.4X$. To predict the cost of maintenance for a car six years old, for example, we merely substitute 6 for X in the equation:

$$Y = 0.77 + 0.4X$$
$$= 0.77 + 0.4(6)$$
$$= 0.77 + 2.4$$
$$= 3.17$$

Chapter 15 – Correlation and regression analysis

In other words, the maintenance will be approximately $3170 for that sixth year. Figure 15.6 illustrates this line of best fit with an obvious direct relationship between the age of the motor vehicle and the maintenance expense. Note that there is little deviation of the estimated points on the regression line from the actual data points, thus reinforcing the strong relationship between the two variables.

Figure 15.6 *Scatter diagram of motor vehicles' age and their corresponding maintenance cost*

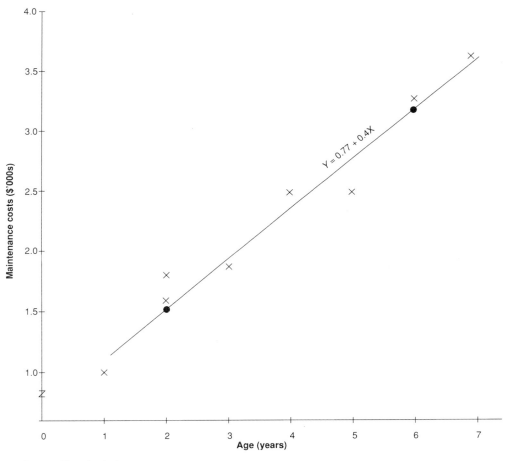

Source: Hypothetical.

Problem: Determine the line of best fit by use of the least-squares method for the data in Table 15.8. In addition, plot appropriate points for a 4% and 7% increase in average weekly earnings.

491

Business mathematics and statistics

Table 15.18 *Average Weekly Earnings (AWE) and Consumer Price Index (CPI) percentage changes, 1983/84 to 1995/96*

Year	AWE (% change) (X)	CPI (% change) (Y)	XY	X^2
1983/84	8.1	6.9	55.89	65.61
1984/85	8.2	4.3	35.26	67.24
1985/86	6.6	8.4	55.44	43.56
1986/87	6.9	9.3	64.17	47.61
1987/88	6.3	7.3	45.99	39.69
1988/89	7.9	7.3	57.67	62.41
1989/90	6.7	8.0	53.60	44.89
1990/91	5.9	5.3	31.27	34.81
1991/92	3.7	1.9	7.03	13.69
1992/93	2.3	1.0	2.30	5.29
1993/94	3.4	1.8	6.12	11.56
1994/95	4.5	3.2	14.40	20.25
1995/96	4.3	4.2	18.06	18.49
	74.8	68.9	447.20	475.10
	$\bar{X} = 5.75$	$\bar{Y} = 5.30$		

Source: Reserve Bank of Australia, *Australian Economic Statistics, 1949–50 to 1994–95*; Australian Bureau of Statistics, *Australian Economic Indicators* (Cat. No. 1350.0).

Solution: First solve for b in the equation $Y = a + bX$:

$$b = \frac{n\Sigma XY - \Sigma X \Sigma Y}{n\Sigma X^2 - (\Sigma X)^2}$$

$$= \frac{13(447.2) - (74.8)(68.9)}{13(475.1) - (74.8)^2}$$

$$= \frac{5813.6 - 5153.72}{6176.3 - 5595.04}$$

$$= \frac{659.88}{581.26}$$

$$= 1.1353$$

$$= 1.14$$

$$a = \bar{Y} - b\bar{X}$$

$$= 5.3 - 1.14(5.75)$$

$$= 5.3 - 6.555$$

$$= -1.255$$

$$= -1.26$$

Chapter 15 – Correlation and regression analysis

Phew, what a painstaking task calculating all these figures! Fortunately, statistical software packages are available which will perform all these calculations for us. However, now you will not only appreciate your computer and software even more but you will also appreciate what the program is doing.

Now, let's get back to the equation just determined, i.e. $Y = -1.26 + 1.14X$, where X is the percentage change in AWE and Y is the percentage change in the CPI. To draw a line on the scatter diagram, first determine and then plot the coordinates—that is, for example, 4% and 7% changes in AWE and the respective CPI results of 3.3 [i.e. $Y = -1.26 + 1.14 (4)$; hence, $Y = 3.3$ when X (AWE) increases by 4%] and a CPI result of 6.72 when AWE increases by 7% [i.e. $Y = -1.26 + 1.14(7)$, hence $Y = 6.72$]. Drawing the straight line through these points provides us with the line of best fit (see Figure 15.7).

Figure 15.7 *Regression line for AWE and CPI percentage changes, 1983/84 to 1995/96 as determined by the least-squares method*

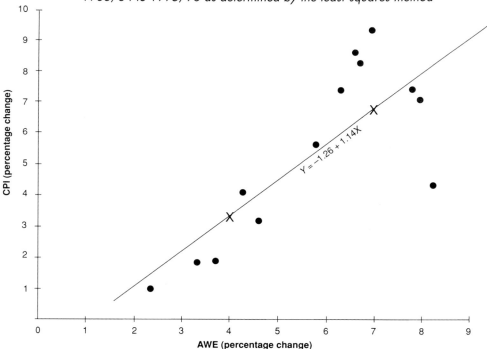

Source: Reserve Bank of Australia, *Australian Economic Statistics 1949–50 to 1994–95*; Australian Bureau of Statistics, *Australian Economic Indicators* (Cat. No. 1350.0).

Although a positive correlation is evident in percentage changes in AWE and the CPI (as measured by the slope, '*b*', bear in mind that other factors might have coexisted with increased average weekly earnings and contributed to movements in the CPI (e.g. foreign investment, government spending, reduced interest rates or a weaker Australian dollar).

The least-squares equation indicates an average relationship between the two variables, with b representing the average amount of change in the dependent variable Y (in our two examples, income and CPI) which can be expected with each unit change in the independent variable X (ages and AWE). To forecast the CPI, for instance, the value of X would be substituted into the equation to determine the appropriate level of CPI that can be expected with a certain change in AWE.

Figure 15.7 and the regression line reveal a fair degree of fluctuation in the actual data above and below the regression line. Obviously, the greater the degree of variation in the plotted points about the regression line, the less reliable the CPI estimates based on the regression-line equation will be. Therefore, from the actual data points and the regression line it appears that the change in AWE is but one of a number of factors that affect the volatile CPI results.

15.4 Competency check

1. What is the purpose of constructing a regression line?
2. (a) Why use the least-squares method to determine the straight-line equation?
   (b) What assumption are we making about the relationship between the two variables when applying the least-squares method?
3. Let's return to the race track to help the manager determine the effect of the outside temperature on race-day attendances. Table 15.19 lists the five randomly selected race meeting days with corresponding attendance numbers.

   (a) Assuming a linear relationship exists, determine the regression line (calculate a and b to one decimal place).
   (b) Interpret the meaning of b (the slope).
   (c) Predict attendance figures when the temperature hits:
      (i) 14°C
      (ii) 21°C

Table 15.19 *Temperature and attendance at race meetings*

Temperature (°C)	Attendance ('000)
15	64
13	64
17	73
19	79
16	70

Source: Hypothetical.

Chapter 15 – Correlation and regression analysis

4. From the data on IQ and examination results in Table 15.20:

   (a) Calculate the least-squares regression line.
   (b) Interpret the meaning of the Y intercept, a, and the slope, b, in this example.
   (c) What exam result (to the nearest per cent) would you expect a student to achieve if her IQ was:
      (i) 118?
      (ii) 105?

Table 15.20 *Intelligence quotients (IQ) and examination results*

IQ	Examination result (%)
100	60
110	70
115	75
125	70
105	55
120	70
90	50
95	55
105	60
110	75

Source: Hypothetical.

5. From the data in Table 15.21, determine the resting pulse rate (to the nearest whole number) that you would expect from someone exercising a daily average of:

   (a) 45 minutes;
   (b) 15 minutes;
   (c) 2.5 hours.

Table 15.21 *Time spent on daily exercises versus resting pulse rate (RPR)*

Daily exercise (minutes)	RPR (pulse per minute)
20	75
30	70
60	70
10	85
100	50
0	90
120	60
160	52
180	48
90	64

Source: Hypothetical.

15.5 Standard error of estimate*

When we examined dispersion to determine the spread of individual data points from the mean, we calculated the standard deviation. We found that the higher the standard deviation from the mean, the greater the spread of the data; therefore, the mean was considered less reliable as a measure of central tendency. The same line of thinking can be applied to the regression line versus the actual data points. Although the least-squares method results in the line that fits the data with the minimum amount of variation, the regression line is not a perfect predictor unless all the observed data points fall on the predicted regression line. As we have seen, the regression line is only one of several tools the forecaster can use and it provides only an approximation or estimate of the dependent value (Y) given the value of the independent variable, X.

If we therefore want to measure the variability of the actual Y values from the predicted Y values, we can use the **standard error of estimate**. Remember that the standard error of estimate of a regression line can be thought of as the standard deviation about the regression line, just as the standard deviation measures the variation about the arithmetic mean. In fact, the standard error of estimate can be used in the same manner as the standard deviation, in that if there is a normal distribution of frequencies, 68% of the items fall within ±1 standard deviation from the mean, 95% of the items fall within ±2 standard deviations and nearly 100% fall within ±3 standard deviations from the mean.

To illustrate the procedure for calculating the standard error of estimate, let's re-examine the ages and respective incomes of 10 individuals (Table 15.22), recalling that the relationship of income to age was described as:

$$Y = 3.0628 + .644X$$

Table 15.22 Age and income of 10 individuals

Age (X)	Income p.a. ($'000) ($Y_a$)	Calculated income ($Y_c = 3.0628 + .644X$)	$Y_a - Y_c$	$(Y_a - Y_c)^2$
17	13	14.0	−1.0	1.00
19	15	15.3	−0.3	0.09
21	16	16.6	−0.6	0.36
23	18	17.9	0.1	0.01
23	20	17.9	2.1	4.41
25	18	19.2	−1.2	1.44
26	21	19.8	1.2	1.44
32	24	23.7	0.3	0.09
37	27	26.9	0.1	0.01
40	28	28.8	−0.8	0.64
263	200	200.1	−0.1	9.49

Source: Hypothetical.

* Optional

Chapter 15 – Correlation and regression analysis

The standard error of estimate is calculated as follows:

1. Substitute each value of X in Table 15.22 into the regression line to calculate the income per individual (e.g. for the 17-year-old, $Y = 3.0628 + .644(17) = 14.0$; for the 26-year-old, $Y = 3.0628 + .644(26) = 19.8$, and so on).
2. Find the deviation of each calculated income (Y_c) from the actual income (Y_a)—that is $Y_a - Y_c$. For example, for the 17-year-old the actual income is 13 and the calculated income is 14; therefore, the deviation is -1.0 (13 – 14). There are deviations between actual and calculated incomes because age is but one criterion affecting income, and the interplay of other determinants (e.g. experience and performance) causes the deviations.
3. Square the deviations and divide the sum of the squared deviations by the total frequency (i.e. $\Sigma(Y_a - Y_c)^2/n$).
4. Find the square root of the sum of the squared deviations divided by the total number of observations. That is:

$$\sqrt{\frac{\Sigma(Y_a - Y_c)^2}{n}}$$

This formula should be used when calculating the standard of estimate for a population.

When determining the standard error for sample data, $n - 2$ is placed in the denominator. That is:

$$\sqrt{\frac{\Sigma(Y_a - Y_c)^2}{n - 2}}$$

Therefore, the standard error of estimate is determined by the formula:

$$Se = \sqrt{\frac{\Sigma(Y_a - Y_c)^2}{n}}$$

Using the values obtained in Table 15.22, we have:

$$Se = \sqrt{\frac{\Sigma(Y_a - Y_c)^2}{n}}$$

$$= \sqrt{\frac{9.49}{10}}$$

$$= \sqrt{.949}$$

$$= .974$$

Note: As we are assuming that the 10 individuals make up the entire population, n is used as the denominator.

The standard error of estimate of .974 (in thousands, or $974) indicates the range of error of the estimates of individual income values. In this case, the estimate appears reasonably accurate since the standard error of $974 is relatively small in comparison to the mean income of $20 000. Also, if income values are normally distributed—

that is, when plotted the data would display a normal bell-shaped curve—then 68% of the values will be within ±1 Se (from the regression line), 95% within ±2 Se and nearly 100% within ±3 Se.

From the regression line of $Y = 3.0628 + .644X$, we see that a 30-year-old would earn an income of approximately $22 400. This $22 400 is the average income expected of a randomly selected individual of 30 years of age according to the regression line. However, to give the estimate precision and to assist in the assessment of its reliability, the standard error of estimate of ±.974 ($974) is included, indicating that there is a 68% likelihood of a 30-year-old's income being between $21 426 and $23 374, a 95% likelihood of that person's income being between $20 452 and $24 348, and nearly a 100% likelihood of his or her realising an income of between $19 478 and $25 322. Remember that such assessments are valid if the data approach a normal distribution. As indicated earlier, the closer the scatter of points to the regression line, the lower the standard of error of estimate and the more useful and precise the regression line for forecasting purposes.

Problem: Determine the standard error of estimate for the CPI changes on the basis of percentage changes in average weekly earnings from Table 15.23. What range contains a 68% chance of the CPI residing if average weekly earnings increase by 4.0%?

Table 15.23 *Average Weekly Earnings (AWE) and Consumer Price Index (CPI) percentage changes, 1983/84 to 1995/96*

Year	AWE (% change) (X)	CPI (% change) (Y_a)	Calculated CPI $Y_c = -1.26 + 1.14X$	$Y_a - Y_c$	$(Y_a - Y_c)^2$
1983/84	8.1	6.9	8.0	−1.1	1.21
1984/85	8.2	4.3	8.1	−3.8	14.44
1985/86	6.6	8.4	6.3	2.1	4.41
1986/87	6.9	9.3	6.6	2.7	7.29
1987/88	6.3	7.3	5.9	1.4	1.96
1988/89	7.9	7.3	7.7	−0.4	0.16
1989/90	6.7	8.0	6.4	1.6	2.56
1990/91	5.9	5.3	5.5	−0.2	0.04
1991/92	3.7	1.9	3.0	−1.1	1.21
1992/93	2.3	1.0	1.4	−0.4	0.16
1993/94	3.4	1.8	2.6	−0.8	0.64
1994/95	4.5	3.2	3.9	−0.7	0.49
1995/96	4.3	4.2	3.6	0.6	0.36
$n = 13$		68.9	69.0	−0.1	34.93

Source: Reserve Bank of Australia, *Australian Economic Statistics, 1949–50 to 1994–95*; Australian Bureau of Statistics, *Australian Economic Indicators* (Cat. No. 1350.0).

Solution: Substitute into the formula:

$$Se = \sqrt{\frac{\Sigma(Y_a - Y_c)^2}{n}}$$

$$= \sqrt{\frac{34.93}{13}}$$

$$= \sqrt{2.687}$$

$$= 1.64$$

The standard error of estimate is 1.6 (i.e. 1.6%).

If AWE is to increase by 4%, substitute this figure into the regression equation of:

$$Y = -1.26 + 1.44X$$
$$= -1.26 + 1.44(4)$$
$$= -1.26 + 4.56$$
$$= 3.3$$
$$= 3.3\%$$

If the data were normally distributed, there would be a 68% likelihood of the CPI falling within ± 1 Se (3.3 ± 1.6 = 1.7 to 4.9) and a 95% chance of its falling within the range of 0.1 to 6.5. It is apparent from this evaluation that the standard error of estimate is large enough to render the regression line of little value since, quite obviously, other factors such as government charges, the depreciation of the Australian dollar and excess demand considerably influenced the CPI. This result, seen in conjunction with the scatter diagram, indicates that the regression line, in this case, would, by itself, be inadequate to rely on for forecasting purposes.

Problem: Determine the standard error of estimate for the sample of nine motor vehicles for which we calculated the regression equation to be $Y_c = 0.77 + 0.4X$.

Table 15.24 *Motor vehicles by age and annual maintenance expenditure*

Age (years) (X)	Maintenance expenditure p.a. ($'000) ($Y_a$)	Calculated maintenance $Y_c = 0.77 + 0.4X$	$Y_a - Y_c$	$(Y_a - Y_c)^2$
6	3.1	3.2	−0.1	0.01
7	3.6	3.6	0.0	0.00
4	2.5	2.4	0.1	0.01
3	1.9	2.0	−0.1	0.01
5	2.5	2.8	−0.3	0.09
6	3.3	3.2	0.1	0.01
2	1.8	1.6	0.0	0.04
2	1.6	1.6	0.0	0.00
1	1.0	1.2	−0.2	0.04
n = 9				0.21

Source: Hypothetical.

Solution: Since we are dealing here with a sample of the population, we should use $n - 2$ in the denominator. Therefore, we have:

$$Se = \sqrt{\frac{\Sigma(Y_a - Y_c)^2}{n - 2}}$$

$$= \sqrt{\frac{0.21}{9 - 2}}$$

$$= \sqrt{\frac{0.21}{7}}$$

$$= 0.173$$

The standard error of the estimate of 0.173 (i.e. $173) represents the measure of variation (which is very slight in this example) around the fitted regression line. This is, of course, reinforced when we note in Table 15.24 just how close the calculated Y_c values are to the actual Y_a values. Therefore, in this case, the regression line's standard error of estimate reinforces the premise of a linear relationship existing with little deviation of calculated values from the actual data points.

15.5 Competency check*

1. What is the importance of the standard error of the estimate?
2. How would you compare the standard error of the estimate with the standard deviation from the mean?
3. For a regression equation, one predicted Y value is 42 with the corresponding standard error of the estimate being 6 units. This means we can be 95% confident that the actual Y value lies between which two limits?
4. (a) Calculate (to the nearest 1000) the standard error of estimate (assume a sample of days were selected) for the race track manager (Table 15.19).
   (b) Interpret the result.
5. (a) Calculate (to the nearest per cent) the standard error of estimate (assume a sample) for IQ versus examination grade results as presented in Table 15.20.
   (b) Interpret the result.

15.6 Summary

A major use of correlation and regression analysis of two variables is to forecast or estimate the values of one variable from known values of the other variable. The estimated variable is referred to as the dependent variable, and the known variable

* Optional

is the independent variable. Regression analysis describes the nature of the relationship between the two variables, while correlation analysis assesses the degree of relationship between the two variables. For our purposes, we assume that the description of correlation between the two sets of data is represented graphically by a straight line. To assist in the analysis, a number of techniques can be used.

The scatter diagram entails plotting the data for the variables and determining whether a relationship does appear to exist between the dependent variable (Y) and the independent variable (X). If a positive correlation occurs, this indicates that a rise or fall in the independent variable results in a rise or fall in the dependent variable. If a line drawn through the plotted points is a near-perfect straight line, it will be considered a perfect correlation. If the drawn line does not follow a strictly linear path, it is an imperfect positive correlation.

If the plotted variables demonstrate that as the independent variable increases, the dependent variable declines (e.g. as the number of break-ins in your neighbourhood increases, the number of insurance companies willing to cover household contents declines), it is considered a negative correlation (as well as indicating that, perhaps, you should move from your neighbourhood). A high negative coefficient of correlation of, say, –0.95 indicates that as the independent variable rises (falls), the dependent variable falls (rises). Because of the strong correlation (very close to –1), we have greater confidence in the reliability of the forecast Y variable. Should a coefficient of correlation of, say, +0.45 exist, this indicates that as the independent variable rises (falls), the dependent variable rises (falls). However, as the correlation is not strong, we cannot make a reliable prediction of the Y variable on the basis of a given value for X.

Correlation analysis describes those techniques used to measure the degree of closeness in the relationship between variables. The statistical determination of the closeness is based on the regression equation. We have examined simple correlation analysis, which is concerned with measuring the relationship between one independent variable and the dependent variable.

The coefficient of correlation (i.e. r) indicates the degree of closeness in the relationship between the two variables. The more r approaches ± 1, the closer the relationship; the closer r is to zero, the less close the relationship.

The product moment formula is used to determine the coefficient of correlation and is defined as:

$$r = \frac{\Sigma(X - \overline{X})(Y - \overline{Y})}{\sqrt{\Sigma(X - \overline{X})^2 \Sigma(Y - \overline{Y})^2}} \quad \text{or} \quad \frac{\Sigma XY - \frac{1}{n}\Sigma X \Sigma Y}{\sqrt{(\Sigma X^2 - \frac{1}{n}(\Sigma X)^2)(\Sigma Y^2 - \frac{1}{n}(\Sigma Y)^2)}}$$

The coefficient of determination is the square of the coefficient of correlation and can be expressed as a percentage. This result indicates the percentage variation in the dependent variable that is directly associated with changes in the independent variable.

To measure the correlation between characteristics which cannot be quantitatively expressed but can be ranked, the rank correlation coefficient is calculated by applying the formula:

$$r_s = 1 - \frac{6\Sigma D^2}{n(n^2 - 1)}$$

When determining whether a relationship exists between two variables (and, if so, to what extent), the use of the above techniques provides an objective, mathematically determined analysis. If there is a distinct linearly defined relationship, this knowledge is thus invaluable for estimating future results. However, when examining any results of correlation analysis, be aware that logic and common sense must also be applied because a high degree of correlation between two variables does not necessarily indicate cause and effect.

If the data illustrate a clearly discernible linear relationship, regression analysis can be used to describe the nature of the relationship between the two variables. The least-squares method is used to indicate how a series of data (the independent variable) tends to change when compared with a given change in another series of data (independent variable).

Once you have calculated a and b values (Y intercept and slope of the line respectively), you can then use this regression equation to forecast values for the Y variable, given the value of the X variable. The straight-line equation is $Y = a + bX$. To solve for a and b, the formulae are:

$$a = \frac{\Sigma X^2 \Sigma Y - \Sigma X \Sigma XY}{n\Sigma X^2 - (\Sigma X)^2} \quad \text{or} \quad \frac{\Sigma Y - b\Sigma X}{n} \quad \text{or} \quad \overline{Y} - b\overline{X}$$

$$b = \frac{n\Sigma XY - \Sigma X \Sigma Y}{n\Sigma X^2 - (\Sigma X)^2}$$

The straight line drawn as a result of the application of the least-squares method is referred to as the regression line—hence the term 'regression analysis'. It is used not only for describing the nature of the relationship between the two variables but also for forecasting purposes. Examination of the constructed line of best fit in conjunction with the scatter diagram will reveal the extent to which the actual data flounders above and below the regression line (i.e. how much they deviate from a strictly linear relationship).

The measure of the degree of variation of the actual data points from the regression line is known as the standard error of estimate, with its formula:

$$Se = \sqrt{\frac{\Sigma(Y_c - Y_a)^2}{n}} \quad \text{for the population and} \quad Se = \sqrt{\frac{\Sigma Y_c - Y_a)^2}{n-2}} \quad \text{for a sample.}$$

The greater the degree of variation by which the plotted points deviate from the regression line, the less reliable your estimate based on the regression line equation.

A university study in the United States[3] found a correlation between the height and income of businessmen. Among 1200 graduates of the University of Pittsburgh's MBA program, individual income rose by an average of $600 for every additional inch (2.54 centimetres) of height. Someone whose height was 6 feet 2 inches (188 centimetres) would, for example, earn $6000 more than someone whose height was 5 feet 4 inches (163 centimetres). In other words, b, the slope, in $Y = a + bX$ was equal to $600 and X was the unit measurement of inches.

The study also considered some women, but found no statistically significant correlation between height and income. The results were not surprising, since innumerable studies have illustrated that tall people enjoy a sharp edge over their shorter colleagues in attaining success. Many argue that this is because taller people are noticed and remembered more than their shorter counterparts, and that promotions in business are basically the result of being noticed.

The application of regression and correlation analysis has been of immeasurable value in all areas, be it business, psychology, sports or medicine. Health science professionals often engage statisticians to work on correlation studies. Recent studies (in the 1990s) include some on the correlation between stress levels and the incidence of cold symptoms and infection; between socially deprived areas and poor health; between gambling expenditure and retail sales; between head injuries and Alzheimer's disease; between fruit consumption and heart disease; and between the storey level in a high-rise building and physical and mental distress. The potential for the application of regression and correlation analysis is boundless, but care as well as logic and much consideration must be exercised in the choice of variables.

15.7 Multiple-choice questions

1. You would expect the correlation between level of education and the rate of unemployment to have:

    (a) a high negative correlation
    (b) a high positive correlation
    (c) a low positive correlation
    (d) no correlation

2. If $r = 0.82$ for one set of paired observations and $r = 0.67$ for another set of paired observations, the strength of the association of the first set of observations compared to the second is:

    (a) the first set of observation's association is 22% stronger than the second set;
    (b) the first set of observation's association is 15% stronger than the second set;
    (c) the first set of observation's association is 50% stronger than the second set;
    (d) none of the above.

3. In the straight-line regression equation $Y = 24.44 + 3.58X$, X represents income earned per year (in $000s) and Y the price of houses purchased by income-earners (in $000s). What does 3.58 represent?

    (a) the correlation of coefficient;
    (b) the coefficient of determination;
    (c) the change in the price of houses purchased with each $1000 change in income;
    (d) the change in the income earned with each $1000 change in house prices;
    (e) none of the above.

4. Based on the regression equation given in question 3, how much would you predict an individual earning $68 000 a year to pay for a house?

    (a) about $268 000
    (b) about $246 000
    (c) about $96 000
    (d) None of the above.

Business mathematics and statistics

5. Using the same regression equation as in question 3, if the house cost $450 000, what was the approximate income earned per year for the home-owner?

   (a) $118 900
   (b) $132 500
   (c) $160 600
   (d) None of the above

15.8 Exercises

1. Why bother with regression and correlation analysis?
2. What is the purpose of a scatter diagram?
3. What is meant by *simple linear regression*?
4. How is the regression equation used in forecasting?
5. Why calculate the standard error of the estimate? What is its purpose?
6. (a) If the coefficient of correlation is +.9, what does this indicate?
   (b) What is the coefficient of determination of the set of data in (a)?
   (c) Interpret what this indicates.
7. Give an example of a spurious correlation (originality counts here).
8. Table 15.25 lists the amount of time spent on daily exercise by 10 individuals, with their respective resting pulse rates (RPR).

   (a) Plot the information in the form of a scatter diagram.
   (b) Determine whether a relationship exists and, if so, what type it is.

Table 15.25 *Time spent on daily exercise versus resting pulse rate (RPR)*

Daily exercise (minutes)	RPR (pulse per minute)
20	75
30	70
60	70
10	85
100	50
0	90
120	60
160	52
180	48
90	64

Source: Hypothetical.

9. You have just been hired as a troubleshooter-cum-consultant for Burke and Wills Coach Line. You randomly select 10 recent trips made throughout the area serviced

Chapter 15 – Correlation and regression analysis

by the coachline and record the number of kilometres travelled as well as the duration of the trips. The results are listed in Table 15.26.

(a) Construct a scatter diagram of this data.
(b) Briefly describe the apparent relationship.

Table 15.26 *Distance travelled by 10 coaches and time taken*

Coach number	Distance (kilometres)	Time (hours)
1	1000	12
2	750	8
3	2000	26
4	250	4
5	500	6
6	1500	20
7	1250	18
8	1000	10
9	800	8
10	1800	25

Source: Hypothetical.

10. Burke and Wills Coach Line are so impressed with your statistical prowess that they now ask you to see what relationship, if any, exists between kilometres travelled and operating costs per vehicle (excluding costs of driver and back-up driver) per kilometre in 1997. Once again, a random sample of 10 coaches was selected and the results are listed in Table 15.27.

(a) Construct a scatter diagram.
(b) How would you describe the relationship, if indeed one exists?

Table 15.27 *Distance travelled and cost per kilometre for Burke and Wills Coach line for 1997*

Coach number	Distance (kilometres) ('000)	Cost per kilometre ($)
1	75	1.40
2	120	1.20
3	80	1.35
4	50	1.50
5	95	1.30
6	150	1.25
7	110	1.20
8	40	1.60
9	85	1.40
10	140	1.10

Source: Hypothetical.

Business mathematics and statistics

11. As a gung-ho troubleshooter for Burke and Wills Coach Line, you have now been commissioned to:
    (a) develop the regression line using the least-squares method (to three decimal places) using Table 15.26;
    (b) explain to management the significance of a and b in the straight-line equation to time and distance travelled;
    (c) estimate how long (to the nearest hour) it will take an average coach to travel:
        (i) 1700 kilometres;
        (ii) 600 kilometres.

12. Now that you have successfully finished that commission, Burke and Wills Coach Line have asked you to do a similar report on the relationship between kilometres travelled and operating cost per kilometre. From Table 15.27:
    (a) Calculate the coefficient of correlation and coefficient of determination and determine the kilometres travelled and operating costs per vehicle (excluding costs of driver and the back-up driver) per kilometre in 1997.
    (b) Calculate the regression line (to three decimal places) using the least-squares method.
    (c) Interpret the meaning of the Y intercept, a, and the slope, b, in this example.
    (d) What cost per kilometre can management expect if the coach travels:
        (i) 60 000 kilometres per annum?
        (ii) 100 000 kilometres per annum?

13. Table 15.28 shows the number of marriages and beehives in Australia. Find the coefficient of correlation and the coefficient of determination (to two decimal places). Interpret the results, including the relationship between being stung and marriage.

Table 15.28 *Marriages and beehives in Australia, 1975/76 to 1988/89*

Year	Beehives ('000)	Marriages ('000)
1975/76	497	110
1976/77	493	105
1977/78	479	103
1978/79	501	104
1979/80	511	109
1980/81	530	114
1981/82	552	117
1982/83	540	115
1983/84	529	109
1984/85	553	115
1985/86	560	115
1986/87	364	113
1987/88	366	110
1988/89	405	114

Source: Australian Bureau of Statistics, *Year Book Australia, 1991* (Cat. No. 1301.0).

14. A conference organiser is asked to forecast the number of participants expected to attend this year (see Table 15.29). The last six conference results indicate that not all those who pay registration fees actually attend, owing to other work commitments and/or for tax purposes (the conference is allowed as a legitimate expense and some people may therefore register but holiday elsewhere on the island).

   (a) Plot a scatter diagram.
   (b) Calculate the least-squares equation for estimating the number of participants attending, on the basis of the number who registered (to three decimal places).
   (c) The October conference for this year has 610 registrants. How many (to the nearest whole number) can the organiser expect to attend?

Table 15.29 *Numbers of registrants and participants*

Registrants	Participants
580	495
550	488
617	501
747	612
480	450
564	510

Source: Hypothetical.

15. Table 15.30 lists a sample of 10 students' allocation of study time outside the classroom, with their respective aggregate grades.

   (a) Plot the scatter diagram.
   (b) Calculate the line of best fit using the least-squares method (to three decimal places).
   (c) Use the line calculated in (b) to estimate the grade for a student spending 21 hours per week studying.
   (d) Comment on the results. (Will these results have any effect on *your* studying habits?)

Table 15.30 Hours studied per week and aggregate grade at end of year for 10 randomly selected students

Hours of study	Aggregate grade (%)
6	36
8	40
11	52
15	57
18	66
22	73
24	61
28	75
32	81
37	80

Source: Hypothetical.

16. Table 15.31 lists the populations of six suburbs, as well as the corresponding sales results for each branch of Norman Loss Pty Ltd in these suburbs. The regional manager wants to know whether a relationship exists between population numbers and sales.

    (a) Construct a scatter diagram and interpret the results.
    (b) Calculate the coefficient of correlation and coeffecient of determination.
    (c) Interpret the results of (b).
    (d) Estimate a regression line (to three decimal places) using the least-squares method.

Table 15.31 Population of six suburbs and corresponding sales of Norman Loss in these suburbs

Suburb	Population ('000)	Sales ($ millions)
1	18	0.75
2	27	1.00
3	14	0.90
4	45	2.20
5	30	1.50
6	34	1.40

Source: Hypothetical.

17. As a salesperson working on a commission only, you are most interested in determining the relationship between the price of the products you are selling and the number of units you have sold. Table 15.32 gives the results you have collected over the past two years.

(a) Calculate the coefficient of determination and interpret the results.
(b) Determine the regression line using the least-squares method (to three decimal places).
(c) If the price of the product is $20, how many units would you expect to sell?

Table 15.32 *Sales of kitchen utensils, 1996/97*

Price ($)	Units sold
5	1500
10	1200
15	1000
20	800
25	400
30	250
35	150
40	50
45	20
50	10

Source: Hypothetical.

18. Lightwait Industries' management is interested in determining whether a relationship exists between the number of years that salespersons have worked for the firm and sales for 1997. The results for the eight randomly selected salespersons (thus a sample) are listed in Table 15.33.

    (a) Determine the coefficient of determination and coefficient of correlation. Interpret the results.
    (b) Use the least-squares method to determine the regression line (to three decimal places).
    (c) Calculate the standard error of estimate and interpret the results.*
    (d) According to the regression equation, what sales (to the nearest $1000) can management expect from staff with six years' experience?

Table 15.33 *Years worked and sales turnover per salesperson for Lightwait Industries*

Years with firm	1997 sales ($'000)
3	250
1	175
9	280
7	240
2	190
4	210
3	190
10	250

Source: Hypothetical.

*Optional

19. With the economic slowdown upsetting profitability, your firm is considering whether it should cut back on advertising expenditure and has asked you to determine whether a strong direct relationship exists between the advertising dollar and sales generation. Therefore, referring to Table 15.34:

   (a) Calculate the least-squares regression equation (to three decimal places).
   (b) What is the standard error (to two decimal places) of estimate (assume this is the population)?*
   (c) What sales amount could be expected from spending $16 000 p.a. on advertising?
   (d) Calculate the coefficient of determination and interpret the results.
   (e) Would you advise cutting back on advertising expenditure in the light of your regression and correlation analysis? Explain briefly.

Table 15.34 Advertising and sales results for your firm

Advertising expenditure ($'000)	Sales ($'000)
8	140
10	165
6	120
14	180
12	185
18	210
15	160

Source: Hypothetical.

20. A major expense for Alien Industries is the maintenance cost of its hydraulic equipment. Table 15.35 lists the annual costs for each of the eight identical machines, which vary only in age.

   (a) Construct a linear regression equation (to three decimal places) using the least-squares method.
   (b) Calculate the standard error (to two decimal places) of estimate (assume this is the population of hydraulic equipment).*
   (c) What expense (to the nearest $1000) can Alien expect for a machine six years old?
   (d) Calculate the coefficient of determination.
   (e) On the basis of this calculation and consideration of possible other factors affecting maintenance expenses, would you use the regression line for forecasting? Explain your answer.

*Optional

Chapter 15 – Correlation and regression analysis

Table 15.35 *Alien Industries' hydraulic equipment and maintenance expense*

Age of equipment (years)	Maintenance expense ($'000)
2	3
5	10
4	6
7	10
3	2
1	1
5	6
4	7

Source: Hypothetical.

21. Opinions on a relationship between aptitude in mathematics (independent variable) and music (dependent variable) are diverse. Table 15.36 lists the results for 17 students.

    (a) Plot the scatter diagram.
    (b) Calculate the line of best fit using the least-squares method. (Calculate each variable to three decimal places.)
    (c) Use the line in (b) to estimate the music result for a student with a maths grade of 80%.
    (d) Calculate the coefficient of correlation and coefficient of determination (to two decimal places). Comment on the reliability and accuracy of your estimate in (c).

Table 15.36 *Maths and music grades for 17 students*

Maths grade (%)	Music grade (%)
56	48
59	51
62	58
63	78
68	62
69	66
72	74
73	76
74	71
76	80
79	81
82	83
85	72
87	90
94	92
96	93
97	96

Source: Hypothetical.

Notes

1. Richard Lyons, 'Run for your life, say men from Harvard', *New York Times*, 26 July 1984, p. 5.
2. Sonya Sandham, 'Drink beer and live longer: study', *Sydney Morning Herald*, 11 June 1996, p. 9.
3. N.R. Kleinfield, 'The taller the executive, the higher the salary, says study', *New York Times*, 10 March 1987, p. 81.

Chapter 16
INDEX NUMBERS

LEARNING OUTCOMES
After studying this chapter you should be able to:
- Define an index number and recognise its pervasiveness throughout the Australian business sector
- Interpret and use a range of index numbers commonly used in the Australian business sector
- Calculate and analyse simple index numbers
- Calculate and analyse simple composite index numbers
- Calculate and analyse weighted index numbers
- Describe the basic structure of the Consumer Price Index (CPI) and perform calculations involving its use
- Describe how to construct index numbers, including shifting base periods and defining in terms of constant dollars (i.e. deflating).

The Dow Jones Index, the All Ordinaries Index, the FTSE-100, the McDonald's Hamburger Index, the Discomfit Index, the Home Affordability Index and the Indexes of Economic Activity all have a common denominator—each one is a statistical device for determining changes over time for a specific variable. In fact, index numbers are used by anyone from statisticians to criminologists, chefs, tour operators, stockbrokers, investment consultants or sportspeople. Certainly, no business person should be without them.

In its simplest form, an **index number** can be thought of as a percentage summarising the changes that occur in prices, production, sales or other variables over time. In other words, it is a convenient way of representing changes in the value of a carefully defined basket of goods and/or services. Therefore, index numbers are widely used and invaluable tools for the business person attempting to assess or make the hard decisions on budgeting, price and/or wage rises, staffing, stock control, product performance, and so on. As you are studying statistics to increase your awareness and understanding of tools available to the decision-maker, we will concentrate on *price, quantity* and *value indexes*.

You may not realise it, but index numbers have a major impact on businesses and consumers alike. This fact is best illustrated by considering the Consumer Price Index (CPI), which measures the changes in retail price levels in Australia over specific periods. However, one cannot live by the CPI alone! Other measures using index numbers include other *price indexes* which summarise changes in prices (e.g. the Retail Price Index, Price Index of Materials Used in House Building, Price Index of Materials Used in the Manufacturing Industry, House Price Index, Export Price Index, Import Price Index, Index of Materials Used in Coal Mining, and Commodity Price Index); *quantity indexes*, which measure changes in physical volume (e.g. factory production and dwelling (housing) commencements); *value indexes*, which measure changes in the product price times quantity (e.g. gross domestic production, and engineering construction activity); *business activity indexes* (the ANZ and ABS economic indicators, for example, are often in index form); and various other indexes (e.g. IQ examination results, the Consumer Confidence Index, Business Confidence Index, Trade Weighted Index, Tourism and Leisure Index, and Award Rates of Pay Index).

A brief exposé of the more interesting indexes is presented for your interest. At the conclusion of most news programs the All Ordinaries Index is mentioned, with its movement actually defined in terms of points gained or lost. The daily movement is related to the previous day's closing All Ordinaries Index amount. Thus, if the All Ordinaries is at 2600 at the beginning of the day's trading and closes up 26 points at the end of the day, it has risen 1% (i.e. 26/2600); similarly, if it loses 13 points, it has closed 0.5% down on the previous day's close (13/2600). Therefore, we are given a snapshot of the weighted average price movements in the stockmarket as a whole. You can, naturally, construct an index of your own investment portfolio in shares and follow the change. You will often find analysts comparing the performance of the All Ordinaries Index to that of other investment options.

The National Bank publishes a Business Confidence Index which provides a guide to business managers' expectations for change in business conditions in upcoming quarters, always using the conditions of the previous quarter as the base. Therefore, zero indicates no change to the business outlook from the conditions prevailing in the previous quarter, while a positive rating indicates expectations of an improvement in business conditions.[1]

In 1996, a new index, the Average Index of Economic Freedom, was published by 11 economic research institutes from around the world, based on 17 factors measuring how freely a country's citizens can engage in economic activities. Somalia took the wooden spoon, with Hong Kong and New Zealand in first and second position and Australia in eighth place. The report was highly critical of the Australian government's high consumption share of GDP and the large and growing transfer sector (i.e. welfare benefits).[2]

Speaking of freedom, what about being able to buy it? Transparency International, a German-based group, has produced a Corruption Index, with New Zealand and Denmark the least corrupted countries (occupying first and second place); Australia is in 10th position. The poll involves 10 sources ranging from statistical data to a survey of businesses; as such, it is more an index of perceptions than an assessment of actual corruption levels.

Finally, the Yellow Pages' *Small Business Index*[3] is an ongoing series of surveys designed to track confidence and behaviour in the small business sector. It is the largest index of such businesses in Australia, specifically focusing on businesses employing 19 or fewer people. The index uses a panel of at least 1200 randomly selected small business proprietors, drawn from all metropolitan (910 businesses) and non-metropolitan regions (290 businesses) who are interviewed by telephone every three months. Quotas are set on geographical location and type of business (e.g. manufacturing, building/construction, wholesale/retail, etc.). Because this is a quota sample, the sample is weighted according to the ABS Register. The primary objectives of the index are to track small business activity overt the past three months and expectations over the next three and 12 months, and to measure overall confidence within the small business community. A further purpose is to provide an independent, objective assessment of proprietors' experience and attitudes on key issues.

Indexes may be simple (sometimes referred to as *simple relative indexes*) or composite. A *simple index* illustrates changes in the price, quantity and/or value of a single commodity, such as a change in the price of a kilogram of steak, the change in the quantity of steak consumed, or a change in Australians' expenditure on steak respectively.

A *composite or aggregate index* also indicates changes in price, quantity and/or value, but this time for a group of items or commodities, some of which are not necessarily expressed in the same unit of measurement. (For example, in the CPI, milk is expressed in terms of litres, coffee in terms of kilograms and shoes in terms of pairs.)

A **weighted aggregate** (or **composite**) **index** is based on the degree of importance that each item contributes to the aggregate, with weights accordingly assigned.

Many of the worked examples, competency tests and exercises in this chapter are based on real Australian statistics to allow you to see possible applications to your workplace.

16.1 Simple index numbers

Simple index numbers can be defined as the ratio or percentage change in the price, quantity or value of a single item in a given period versus the price, quantity or value of the same item in another period (the **base period**). The base period is defined as the period with which subsequent periods are compared (the reference period), with the base usually given a value of 100.

Simple price index

Price indexes summarise the change in the price of an item(s), usually over time. It can therefore be thought of as providing information on the movement in the price level of items included within the index. For example, let us assume we are interested in a simple price index for milk, which cost $1.10 a litre in 1992 and $1.36 in 1997.

The formula for a simple price index is: $\dfrac{P_g}{P_b} \times 100$

where P_g = price of the product in the given year
P_b = price of the same product in the base year.

Business mathematics and statistics

Note that the base period is always in the denominator and that you multiply the fraction by 100 to convert it to an index or percentage relative, although by convention the per cent symbol is dropped.

Therefore, in calculating the change in the price of milk over the five years in question, we have:

$$\text{1997 simple price index} = \frac{\text{Price of milk in given period (1997)}}{\text{Price of milk in base period (1992)}} \times 100$$

$$= \frac{1.36}{1.10} \times 100$$

$$= 123.64$$

This means that milk increased in price by 23.64% over the period under consideration (in this case, the five-year period from 1992 to 1997).

The price index can also be calculated with a different base year. If, for instance, you decided that 1997 should be the base period, the price index would be:

$$\text{Simple price index} = \frac{\text{Price of milk in 1992}}{\text{Price of milk in 1997}} \times 100$$

$$= \frac{1.10}{1.36}$$

$$= 80.88$$

Since the price index is less than 100, this indicates that milk in 1992 cost less than compared to the base (or 100). In this case, milk in 1992 cost 19.12% less than in 1997. It is incorrect to say that milk in 1997 cost 19.12% more than in 1992, because this statement has made 1992 the base when, in fact, it is 1997.

Problem: The typical rent for a two-bedroom unit for the years 1990–97 is listed in Table 16.1. Calculate the simple price index for all years, using 1992 as the base period.

Table 16.1 *Rent for two-bedroom units in Fernleigh, 1990–97*

Year	Rent ($)
1990	140
1991	150
1992	160
1993	175
1994	185
1995	190
1996	185
1997	195

Source: Hypothetical.

Solution:

$$1990 \text{ simple price index} = \frac{1990 \text{ price}}{1992 \text{ price}} \times 100$$
$$= \frac{140}{160} \times 100$$
$$= 87.5$$

Thus, rent in 1990 was about 12.5% less than in the base year, 1992; or rent in 1990 was 87.5% of the rent charged in 1992.

$$1991 \text{ simple price index} = \frac{1991 \text{ price}}{1992 \text{ price}} \times 100$$
$$= \frac{150}{160} \times 100$$
$$= 93.75$$

Rent in 1991 was about 6% less than in the base period; or rent in 1991 was about 94% of the rent charged in 1992.

$$1993 \text{ simple price index} = \frac{1993 \text{ price}}{1992 \text{ price}} \times 100$$
$$= \frac{175}{160} \times 100$$
$$= 109.375$$

Rent in 1993 was about 9% above that of 1992.

Is it correct to say that rent in 1993 was about 15.5% higher than in 1991 (i.e. 109.375 − 93.75)? Why?

$$1994 \text{ simple price index} = \frac{1994 \text{ price}}{1992 \text{ price}} \times 100$$
$$= \frac{185}{160} \times 100$$
$$= 115.625$$

Rent in 1994 was about 15.5% more than in 1992.

$$1995 \text{ simple price index} = \frac{1995 \text{ price}}{1992 \text{ price}} \times 100$$
$$= \frac{190}{160} \times 100$$
$$= 118.75$$

Rent for a two-bedroom unit had increased by approximately 19% compared with the 1992 base period.

$$1996 \text{ simple price index} = \frac{1996 \text{ price}}{1992 \text{ price}} \times 100$$

$$= \frac{185}{160} \times 100$$

$$= 115.625$$

Rent in 1996 was 15.6% higher than in 1992 (the base period).

Is it correct to say that rent in 1996 was about 3% less than in 1995 (118.75 − 115.625)?

$$1997 \text{ simple price index} = \frac{1997 \text{ price}}{1992 \text{ price}} \times 100$$

$$= \frac{195}{160} \times 100$$

$$= 121.875$$

The 1997 rent for a two-bedroom unit was about 22% greater than the base period.

Simple quantity index

Comparisons are frequently made between quantities of products (e.g. production) in index form. These indexes are called **quantity relatives**. The method used to calculate them is similar to that used for the price relatives, except that the percentage change of quantity produced in the given period is compared with the base period. The formula for calculating a simple **quantity index** is:

$$\text{Simple quantity index} = \frac{\text{Quantity in given period}}{\text{Quantity in base period}} \times 100$$

$$= \frac{Q_g}{Q_b} \times 100$$

Problem: Calculate the simple quantity indexes for the three articles listed in Table 16.2 for 1991/92 and 1994/95, using 1991/92 as the base.

Table 16.2 *Production of articles for 1991/92 and 1994/95*

Article	1991/92	1994/95
Beer (ml)	1862	1778
Confectionery ('000 tonnes)	164	182
Cars and station wagons ('000)	269	301

Source: Australian Bureau of Statistics, *Year Book Australia, 1996* (Cat. No. 1301.0).

Solution:
Beer:

$$\text{1994/95 simple quantity index} = \frac{\text{1994/95 quantity}}{\text{1991/92 quantity}} \times 100$$

$$= \frac{1788}{1862} \times 100$$

$$= 96.03$$

There was a decline of 4% (i.e. 100 – 96.03) in beer production in 1994/95 versus the base period, perhaps as a result of changes in the pattern of consumption (with wine increasingly in favour) and Australians becoming more health-conscious.

Confectionery:

$$\text{1994/95 simple quantity index} = \frac{\text{1994/95 quantity}}{\text{1991/92 quantity}} \times 100$$

$$= \frac{182}{164} \times 100$$

$$= 110.98$$

The sweet tooth prevailed, as there was an increase of 10.98% in the production of confectionery in 1989/90 compared with the base period.

Cars and station wagons:

$$\text{1994/95 simple quantity index} = \frac{\text{1994/95 quantity}}{\text{1991/92 quantity}} \times 100$$

$$= \frac{301}{269} \times 100$$

$$= 111.9$$

Production of cars and station wagons rose by about 11.9% in 1994/95 compared with the base period of 1991/92. If you remember, 1994/95 was a time of steady business and consumer confidence, while 1991/92 was a period of high unemployment and deep economic gloom.

Simple value index

We should all be aware that price × quantity is equal to the value of sales. The comparison between values of sales over a given time period is therefore called *value relative*. This is simply the total value of a product(s) (i.e. quantity × price) sold in the given period, compared with the value of the same product(s) sold in another period (the base period). The formula is:

$$\text{Simple value index} = \frac{\text{Value in given period}}{\text{Value in base period}} \times 100$$

$$= \frac{V_g}{V_b} \times 100$$

Business mathematics and statistics

For example, let's calculate the simple value index of turnover in the Australian retail industry for each of the years listed in Table 16.3, using 1991/92 as the base period.

Table 16.3 *Turnover in the Australian retail industry, 1991/92 to 1995/96*

Year	Turnover ($ billion)
1991/92	98
1992/93	100
1993/94	104
1994/95	112
1995/96	121

Source: Australian Bureau of Statistics, *Year Book Australia, 1996* (Cat. No. 1301.0).

$$1992/93 \text{ simple quantity index} = \frac{1992/93 \text{ value}}{1991/92 \text{ value}} \times 100$$
$$= \frac{100}{98} \times 100$$
$$= 102.00$$

(An increase of 2% in turnover compared with the base period.)

$$1993/94 \text{ simple quantity index} = \frac{1993/94 \text{ value}}{1991/92 \text{ value}} \times 100$$
$$= \frac{104}{98} \times 100$$
$$= 106.1$$

(A rise of 6.1% compared with the base period.) Is it correct to say that from 1992/93 to 1993/94, retail sales increased by about 4.1%? If so, why? If not, why not?

$$1994/95 \text{ simple quantity index} = \frac{1994/95 \text{ value}}{1991/92 \text{ value}} \times 100$$
$$= \frac{112}{98} \times 100$$
$$= 114.3$$

The 1994/95 financial year's retail sales were 14.3% above those of the 1991/92 base year.

$$1995/96 \text{ simple quantity index} = \frac{1995/96 \text{ value}}{1991/92 \text{ value}} \times 100$$
$$= \frac{121}{98} \times 100$$
$$= 123.5$$

By 1995/95, retail sales were 23.5% greater than four years earlier (the 1991/92 base year).

The results for retailing look healthy. However, what must be considered in conjunction with turnover to get a better idea whether, in fact, the number of unit sales had risen (or was it merely price increases, or a combination of both)? You must look at the inflation rate. Interestingly, over the period from 1991/92 to 1995/96, inflation was 10.6%; which means that just under half of the rise in retail turnover was a result of inflationary pressures. Also remember that negative economic growth was recorded during the 1991/92 financial year, which stifled retail sales at that time.

16.1 Competency check

1. What is an *index number*?

2. (a) What does a *simple price index* measure?
   (b) What does a *simple quantity index* measure?
   (c) What does a *simple value index* indicate?

3. A standard video recorder cost $750 in 1995, $690 in 1996 and $650 in 1997.

   (a) Find the simple price indexes using 1995 as your base period. Briefly interpret the results.
   (b) Find the simple price indexes using 1997 as your base period. Interpret the results.

4. A high-season economy return fare from Sydney to Los Angeles cost $1250 in 1994, $1400 in 1995, $1475 in 1996 and $1525 in 1997.

   (a) Find the simple price indexes using 1994 as the base period. Interpret the results.
   (b) Find the simple price indexes using 1996 as the base period. Interpret the results.

5. Table 16.4 lists the number of dwelling units approved in new residential buildings in Australia between 1993/94 and 1995/96. Construct simple quantity indexes (to one decimal place) using:

   (a) 1993/94 as the base period. Interpret the results.
   (b) 1995/96 as the base period. Interpret the results.

Table 16.4 *Dwelling units approved in new residential buildings, Australia, 1993/94 to 1995/96*

Year	Dwelling units ('000)
1993/94	189
1994/95	171
1995/96	125

Source: Australian Bureau of Statistics, *Building Approvals, September 1996* (Cat. No. 8731.0).

16.2 Simple aggregate (composite) index numbers

Our previous examples of price, quantity and value indexes related only to single items or to an industry as a whole. However, in practice, index numbers are commonly used to assess the movement of a group of commodities or items (e.g. retail prices or exports). The **simple** (sometimes referred to as *unweighted*) **aggregate** (or **composite**) **index** method obtains a single value or index which represents the sum of the values of a group of items for a given year divided by the sum of values for the same group during the base period, with this ratio multiplied by 100. Since we are simply adding the values of every item, we are assuming that each is of equal importance to the other.

Simple aggregate price index

Let's assume that we have been asked to calculate an index of food prices for 1991/92 and 1996/97 using 1991/92 as the base period. Assume that the food prices are as listed in Table 16.5.

Table 16.5 Prices of selected foodstuffs, 1991/92 and 1996/97

Item	Price 1991/92 ($)	Price 1996/97 ($)
Bread (loaf)	1.21	1.85
Rump steak (kg)	7.45	7.10
Butter (kg)	3.50	3.70
Soft drinks (litre)	1.35	1.35
Caviar (kg)	30.00	40.00
Poultry (kg)	2.00	4.00
Potatoes (kg)	0.75	0.90

Source: Author's shopping list.

If we add the 1996/97 prices and divide that sum by the total of the 1991/92 prices (the base period), we obtain:

$$\text{Simple aggregate price index} = \frac{\text{Sum of prices in given year}}{\text{Sum of prices in base year}} \times 100$$

$$\text{Simple aggregate price index} = \frac{1.85 + 7.10 + 3.70 + 1.35 + 40.00 + 4.00 + 0.90}{1.21 + 7.45 + 3.50 + 1.35 + 30.00 + 2.00 + 0.75} \times 100$$

$$= \frac{58.90}{46.26} \times 100$$

$$= 127.32$$

Therefore, the composite 1996/97 prices were 127.32% of what they were in 1991/92. In other words, the prices of the items in this basket of food increased by 27.32% over the given period.

Obviously, the advantages of using this method are simplicity and clarity, but its failings far outweigh its usefulness. There is no acknowledgment of the relative importance of the various commodities comprising the index—that is, no indication of how much of each item was purchased. This is evidenced by the fact that a price change in a rarely purchased item (e.g. caviar) has the same effect as a similar price change in an item that is frequently consumed (e.g. bread). Furthermore, in this example, the prices of different units of measurement (litres and kilograms) were added together.

To overcome these major problems, a weighted aggregate index is used. This index permits the relative importance of examined commodities to be calculated and included in the final composite index result (found in Section 16.3).

Simple (unweighted) average of relative index

Since businesses often deal in different units of measurement (e.g. tonnes, kilograms and litres), these varying quantities cannot be added together. Instead, the commodity price or quantity (depending on whether you are calculating a price or quantity index) must be expressed as a percentage of its price or quantity in the base period. The formula for calculating the simple average of relative price index is:

$$\text{Simple price relative index} = \frac{\Sigma\left(\frac{P_g}{P_b} \times 100\right)}{n}$$

where P_g = relative price in the given year
P_b = relative price in the base year
n = number of items.

The formula for calculating the simple average of relative quantity index is:

$$\text{Simple quantity relative index} = \frac{\Sigma\left(\frac{Q_g}{Q_b} \times 100\right)}{n}$$

where Q_g = relative quantity in the given year
Q_b = relative quantity in the base year
n = number of items.

In other words, when using this method you first calculate the relative price changes (i.e. the ratio of present price to base period price) per individual commodity and then obtain the index by averaging these price relatives for each period. Let's apply this method to the basket of foodstuffs cited earlier (see Table 16.6) using 1991/92 as the base period.

$$\begin{aligned}\text{Simple price relative index} &= \frac{\Sigma\left(\frac{P_g}{P_b} \times 100\right)}{n} \\ &= \frac{907.2}{7} \\ &= 129.6\end{aligned}$$

Business mathematics and statistics

Table 16.6 Prices of selected foodstuffs, 1991/92 and 1996/97

Item	Price		$\dfrac{P_g}{P_b} \times 100$
	1991/92 ($) ($P_b$)	1996/97 ($) ($P_g$)	
Bread (loaf)	1.21	1.85	152.9
Rump steak (kg)	7.45	7.10	95.3
Butter (kg)	3.50	3.70	105.7
Soft drinks (litre)	1.35	1.35	100.0
Caviar (kg)	30.00	40.00	133.3
Poultry (kg)	2.00	4.00	200.0
Potatoes (kg)	0.75	0.90	120.0
Total			907.2
Index	100	129.6	

The **relative index** is an improvement on the simple aggregate price index, since the units of measurement are accounted for. Furthermore, the figure of 129.6 represents the average of price changes for each of the items included in the food basket, while the simple price index represents the change in the price of the entire basket of food. Note that poultry had the largest percentage price change and therefore influenced the final index more than the price relatives for bread, caviar or potatoes. Thus, since weights are not considered, such an index is inappropriate more often than not.

Problem: For the production listed in Table 16.7:

(a) Calculate the simple quantity index using 1992/93 as the base period.
(b) Calculate the quantity (unweighted) relative index.
(c) Briefly discuss which of the two methods is more appropriate.

Table 16.7 Australian production of selected commodities, 1992/93 and 1994/95

Commodity	Quantity	
	1992/93	1994/95
Gas (terajoules)	568 820	622 047
Electricity (mil. kWh)	159 872	165 063
Yarn wool (tonnes)	18 167	23 093
Bricks (millions)	1 722	1 860
Beer (millilitres)	1 805	1 788

Source: Australian Bureau of Statistics, *Year Book Australia, 1994, 1996* (Cat. No. 1301.0).

Chapter 16 – Index numbers

Solution:

(a) Simple quantity index $= \dfrac{Q_g}{Q_b} \times 100$

$= \dfrac{622\,047 + 165\,063 + 23\,093 + 1860 + 1788}{568\,820 + 159\,872 + 18\,167 + 1722 + 1805} \times 100$

$= \dfrac{813\,851}{750\,386} \times 100$

$= 108.46$

Thus there was an 8.46% increase in production according to this method. However, because different units of measurement (e.g. millilitres, terajoules, tonnes, and millions of kilowatt hours) were added together, such calculations are unacceptable and therefore invalid.

(b) The above commodities are expressed in different units, and it is therefore more accurate to take an average of the relative quantities. That is, determine $\dfrac{Q_g}{Q_p} \times 100$ for each of the commodities.

Commodity	$\dfrac{Q_g}{Q_b} \times 100$
Gas	$\dfrac{622\,047}{568\,820} \times 100 = 109.36$
Electricity	$\dfrac{165\,063}{159\,872} \times 100 = 103.25$
Wool	$\dfrac{23\,093}{18\,167} \times 100 = 127.12$
Bricks	$\dfrac{1860}{1722} \times 100 = 108.01$
Beer	$\dfrac{1788}{1805} \times 100 = 99.06$
Total	546.8

Now apply the formula:

Simple quantity relative index $= \dfrac{\Sigma\left(\dfrac{Q_g}{Q_b} \times 100\right)}{n}$

$= \dfrac{546.8}{5}$

$= 109.36$

Note the difference between the simple aggregate quantity index and the simple quantity relative index. The latter's rise of 9.36% in 1994/95 over the base period of 1992/93 is a difference of about 10.6% (that is, a rise of 9.36% is 10.6% greater than a rise of 8.46%, i.e. 9.36/8.46). Remember that when different units of measurement are presented, one cannot just add them up. Using the relative quantity index is an improvement, because the quantity of each commodity is expressed relative to its production in the base year. However, the lack of any weights indicating the relative importance of each one still severely detracts from its usefulness.

16.2 Competency check

1. What is the difference between a *simple aggregate (composite) price index* and a *simple price index*?

2. Although calculating the aggregate price index is simple, it has some problems which outweigh its usefulness. What are the major failings of using such an index?

3. When is the average of relative index useful?

4. From Table 16.8, using 1992/93 as the base period, calculate:

    (a) the simple price index;
    (b) the simple unweighted relative price index.

Table 16.8 *Prices of selected foodstuffs, 1992/93 and 1996/97*

Item	Price per kg	
	1992/93 ($)	1996/97 ($)
Apples	1.40	2.30
Bananas	1.49	1.72
Peaches	3.50	4.40
Strawberries	6.00	8.00

Source: Hypothetical.

5. Table 16.9 contains data on the production of articles in Australian manufacturing establishments.

    (a) Calculate the unweighted average of relative quantity index using 1992/93 as the base period.
    (b) Why was there no point in asking you to calculate the simple aggregate quantity index?

Chapter 16 – Index numbers

Table 16.9 *Quantities of certain articles produced by Australian manufacturers, 1992/93 and 1994/95*

Article	Quantity	
	1992/93	1994/95
Domestic refrigerators ('000)	393	408
Textile floor coverings ('000 sq. m)	42 106	47 258
Chocolate (tonnes)	105 681	109 709
Tobacco and alcohol (tonnes)	24 001	23 083

Source: Australian Bureau of Statistics, *Year Book Australia, 1996* (Cat. No. 1301.0).

16.3 Weighted aggregate (or composite) index

Weighted aggregate price index

In order to attribute the appropriate importance to each of the items examined, some reasonable weighting plan must be followed. For example, examine the selection of certain manufactured commodities in Table 16.10.

Table 16.10 *Price and production of certain manufactured commodities, 1991/92 and 1996/97: calculation of a Laspeyres index*

Item	P_b 1991/92 Price ($)	Q_b 1991/92 Quantity	P_g 1996/97 Price ($)	Q_g 1996/97 Quantity	$P_b Q_b$	$P_g Q_b$
(1)	(2)	(3)	(4)	(5)	(6) = (2) × (3)	(7) = (4) × (3)
Washing machines	460	220	550	294	101 200	121 000
Vacuum cleaners	103	175	180	202	18 025	31 500
Electric fans	43	598	75	694	25 714	44 850
Television sets	570	375	625	250	213 750	234 375
Total					358 689	431 725

Source: Hypothetical.

There are two principal methods (and hence, formulae) for calculating weighted price and quantity indexes: the *Laspeyres* formula and the *Paasche* formula. The former uses the base year's consumption as the quantity weights, while the latter uses the current-period consumption as the quantity weights.

The **Laspeyres method** is more widely used because it needs quantity measures for only one period (i.e. the base period's consumption quantity). For example, in the case of manufactured goods (see Table 16.10), to calculate the weighted aggregate price index using 1991/92 as the base year:

1. Multiply the price per item in the base year (column 2) by the quantity produced in the base year (column 3).

2. Find the total spent during the base period by finding the sum of these items (column 6).
3. Multiply the price per commodity in the current (or given) period (column 4) by the respective production in the base period (column 3).
4. Find the sum of these values (column 7).
5. Divide the current year's sum (column 7) by the base year's sum (column 6) and multiply this figure by 100 to convert it to index form.

Laspeyres method:

$$\text{Weighted aggregate price index} = \frac{\Sigma(P_g Q_b)}{\Sigma(P_b Q_b)} \times 100$$

where P_b = price of the commodity in the base year
Q_b = quantity of the commodity in the base year
P_g = price of the commodity in the given year.

Therefore, using the Laspeyres method, we have:

$$\begin{aligned}
\text{Weighted aggregate price index} &= \frac{\Sigma(P_g Q_b)}{\Sigma(P_b Q_b)} \times 100 \\
&= \frac{(550 \times 220) + (180 \times 175) + (75 \times 598) + (625 \times 375)}{(460 \times 220) + (103 \times 175) + (43 \times 598) + (570 \times 375)} \times 100 \\
&= \frac{121\,000 + 31\,500 + 44\,850 + 234\,375}{101\,200 + 18\,025 + 25\,714 + 213\,750} \times 100 \\
&= \frac{431\,725}{358\,689} \times 100 \\
&= 120.36
\end{aligned}$$

In other words, during the base period of 1991/92 the items produced cost $358 689; the same items in the same quantities would have cost $431 725 in 1996/97, an increase of 20.36%. Therefore, there was an increase of 20.36% in the price level. It is most important to remember that when using the Laspeyres method to calculate a **weighted price index**, you must always use only the base year quantity as your weight. Why? Because the objective of the Laspeyres weighted price index is to determine the change in cost of buying, in the current year, the same products and same quantity of each product as you bought in the base period. In other words, you are finding out how much more it costs today to buy the same quantity of goods bought in the base year (irrespective of the actual amount bought during the current period).

The **Paasche formula**,* on the other hand, uses the current consumption pattern as weights (Table 16.11). In other words, the results indicate how much was actually spent in the current year compared to what would have been spent in other years had we purchased the same quantity of goods as in the current year. Therefore, to calculate the weighted aggregate price index using the Paasche formula:

* Optional.

Chapter 16 – Index numbers

1. Multiply the current year's price times the production for each item during the current year (column 4 times column 5) to derive column 6.
2. Find the sum of the items listed in column 6.
3. Multiply the price in the base period (column 2) times the quantity produced in the current period (column 5) to derive column 7.
4. Find the sum of the items listed in column 7.
5. Divide the total actual expenditure of the current year (column 6) by the sum of what would have been spent in the base year had the same quantities as in the current year been purchased (column 7) multiplied by 100 to convert to an index.

Paasche method:

$$\text{Weighted aggregate price index} = \frac{\Sigma(P_g Q_g)}{\Sigma(P_b Q_g)} \times 100$$

Table 16.11 *Price and production of certain manufactured commodities, 1991/92 and 1996/97: calculation of a Paasche index*

Item (1)	P_b 1991/92 Price ($) (2)	Q_b 1991/92 Quantity (3)	P_g 1996/97 Price ($) (4)	Q_g 1996/97 Quantity (5)	$P_g Q_g$ (6) = (4) × (5)	$P_b Q_g$ (7) = (2) × (5)
Washing machines	460	220	550	294	161 700	135 240
Vacuum cleaners	103	175	180	202	36 360	20 806
Electric fans	43	598	75	694	52 050	29 842
Television sets	570	375	625	250	156 250	142 500
Total					406 360	328 388

Source: Hypothetical.

Therefore, we have:

$$\begin{aligned}
\text{Weighted aggregate price index} &= \frac{\Sigma(P_g Q_g)}{\Sigma(P_b Q_g)} \times 100 \\
&= \frac{(550 \times 294) + (180 \times 202) + (75 \times 694) + (625 \times 250)}{(460 \times 294) + (103 \times 202) + (43 \times 694) + (570 \times 250)} \times 100 \\
&= \frac{161\,700 + 36\,360 + 52\,050 + 156\,250}{135\,240 + 20\,806 + 29\,842 + 142\,500} \times 100 \\
&= \frac{406\,360}{328\,388} \times 100 \\
&= 123.74
\end{aligned}$$

Thus, the actual cost of the commodities produced during the current year, 1996/97, was $406 360. Had these same commodities been purchased in 1991/92 in identical quantities, they would have cost $328 388. Therefore, there was an increase of 23.74% in the price level.

The Laspeyres method and Paasche methods provided similar results in this example. However, they can vary widely depending on the number of items covered, as well as on the differences in the weights applied. Usually, you will find widening differences in results between the Laspeyres and Paasche methods the greater the number of years between the index's base year and the current year.

Which of the two methods should you use? There is no clearcut answer. Certainly, the Laspeyres method is the most commonly used for the following reasons:

1. Only one set of weights is needed and it can be applied to all periods. In other words, if the results for 1997/98 were available, we could easily compare the index for that year with the index for 1996/97 or even 1992/93.
2. If you are trying to discover how much more it costs you today to buy the same goods as you bought previously, this method must be applied. It indicates the change in the cost of living as defined by the items you included in your basket of goods during the base period.

However, the Laspeyres method is imperfect in that it does not allow for the following facts:

1. Price rises change our consumption pattern and therefore the weights. You could say that the Laspeyres method does not really reflect the changes in the price level, since we may no longer be buying the same products we bought during the base period.
2. Changes occur in consumption due to changing consumer needs or tastes.

The Paasche method, on the other hand, has this major benefit: it reflects the changing pattern of consumption and prices, since it uses current prices and current quantities. With this method we should find that rising prices from the base period to the current period resulted in reduced consumption of those goods with the largest increases.

The major disadvantages of this method are as follows:

1. You have to collect quantity measures for each year examined. This makes it far more expensive to construct than a Laspeyres index.
2. Comparisons between indexes of different periods are difficult. This is because the quantity measures of one index period are usually different from the quantity measures of another index period (since we always take the current period's quantity as the weights).

Problem: Calculate the weighted aggregate price index for appliance sales of the All-purpose Company (see Table 16.12) using:

(a) the Laspeyres method and 1992/93 as the base period;
(b) the Paasche method.*

---

* Optional.

Chapter 16 – Index numbers

Table 16.12 *Appliance sales for the All-purpose Company, 1992/93 and 1996/97*

Item	1992/93 Price($) ($P_b$)	1992/93 Quantity ('000) (Q_b)	1996/97 Price($) ($P_g$)	1996/97 Quantity ('000) (Q_g)
Blenders	77	120	90	124
Juicers	36	50	38	63
Toasters	39	140	44	135
Woks	62	40	64	100

Source: Hypothetical.

Solution:

(a) *Laspeyres method:*

Weighted aggregate price index

$$= \frac{\Sigma(P_g Q_b)}{\Sigma(P_b Q_b)} \times 100$$

$$= \frac{(90 \times 120) + (38 \times 50) + (44 \times 140) + (64 \times 40)}{(77 \times 120) + (36 \times 50) + (39 \times 140) + (62 \times 40)} \times 100$$

$$= \frac{10\ 800 + 1900 + 6160 + 2560}{9240 + 1800 + 5460 + 2480} \times 100$$

$$= \frac{21\ 420}{18\ 980} \times 100$$

$$= 112.86$$

In other words, if the All-purpose Company had sold the same number of units in 1996/97 as it did during the base period of 1992/93, sales turnover would have been 12.86% higher in 1996/97 as compared to the base.

(b) *Paasche method:**

Weighted aggregate price index

$$= \frac{\Sigma(P_g Q_g)}{\Sigma(P_b Q_g)} \times 100$$

$$= \frac{(90 \times 124) + (38 \times 63) + (44 \times 135) + (64 \times 100)}{(77 \times 124) + (36 \times 63) + (39 \times 135) + (62 \times 100)} \times 100$$

$$= \frac{11\ 160 + 2394 + 5940 + 6400}{9548 + 2268 + 5265 + 6200} \times 100$$

$$= \frac{25\ 894}{23\ 281} \times 100$$

$$= 111.22$$

Using the Paasche method, we find that the sales revenue received in 1996/97 was 11.22% greater than if the same number of appliances as sold in 1996/97 had been sold in 1992/93 given the prices existing in 1992/93.

* Optional.

Business mathematics and statistics

Weighted average of price relative

As we have seen in previous examples, a composite index is often composed of a variety of unit measurements, and when you are determining the weighted price or quantity relative, these cannot be added up. Instead, when calculating the weighted price relative, use the value of, for example, consumption (i.e. price multiplied by quantity consumed), since this is defined in dollar terms, hence permitting total consumption to be defined in terms of total dollars spent. For example, determine the weighted average price relative index for the production of certain manufactured goods (see Table 16.13).

Table 16.13 *Price and production of certain manufactured commodities, 1991/92 and 1996/97*

Item	1991/92 Price ($)	1996/97 Price ($)	$\dfrac{P_g}{P_b}$	1991/92 Quantity	1991/92 $P_b \times Q_b$	$\dfrac{P_g}{P_b} \times w \times 100$
	(P_b)	(P_g)		(Q_b)	(w)	
(1)	(2)	(3)	(3) ÷ (2) = (4)	(5)	(6) = (2) × (5)	(7) = (4) × (6) × 100
Washing machines	460	550	1.20	220	101 200	12 144 000
Vacuum cleaners	103	180	1.75	175	18 025	3 154 375
Electric fans	43	75	1.74	598	25 714	4 474 236
Television sets	570	625	1.10	375	213 750	23 512 500
Total					358 689	43 285 111

Source: Hypothetical.

1. Define each separate item as a price relative (i.e. compare the given year with the base year—column 4).
2. The weight selected must permit the addition of the units measured. Determine the value, or in this case, the amount of manufactured goods sold for each item during the base period; this value will be the respective weight per item (column 6).
3. Sum the results (e.g. 358 689).
4. Calculate the weighted average of relative prices by dividing the weighted percentage relative total (column 7) by the total weight (which in this case is column 6, the base year's value).
5. Multiply this result by 100 to get it into index form. Thus, we have the following:

$$\text{Weighted average price relative index} = \frac{\Sigma\left(\dfrac{P_g}{P_b} \times w \times 100\right)}{\Sigma w}$$

where w = weights (i.e. value in above case; however, if units are the same throughout, then the quantity will act as the weight).

Chapter 16 – Index numbers

$$\text{Weighted average price relative index} = \frac{\Sigma\left(\frac{P_g}{P_b} \times w \times 100\right)}{\Sigma w}$$

$$= \frac{43\ 285\ 111}{358\ 689}$$

$$= 120.68$$

Therefore, prices have increased an average of 20.68%, with television sets showing the lowest price relative increase (10%) and vacuum cleaners the highest (75%). Thus, Table 16.13 reveals the price index for each commodity (column 4), as well as the composite index. The composite index should provide the same answer as the weighted aggregate method (120.36), with the minor difference due to rounding off (in column 4).

Problem: Calculate the weighted average relative price index for the All-purpose Company (see Table 16.14).

Table 16.14 *Appliance sales for the All-purpose Company, 1992/93 and 1996/97*

Item (1)	1992/93 Price ($) (P_b) (2)	1996/97 Price ($) (P_g) (3)	$\frac{P_g}{P_b}$ (3) ÷ (2) = (4)	1992/93 Quantity (Q_b) (5)	1992/93 $P_b \times Q_b$ (w) (6) = (2) × (5)	$\frac{P_g}{P_b} \times w \times 100$ (7) = (4) × (6) × 100
Blenders	77	90	1.17	120	9 240	1 081 080
Juicers	36	38	1.06	50	1 800	190 800
Toasters	39	44	1.13	140	5 460	616 980
Woks	62	64	1.03	40	2 480	255 440
Total					18 980	2 144 300

Source: Hypothetical.

Solution:

$$\text{Weighted average price relative index} = \frac{\Sigma\left(\frac{P_g}{P_b} \times w \times 100\right)}{\Sigma w}$$

$$= \frac{2\ 144\ 300}{18\ 980}$$

$$= 112.98$$

Therefore, there is a 12.98% rise in the weighted average of price relative index, with blenders showing the largest price relative increase (17%) and woks the least (3%).

To summarise, the weighted average of relative index reveals the relative price changes for each individual component as well as the aggregate change.

The weights must be values (P × Q), usually base year values. Furthermore, since available data are often in the form of dollar values only, such values must act as weights.

Finally, it is possible to apply given year weights (the Paasche method) rather than base year weights (the Laspeyres method); but as this is rarely done, since results would not match any weighted composite index, an example has not been provided.

The Consumer Price Index

The most frequently used weighted average of price relative index exerting the most influence on the typical Australian is, of course, the **Consumer Price Index**. According to the Australian Bureau of Statistics,[4] the CPI measures the change in the cost of purchasing a constant basket of retail goods and services representative of the purchases made by a particular population group in a specified time period.

The present composition of the CPI basket is based on the estimated pattern of household expenditure in 1988, with the information on Australian spending habits derived from a multi-stage area sampling of approximately 8400 households in dwellings and caravan parks throughout Australia. The pattern is expected to change in 1998 as the results of the most recent Household Expenditure Survey (1993–94, published in 1996) are incorporated. In addition, CPI reviews are usually carried out at approximately five-year intervals which involves revising the list of items and their respective weights. The latest revision was in 1992, with the next expected in the March quarter of 1998.

The CPI is categorised into eight expenditure groups and subdivided into 105 subgroups which are classed into expenditure areas. Naturally, not every item purchased by householders can be included, but price movements in those items represented are expected to mirror similar price changes in related items not included.

When the CPI results are released by the ABS each quarter, a headline CPI and an underlying CPI are publicly discussed. The headline inflation rate is the CPI as published, while the underlying inflation is the Treasury's definition of inflation which discounts various factors impacting on the CPI, such as seasonal changes, government-initiated increases (such as increases in sales tax) and the effect of exchange rate variation. The rationale behind this is that the Treasury believes inflation should measure price level changes directly resulting from imbalances in supply and demand, rather than from external influences such as government charges.

Weighting of Consumer Price Index items
The household expenditure estimates not only provide the information needed to select representative items for CPI inclusion but also reveal the degree of importance of each item in the total spending by consumers.

Within each expenditure class, weights are calculated for each item. For example, within the 'cakes and biscuits' expenditure class, various sizes, qualities and types of cakes and biscuits are priced. The underlying weights for each expenditure class are fixed. However, to obtain a true representation of people's expenditure patterns, the weights are revised periodically, usually at five-year intervals from the results of the Household Expenditure Survey. The weights prevailing in early 1997 were as follows:

Table 16.15 *Consumer Price Index (CPI): weighted average of capital cities*

Group	Percentage contribution to the group's CPI
Food	18.324
Clothing	6.264
Housing	15.900
Household equipment and operation	18.370
Transportation	15.967
Tobacco and alcohol	7.475
Health and personal care	6.850
Recreation and education	10.850
Total all groups	100.00

Source: Australian Bureau of Statistics, *Year Book Australia, 1997* (Cat. No. 1301.0).

Therefore, of every dollar spent by the average consumer, household equipment consumes the largest share of total expenditure at 18.370%, while we sent a greater percentage of our spending on transportation than we do on housing. This can be explained by the fact that approximately 40% of Australian households have fully paid for their homes, thereby reducing the average spending on housing.

Collecting prices
Price information is collected from retail outlets where Australian householders normally purchase goods and services. Items such as rail fares, electricity and gas charges, telephone charges and local government rates are collected from the authorities concerned. In total, about 100 000 separate price quotations are collected quarterly in capital cities only. Perishables such as fruit and vegetables are collected weekly, fresh fish fortnightly, and fresh meat, bread, cigarettes, tobacco and alcohol monthly. Most other items are collected once during the quarter, normally by the middle of that period.

Calculating the Consumer Price Index
The first step is to calculate an average price for each of the individual items (the weighted average price relative index) which is derived from the average of prices charged by the sampled outlets carrying the said goods or services. An example of the calculation for the expenditure class of bread is shown in Table 16.16.

Each expenditure class is calculated the same way. The 2.7% in Table 16.16 indicates that to purchase the same quantity and quality of bread in the current quarter as was purchased in the previous quarter would cost 2.7% more.

After the average price changes for each expenditure class are calculated, the calculated price changes are combined to obtain an average price change per subgroup group, and finally for 'all groups'.

For example, Table 16.17 shows that cereal products have a price index of 101.713, or an increase in price of 1.7%, using the previous quarter as the reference period. The same procedure applies to the calculation of index numbers for all other subgroups and 'all groups'.

Table 16.16 *Expenditure class: bread*

Specification	Percentage weight (quantity)	Average price at		Percentage weight × average price	
		Previous quarter (cents)	Current quarter (cents)	Previous quarter (cents)	Current quarter (cents)
Type A	50	93.4	95.9	4670	4795
Type B	30	99.8	102.5	2994	3075
Type C	15	99.0	101.7	1485	1526
Type D	5	112.0	115.0	560	575
Total	100			9709	9971

The average price change for the expenditure class equals 9971 ÷ 9709 = 1.027 (or an increase of 2.7%).

Source: Australian Bureau of Statistics, *A Guide to the Consumer Price Index* (Cat. No. 6440.0). Commonwealth of Australia copyright, reproduced by permission.

Table 16.17 *Calculation of cereal products index*

Cereal products	Percentage weight (1)	Price change from previous quarter (2)	Price change × weight (3) = (1) × (2)
Bread	38.6	1.027	39.642
Cakes and biscuits	44.4	1.009	44.800
Breakfast cereals	10.4	1.014	10.546
Other cereal products	6.6	1.019	6.725
Total	100.0		101.713

Source: Australian Bureau of Statistics, *A Guide to the Consumer Price Index* (Cat. No. 6440.0). Commonwealth of Australia copyright, reproduced by permission.

Uses of the Consumer Price Index

Australia's inflation rate is defined as being measured by movements in the CPI. Increases in unemployment and pension benefits, rental or lease charges, and council rates, as well as semi-annual increases in tobacco, alcohol and petrol prices, are often based on movements in the CPI. Until the introduction of a more flexible wage system in March 1987, wages had been indexed according to the CPI since approximately 1976, except for a couple of brief periods of collective bargaining and later during an effective freeze on wages. You can be sure that the CPI result is never far away from the minds of workers and consumers because of its impact on the purchasing power of the dollar.

Australia's ability to combat inflation compared with that of several leading Western countries is illustrated in Table 16.18 by using the CPIs of the respective countries. It should be pointed out that Australia's CPI is published quarterly, while in most western countries, a measure of inflation is available monthly.

Table 16.18 indicates that Italy had the greatest percentage rise in prices over the base period, and Japan had the lowest. However, this does not necessarily imply that goods and services are more expensive in Italy than in any of the other countries selected. Remember that the percentage change in price is related to the cost of purchasing the basket of goods and services in the country in question during the base year, and that the cost in the base period in Italy may well have been below that of any other country; therefore, in spite of having the largest percentage increase in prices, Italy may not be the most expensive place to live. In fact, you will probably find that Japan is the most expensive country to live, even though it had the lowest level of inflation since the base period. Also, note that the inflation rate in Australia since the base period is below the major seven of the OECD (Organisation of Economic Cooperation and Development—comprised of 25 Western countries) average, while in the 1970s and 1980s our rate was often double the OECD average.

Table 16.18 *Consumer Price Index comparison, 1990/91 to 1994/95 (1990 = 100)*

Period	United States	Japan	France	Italy	United Kingdom	Canada	OECD Major 7	Australia
1990/91	102.5	101.8	101.7	103.3	103.6	103.2	102.4	102.0
1991/92	105.8	104.3	104.6	109.4	108.0	106.4	106.1	104.0
1992/93	109.1	105.6	106.7	114.4	110.6	108.2	109.1	105.0
1993/94	111.9	106.9	108.8	119.1	112.9	109.3	111.8	107.0
1994/95	115.1	107.2	110.6	124.4	116.2	110.5	114.4	110.4

Source: Australian Bureau of Statistics, *Australian Economic Indicators, September 1996* (Cat. No. 1350.0). Commonwealth of Australia copyright, reproduced by permission.

Before finally leaving the CPI as it is measured in Australia, it should be recognised that as a measure of inflation it is imperfect because it is based on a fixed basket of goods which does not allow for consumers to reduce their costs by substituting lower-priced products for ones whose relative prices have risen. In other words, the index assumes that householders' buying habits remain constant in spite of price changes. Likewise, the introduction of new products or a reduction in the use of existing items, such as cigarettes, suggests that the fixed weights and prices measured are no longer representative of actual expenditure patterns. Also what about increases in product quality which are difficult to measure? If the price increases by 10% as a result of quality improvement extending the life of the item by 20%, is that inflationary? The weights themselves are extracted from Household Expenditure Surveys which are only given to those working, aged under 65 and living in capital cities. Thus, how accurately does the CPI reflect inflation Australia-wide?

Weighted aggregate quantity index*

Although not as common as the aggregate price index, an aggregate quantity is useful for measuring changes in actual output at times of high inflation. The and Paasche methods are once again used, but with the w

* Optional.

When we calculated the weighted aggregate price index using the Laspeyres method, the base year quantity was used as the weight (Table 16.19, column 3); when we calculate the weighted aggregate quantity index, the price of the base year is applied as the weight (column 2) for both the given and base period. Therefore, constant prices are applied as the weighting factor for quantity indexes. For example, let's look at the list of manufactured goods and determine how to calculate this index by the Laspeyres and Paasche methods.

Table 16.19 *Price and production of certain manufactured commodities, 1991/92 and 1996/97: calculation of a quarterly index using the Laspeyres method*

Item	1991/92 Price ($) ($P_b$)	1991/92 Quantity ('000) (Q_b)	1996/97 Price ($) ($P_g$)	1996/97 Quantity ('000) (Q_g)
Washing machines	460	220	550	294
Vacuum cleaners	103	175	180	202
Electrical fans	43	598	75	694
Television sets	570	375	625	250

Source: Hypothetical.

Problem: Calculate the weighted aggregate quantity index for 1996/97, using 1991/92 as the base.

Solution:

1. *Laspeyres method*

Weighted aggregate quantity index

$$= \frac{\Sigma(P_b Q_g)}{\Sigma(P_b Q_b)} \times 100$$

$$= \frac{(460 \times 294) + (103 \times 202) + (43 \times 694) + (570 \times 250)}{(460 \times 220) + (103 \times 175) + (43 \times 598) + (570 \times 375)} \times 100$$

$$= \frac{135\,240 + 20\,806 + 29\,842 + 142\,500}{101\,200 + 18\,025 + 25\,714 + 213\,750} \times 100$$

$$= \frac{328\,388}{358\,689} \times 100$$

8.45% reduction in the volume of sales compared 1991/92.

2. *Paasche method** (Table 16.20)

Weighted aggregate quantity index

$$= \frac{\Sigma(P_g Q_g)}{\Sigma(P_g Q_b)} \times 100$$

$$= \frac{(550 \times 294) + (180 \times 202) + (75 \times 694) + (625 \times 250)}{(550 \times 220) + (180 \times 175) + (75 \times 598) + (625 \times 375)} \times 100$$

$$= \frac{161\,700 + 36\,360 + 52\,050 + 156\,250}{121\,000 + 31\,500 + 44\,850 + 234\,375} \times 100$$

$$= \frac{406\,360}{431\,725} \times 100$$

$$= 94.1$$

The volume of sales was approximately 6% less in 1996/97 compared with 1991/92.

Table 16.20 *Price and production of certain manufactured commodities, 1991/92 and 1996/97: calculation of a quantity index using the Paasche method**

Item	P_b 1991/92 Price ($)	Q_b 1991/92 Quantity ('000)	P_g 1996/97 Price ($)	Q_g 1996/97 Quantity ('000)	$P_g Q_g$	$P_g Q_b$
(1)	(2)	(3)	(4)	(5)	(6) = (4) × (5)	(7) = (4) × (3)
Washing machines	460	220	550	294	161 700	121 000
Vacuum cleaners	103	175	180	202	36 360	31 500
Electric fans	43	598	75	694	52 050	44 850
Television sets	570	375	625	250	156 250	234 375
Total					406 360	431 725

Source: Hypothetical.

Weighted average of quantity relative*

The weighted average of quantity relative reveals the quantity relatives between different periods, as well as a weighted relative index. The selection of weights depends on the type of quantity changes we are measuring. If we are interested in sales (dollar) turnover, then the dollar value will be applied as the weight. For example, see Table 16.21.

* Optional.

Table 16.21 *Appliance sales for the All-purpose Company, 1992/93 and 1996/97*

Item	1992/93 Quantity (Q_b)	1996/97 Quantity (Q_g)	$\dfrac{Q_g}{Q_b}$	1992/93 Price (P_b)	1992/93 $(P_b \times Q_b)$ (w)	$\dfrac{Q_g}{Q_b} \times w \times 100$
(1)	(2)	(3)	(3) ÷ (2) = (4)	(5)	(6) = (5) × (2)	(7) = (4) × (6) × 100
Blenders	120	124	1.033	77	9 240	954 492
Juicers	50	63	1.26	36	1 800	226 800
Toasters	140	135	0.964	39	5 460	526 344
Woks	40	100	2.50	62	2 480	620 000
Total					18 980	2 327 636

Source: Hypothetical.

The weighted average of quantity relative is calculated in the same fashion as the price relative, except that this time given year quantities of each item are compared with base year quantities, thus providing a quantity relative (column 4). To calculate its value:

$$\text{Weighted average quantity relative} = \dfrac{\Sigma\left(\dfrac{Q_g}{Q_b} \times w \times 100\right)}{\Sigma w}$$

$$= \dfrac{2\,327\,636}{18\,980}$$

$$= 122.6$$

The woks had the greatest change in relative terms as their sales rose 150%, while toasters showed a decline in relative terms. Furthermore, the result is the same as for the aggregate quantity index (allowing for rounding off), which will always be the case as long as dollar values are used as the weights and the Laspeyres method is applied.

Weighted aggregate value index*

We have all probably used a value index without recognising it as such. For instance, when your sales division's monthly result is compared with that of the same month last year, you are in fact using a value index. For example, if in 1996/97 your department sold 7000 pairs of shoes at an average price of $125 per pair compared with the 1994/95 result of 6500 pairs at $90 per pair, the weighted aggregate value index (using 1989/90 as the base) would be:

$$\text{Value index} = \dfrac{7000 \times 125}{6500 \times 90} \times 100$$

$$= \dfrac{875\,000}{585\,000} \times 100$$

$$= 149.57$$

*Optional

In other words, there was a 49.57% increase in total dollar sales in 1996/97 compared with 1994/95.

The value index is referred to as a weighted index, since both price and quantity are necessary in its calculation and are both considered as weights. This is also, perhaps, the basic disadvantage of the value index: because it is affected by both price and quantity changes, it does not distinguish between the effects of either price or quantity. This does not detract, however, from the fact that this index is the simplest to calculate and the easiest to understand. When, for example, businesses forecast sales results, they are in fact looking at anticipated changes in prices and unit sales to assess future sales value more accurately. In other words, the principle of a value index is applied.

16.3 Competency check

1. What is the main benefit of using a weighted aggregate price index rather than a simple aggregate price index?
2. To calculate a weighted aggregate index, the Laspeyres and Paasche methods are used, the former being more commonly applied. What is the difference between the two measures?
3. What does a weighted aggregate price index indicate if you use the Laspeyres method?
4. Mrs Fragola sells various fruits from her homestead site at the prices and quantities listed in Table 16.22. Calculate to two decimal places the weighted average price relative index.

Table 16.22 Mrs Fragola's fruit patch, 1992 and 1997

Fruit	Price per kg		Quantity	
	1992 ($)	1997 ($)	1992 (kg)	1997 (kg)
Apples	1.40	1.60	1150	1090
Bananas	1.09	1.39	1700	1400
Peaches	2.40	2.80	2500	2700
Strawberries	4.50	5.60	800	940

Source: Hypothetical.

5. Referring to Mrs Fragola's fruit patch (Table 16.22), calculate:
   (a) the weighted aggregate price index using the Laspeyres method using 1992 as the base period;
   (b) the weighted aggregate quantity index using this same method;
   (c) the weighted aggregate value index.

16.4 Index number construction

Having examined the various types of index numbers and their respective calculations, let's look at the procedure and the inherent problems involved in index number construction.

1. *Purpose.* A clear and precise definition of your purpose is necessary, as it has a definite bearing upon your choice of data and the method used.
2. *Availability and compatibility of data.* Is the information available for the time period you are investigating? There may be a lack of uniformity in the methods used—for example, Commonwealth Employment Service figures as compared with Australian Bureau of Statistics unemployment figures.
3. *Selection of items.* Is the number of items adequate? Is it a representative sample?
4. *Choice of base period.* Is it a fairly normal period, divorced from any abnormal influences? Is it a fairly recent period? Is it a base period commonly used by other series with which you are likely to compare your data?
5. *Choice of weights.* A method of combining, for example, confectionery and diamonds in the CPI or pig iron with paints in industrial production must be determined.
6. *Choice of index number formula.* Laspeyres versus Paasche. It is said there are well over 150 different formulae devised for index numbers.
7. *Method of construction.* Is it simple unweighted? Weighted aggregate? Weighted relative?

Limitations in the use of index numbers

Although you may follow each of the above steps to overcome any problems, you should realise that there are limitations in the use of index numbers. These include the following:

1. Sampling errors are unavoidable, since an index number is commonly based on a sample.
2. The greater the time difference between the current period and the base period, the greater the likelihood of bias in the result.
3. It may be impossible to take account of all quality changes.
4. The base year selection may be arbitrary and inappropriate.
5. Weightings may not be universally applicable. (Consider the weightings for the CPI.)
6. Data to construct the desired index may not be available.
7. Certain index numbers may be incompatible with other useful indexes.

Selection of the base period

Nominating an appropriate base period is vital to the construction of a relevant, useful and valid index number, because all subsequent and previous periods will be compared with it. Therefore, several criteria must be applied to its selection, including the following:

1. A normal period should be chosen (i.e. a time when the prices of the items used were normal and not affected by erratic, abnormal or cyclical variations).

Chapter 16 – Index numbers

2. A fairly recent period is necessary because commodities in a certain basket may change (e.g. the content of the CPI) and the product quality itself may change, thus making comparisons with a base period of yesteryear less meaningful.

Shifting the base period

The Consumer Price Index, the Import Price Index and the Price Index of Materials Used in House Building all had their base years altered during the 1980s because the previous base period was too long ago (e.g. for the CPI and Import Price Index it was 1966/67). Therefore, the base period may be altered if:

- the original base period is too long ago;
- the two indexes being compared are incompatible because of different base periods.

For example, Table 16.23 lists the CPI for 1986/87 to 1995/96, with a base year of 1966/67 (column 2). As the index is now less relevant (as 30 years have passed), let's shift the base to 1989/90. To shift a base period, let the newly selected base equal 100 and relate all other periods to this new base. In the above example, this is accomplished by dividing the outdated CPI indexed results by the CPI results corresponding to the newly selected base period—in this case, 1989/90's 636.3 CPI result. You divide by 636.3 in converting all preceding and successive years, as the relativities between the years must be kept constant. The resulting decimal is multiplied by 100 to convert it to index form.

Thus, in this case we find that inflation of 18.7% occurred between 1989/90 and 1996/97; but if we had not altered the base period from 1966/67, we would have had to say that inflation of 655.2% occurred between 1966/67 and 1995/96.

Table 16.23 *Australia's Consumer Price Index, 1986/87 to 1995/96*

Year	CPI (1966/67 = 100)	Revised CPI (1989/90 = 100)
1986/87	510.8	80.3 (510.8/636.3 × 100)
1987/88	548.1	86.1 (548.1/636.3 × 100)
1988/89	588.1	92.4 (588.1/636.3 × 100)
1989/90	636.3	100.0 (636.3/636.3 × 100)
1990/91	670.0	105.3 (670.0/636.3 × 100)
1991/92	682.7	107.3 (682.7/636.3 × 100)
1992/93	689.7	108.4 (689.7/636.3 × 100)
1993/94	701.8	110.3 (701.8/636.3 × 100)
1994/95	724.7	113.9 (724.7/636.3 × 100)
1995/96	755.2	118.7 (755.2/636.3 × 100)

Source: Australian Bureau of Statistics, *Consumer Price Index* (Cat. No. 6401.0).

Problem: The importance of the Export Price Index has been emphasised by many Australian economists, as it indicates the rate of change in prices received for Australian exports. Table 16.24 shows Australia's Export Price Index with 1974/75 as the base period. Shift the base to 1989/90 and comment on the findings.

Table 16.24 Australia's Export Price Index, 1986/87 to 1995/96

Year	Export Price Index (1974/75 = 100)	Revised Export Price Index (1989/90 = 100)
1986/87	243	81.5 (243/298 × 100)
1987/88	265	88.9 (265/298 × 100)
1988/89	283	95.0 (283/298 × 100)
1989/90	298	100.0 (298/298 × 100)
1990/91	283	95.0 (283/298 × 100)
1991/92	267	89.6 (267/298 × 100)
1992/93	279	93.6 (279/298 × 100)
1993/94	274	91.9 (274/298 × 100)
1994/95	283	95.0 (283/298 × 100)
1995/96	287	96.3 (287/298 × 100)

Source: Australian Bureau of Statistics, *Export Price Index* (Cat. No. 6405.0).

Solution: Since 1989/90 has been nominated as the new base period, divide all other periods by the Export Price Index corresponding to 1989/90 (i.e. 298) and multiply the decimal result by 100 to convert into index form. Thus, we find that prices of Australian exports decreased by about 4% from 1989/90 to 1995/96. This should now be compared with Australia's Import Price Index to determine how our terms of trade have progressed or retrogressed.

The second reason given for shifting a base period was the comparison of data with incompatible base periods. For example, let's assume you want to determine whether average weekly earnings for all employees (Table 16.25) have increased at a higher rate than inflation, thereby resulting in an increase in employees' purchasing power. You are confronted with indexes of two different base periods and must shift one or both to allow a comparison between the two economic indicators.

Table 16.25 Consumer Price Index (CPI) and Average Weekly Earnings (AWE) Index, 1986/87 to 1995/96

Year	CPI (1989/90 = 100)	AWE (1981/82 = 100)	Revised AWE (1989/90 = 100)
1986/87	80.3	140.2	81.7 (140.2/171.7 × 100)
1987/88	86.1	149.1	86.8 (149.1/171.7 × 100)
1988/89	92.1	160.9	93.7 (160.9/171.7 × 100)
1989/90	100.0	171.7	100.0 (171.7/171.7 × 100)
1990/91	105.3	181.8	105.9 (181.8/171.7 × 100)
1991/92	107.3	188.5	109.8 (188.5/171.7 × 100)
1992/93	108.4	192.8	112.3 (192.8/171.7 × 100)
1993/94	110.3	199.4	116.1 (199.4/171.7 × 100)
1994/95	113.9	207.4	120.8 (207.4/171.7 × 100)
1995/96	118.7	215.3	125.4 (215.3/171.7 × 100)

Source: Australian Bureau of Statistics, *Consumer Price Index* (Cat. No. 6401.0) and *Average Weekly Earnings* (Cat. No. 6304.0).

As you are interested in how earnings have fared in comparison with inflation since 1989/90, the AWE can be altered to a base year of 1989/90 from the present 1981/82 base period. In order to accomplish this, let the 1989/90 period equal 100 as it is the new base year. According to the 1981/82 base year result, the 1989/90 AWE result was 171.7; to convert this to 100, divide by 171.7 and multiply by 100. The same technique must be followed in converting preceding and succeeding years to the new base year of 1989/90 as the relativities between the years must remain constant; each successive year must therefore be divided by 171.7 and multiplied by 100. The revised AWE is shown in column 4, together with the necessary calculations. This operation allows a direct comparison with the necessary calculations. That is, average weekly earnings have risen 25.4%, while the CPI has risen by 18.7% between 1989/90 and 1995/96; hence, there has been a real increase in average weekly earnings of approximately 5.6% (i.e. 125.4/118.7 × 100 = 105.6%). You could just as easily have nominated 1991/92 as the base period and would have then had to convert both the CPI and AWE indexes to a 1991/92 base. The decision would depend on the objective of the exercise.

Uses of index numbers

Index numbers have the following uses:

1. They measure change, be it in prices (price index), quantity (production index) or value (value index). Their application is common in the workplace and in government.
2. They permit the user to examine quickly trends for further analysis by examining in one figure the composite results of a group of items.
3. They may prove invaluable to business and government sectors for budgeting and forecasting.
4. They may be applied as a deflator—that is, a device that adjusts current dollars by a price index and hence expresses a result in constant or base year dollars.

This last use is of immeasurable benefit when you contemplate approaching your boss for a raise and you need some ammunition in the form of the downward trend in your real wages.

Deflation permits businesspeople to determine whether their sales performance has improved in real terms (i.e. whether the percentage increase in sales has exceeded the percentage increase in the inflation rate). Some examples follow. Table 16.26 lists the value of Australia's retail sales (excluding motor vehicles) for the years 1993/94 to 1995/96. Looking at the actual dollar results (column 2), we find that there were increases of 7.5% from 1993/94 to 1994/95 and 7.6% from 1994/95 to 1995/96. (This calculation is called a *linked* relative, as the percentage change in this case is taken from successive periods, with the preceding period seen as the base.)

For example, the 1995/96 retail sales result divided by the 1994/95 retail sales result (120 744/112 244 × 100) indicates a 7.6% increase in sales. Similarly, the 1994/95 result divided by the 1993/94 result (112 244/104 424 × 100) indicates a 7.5% rise in sales in money terms. For the period from 1993/94 to 1995/96 there was a 15.6% increase in retail dollar sales (120 744/104 424 × 100 = 115.6).

Before breaking out the bubbly over the buoyant state of the economy, we must examine the price level and relate it to retail sales. This is necessary since, if the

Table 16.26 *Value of Australia's retail sales (excluding motor vehicles), 1993/94 to 1995/96*

Year	Retail sales ($ million)	CPI (1989/90 = 100)	Retail sales (1989/90 constant $)
1993/94	104 424	110.3	94 673 (104 424/110.3 × 100)
1994/95	112 244	113.9	98 546 (112 244/113.9 × 100)
1995/96	120 744	118.7	101 722 (120 744/118.7 × 100)

Source: Australian Bureau of Statistics, *Retail Sales* (Cat. No. 8501.0) and *Consumer Price Index* (Cat. No. 6401.0).

price level (i.e. inflation) has risen by, for example, over 16% during the period, then retail sales in real terms (after taking inflation into account) will have shown a decline.

To deflate the retail sales, which are given in current dollar terms, we divide these data by the price index and multiply the fraction by 100 to convert it into constant dollar terms. In the above example, divide the actual or current retail sales by the corresponding CPI (with a base of 1989/90 = 100), thereby taking into account the effect of inflation on sales. The results will therefore be the real sales in terms of 1989/90 dollars (column 4).

From our compilation we see that retail sales (in constant 1989/90 dollars) increased by approximately 4.1% from 1993/94 to 1994/95 (98 546/94 673 × 100) and by 3.2% from 1994/95 to 1995/96 (101 722/98 546 × 100). Therefore, a real increase of 7.4% was realised over the period from 1993/94 to 1995/96 (101 722/94 673 × 100) versus the 15.6% increase in current dollar terms. In other words, the importance of considering inflation's impact on your results can lead to a very different picture of your performance.

16.4 Competency check

1. (a) What factors should you consider when deciding on which year to use as the base period?
   (b) Why is the selection of an appropriate base year considered vital?

2. Assume that any increases in rents on industrial premises are aligned with changes in the CPI. If at the end of 1993/94 the rent for the warehouse was $75 per square metre, what was the rent (to the nearest dollar) in July 1996? (Use Table 16.26 to calculate the CPI increases for each of the last two years.)

3. In 1995/96, sales for the Joggers' Jogger totalled $1 265 000 compared to $1 194 000 in 1994/95. What were sales in each of these two years in constant 1989/90 dollars? (*Hint:* The total sales must be CPI-adjusted, with 1989/90 as the base for the CPI.)

4. Table 16.27 lists the price index of materials used in house building for three capital cities.

Chapter 16 – Index numbers

(a) Is it possible to determine from these data which capital city is the most expensive in which to build a house? Briefly explain.
(b) Assume the base year is considered antiquated, so that relevant comparison with recent times is difficult. Shift the base for each of the three capital cities to 1991/92 (to one decimal place) and briefly evaluate the results. Which city has experienced the largest percentage increase in house building materials?

Table 16.27 *Price index of house building materials for three capital cities, 1991/92 to 1994/95 (1966/67 = 100)*

Year	Sydney	Brisbane	Perth
1991/92	749.3	783.2	731.2
1992/93	745.1	794.1	734.7
1993/94	748.6	802.9	744.4
1994/95	771.7	823.3	765.3

Source: Australian Bureau of Statistics, *Year Book Australia, 1996* (Cat. No. 1301.0) and *Price Index of Materials Used on House Building, State Capital Cities* (Cat. No. 6408.0).

5. Table 16.28 shows the price indexes for materials used in house building and materials used in building other than house building. Since the data are of two different base periods, alter both to the base period of 1991/92. Briefly evaluate the result.

Table 16.28 *Price indexes of house building materials and materials used in building other than house building, 1991/92 to 1995/96*

Year	House building (1985/86 = 100)	Other than house building (1979/80 = 100)
1991/92	142.5	256.1
1992/93	145.2	257.7
1993/94	152.2	258.2
1994/95	156.7	262.0
1995/96	157.0	267.6

Source: Australian Bureau of Statistics, *Price Index of Materials Used in House Building* (Cat. No. 6408.0) and *Price Index of Materials Used in Building Other Than House Building* (Cat. No. 6407.0).

16.5 Summary

Summarising in one figure the relative comparison between groups of related items at different points in time is easily accomplished by using index numbers. You read about it or hear it daily in relation to the All Ordinaries Index or the Dow Jones Index. For example, over the 1995/96 financial year the All Ordinaries moved from 2017.0

Business mathematics and statistics

to 2242.1, an increase of 11.2% during the calendar year. Thus, it far outperformed the Gold Index, which rose from 1867.5 to 1978.1 (an increase of 5.9%) during the same period.

1. An index number in its most basic form is simply a percentage relative that indicates the relationship between values over time. The reference period, called the base period, is given the value of 100. In fact, calculated index numbers are actually percentages, but by convention the per cent symbol is dropped.

2. The use of index numbers by business and government is widespread. Because it permits information to be summarised in a single number, comparisons are simpler and easily understood. For example, a change of 8.5% in sales in a multimillion dollar company's annual results is more comprehensible than a change of $2 475 647.65. What does this latter figure mean? How does it compare with last year's results? Converting the result to its index number equivalent makes much more sense, as it simplifies the data and allows for comparisons.

3. Index numbers are invaluable in determining real sales or income results (by discounting for inflation). You will see in the next chapter that index numbers are also used in expressing seasonal changes and, as a result, can be used for forecasting purposes.

4. Most indexes are classified in any of three forms—price, quantity or value indexes.

    (a) The price index illustrates the relative change in the price of goods and/or services (such as the Export Price Index or the Consumer Price Index).
    (b) The quantity index indicates the relative change in physical volume—for example, motor vehicle production this year versus last year.
    (c) The value index incorporates changes in both price and quantity and is used to express changes in, for example, total costs or sales turnover.

5. There are simple, simple aggregate and weighted aggregate indexes.

    (a) A simple index expresses the percentage change in the price, quantity or value of a commodity in a given period versus the price, quantity or value of the same commodity in another period.
    (b) A simple aggregate index expresses the percentage change over a period of time in price, quantity or value for a group of items.
    (c) In a simple average of relative index the current price or quantity is compared with the base price or quantity for each commodity, and the resulting relatives are added with the sum divided by the number of items listed times 100. Such an index assists in overcoming the problems of using different measurements for individual items (e.g. kilograms, litres, tonnes).
    (d) A weighted aggregate index allocates weights to each item contributing to the composite.
    (e) In a weighted average of relative index, the weights attached to each item are generally measured in dollar terms, such as consumption or production expenditure.

6. In constructing price indexes, the most commonly used form of aggregate index number is the Laspeyres method. According to this method, the quantities consumed during the base period are the weights applied to other periods. The Paasche method, on the other hand, applies current quantities consumed as the weights. Therefore, the Laspeyres method is applied when we want to learn how much it would cost to buy the base year's basket of goods at current prices. The Paasche index, however, indicates how much it would cost to buy the current period's basket of goods at the prices that prevailed in some previous period.

   The Laspeyres method needs quantity measures for only one period (that of the base period). This makes comparisons easier and less costly than with the Paasche method, which demands annual changes in weights because it reflects quantities consumed in the current period. However, the Laspeyres method, in determining how much more it costs today to buy the same basket of goods as we bought yesteryear, does not allow for changes in the consumption pattern, which may certainly alter with fluctuations in prices. Therefore, in this case, the Paasche method better reflects the current pattern of consumption and prices (if this is what you want to measure).

7. If index numbers are to be relevant, it is necessary to keep in mind certain procedures which may include:

   (a) ensuring that base year selection is in accord with the need for a normal, not distant period and one used by other series;

   (b) selecting appropriate weights, the problem being, of course, that over time the degree of importance can change, hence the need for Household Expenditure Surveys every five years to determine the weights applying for the CPI;

   (c) determining what products should be included in the basket and whether they are representative.

Simple index numbers

Simple index numbers indicate the ratio or percentage change of the price, quantity or value of a commodity in a given period versus the price, quantity or value of the same commodity in another period (the base). The following formulae are used:

- Simple price index = $\dfrac{P_g}{P_b} \times 100$

- Simple quantity index = $\dfrac{Q_g}{Q_b} \times 100$

- Simple value index = $\dfrac{V_g}{V_b} \times 100$

Simple aggregate (or composite) index

A simple aggregate index is used to determine the movement of several prices and/or quantities. The prices or quantities of the commodities in a given year are totalled and expressed as a percentage of the base year result. The following formulae are used:

Business mathematics and statistics

- Simple aggregate price index $= \dfrac{\Sigma P_g}{\Sigma P_b} \times 100$

- Simple aggregate quantity index $= \dfrac{\Sigma Q_g}{\Sigma Q_b} \times 100$

The formulae for calculating averages (for relatives) are:

- Simple price relative index $= \dfrac{\Sigma\left(\dfrac{P_g}{P_b} \times 100\right)}{n}$

- Simple quantity relative index $= \dfrac{\Sigma\left(\dfrac{Q_g}{Q_b} \times 100\right)}{n}$

Weighted aggregate (or composite) index

A weighted aggregate index is used in order to attribute the appropriate importance to each of the items being examined. The following formulae are used:

1. *Laspeyres method:*

    - Weighted aggregate price index $= \dfrac{\Sigma(P_g Q_b)}{\Sigma(P_b Q_b)} \times 100$

    - Weighted aggregate quantity index $= \dfrac{\Sigma(P_b Q_g)}{\Sigma(P_b Q_b)} \times 100$

2. *Paasche's method:* *

    - Weighted aggregate price index $= \dfrac{\Sigma(P_g Q_g)}{\Sigma(P_b Q_g)} \times 100$

    - Weighted aggregate quantity index $= \dfrac{\Sigma(P_g Q_g)}{\Sigma(P_g Q_b)} \times 100$

    - Weighted aggregate value index $= \dfrac{\Sigma(P_g Q_g)}{\Sigma(P_b Q_b)} \times 100$

The formulae for calculating a weighted average price relative index and a weighted average quantity relative index are:

- Weighted average price relative index $= \dfrac{\Sigma\left(\dfrac{P_g}{P_b} \times w \times 100\right)}{\Sigma w}$

- Weighted average quantity relative index $= \dfrac{\Sigma\left(\dfrac{Q_g}{Q_b} \times w \times 100\right)}{\Sigma w}$

* Optional.

Chapter 16 – Index numbers

The Australian Stock Exchange Indices

The Australian Stock Exchange (ASX) publishes several indexes, including the ASX series of price and accumulation indexes. Since you are probably most familiar with the ASX price indexes, including the All Ordinaries Index, a brief explanation of their workings follows.

How the ASX price indexes work

The ASX share price indexes summarise price movements on the Australian stockmarket by following the daily change in the aggregate market values (AMV) of the companies covered by each index. For each company, its market value (or market capitalisation) is the number of listed shares multiplied by their market price. These company market values are added to obtain the AMV for each index portfolio and the price indexes are calculated as follows:

$$\text{Today's closing Price index} = \text{Yesterday's closing index} \times \frac{\text{Today's closing AMV}}{\text{Today's Start-of-Day AMV}}$$

The value of each security in the AMV calculation is the last sale price recorded by SEATS (the Stock Exchange Automated Trading System), modified by any subsequent higher bids or lower offers made before normal trading ends at 16:00 hours (EST).

The ASX indexes are market capitalisation weighted, with the larger companies having proportionately more influence on the indexes than the smaller companies. Some comparable overseas indexes are the New York Stock Exchange Composite Index, the Financial Times Actuaries Indices (e.g. FTSE-100) and the Topix Index (Japan).

Market coverage

The index sample changes slightly from month to month and currently consists of approximately the top 322 listed companies (based on market capitalisation), although companies which are not traded regularly will normally be excluded. These 322 companies account for about 93% of listed Australian/Papua New Guinea equities by market capitalisation. To maintain this coverage, the sample is updated throughout the year.

We have all undoubtedly heard of or seen reference to the All Ordinaries Index. This index is divided into 23 industry sectors, as seen in Table 16.29. Note the share of each industry sector and the number of companies included in each index. The all industrials' share of the All Ordinaries Index is currently 66.8%, and approximately 73% of all companies listed in the index were from this sector at 30 June 1996. Interestingly, the banks and finance share of the All Ordinaries Index is 15.07% (see column 2, no. 16) even though it comprises only 4% of the companies in the index.

Table 16.29 All Ordinaries Index as of 30 June 1996

No.	Group	Industry groups as % of All Ordinaries Index, 30/6/96	Companies in each index, 30/6/96
1.	Gold	5.86	45
2.	Other metals	8.20	24
3.	Solid fuels	0.39	2
27.	**ALL MINING**	**14.45**	**71**
4.	Oil and gas	2.86	11
5.	Diversified resources	15.89	5
28.	**ALL RESOURCES**	**33.20**	**87**
6.	Developers and contractors	2.82	12
7.	Building materials	4.20	10
8.	Alcohol and tobacco	2.06	5
9.	Food and household goods	3.58	8
10.	Chemicals	1.35	5
11.	Engineering	0.68	8
12.	Paper and packaging	2.79	4
13.	Retail	3.56	11
14.	Transport	3.05	8
15.	Media	9.48	15
16.	Banks and Finance	15.07	13
17.	Insurance	1.82	8
18.	Entrepreneurial investors	0.50	2
19.	Investment and financial services	1.49	23
20.	Property trusts	4.05	28
21.	Miscellaneous services	1.99	28
22.	Miscellaneous industrials	1.79	20
23.	Diversified industrials	3.65	12
24.	Tourism & Leisure	2.71	15
29.	**ALL INDUSTRIALS**	**66.80**	**235**
30.	**ALL ORDINARIES**	**100.0**	**322**
26.	Twenty leaders	53.90	20
31.	Fifty leaders	72.91	50

Source: The Australian Stock Exchange Index Services, Companies on the Australian Stock Exchange Indices, 30 June 1996. *Monthly Index Analysis.*

Chapter 16 – Index numbers

16.6 Multiple-choice questions

1. Index numbers permit all of the following except:

   (a) rates of change to be measured;
   (b) a summary of a numbers of items in one composite figure to be calculated;
   (c) elimination of sampling errors;
   (d) budgeting and forecasting to be undertaken by businesses and government;
   (e) its use as a deflator, i.e. to take account of inflation.

2. In constructing index numbers, you must:

   (a) use appropriate weights;
   (b) ensure that the items are adequate in number and representative;
   (c) be clear on the purpose of the index number you are constructing;
   (d) be sure that data are available for the periods you are investigating;
   (e) all of the above.

3. On 1 July 1995, you were appointed as a supervisor on a salary of $50 000 p.a. with your salary indexed according to the inflation rate. What was your salary on 1 July 1996? (See Table 16.26 for the CPI.)

   (a) $59 350   (b) $52 107
   (c) $52 400   (d) $54 200

4. At the end of June 1996, your investment property was valued at $245 000 after having purchased it for $215 000 in June 1993. What was the real property value of your property at the end of the 1995/96 financial year? (*Hint:* Refer to Table 16.25 for the CPI and deflate the property value by the rise in the CPI over the period.)

   (a) $290 815   (b) $265 580
   (c) $223 741   (d) $206 403
   (e) $222 121

5. Referring to Table 16.18, which country experienced the lowest rate of inflation in 1993/94?

   (a) Japan   (b) Canada
   (c) Australia   (d) France

16.7 Exercises

1. How would you distinguish between a price index and a quantity index?

2. Briefly explain what the Consumer Price Index measures.

3. The CPI in June 1994/95 was 113.9 and 118.7 in June 1995/96. What does this mean?

4. If you were given two indexes with two different base years, how would you compare the two?

Business mathematics and statistics

5. In 1994/95, there were approximately 177 000 dwelling commencements in Australia, the highest on record. Should 1994/95 therefore be considered as the new base period when examining the rate of change in residential building activity? Briefly explain your answer.

6. Many private rental tenants face increases in their rents according to the movement in the CPI. Using Table 16.26, what rent (to the nearest dollar) would the tenant be paying on 30 June 1996 if the rent has been indexed according to the CPI since 30 June 1994 when the rent was $180 a week?

7. Table 16.30 lists the value (in millions of dollars) of travel revenue received in Australia (as per the balance of payments) resulting from overseas visitors coming here. Calculate the simple value index for each period using 1992/93 as the base period. Interpret the results.

Table 16.30 *Travel credit in Australia, 1992/93 to 1994/95*

Year	Travel credit ($ million)
1992/93	6417
1993/94	7502
1994/95	8879

Source: Australian Bureau of Statistics, *Balance of Payments* (Cat. No. 5301.0).

8. Table 16.31 lists the value (in millions of dollars) of travel expenditure debited to Australia's balance of payments, by Australians holidaying overseas.

   (a) Calculate the simple value index for each period using 1992/93 as the base period. Interpret the results.
   (b) What are your observations of the trend in changes in travel credits and travel debits over the period examined?

Table 16.31 *Travel debit for Australia, 1992/93 to 1994/95*

Year	Travel debit ($ million)
1992/93	4985
1993/94	5157
1994/95	5774

Source: Australian Bureau of Statistics, *Balance of Payments* (Cat. No. 5301.0).

9. The sales results for four different divisions of a multinational firm are set out in Table 16.32. Calculate an unweighted average of relatives index (to one decimal place) for each year, using 1994 as the base period.

Chapter 16 – Index numbers

Table 16.32 *Sales by four divisions of B-Wear Pty Ltd, 1994–97 ($ million)*

Division	Year			
	1994	1995	1996	1997
Graphics	14.5	14.6	15.0	16.0
Fisheries	3.4	4.6	6.0	8.0
Wines and spirits	9.2	10.1	10.5	11.4
Sporting goods	2.7	3.1	3.5	4.1

Source: Hypothetical.

10. The average weekly earnings for all adult workers for the years 1992/93 to 1995/96 can be found in Table 16.33 along with the CPI for the same years (with 1989/90 as the base year). Calculate (to the nearest dollar) the real average weekly earnings (i.e. after taking account of inflation using the existing base year) for each of the four years.

Table 16.33 *Average Weekly Earnings (AWE) for full-time working adults and the Consumer Price Index (CPI), 1992/93 to 1995/96 (1989/90 = 100)*

Period	AWE ($)	CPI
1992/93	625	108.4
1993/94	646	110.3
1994/95	675	113.9
1995/96	704	118.7

Source: Australian Bureau of Statistics, *Australian Economic Indicators* (Cat. No. 1350.0).

11. Interpret the results found in question 10.

12. From the data in Table 16.34 calculate (to one decimal place):

    (a) the simple aggregate price index for 1997 using 1992 as the base year;
    (b) the simple aggregate quantity index for 1997 using 1992 as the base year.

Table 16.34 *Annual dairy product consumption and pricing, 1992 and 1997*

Commodity	1992		1997	
	Price ($)	Quantity	Price ($)	Quantity
Cheese (kg)	4.20	7	6.00	6
Milk (litre)	0.74	102	0.91	105
Eggs (dozen)	1.55	26	1.80	16
Yoghurt (kg)	1.40	20	1.75	12
Butter (kg)	1.40	2	2.00	2

Source: Hypothetical.

13. (a) Referring to Table 16.34, calculate the weighted aggregate price index for 1997 using 1992 as the base year (i.e. use the Laspeyres method).
    (b) Briefly interpret the results.

14. In evaluating the possibility of buying a small whitegoods business, information (Table 16.35) on a range of items was collected.

    (a) Calculate the unweighted average of relative price index (to the nearest whole number) using 1992 as the base period.
    (b) Calculate the Laspeyres price index using 1992 as the base period.
    (c) Calculate the weighted average of relative price index.
    (d) What do the answers in parts (b) and (c) suggest?

Table 16.35 *Small business sales information, 1992 and 1997*

Item	1992 Price ($)	1997 Price ($)	1992 Quantity
Washing machines	500	720	50
Dishwashers	600	875	27
Television sets	550	650	100
Microwave ovens	525	625	120

Source: Hypothetical.

15. Using Table 16.36, determine the weighted aggregate quantity index for the appliance section of the All-purpose Company, using:

    (a) the Laspeyres method;
    (b) the Paasche method*.

Table 16.36 *Appliance sales for the All-purpose Company, 1992/93 and 1996/97*

Item	1992/93		1996/97	
	Price ($) ($P_b$)	Quantity ('000) (Q_b)	Price ($) ($P_g$)	Quantity ('000) (Q_g)
Blenders	77	120	90	124
Juicers	36	50	38	63
Toasters	39	140	44	135
Woks	62	40	64	100

Source: Hypothetical.

16. Convert the GDP data in Table 16.37 into index form (to one decimal place), using 1990/91 as the base, and interpret the results.

*Optional

Chapter 16 – Index numbers

Table 16.37 *Gross Domestic Product (GDP) and the Consumer Price Index (CPI), 1990/91 to 1994/95*

Year	GDP ($ billion)	CPI (1989/90 = 100)
1990/91	379.0	107.3
1991/92	387.2	108.4
1992/93	405.8	110.3
1993/94	430.4	113.9
1994/95	455.6	118.7

Source: Australian Bureau of Statistics, *Consumer Price Index* (Cat. No. 6401.0) and *Australian National Accounts* (Cat. No. 5206.0).

17. Referring to Table 16.37, deflate the current dollar GDP at constant 1990/91 prices (to the nearest $100 million) and evaluate. *(Hint:* First alter the CPI base to a base of 1990/91, to one decimal point, then deflate the current GDP by the revised CPI result.)

18. (a) Index the adult full-time working males' and females' wages in Table 16.38 (to the nearest per cent) with a base year of 1991/92 and comment on the results.

    (b) Using Table 16.38, calculate the real average weekly earnings (to the nearest dollar) using 1991/92 as the CPI base period for:

    (i) males;
    (ii) females.

    (c) Interpret the results.

Table 16.38 *Average Weekly Earnings for adult male and female employees, 1991/92 to 1995/96*

Year	Male ($)	Female ($)	CPI (1989/90 = 100)
1991/92	656	528	107.3
1992/93	673	538	108.4
1993/94	696	556	110.3
1994/95	729	578	113.9
1995/96	762	600	118.7

Notes: (a) Rounded to the nearest dollar.
(b) Non-managerial employees over 21 years of age.

Source: Australian Bureau of Statistics, *Average Weekly Earnings, States and Australia* (Cat. No. 6302.0) and *Consumer Price Index* (Cat. No. 6401.0). Commonwealth of Australia copyright, reproduced by permission.

19. Table 16.39 shows the prices and quantities of tennis equipment sold by a sports shop. Calculate to two decimal places:
    (a) the weighted aggregate price index using the Laspeyres method and 1992 as the base period;
    (b) the weighted aggregate price index using the Paasche method.*

Table 16.39 Tennis equipment and clothing sales results, 1992 and 1997

Item	Price		Quantity	
	1992 ($)	1997 ($)	1992	1997
Racquets	86.00	102.00	605	700
Balls	5.50	6.50	1150	1400
Bags	24.00	31.00	140	200
Shorts	22.50	26.00	380	420
Tops	17.50	18.00	620	650
Skirts	24.00	28.00	655	700
Shirts	15.50	20.00	400	440

20. To earn some extra spending money during these trying times, Mrs Zucchero bakes from home and sells the fruits of her labour to various eateries. Table 16.40 shows her costs and the quantities purchased over the past two years. With 1995 as the base period, determine (to two decimal places):

    (a) the weighted aggregate price index using the Laspeyres method;
    (b) the weighted aggregate quantity index using the Laspeyres method;
    (c) the weighted aggregate value index.
    (d) Interpret parts (a) and (c).

Table 16.40 Mrs Zucchero's shopping list, 1995 and 1997

Item	Price per kg		Quantity	
	1995 ($)	1997 ($)	1995 (kg)	1997 (kg)
Assorted fruits	2.10	2.75	252	300
Baking powder	2.40	2.52	420	432
Cocoa	1.85	2.20	190	240
Cream	2.60	2.77	350	400
Eggs (per dozen)	1.50	1.67	920	1040
Flour	0.80	0.84	2000	2400
Sugar	0.60	0.70	245	260

* Optional.

Notes

1. National Bank, *Decision*, June 1995, Vol. 7, No. 2, p. 4.
2. Nina Field, 'Australia improves but needs cuts in Canberra, says IPA', *Australian Financial Review*, 16 January 1996, p. 9.
3. Yellow Pages Australia, *Small Business Index*™, August 1996 Report.
4. Australian Bureau of Statistics, *A Guide to the Consumer Price Index* (Cat. No. 6440.0).
5. The Australian Exchange Index Services, *Companies on the Australian Stock Exchange Indices*, 30 June 1996.

Chapter 17
TIME SERIES

LEARNING OUTCOMES
After studying this chapter you should be able to:
- Describe time series and explain their use in business
- Identify and explain the influence of each of the four components in a time series: secular trend, and cyclical, seasonal and erratic variations on business activity
- Evaluate secular-trend results and forecast using the:
  (a) freehand graphic method;
  (b) semi-average method;
  (c) moving average method;
  (d) least-squares method; and
  (e) exponential smoothing method
- Calculate and use a seasonal index.

An efficiently run business operates on the principle of learning from the past, but more important, a business must have some idea of what lies ahead in order to plan and make competent decisions. The increased complexity of business operations has made it necessary for all businesses to be more aware of the importance of keeping statistical records. Accuracy in examining past and present trends and in forecasting future trends is mandatory for any firm aiming to improve its overall efficiency, and hence its profitability. An invaluable tool to assist the firm in realising its aim is time-series analysis. A **time series** is a set of statistical observations arranged in chronological sequence, normally at equal time intervals (such as monthly or quarterly) over relatively long periods, with the quality of management forecasts very much a function of the information gathered over time—that is, the time series.

For example, at your workplace you have probably come across time-series information in the form of daily, weekly, fortnightly, monthly, quarterly or annual results concerning production, sales, costs, wages, stock, absenteeism and the like.

Such data are represented visually by a graph with the independent variable, the period of time (e.g. days, weeks, months, quarters, years), on the x axis and the dependent variable—that is, the value of what you are measuring over the time interval (e.g. sales or profit)—on the y axis. An evaluation of such a series of data may indicate a persistent trend in the results which may improve management's understanding of past and current changes, thus enabling it to formulate more accurate projections. In other words, if the data collected over regular time intervals illustrate a distinct pattern, this may help you to estimate future results. Therefore, time-series analysis improves the user's ability to cope with the uncertainty of the future.

17.1 Time series components

As suggested above, any group of statistical data collected at regular intervals is referred to as time series. There are four movements or components which are involved in time series analysis and influence businesses, industries and the general economy. The long-term factors are the *secular trend* and *cyclical fluctuations*. The former describes a long-term pattern, normally represented as a line graph; the latter is a recurring upward and downward fluctuation lasting from three to nine years. Seasonal variations are also upward and downward movements, but the cycle is completed within a year (and is thus considered short-term) and then repeats itself. Since seasonal movements exhibit regular periodic patterns, they can be anticipated and applied to forecasting techniques. Finally, the results of a time series may be affected by random, unpredictable factors such as the Iraqi invasion of Kuwait in 1991 and its immediate impact on petrol prices around the world. However, once we extend the time-frame to a year, or a business cycle, or the demarcation point of a trend, this erratic or irregular movement's effect on the results of that month or year will be eliminated.

Secular trend

Secular trend describes a relatively persistent long-term movement, indicating the general trend of the change of values, which may be upward, downward or in both directions. You may think of the secular trend as the underlying long-term change in the average level of the variable per unit of time. The period considered should be from 10 to 20 years, long enough to permit any persistent pattern to emerge. By evaluating a long-term result, we are able to eliminate the seasonal and cyclical influences as well as the sudden upward or downward movements due, for example, to a brief drought or a sudden one-day sharemarket collapse.

The long-term or secular trends of business and industry are primarily the consequence of the degree of population growth, capital investment, technological progress and large-scale shifts in consumer tastes. For example, black and white televisions, slide rules, manual typewriters, steel, computers and overseas travel, to name a few products, have all produced dramatic secular trend results. When the data are graphed, the secular trend is usually represented by a straight line or smooth curve. Figure 17.1 illustrates a pressing problem confronting Australia—its deteriorating terms of trade (the movement in Australia's export prices versus import prices). The trend until 1987/88 was unmistakably downward and inhibited a more rapid recovery in Australia's balance of payments.

After an improvement, which continued for about two years, the trend was downward again until 1993 when it finally began to climb. You will find that most well-run organisations are heavily dependent on long-term forecasting as their planning cycles expand in the light of major capital expenditure and extended lead times in realising positive returns.

Cyclical variation

Cyclical variations indicate the expansion and contraction of business activity, describing the tendency for a particular firm, industry or economy to recover from a recession, become prosperous, contract again, recover, contract and so on. There is no available mathematical model capable of predicting precisely when business activity will expand, by how much and for how long. In other words, the duration of a future expansion or contraction is unknown, as is the range (the difference between the peak and the trough) of the change.

The factors responsible for cyclical fluctuations are numerous, but include variations in the level of investment (in the early to mid-1980s the low level of investment,

Figure 17.1 *Australia's terms of trade: secular trend (Index 1989/90 = 100)*

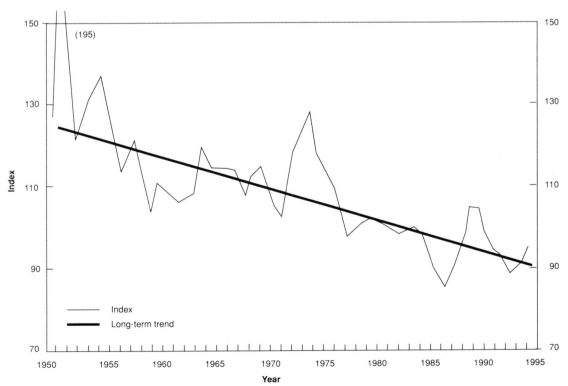

Source: Reserve Bank of Australia, *Australian Economic Statistics, 1949–50 to 1994–95*, June 1996. Commonwealth of Australia copyright, reproduced by permission.

especially in plant and equipment, made a large contribution to Australia's trade difficulties), production, consumption (e.g. consumers lacking confidence reduce their spending, and this may lead to an economic slowdown), government monetary policy (e.g. the level of interest rates) and fiscal policy (government expenditure and receipts, i.e. the budget). Nature (e.g. the drought in New South Wales from 1989 to early 1992) and major changes in population (since 1945 over 5.5 million migrants have arrived in Australia, contributing well over half the increase in the workforce and approximately 40% of the population increase) also play an important part. You have probably heard reference to the building cycle: Figure 17.2 illustrates the variations in the duration and intensity of the expansions and contractions of dwelling completions in Australia. You will note that cyclical movements do not follow any definite trend but move in a somewhat unpredictable fashion.

Seasonal variation

Seasonal variations represent repeating periodic movements and can be thought of as a type of pattern formed within a calendar year, repeating itself year after year. Changes in the level of business and economic activities are often a direct result of repetitive yearly patterns of seasonal variations in weather and holidays. For example, sales of certain foodstuffs, such as avocados, mangoes and ice-cream, and sales of jumpers, air-conditioners, airline tickets, toys and flowers are all seasonal and therefore considerably influence the industries concerned. Figure 17.3 illustrates the seasonal variation in the number of foreign tourists visiting Australia. Note how each quarter of each year has surpassed the same quarter of the previous year and how Australia's summer months are favoured by the overseas tourists.

Figure 17.2 *Dwelling completions in Australia, 1976–95: cyclical variations*

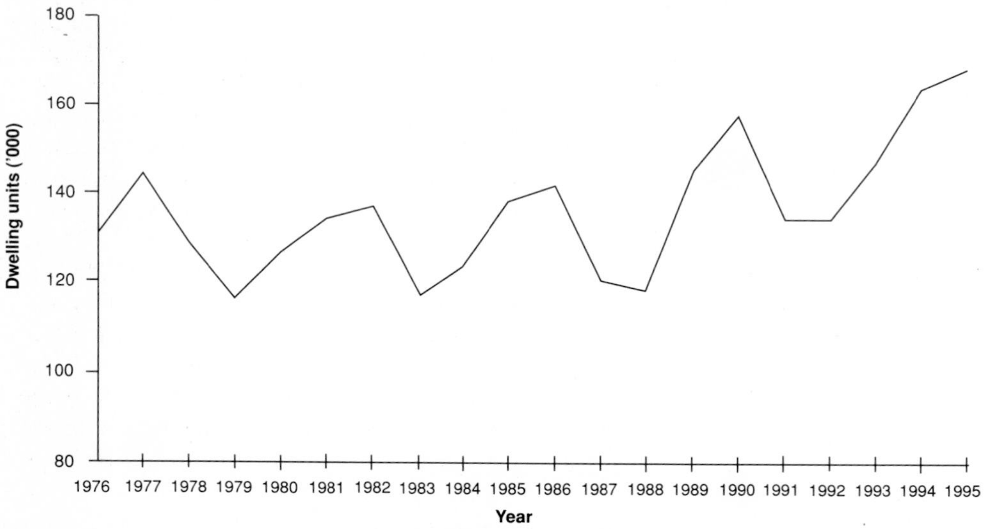

Source: Australian Bureau of Statistics, *Australian Economic Indicators* (Cat. No. 1350.0).

Chapter 17 – Time series

Figure 17.3 *Visitors to Australia, 1991–95: seasonal variations*

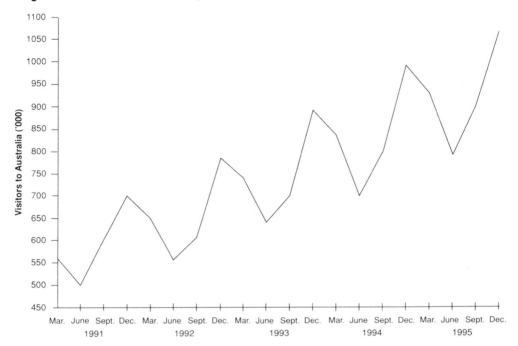

Source: Australian Bureau of Statistics, *Australian Economic Indicators* (Cat. No. 1350.0).

As a result of these seasonal variations, companies in the tourism industry can, for example, offer discount incentives during the off-peak season or even for certain days of the week (as airlines and many hotels have done). Furthermore, a recurring movement enhances a company's ability to forecast and allocate resources accordingly. Keep in mind the fact that seasonal variations may be daily (such as Telstra's charges, which are lower during certain periods of reduced use); weekly (e.g. on Tuesdays tickets are half-price in several cinemas in Sydney); monthly (prices of international airline tickets and cruises depend on the time of year, e.g. low, shoulder or high season); or quarterly (e.g. the CPI is normally highest in the December quarter and lowest in the March quarter).

Irregular or erratic variation

Irregular variations are sometimes referred to as **residual variations** since, by definition, they represent what is left over in an economic time series after trend, cyclical and seasonal elements have been accounted for. The forces that make business erratic are numerous and random. They may include events such as a snap election, a sudden sharemarket collapse, exchange rate variations, strikes, natural disasters (e.g. floods and cyclones) and wars. The US stockmarket reacted in 1986 when then President Ronald Reagan had to have a sunspot removed from his nose. These irregular movements are obviously of varying length and influence, but they are shorter than cycles and, often, shorter than seasonal variations (see Figure 17.4).

Business mathematics and statistics

Figure 17.4 *US cents/A$: irregular variations*

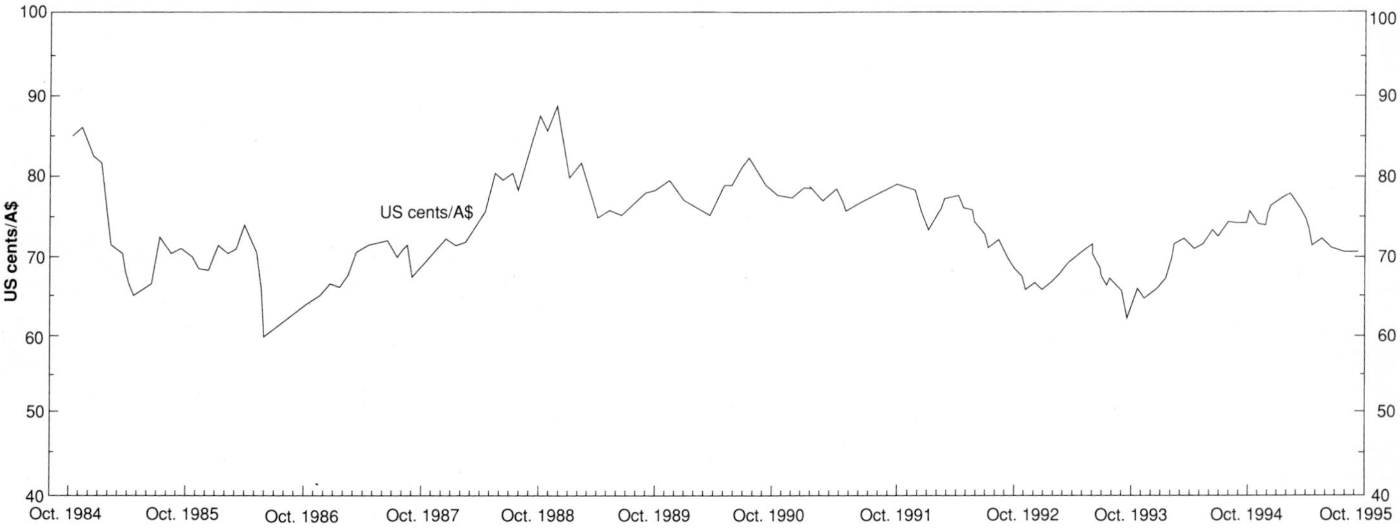

Source: The Treasury, *Economic Round-up*, Autumn 1991, p. 51; Reserve Bank of Australia, *Bulletin*. Commonwealth of Australia copyright, reproduced by permission.

In most instances, time series tends to incorporate several of these four components rather than merely exhibiting one or another.

17.1 Competency check

1. What is the purpose of time-series analysis?

2. If you were the newly appointed sales manager of white goods responsible for a particular region, what different sets of time-series data would you be interested in collecting for analysis?

3. What are the four movements in a time series? Explain the kind of change, over time, to which each applies.

4. Provide an example of each time-series component (other than those suggested in the text).

5. (a) If you were evaluating the behaviour of computer prices from 1982 to 1997, which time-series component would best explain the data?
   (b) Which component would you use to describe the effect of Wimbledon and the US Open on the sales of tennis racquets?
   (c) Which component best describes the effect of a hailstorm on the demand for panelbeaters?
   (d) Which component assists in describing the effect of a reduced level of business and consumer confidence on general business activity?

17.2 Secular trend movements

Secular trend

An underlying trend of growth or decline over a long period of time may be graphically represented by a straight line or smooth curve and is referred to as the *secular trend*. An example of an upward growth trend is Sydney real estate values, which show an average of approximately 2% increase in real terms per annum over an extended period. A secular trend in the opposite direction can be seen in the consumption of sugar or meat per capita. Over this long period the trend is affected by changing population size, tastes (e.g. an increased awareness of the benefits of a healthy diet), technological change (e.g. substitutes for sugar) and improvements in productivity (e.g. solar energy).

If you were to examine the real GDP over the past 20 years there would be an underlying pattern of growth, graphically illustrated by an almost-straight line, indicating an approximate constant per annum growth amount. You may disagree and claim that there were recessions over those 20 years. You would be correct, but remember that the secular trend spans a period long enough to balance the high and low cyclical and seasonal influences. In other words, think of the secular trend as an average value for the series of data, cutting between the highs and lows of the actual values. Getting back to our real GDP, its growth could be attributed to the continued but gradual increase in Australia's population, technological change (providing new and improved products, and hence business activity, which is what the GDP measures), and improved productivity due to the level of capital investment and increased educational attainment by the workforce.

An understanding of secular trends is useful for a number of reasons, including:

- it allows the user to present a historical pattern of results; and
- on the basis of past patterns, the user can project or forecast future patterns or trends.

There are four methods we will use for isolating secular trends, each with basic advantages and disadvantages. One common feature of three of the methods is the assumption that, when the long-term trend is plotted, a straight line will be formed. This results from the selection of a period of 10 to 20 years which permits the elimination of cyclical, irregular and seasonal influences. If, however, the data does not approach a linear trend, one may measure non-linear trends (i.e. trend curves) by using the second degree parabola (which may, for example, show the number of Australians travelling by coach, which until the early to mid-1980s declined, then rose until several fatal bus crashes occurred and the aviation industry was deregulated).

If, instead of illustrating a constant absolute amount of growth, the data portrayed a constant rate of growth per annum, an exponential trend curve would be preferred. For instance, up until the October 1987 sharemarket crash, several public companies had realised exponential growth in share price, often over the life of their share register. Finally, if regular cyclical fluctuations are apparent, then measuring the moving average trend will smooth out these variations.

The first step in any secular-trend evaluation should be graphing the data to discover whether an obvious persistent underlying pattern does, in fact, exist. If a linear trend is evident after the time series has been graphed, one of the four measures available to measure secular trend should be chosen and a straight line should be drawn showing the long-term gradual direction of the time series.

Freehand graphic method

The freehand graphic method is the simplest and fastest method of determining the secular trend. It entails:

1. plotting the time-series data onto a graph;
2. examining the direction of the trend;
3. constructing a straight line which you, the analyst, think is most appropriate for the trend's direction.

Figures 17.5 and 17.6 illustrate the freehand graphic method (data are from Tables 17.1 and 17.2). The line that has been drawn through the data in both figures represents the trend. If you are thoroughly familiar, for example, with a particular firm or industry and the prevailing market conditions, this method might, in some cases, provide a better description of the trend than would a mathematical equation. In other words,

Figure 17.5 *Sales turnover for Green Thumbs Nurseries, June 1983–97: freehand graphic method*

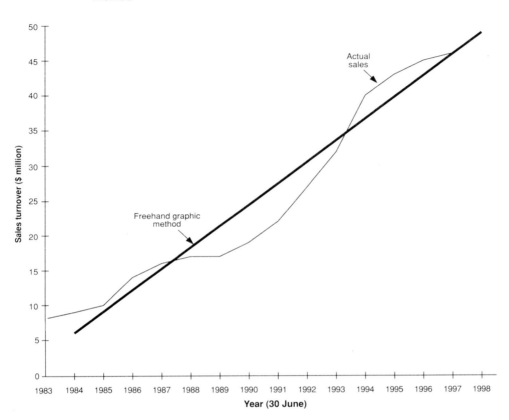

Source: Hypothetical.

lines fitted by this method could prove a close approximation to a mathematically fitted line, but the latter would be far more complicated to determine. To predict the level of sales anticipated in 1998, simply extend the trend line and read the result corresponding to where 1998 would lie on the x axis. Similarly, to predict the amount of imports for 1998, extend the trend line and read the result from the point corresponding to 1998.

The freehand graphic method has the disadvantage that the result is subjective and someone unfamiliar with this technique and/or the operating characteristics of the firm will find it difficult to draw the 'true' trend line accurately. For instance, in the above examples (of sales turnover and imports), you may interpret the two trends differently from the way in which the trend has been depicted. Therefore, your forecasts will differ from mine. This should not, however, detract from the fact that this method allows one to determine whether an obvious linear trend does persist.

Figure 17.6 Imports of goods and services into Australia ($ billion), 1984–96: freehand graphic method

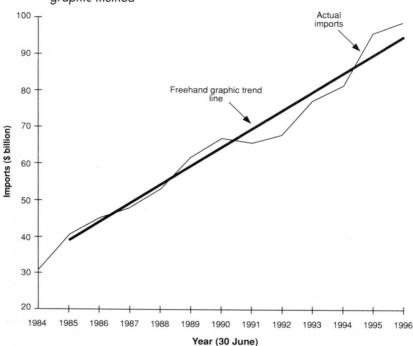

Source: Australian Bureau of Statistics, *Balance of Payments* (Cat. No. 5301.0).

Semi-average method

The semi-average is the easiest method of finding a straight-line trend without having to make a subjective judgment. The procedure is as follows:

1. The original data are divided into two equal groups, with the arithmetic mean of each group calculated. Should the data contain an odd number of years, the middle year is disregarded.
2. The mean of each group is then centred in the middle period of each of the two groups. These results are then plotted on the graph, with a straight line drawn through the two means. As each group has an equal number of years, the mean is plotted at the midpoint of the respective periods. This device is based on the premise that the mean of the original data of a given period equals the trend values of the period. To predict future results, the trend line has to be extended with the forecast result obtained by observing the year in question on the x axis and the corresponding value on the y axis.

Tables 17.1 and 17.2 and Figures 17.7 and 17.8 illustrate the semi-average method for the data on sales turnover and imports. Note that Table 17.1, with seven years in each of the two groups, centres the semi-total and semi-average corresponding to

Chapter 17 – Time series

the fourth (or middle) year of each group (i.e. the centre of the group), whilst in Table 17.2, with six years in each group, the semi-total and semi-average must be centred between the third and fourth year in each group and similarly plotted.

Table 17.1 *Sales turnover for Green Thumbs Nurseries, 30 June 1983–97*

Year (30 June)	Sales turnover ($ million)	Semi-totals	Semi-average
1983	8		
1984	9		
1985	10		
1986	14	91	13
1987	16		
1988	17		
1989	17		
1990	19	←(ignore middle year)	
1991	22		
1992	27		
1993	32		
1994	40	255	36.43
1995	43		
1996	45		
1997	46		

Source: Hypothetical.

Table 17.2 *Imports of goods and services into Australia, 30 June 1984–96*

Year (30 June)	Imports ($ billion)	Semi-totals	Semi-average
1984	31		
1985	40		
1986	46		
		279	46.5
1987	48		
1988	53		
1989	61		
1990	67	←(ignore middle year)	
1991	66		
1992	68		
1993	78		
		489	81.5
1994	81		
1995	96		
1996	100		

Source: Australian Bureau of Statistics, *Balance of Payments* (Cat. No. 5301.0). Commonwealth of Australia copyright, reproduced by permission.

Figure 17.7 *Sales turnover for Green Thumbs Nurseries, 1983–97: semi-average method*

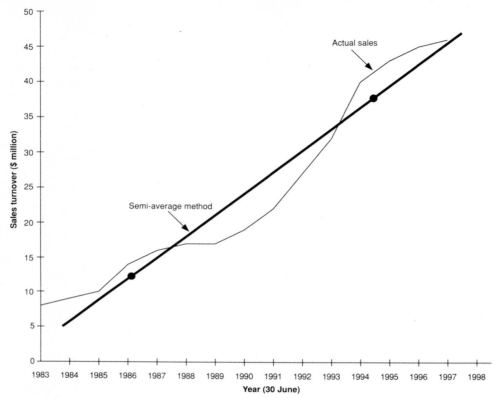

Source: Hypothetical.

Although the semi-average method is simple and objective, its use of the arithmetic mean, which is greatly affected by extremes, renders the semi-average trend less accurate if unusual events occur, such as a sharemarket collapse, massive industrial unrest, trade restrictions and/or trade wars, which may influence the result. Also, the selection of the number of periods under consideration has a direct bearing on the slope of the line and, therefore, on the trend results. If, for example, only the last 10 years of sales turnover or imports had been considered, the steepness (slope) of the trend line would have been more pronounced, and forecast results would differ. Finally, it must not be forgotten that this method is appropriate only when fitting a straight-line trend.

Least-squares method

You may recall from Chapter 15 the use of regression analysis and the application of the least-squares straight-line equation. The straight-line equation is also used in secular trend analysis. However, rather than the straight-line equation revealing how a change in a series of data tends to compare with a given change in another series of data (in other words, in regression analysis, it describes the average relationship

Figure 17.8 *Imports of goods and services into Australia ($ billion), 1984–96: semi-average method*

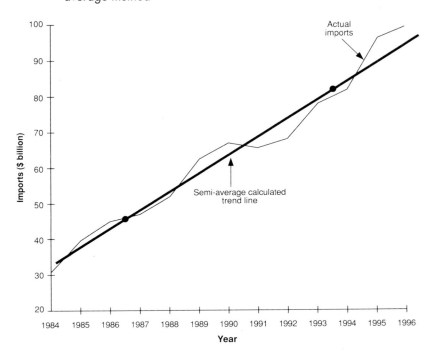

Source: Australian Bureau of Statistics, *Balance of Payments* (Cat. No. 5301.0).

between the variables of X and Y), in trend analysis, the straight-line equation expresses the annual change in a series of data.

The **least-squares method** is the method most widely used to evaluate the straight-line secular trend. The term 'least squares' is derived from the fact that the sum of the squares of the deviations of the original values, Y, from the estimated trend values (Y_c) is less than the sum of the squares of the deviations of the original values from any other straight line used to determine the trend. Thus, the term 'least-squares method' provides us with the line of best fit. To calculate a straight-line trend using this method, the straight-line equation is employed:

$$Y_c = a + bX$$

where X = time periods
  Y_c = computed trend value for a given time period
  a = value of Y when $X = 0$ (the value of the height of the straight line at the point it intersects the vertical Y axis, also known as the Y intercept)
  b = slope of the straight line (the amount that Y changes for a unit change in X, e.g. year).

Once the values of the straight-line equation for the data are known, the trend values can be estimated. Let's first look at sales turnover and its graphic representation with the least-squares method (see Table 17.3 and Figure 17.9).

Table 17.3 *Sales turnover for Green Thumbs Nurseries, 30 June 1983–97*

Year (30 June)	Sales turnover ($ million) (Y)	Deviation in years (X)	XY	X^2	Y_c
1983	8	−7	−56	49	3.75
1984	9	−6	−54	36	6.69
1985	10	−5	−50	25	9.63
1986	14	−4	−56	16	12.57
1987	16	−3	−48	9	15.51
1988	17	−2	−34	4	18.45
1989	17	−1	−17	1	21.39
1990	19	0	0	0	24.33
1991	22	1	22	1	27.27
1992	27	2	54	4	30.21
1993	32	3	96	9	33.15
1994	40	4	160	16	36.09
1995	43	5	215	25	39.03
1996	45	6	270	36	41.97
1997	46	7	322	49	44.91
	365	0	824	280	364.95

Source: Hypothetical.

Column 3 in Table 17.3, denoted X, refers to the time periods of the data. However, instead of having to work with four-digit figures of, say, 1989, deviations of the periods (years in this case) from the centre or middle period of the data are substituted. In this case, 15 years are examined, making the eighth year the middle period. The mean has two mathematical properties: the algebraic sum of the deviations of individual values from the mean is zero, and the sum of the squared deviations of the individual values from the mean is minimised. As the least-squares method is applied to calculate a representative value for a series of data and uses the arithmetic mean of the examined data as its basis, the mathematical properties of the mean can therefore be applied—that is, the sum of the deviations from the mean (1990) equals zero. Zero is assigned to this X value, which is the middle of the periods included in the time series and this period (e.g. 1990) becomes the point of origin. In other words, we have shifted the origin from the beginning of the time period (1983) to the middle of the time period (1990).

The origin must always be selected on the basis of ensuring that the deviations from it are equal to zero. The years preceding the origin are assigned negative numerical values and those succeeding it are assigned positive values. The formulae for calculating the values of a and b are:

and
$$a = \frac{\Sigma Y}{n}$$

$$b = \frac{\Sigma XY}{\Sigma X^2}$$

where n = number of periods in the data
  Y = dependent variable
  X = time deviation from the middle period which has been selected as the origin.

Therefore:
$$a = \frac{\Sigma Y}{n}$$
$$= \frac{365}{15}$$
$$= 24.3\overline{3}$$

and
$$b = \frac{\Sigma XY}{\Sigma X^2}$$
$$= \frac{824}{280}$$
$$= 2.94$$

By now substituting the calculations undertaken in Table 17.3 into the straight-line equation, we find:

$$Y_c = a + bX$$
$$= 24.33 + 2.94X \text{ with the origin: 30 June 1990}$$
$$X \text{ unit: 1 year}$$
$$Y \text{ unit: \$1 million in company sales}$$

Thus, a, or 24.33, really represents the mean sales turnover for the entire period, while b is the annual change in sales. The long-term trend estimates of sales can now be determined (column 6) by substituting values from column 3 for X. For example, the estimated trend for 1994 is found by substituting 4 for X:

$$Y_c = 24.33 + 2.94X$$
$$Y_{1994} = 24.33 + 2.94(4)$$
$$= 24.33 + 11.76$$
$$= 36.09$$

All the trend values can be calculated the same way, by substituting the appropriate X value. However, a simpler way to determine successive estimated trend values would be to add the constant b (2.94) to each preceding trend value—for example, 1992 is 27.27 + 2.94 which equals 30.21, while 1993 is equal to this 1992 result of 30.21

plus the constant 2.94 or 33.15. Furthermore, to forecast sales turnover for 2000, for example, we merely substitute into the formula:

$$Y_c = 24.33 + 2.94X$$
$$Y_{2000} = 24.33 + 2.94(10)$$
$$= 24.33 + 29.4$$
$$= 53.73$$

In this case, X (i.e. the year's deviation from the origin) was 10, since the nominated period of 2000 is 10 years after the origin at 1990. Therefore, if sales turnover, indeed, illustrated an annual constant change, and hence a linear trend, we would expect sales in 2000 to reach approximately $54 million.

As the shortest distance between two points is a straight line and the least-squares method is based on the straight-line equation, if you wanted to graph the results you would need only to plot two estimated trend values of two years and connect the points (see Figure

Figure 17.9 *Sales turnover for Green Thumbs Nurseries, 1983–97: least-squares method*

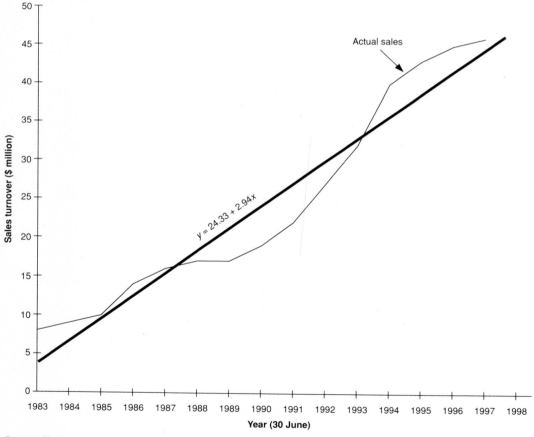

Source: Hypothetical.

Chapter 17 – Time series

17.9). Theoretically, the results we calculated for the trend values of sales indicate the sales turnover resulting from the secular trend alone—that is, after eliminating cyclical and/or irregular variations (due to the time-frame extending beyond such variations). It is estimated that sales for 1986, for example, would be $12.57 million without these influences. In fact, actual sales were $14 million, which means that the cyclical and irregular movements forced sales above their long-term trend line.

Problem: Calculate the trend-line equation and graphically represent the imports of goods and services into Australia (see Table 17.4 and Figure 17.10).

Table 17.4 *Imports of goods and services into Australia, 30 June 1984–96*

Year (30 June)	Imports ($ billion) (Y)	Deviation in years (X)	XY	X^2	Y_c
1984	31	−6	−186	36	32.4
1985	40	−5	−200	25	37.7
1986	46	−4	−184	16	43.0
1987	48	−3	−144	9	48.3
1988	53	−2	−106	4	53.6
1989	61	−1	−61	1	58.9
1990	67	0	0	0	64.2
1991	66	1	66	1	69.5
1992	68	2	136	4	74.8
1993	78	3	234	9	80.1
1994	81	4	324	16	85.4
1995	96	5	480	25	90.7
1996	100	6	600	36	96.0
	835	0	959	182	834.6

Source: Australian Bureau of Statistics, *Balance of Payments* (Cat. No. 5301.0). Commonwealth of Australia copyright, reproduced by permission.

Solution: To solve for *a* and *b* in the equation $Y_c = a + bX$

$$a = \frac{\Sigma Y}{n}$$
$$= \frac{835}{13}$$
$$= 64.2$$

and

$$b = \frac{\Sigma XY}{\Sigma X^2}$$
$$= \frac{959}{182}$$
$$= 5.3$$

Business mathematics and statistics

Therefore $Y_c = 64.2 + 5.3X$ with the origin 1 January 1990 and with units of X expressed in terms of years and Y in billions of dollars. Examining the estimated trend results versus the actual imports, we find that although at times the estimate is close to the actual, the dramatic rise in imports in latter years resulted in the fitting of a straight line, being not the ideal or appropriate solution.

Figure 17.10 *Imports of goods and services into Australia, 1984–96: least-squares method*

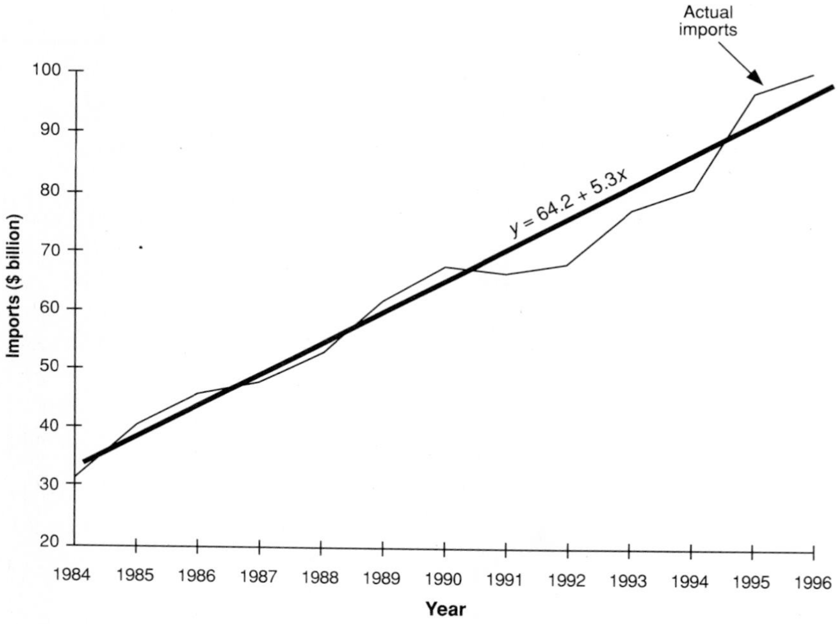

Source: Australian Bureau of Statistics, *Balance of Payments* (Cat. No. 5301.0).

Thus far, only odd-number years were considered when applying the least-squares method, which made life easy since selection of the middle period was simple, as the middle period was nominated as the origin and zero was assigned to its X value. If you recall, this was necessary in order to satisfy the condition that all the X values above and below the zero point equalled zero. In Table 17.5 there is an even number of periods, with the midpoint or point of origin at 1 January 1991. By applying a value of 1 to each six-monthly period, X values of -1 and $+1$ can be assigned to 30 June 1990 and 30 June 1991 respectively. The X value for 1993 must be $+5$, as it is four six-monthly intervals from 30 June 1991; all the other periods are treated similarly.

To solve for a and b in the equation $Y_c = a + bX$:

$$a = \frac{\Sigma Y}{n}$$

$$= \frac{357}{14}$$

$$= 25.5$$

and

$$b = \frac{\Sigma XY}{\Sigma X^2}$$

$$= \frac{1403}{910}$$

$$= 1.54$$

Hence, $Y_c = 25.5 + 1.54X$ for an origin of 1 January 1991, with each X unit equal to a six-month period. The graphic representation is shown in Figure 17.11.

Table 17.5 *Sales turnover for Green Thumbs Nurseries, 30 June 1984–97*

Year (30 June)	Sales turnover ($ million) (Y)	Deviation in years (X)	XY	X^2	Y_c
1984	9	−13	−117	169	5.48
1985	10	−11	−110	121	8.56
1986	14	−9	−126	81	11.64
1987	16	−7	−112	49	14.72
1988	17	−5	−85	25	17.80
1989	17	−3	−51	9	20.88
1990	19	−1	−19	1	23.96
1991	22	1	22	1	27.04
1992	27	3	81	9	30.12
1993	32	5	160	25	33.20
1994	40	7	280	49	36.28
1995	43	9	387	81	39.36
1996	45	11	495	121	42.44
1997	46	13	598	169	45.52
	357	0	1403	910	357.00

Source: Hypothetical.

To forecast sales for 2000, the X value would be 19 (as it is three years from 1997 when X is equal to 13); thus:

$$Y_c = 25.5 + 1.54X$$
$$Y_{2000} = 25.5 + 1.54(19)$$
$$= 25.5 + 29.26$$
$$= 54.76$$

Business mathematics and statistics

This result is approximately $1 million more than the forecast of 53.73 (see p. 576) because the latter result included the 1984 sales results of $9 million, which contributed to a smaller a and b.

Figure 17.11 *Sales turnover for Green Thumbs Nurseries, 1984–97: least-squares method*

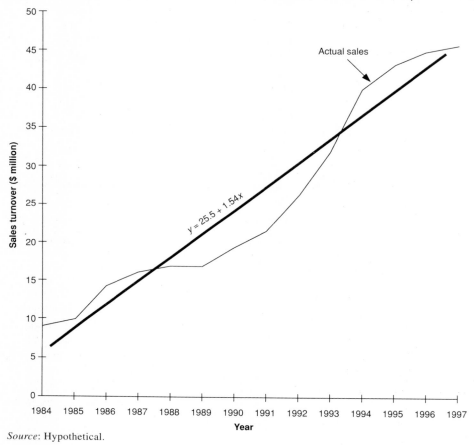

Source: Hypothetical.

In review, the *least-squares method* has two main advantages:

1. It expresses the secular trend in a mathematical formula which permits extrapolation into the past, present and future.
2. Its calculated results are objective (although the subjective selection of different periods can vary the results dramatically).

The disadvantage of this method is that it is based on the assumption that the data follow a straight-line trend, which—as we found in a number of data examined—is seldom the case. This method is also far more time-consuming than the other techniques examined, as it demands far more mathematical calculations.

We have now completed considerations of the methods used to calculate secular-trend estimates. It should be fairly obvious that the selection of the appropriate method and period of years will affect the degree of accuracy of the estimates. Variables affecting the secular trend, such as technology, consumer taste and population, may also alter the long-term industry trend and must therefore be taken into account when forecasting future trends.

In deciding how to select the appropriate method for secular-trend analysis, we must first understand the reasons for the analysis. If management is interested in the historical perspective of the company over a definite period, it might need only the *freehand trend line* drawn by an experienced statistician. However, the *semi-average method* should be used by those who are not trained to construct a freehand graphic representation. If a good fit to the data is necessary, the *least-squares method* should be applied. Since we use trend analysis for comparing trends of various groups of data and since b represents the slope of the line and, therefore, the trend's direction and magnitude, the least-squares method would be the most appropriate. This method would also be used for forecasting results—but remember that other variables must be considered in forecasting activity (e.g. deregulation pervading an increased number of business areas such as financial systems, airlines and telecommunications, and major taxation changes). Remember also that substantial variations in the values near the end of a time series can alter considerably the values of a and b when they are calculated by the least-squares method and so here, too, care must be applied when forecasting.

Moving average method

When the data and the corresponding graphic representation are examined, a straight-line description may be considered inappropriate. Furthermore, often inspection of any underlying regularity in a time series may be difficult to decipher due to erratic and short-term fluctuations. Therefore, to eliminate such short-term variations, the moving average method is often used. The **moving average method** allows us to describe a smooth curve that will minimise the effect of the cyclical, seasonal and irregular variations, thereby indicating the data's general trend. It does this by literally calculating successive averages while moving along the time series.

To select the appropriate number of periods to be averaged, the number of periods required to eliminate the fluctuations (i.e. cyclical, seasonal and irregular) must be considered. For example, quarterly fluctuations require a quarterly moving average. The same rationale applies to monthly fluctuations when a 12-term moving average is used (e.g. used frequently for economic time series which are collected on a monthly basis—such as unemployment data). In other words, what you are doing is basically replacing each observation (e.g. actual sales turnover) with an average of the replaced observation and some adjacent observations.

However, when examining yearly periods it is not so simple to select the appropriate number of periods, as cyclical fluctuations vary considerably in duration. Therefore, a period equal to at least the length of one business cycle must be selected in order to eliminate its effect on the secular trend. Probably the best way to decide on the length of the cyclical period is to determine from the graph the period between peaks and/or troughs.

If, on the other hand, you want to remove the impact of seasonality on the collected data and a daily time series is being examined, apply a seven-item moving average. Likewise, to remove quarterly or monthly seasonality, use a four-item and a 12-item moving average respectively. Let's have a look at a three-year moving average for sales turnover (see Table 17.6).

Table 17.6 *Sales turnover for Green Thumbs Nurseries, 30 June 1983–97: three-year moving average*

Year (30 June)	Sales turnover ($ million)	Three-year moving total	Three-year moving average
1983	8		
1984	9	27	9.0
1985	10	33	11.0
1986	14	40	13.3
1987	16	47	15.7
1988	17	50	16.7
1989	17	53	17.7
1990	19	58	19.3
1991	22	68	22.7
1992	27	81	27.0
1993	32	99	33.0
1994	40	115	38.3
1995	43	128	42.7
1996	45	134	44.7
1997	46		

Source: Hypothetical.

The moving average technique consists in successively averaging the data in the cycle selected. The cycle represents the number of years (quarters, months, weeks, days) to be averaged; the moving average is calculated by adding the values before and after the value for the period for which the average is being obtained, and then dividing this sum by the number of periods in the cycle.

For example, to calculate a three-year moving average:

1. The first three numbers (8, 9 and 10) are added, with the total entered in column 3 next to the middle year of that group (in this case, 1984).
2. The first number (8) is then replaced by the next number in the column of figures (14) and the procedure is repeated until the entire series has been similarly included.

Thus the first, second and third periods are added and centred in the middle period, followed by the second, third and fourth, third, fourth and fifth, and so on. Each total is subsequently divided by three to arrive at the three-year moving average found in column 4. This result can now be fitted onto the graph where the original points were

Chapter 17 – Time series

plotted (see Figure 17.12). Observe Table 17.6 and the graphing of the result on Figure 17.12 and note how closely the three-year moving average fits the original trend.

Figure 17.12 *Sales turnover for Green Thumbs Nurseries, 1983–97: three-year moving average*

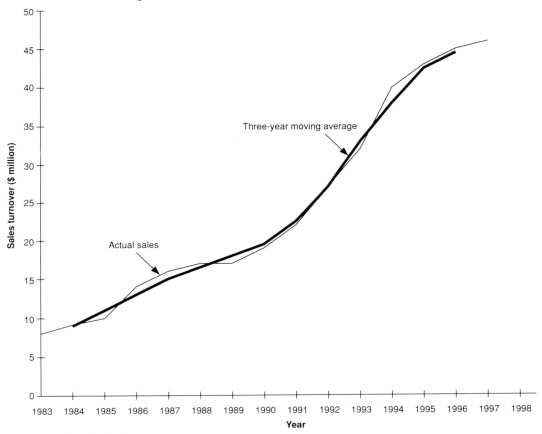

Source: Hypothetical.

If you now apply a five-year moving average to the same data (see Table 17.7 and Figure 17.13), you find that the curve eliminates to a greater extent the fluctuations of the original data; hence, as in the three-year moving average, the slope of the line changes from year to year, maintaining a course above the troughs and below the peaks of the actual sales turnover.

Problem: Table 17.8 lists imports of goods and services into Australia for the period 1984–96. Calculate a three-year moving average (checking your answers against the entries in column 4) and plot your results (the same as Figure 17.14).

Business mathematics and statistics

Table 17.7 *Sales turnover for Green Thumbs Nurseries, 30 June 1983–97: five-year moving average*

Year (30 June)	Sales turnover ($ million)	Five-year moving total	Five-year moving average
1983	8		
1984	9		
1985	10	57	11.4
1986	14	66	13.2
1987	16	74	14.8
1988	17	83	16.6
1989	17	91	18.2
1990	19	102	20.4
1991	22	117	23.4
1992	27	140	28.0
1993	32	164	32.8
1994	40	187	37.4
1995	43	206	41.2
1996	45		
1997	46		

Source: Hypothetical.

Solution:

Table 17.8 *Imports of goods and services into Australia, 30 June 1984–96: three-year moving average*

Year (30 June)	Imports ($ billion)	Three-year moving total	Three-year moving average
1984	31		
1985	40	117	39.0
1986	46	134	44.7
1987	48	147	49.0
1988	53	162	54.0
1989	61	181	60.3
1990	67	194	64.7
1991	66	201	67.0
1992	68	212	70.7
1993	78	227	75.7
1994	81	255	85.0
1995	96	277	92.3
1996	100		

Source: Australian Bureau of Statistics, *Balance of Payments* (Cat. No. 5301.0). Commonwealth of Australia copyright, reproduced by permission.

Chapter 17 – Time series

Figure 17.13 *Sales turnover for Green Thumbs Nurseries, 1983–97: five-year moving average*

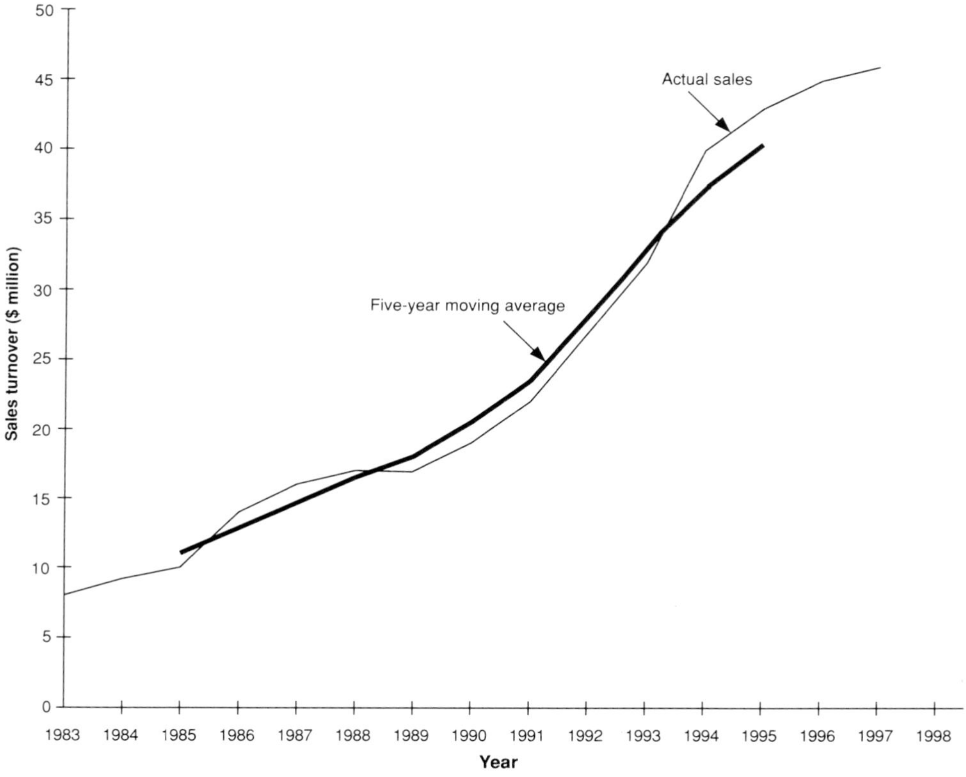

Source: Hypothetical.

Problem: Calculate a five-year moving average of imports and plot your results (see Table 17.9 and Figure 17.15).

Solution:

Table 17.9 *Imports of goods and services into Australia, 30 June 1984–96: five-year moving average*

Year (30 June)	Imports ($ billion)	Five-year moving total	Five-year moving average
1984	31		
1985	40		
1986	42	218	43.6
1987	48	248	49.6

(continued)

Business mathematics and statistics

Table 17.9 *(continued)*

Year (30 June)	Imports ($ billion)	Five-year moving total	Five-year moving average
1988	53	275	55.0
1989	61	295	59.0
1990	67	315	63.0
1991	66	340	68.0
1992	68	360	72.0
1993	78	389	77.8
1994	81	423	84.6
1995	96		
1996	100		

Source: Australian Bureau of Statistics, *Balance of Payments* (Cat. No. 5301.0). Commonwealth of Australia copyright, reproduced by permission.

Figure 17.14 *Imports of goods and services into Australia, 1984–96: three-year moving average*

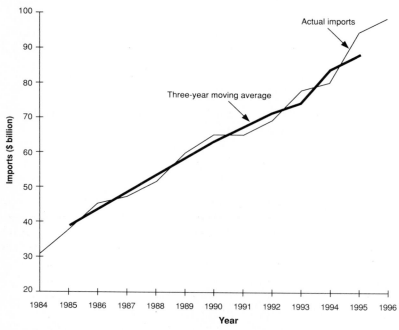

Source: Australian Bureau of Statistics, *Balance of Payments* (Cat. No. 5301.0).

Observe that when Australia was in the midst of a recession (in 1991), import growth was stifled, but once the economy turned around (after 1993) as result of a combination of budget stimulus, the end of the drought, lower interest rates, lower inflation, improved company profits and a steady worldwide economic recovery, production, investment and employment rose with increased imports a by-product. In other words, cyclical, irregular and/or seasonal forces have contributed to fluctuations in Australia's annual import bill and these have been removed or partially eliminated by calculating the moving average.

Figure 17.15 *Imports of goods and services into Australia, 1984–96: five-year moving average*

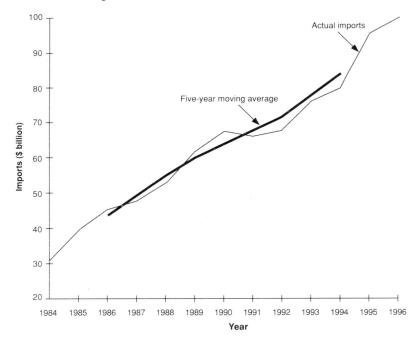

Source: Australian Bureau of Statistics, *Balance of Payments* (Cat. No. 5301.0).

Until now we have examined only odd-number periods (in this case, years). We will now examine even-number periods—for example, a four-year moving average for imports (see Table 17.10). Whenever an even number of periods is included in a moving average, the centre point of each group will be between the second and third periods of the group. It is therefore necessary to shift the moving averages so that they coincide with the periods (years) rather than resting between periods. Column 4 in Table 17.10 (the four-year moving total) is obtained by the identical method used for calculating the moving total for odd-number periods. Note, however, that the results are centred between years. As we want to realign the values in line with the years, a two-period moving average of the even-period moving averages must be calculated. This is done by taking a two-period moving average by dividing the two-period moving total by eight, since we have added two four-quarter totals (see column 5 in Table 17.10).

A graphic representation of the results is shown in Figure 17.16. Observe the moving average trend over the three-year, four-year and five-year cycles and note how, with each cyclical increment, more fluctuations in the original data are removed.

In other words, the moving average is much smoother than the original series and for this reason, moving average is referred to as a *smoothing* technique. As we observed, the smoothing effect increased as the number of items in the moving average rose.

Table 17.10 *Sales turnover for Green Thumbs Nurseries, 30 June 1983–97: four-year moving average*

Year (30 June)	Sales turnover ($ million)	Four-year moving total	Two four-year central moving total	Four-year central moving total
1983	8			
1984	9			
		41		
1985	10		90	11.2
		49		
1986	14		106	13.2
		57		
1987	16		121	15.1
		64		
1988	17		133	16.6
		69		
1989	17		144	18.0
		75		
1990	19		160	20.0
		85		
1991	22		185	23.1
		100		
1992	27		221	27.6
		121		
1993	32		263	32.9
		142		
1994	40		302	37.8
		160		
1995	43		334	41.8
		174		
1996	45			
1997	46			

Source: Hypothetical.

Figure 17.16 *Sales turnover of Green Thumbs Nurseries, 1983–97: four-year moving average*

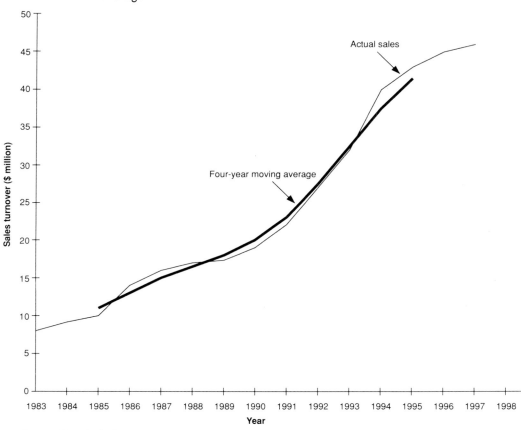

Source: Hypothetical.

In summary, a moving average trend should be used when eliminating erratic and short-term movements in time series and when a non-linear trend seems apparent in regular cyclical or seasonal fluctuations. By applying moving averages over the prevailing cycle, you are really averaging the results within the term of the cycle, hence removing sharp upward or downward movements; in other words, smoothing movements in the time series. However, there is no clear-cut rule on how many items to include in the moving average. If we use too many items, any discernible pattern will be lost. It is also obvious from our study of moving averages that the trend cannot be brought up to date.

Furthermore, the moving average is calculated by the mean and since the mean is affected by extreme values, the moving average will be distorted by any unusual events occurring during the time under consideration (e.g. drought, sharemarket crash or trade war). Thus, this method is most appropriate if fluctuations are regular in period and amplitude. If the cycle varies in either the former or the latter, moving averages will not remove all of the cyclical or seasonal variations from the data, thereby rendering it of little use. When considering sales, three-year, four-year and five-year cycles were chosen to illustrate the use and manipulation of the moving average technique. Perhaps on the basis of the available data, you might have felt the average length of the cycle was six years, in which case your calculated trend would be marginally different, illustrating that the moving average can be considered partially subjective if an obviously regular cyclical pattern is not evident.

In fact, when moving averages are used for forecasting results, the moving average forecast for any period is usually the value of the moving average for the preceding period. Furthermore, if you want to forecast for later periods, that same last moving average value will apply.

The major problem of a moving average forecast is evident. What justification is there for applying only the observations of the last three to five periods (which we are averaging) to forecast future results, thereby completely disregarding all other observations? In addition, the larger the number of terms employed in calculating the moving average, the smoother the trend line becomes and the less realistic the resulting movements when compared to the original data.

Exponential smoothing

Another smoothing technique that is sometimes applied because it is a simple and inexpensive forecasting tool is **exponential smoothing**. You can find this method being used in inventory management. The technique applies to stationary data—in other words, data that fluctuate about a constant level rather than having a trend. Just as its name implies, exponential smoothing is derived by smoothing random variations to create a new smoothed series. The forecast is then the most recent smoothed value. For example, Green Thumbs Nurseries' smoothed series is determined as shown in Table 17.11 and its graphic representation is as shown in Figure 17.17.

The data series is smoothed using the formula:

New smoothed value =

$(1 - \alpha) \times$ Previous smoothed value $+ (\alpha \times$ Most recent actual value)

or $\quad \Sigma_t = (1 - \alpha)\Sigma_{t-1} + \alpha Y_t$

where Σ = value of the exponentially smoothed series being calculated in time period t
Σ_{t-1} = value of the exponentially smoothed series already calculated in time period $t - 1$
Y_t = observed value of the time series in period t
α = subjectively assigned weight or smoothing coefficient.

Table 17.11 *Sales turnover for Green Thumbs Nurseries, 30 June 1983–97: exponential smoothing*

Year (30 June)	Actual sales turnover ($ million)	Smoothed series (using $\alpha = .6$) $\Sigma_t = (1-\alpha)\Sigma_{t-1} + \alpha Y_t$
1983	8	8
1984	9	8.6 (0.4 × 8 + 0.6 × 9)
1985	10	9.44 (0.4 × 8.6 + 0.6 × 10)
1986	14	12.18 (0.4 × 9.44 + 0.6 × 14)
1987	16	14.47 (0.4 × 12.18 + 0.6 × 16)
1988	17	15.99 (0.4 × 14.47 + 0.6 × 17)
1989	17	16.60 (0.4 × 15.99 + 0.6 × 17)
1990	19	18.04 (0.4 × 16.6 + 0.6 × 19)
1991	22	20.42 (0.4 × 18.04 + 0.6 × 22)
1992	27	24.37 (0.4 × 20.42 + 0.6 × 27)
1993	32	28.95 (0.4 × 24.37 + 0.6 × 32)
1994	40	35.58 (0.4 × 28.95 + 0.6 × 40)
1995	43	40.03 (0.4 × 35.58 + 0.6 × 43)
1996	45	43.01 (0.4 × 40.03 + 0.6 × 45)
1997	46	44.80 (0.4 × 43.01 + 0.6 × 46)

Source: Hypothetical.

The equation is applied to the series one number at a time so that a smoothed series is built up. The value of α (alpha), known as the *smoothing constant*, must be within the range of 0 to 1 but its exact value is decided by the forecaster. A value close to 1 indicates that heavy weighting has been given to the most recent actual value and light weighting to the rest of the series (represented by the previous smoothed value). This would be desirable if the future was determined mainly by very recent values. A value close to 0 gives little weighting to the recent value. This would be desirable if there were large random fluctuations which could be averaged out by spreading the weighting over many past values.

If you think about it, the use of exponential smoothing appears more reasonable than the moving average, as the latter gives equal weight to all observations in determining the forecast, while the former may make the more recent observations more important in forecasting (as we did in the above example).

If you wanted to interpret the above formula and results, you could say that the smoothed value of 24.37 in 1992, for example, is the weighted average of the smoothed value for the previous period (i.e. 20.42 × 0.4) and the actual value in 1992 (27 × 0.6). The forecast for any future periods is simply the exponentially smoothed value of the last period (in this case in 1997, it was 44.8). Therefore, if you were asked to forecast the sales turnover for 1999, it would be 44.8.

There is no set rule that provides the best smoothing constant for forecasting; therefore, a problem in its application is the arbitrariness of the nomination of this constant. However, if there is a great deal of random variation in the original data, you can usually use a low smoothing constant to suppress this variation in the smoothed series. Typically smoothing constants are in the range of 0.2 to 0.8.

In review, the exponential smoothing technique has three basic advantages:

1. It is simple to calculate and once α (alpha) is chosen, a computer can make forecasts without intervention by the forecaster.
2. It allows the importance of recent or past values to be allocated appropriate weights in terms of alpha.
3. It incorporates all previous values in each calculation.

The obvious disadvantage is that the determination of the smoothing constant is left to the discretion of the forecaster.

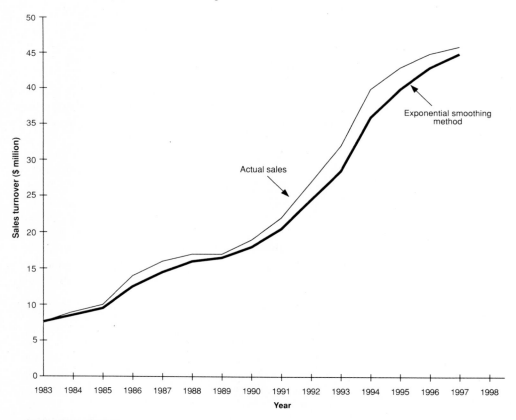

Figure 17.17 *Sales turnover for Green Thumbs Nurseries, 1983–97: exponential smoothing method*

Source: Hypothetical.

Chapter 17 – Time series

17.2 Competency check

1. The freehand graphic method, the semi-average method and the least-squares method of determining the secular trend are similar in what respect(s)?

2. Although the semi-average method is simple and objective, the least-squares method is considered the more preferred option in evaluating the straight-line secular trend. Why is this the case?

3. In Table 17.12, the annual sales of Drivem Power Tools is listed for 1989–97.

    (a) Plot the data and sketch a trend line using the freehand graphic method.
    (b) Forecast sales for 2003 using this method.

    Table 17.12 *Annual sales for Drivem Power Tools, 30 June 1989–97*

Year (30 June)	Sales ($ million)
1989	1
1990	3
1991	6
1992	7
1993	10
1994	14
1995	15
1996	15
1997	16

    Source: Hypothetical.

4. Management has now requested that you estimate sales for 2003 using the least-squares method.

    (a) Find the linear equation that describes the trend in sales.
    (b) Estimate the sales results for 2003.
    (c) Do you believe a trend line best describes the data in the light of actual sales results? Explain briefly.

5. (a) Exponentially smooth the data in Table 17.12, with a smoothing constant of 0.2.
    (b) Now apply a smoothing constant of 0.6 and, after exponentially smoothing out the time series, comment on the difference in results when applying a smoothing constant of 0.2 and 0.6.

17.3 Seasonal variation*

When a time series is classified into periods of less than one year—for example, weeks, months or quarters—there may be a repetitive periodic movement which is referred to as a *seasonal fluctuation* or *variation*. Such variation is evident, for example, in retail sales of certain types of clothing, soft drinks and ice-cream. Seasonal variations need not necessarily occur in a year, but we may see daily fluctuations such as in the use of telecommunications, public transport, public utilities and the like; weekly variations in supermarket sales; and monthly variations such as in unemployment statistics, trade statistics or retail sales.

Just as estimates of secular trends are needed for long-term planning, estimates of seasonal variations are important for short-term planning. Management must be aware of any seasonal variation in order to ensure continuity of supply to meet demand but without, at the same time, being left with excessive inventory. An index must be constructed which can define, for example, each month's sales in relation to the year's results as a whole, hence improving short-term planning and enhancing the likelihood of cost efficiency and profitability. In other words, measuring seasonal variation may serve any or all of the following purposes:

1. To help understand seasonal patterns. For example, October sales are 5% down on September results and you want to discover whether the decline is more or less than the expected seasonal decline.
2. To examine existing patterns in order to project future results. Seasonal variation patterns are important for short-term planning. For example, you may be uncertain about the number of casual staff to hire during certain times of the year. Evaluating the seasonal variation results will help to minimise the risk of a shortage or surplus of staff.
3. To deseasonalise the time series. We can measure the seasonal impact on the prevailing results and allow for these in order to reveal any cyclical pattern. If you examine statistical publications by the Australian Bureau of Statistics, you will find that most are seasonally adjusted (e.g. unemployment, exports and imports, retail sales, production, construction activity, and so on).

A number of methods of calculating a seasonal index are available. Two methods—the simple average and the ratio to moving average—are discussed below.

Simple average method

The **simple average method** is the simplest method of calculating the seasonal index and operates under the (not necessarily valid) assumption that secular trend has little effect on time series and that cyclical and irregular fluctuations are eliminated by being fairly well balanced over the time series with, for example, peaks and troughs in the case of cyclical movements cancelling each other over time.

To calculate the seasonal index:

1. Calculate the arithmetic mean for each period (e.g. quarter, month) for all years, regarding the calculated mean as typical for each of the periods.

* Optional

2. Calculate the seasonal index from the averages of each period, with the index normally expressed in percentage form. The total of the percentages is 400% for a quarterly seasonal index and 1200% for a monthly seasonal index.

Let's determine the seasonal index of the data on overseas visitors to Australia contained in Table 17.13.

1. Find the average number of visitors for each quarter. For example, to calculate the average visitors for the March quarter, find the total of the March quarter results (3755) and divide by the number of March quarters in the series (5). Therefore, $3755 \div 5 = 751$.
2. The seasonal index is then calculated by determining each respective quarterly average in proportion to the mean average of all quarters (see Table 17.14). For example, the mean overseas visitors for June (641) is divided by the mean average for the four quarters (753) and multiplied by 100. That is:

$$\text{June seasonal index} = \frac{641}{753} \times 100$$
$$= 85.1$$

The seasonal index of the total number of visitors to Australia clearly illustrates that the quietest period is the June quarter, while the December quarter is the busiest. In fact, there are normally almost 40% more visitors in the latter quarter than in the former quarter. As tourism is now the greatest generator of foreign exchange for Australia, taking into account the seasonal variation is of immeasurable help to all businesses relying on the foreign tourist trade. The ability to measure the seasonal variation in this trade will help them with planning, inventory control, staffing levels and, ultimately, profitability.

Table 17.13 *Total number of visitors ('000s) to Australia, 1991–95: quarterly basis*

Year	Quarter				Total
	March	June	September	December	
1991	559	502	606	707	2 374
1992	652	556	608	787	2 603
1993	746	644	708	898	2 996
1994	862	703	800	997	3 362
1995	936	800	903	1 086	3 725
Total	3 755	3 205	3 625	4 475	15 060
Mean average	751	641	725	895	3 012

Source: Australian Bureau of Statistics, *Australian Economic Indicators* (Cat. No. 1350.0).

Table 17.14 Total number of visitors to Australia, 1991–95: quarterly basis and seasonal index

Quarter	Average number of visitors for quarter ('000)	Seasonal index (%)
March	751	99.7
June	641	85.1
September	725	96.3
December	895	118.9
Total	3012	400
Mean average	753	100

Source: Australian Bureau of Statistics, *Australian Economic Indicators* (Cat. No. 1350.0).

Problem: Thus far, we have examined quarterly indexes, but monthly indexes are just as common. Let's have a look at the wine sales from winemakers of the Besoffen Valley for the period 1993–97.

Table 17.15 Sales by Besoffen Valley winemakers (millions of litres), 1993–97

Month	1993	1994	1995	1996	1997	Total	Mean
January	24.2	16.9	17.6	18.4	20.0	97.1	19.4
February	23.4	20.0	22.1	21.8	21.7	109.0	21.8
March	23.4	23.5	25.7	26.5	26.2	125.3	25.1
April	26.4	21.7	22.2	22.4	24.5	117.2	23.4
May	24.7	25.6	26.9	28.1	24.6	129.9	26.0
June	23.4	23.9	23.2	25.5	26.5	122.5	24.5
July	24.7	25.5	27.8	24.9	30.7	133.6	26.7
August	23.2	31.0	35.4	33.4	34.6	157.6	31.5
September	24.3	24.5	23.2	24.1	26.7	122.8	24.6
October	22.6	23.1	25.1	28.4	26.2	125.4	25.1
November	22.9	29.7	30.0	33.5	32.1	148.2	29.6
December	23.2	34.4	35.7	37.4	38.9	169.6	33.9

Source: Hypothetical.

Solution:
1. The mechanics are identical for a monthly index and a quarterly index, with the arithmetic mean of each month's results (over the five years) calculated (last column in Table 17.15).
2. The arithmetic mean of the monthly mean sales must then be calculated (e.g. 25.96 from Table 17.16).
3. Each individual month's mean average is taken as a proportion to the mean average for all months (i.e. 25.96) multiplied by 100, with the result being the seasonal index for the respective months (e.g. April—23.4/25.96 × 100 = 90.1).

Table 17.16 *Sales by Besoffen Valley winemakers (millions of litres) and seasonal index*

Month	Mean sales	Seasonal index
January	19.4	74.7
February	21.8	84.0
March	25.1	96.7
April	23.4	90.1
May	26.0	100.1
June	24.5	94.4
July	26.7	102.8
August	31.5	121.3
September	24.6	94.7
October	25.1	96.7
November	29.6	114.0
December	33.9	130.5
Total	311.6	1200.0
Mean average	25.96	100.0

Source: Hypothetical.

Interestingly, only three months' sales exceed the monthly average by more than a marginal rate, these being August, November and December. Winemakers, aware of the seasonal demand, can ensure that at these particular times stock is ample to satisfy demand. During particularly quiet times, such as January and February, the required stock on hand is reduced; this will allow the winemakers to save on warehousing and wastage.

Ratio-to-moving average

The simple average method operates under the assumption that the upward and downward phases of cyclical and trend fluctuations balance out, hence permitting a seasonal index to be determined, unaffected by trend or cyclical variations. However, this is not the case in most business and economic data, since there is usually a degree of upward or downward bias in the direction of the trend or cyclical fluctuations, subsequently placing similar directive bias on the seasonal index for particular periods. To remove such bias, the moving average method may be applied. Previously we applied the moving average method to determine a non-linear trend with the period equal to the length of the cycle, be it three, four or five years. When calculating seasonal variation, the time-frame is always one year—that is, four quarters for quarterly data and 12 months for monthly data. The application of the moving average will smooth out seasonal variations as long as the period selected for calculating the moving average (e.g. month) is equivalent to the average duration of the variations within the time series (e.g. monthly).

The process of calculating a moving average over a one-year period will eliminate the effect of seasonal variations, since by definition such variations are periodic changes

Business mathematics and statistics

which repeat themselves each year and irregular variations (for the most part, since such variations are normally of short duration). Therefore, moving average results can be thought of as the values that would exist without seasonal or irregular influences, thus values directly affected only by cyclical and trend variations. However, since cyclical fluctuations may vary considerably in duration, moving averages may also remove a degree of the cyclical influence on the data.

Let's have another look at the overseas visitors to Australia, but this time let's determine the seasonal index with the ratio-to-moving average method. Table 17.17 lists the period (column 1) and actual number of visitors (column 2). To determine the seasonal index, the following steps should be followed:

1. Since the data are provided on a quarterly basis, a four-quarter moving average must be calculated—that is, add the first four quarters' sales and centre the result between the second and third quarter in column 3, with the third quarter's moving total centred between the third and fourth quarter, and so on. Thereafter, to complete this column, add the next quarter's sales and subtract the same quarter's result of the previous year. For example, for September 1992 the moving average of June, September and December 1992 and March of 1993 must be added and then centred between the September and December quarters in column 3 (i.e. 556 + 608 + 787 + 746 = 2697). However, instead of having to go through this exercise with each month, it is easier to take the difference between March 1993 and March 1992 (746 − 652 = 94), which is added to the previous quarter's moving total (June 1992 at 2603) with a result of 2697, but this time without having to add four four-digit numbers.
2. After completing column 3, add the four-quarter moving totals in consecutive pairs (column 4) with the total now opposite the original values of column 2.
3. A four-quarter moving average is calculated by dividing the two four-quarter moving totals (column 4) by eight (since two four-quarter moving totals are equivalent to eight quarters being considered).
4. Find the ratio of each quarter's retail sales to the corresponding four-quarter moving average (column 6). For example, the ratio for September 1994 is determined by $(800 \div 849.8) \times 100 = 94.1$.
5. Arrange these ratios in another table (Table 17.18) in columns by periods and calculate the average, which is usually in the form of the mean, median or modified mean. The modified mean is calculated by excluding extreme ratios in a set of data and calculating the mean of the remaining ratios. This is done to eliminate extreme cyclical and irregular variations. The degree of modification is at the discretion of the analyst. For example, on a quarterly basis, the highest and lowest may be excluded with the mean of the two middle values calculated. This procedure would have to be followed throughout the particular set of data. If 11 years were being considered, the three highest and three lowest values could be excluded, for example, with the mean calculated for the remaining five years' ratios. In Table 17.18, quarterly results are in the form of the mean rather than the modified mean.
6. Calculate the seasonal index from which the average has been selected (e.g. mean, modified mean or median). The four quarters' total seasonal index should equal

Table 17.17 *Calculation of seasonal indexes for the quarterly number of visitors to Australia using four-quarter moving averages for the period 1991–95*

Year and quarter		Overseas visitors ('000)	Four-quarter moving total	Two four-quarter centred moving total	Four-quarter centred moving average	Ratio of actual sales to moving average
(1)		(2)	(3)	(4)	(5) = (4) ÷ 8	(6) = (2) ÷ (5) × 100
1991	M	559				
	J	502				
			2374			
	S	606		4841	605.1	100.1
			2467			
	D	707		4988	623.5	113.4
			2521			
1992	M	652		5044	630.5	103.4
			2523			
	J	556		5126	640.8	86.8
			2603			
	S	608		5300	662.5	91.8
			2697			
	D	787		5482	685.2	114.9
			2785			
1993	M	746		5670	708.8	105.2
			2885			
	J	644		5881	735.1	87.6
			2996			
	S	708		6108	763.5	92.7
			3112			
	D	898		6283	785.4	114.3
			3171			
1994	M	862		6434	804.2	107.2
			3263			
	J	703		6625	828.1	84.9
			3362			
	S	800		6798	849.8	94.1
			3436			
	D	997		6969	871.1	114.5
			3533			
1995	M	936		7169	896.1	104.5
			3636			
	J	800		7361	920.1	86.9
			3725			
	S	903				
	D	1086				

Source: Australian Bureau of Statistics, *Australian Economic Indicators* (Cat. No. 1350.0).

400 (as a 12-month seasonal index should equal 1200). Sometimes, because of centring and rounding-off processes, the total may be more or less than that required, in which case an adjustment factor must be applied (as exemplified in Table 17.19) to determine the seasonal index.

Table 17.18 Quarterly seasonal index of number of visitors to Australia based on the mean for the period 1991–95

Year	Quarter			
	March	June	September	December
1991			100.1	113.4
1992	103.4	86.8	91.8	114.9
1993	105.2	87.6	92.7	114.3
1994	107.2	84.9	94.1	114.5
1995	104.5	86.9		
Mean	105.1	86.6	94.7	114.3

Table 17.19 Calculation of seasonal index of visitors to Australia with seasonal adjustment factor

Quarter (1)	Unadjusted seasonal index (2)	Adjusted seasonal index Col. 2 × 0.9983 (3)
March	105.1	104.9
June	86.6	86.5
September	94.7	94.5
December	114.3	114.1
Total	400.7	400.0
Average	100.175	100.0

Note: Adjustment factor $= \dfrac{400}{400.7} = 0.9983$.

The graphic representation of the original data and the seasonally adjusted indexes (Figure 17.18) depicts a smooth upward-sloping curve illustrating how many visitors would have visited Australia without the influence of seasonal variations.

On the basis of the original and seasonally adjusted data, it is obvious that there is a periodic pattern of overseas visitors touring Australia. Understanding this would greatly help planning and forecasting.

Comparing Table 17.19 with Table 17.14 (calculating the seasonal index for the same data but using the simple average method), we find similar results as extremes were absent. Had extremes been present in the data, the two indexes would have differed according to the severity of the extremes.

Chapter 17 – Time series

Figure 17.18 *Quarterly results of the number of visitors to Australia, 1991–95: four-quarter moving average method*

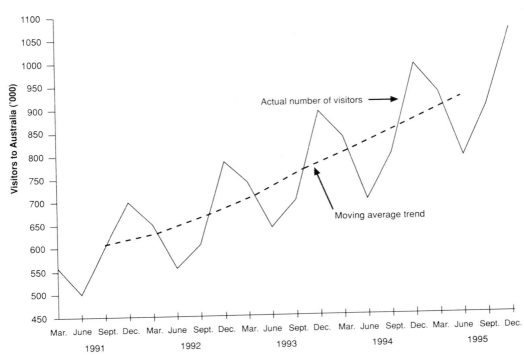

Source: Australian Bureau of Statistics, *Australian Economic Indicators* (Cat. No. 1350.0).

Problem: Determine (from Table 17.20) the seasonal index of sales of wine by Besoffen Valley winemakers using the ratio-to-moving average method.

Solution:
1. A 12-month moving average is determined by using the same principles as for a four-quarter moving average—that is, find the 12-month moving total by centring the sum of successive 12-monthly sales results between the sixth and seventh months.
2. Find the sum of all consecutive pairs of the 12-month moving total, in line with the original sales.
3. Determine the 12-month moving average by dividing the two 12-month moving total (column 4) by 24 (i.e. there are 2 × 12 monthly totals = 24).
4. Find the ratio of actual sales to the corresponding moving average (column 2 divided by column 5 × 100).
5. Construct another table (Table 17.21) listing the calculated ratios per month. Determine the mean or median. Do the total indexes for the 12 months equal 1200? If so, these indexes are the seasonal indexes. If not, an

adjustment factor must be applied. In the case of wine sales, the mean is less than 1200 while the median is greater than 1200. Thus, Tables 17.22 and 17.23 illustrate the adjustment for the mean and median respectively. For example, the mean totalled 1190.9 instead of 1200. The respective months' ratios results must be multiplied by an adjustment factor for the seasonal indices to total 1200. This factor is found by dividing 1200 (the desired total) by the actual total (1190.9). Thus each month's mean result (e.g. August's 120.3) is multiplied by 1.007641, with the final indexes (e.g. August's 121.2) in column 3 of Table 17.22.

Table 17.20 Sales by Besoffen Valley winemakers (millions of litres), 1993–97: 12-month moving average method

Year and month (1)	Sales (2)	Twelve-month moving total (3)	Two 12-month centred moving total (4)	Twelve-month centred moving average (5) = (4) ÷ 24	Ratio of actual sales to moving average (6) = (2) ÷ (5) × 100
1993 J	24.2				
F	23.4				
M	23.4				
A	26.4				
M	24.7				
J	23.4				
		286.4			
J	24.7		565.5	23.6	104.7
		279.1			
A	23.2		554.8	23.1	100.4
		275.7			
S	24.3		551.5	23.0	105.7
		275.8			
O	22.6		546.9	22.8	99.1
		271.1			
N	22.9		543.1	22.6	101.3
		272.0			
D	23.2		544.5	22.7	102.2
		272.5			
1994 J	16.9		545.8	22.7	74.4
		273.3			
F	20.0		554.4	23.1	86.6
		281.1			
M	23.5		562.4	23.4	100.4
		281.3			
A	21.7		563.1	23.5	92.3
		281.8			
M	25.6		570.4	23.8	107.6
		288.6			
J	23.9		588.4	24.5	97.6
		299.8			

Table 17.20 *(continued)*

Year and month (1)	Sales (2)	Twelve-month moving total (3)	Two 12-month centred moving total (4)	Twelve-month centred moving average (5) = (4) ÷ 24	Ratio of actual sales to moving average (6) = (2) ÷ (5) × 100
J	25.5		600.3	25.0	102.0
		300.5			
A	31.0		603.1	25.1	123.5
		302.6			
S	24.5		607.4	25.3	96.8
		304.8			
O	23.1		610.1	25.4	90.9
		305.3			
N	29.7		611.9	25.5	116.5
		306.6			
D	34.4		612.5	25.5	134.9
		305.9			
1995 J	17.6		614.1	25.6	68.8
		308.2			
F	22.1		620.8	25.9	85.3
		312.6			
M	25.7		623.9	26.0	98.8
		311.3			
A	22.2		624.6	26.0	85.4
		313.3			
M	26.9		626.9	26.1	103.1
		313.6			
J	23.2		628.5	26.2	88.5
		314.9			
J	27.8		630.6	26.3	105.7
		315.7			
A	35.4		631.1	26.3	134.6
		315.4			
S	23.2		631.6	26.3	88.2
		316.2			
O	25.1		632.6	26.4	95.1
		316.4			
N	30.0		634.0	26.4	113.6
		317.6			
D	35.7		637.5	26.6	134.2
		319.9			
1996 J	18.4		636.9	26.5	69.4
		317.0			
F	21.8		632.0	26.3	82.9
		315.0			
M	26.5		630.9	26.3	100.8
		315.9			

Table 17.20 (continued)

Year and month (1)	Sales (2)	Twelve-month moving total (3)	Two 12-month centred moving total (4)	Twelve-month centred moving average (5) = (4) ÷ 24	Ratio of actual sales to moving average (6) = (2) ÷ (5) × 100
A	22.4		635.1	26.5	84.5
		319.2			
M	28.1		641.9	26.7	105.2
		322.7			
J	25.5		647.1	27.0	94.4
		324.4			
J	24.9		650.4	27.1	91.9
		326.0			
A	33.4		651.9	27.2	122.8
		325.9			
S	24.1		651.5	27.1	88.9
		325.6			
O	28.4		653.3	27.2	104.4
		327.7			
N	33.5		651.9	27.2	123.2
		324.2			
D	37.4		649.4	27.1	138.0
		325.2			
1997 J	20.0		656.2	27.3	73.3
		331.0			
F	21.7		663.2	27.6	78.6
		332.2			
M	26.2		667.0	27.8	94.2
		334.8			
A	24.5		667.4	27.8	88.1
		332.6			
M	24.6		663.8	27.7	88.8
		331.2			
J	26.5		663.9	27.7	95.7
		332.7			
J	30.7				
A	34.6				
S	26.7				
O	26.2				
N	32.1				
D	38.9				

Source: Hypothetical.

Table 17.21 *Calculation of seasonal index of Besoffen Valley wine sales for 1993–97: 12-month moving average method*

Year	Month											
	Jan.	Feb.	Mar.	Apr.	May	June	July	Aug.	Sept.	Oct.	Nov.	Dec.
1993							104.7	100.4	105.7	99.1	101.3	102.2
1994	74.4	86.6	100.4	92.3	107.6	97.6	102.0	123.5	96.8	90.9	116.5	134.9
1995	68.8	85.3	98.8	85.4	103.1	88.5	105.7	134.6	88.2	95.1	113.6	134.2
1996	69.4	82.9	100.8	84.5	105.2	94.4	91.9	122.8	88.9	104.4	123.2	138.0
1997	73.3	78.6	94.2	88.1	88.8	95.7						
Total	285.9	333.4	394.2	350.3	404.7	376.2	404.3	481.3	379.6	389.5	454.6	509.3
Mean	71.5	83.4	98.6	87.6	101.2	94.0	101.1	120.3	94.9	97.4	113.6	127.3
Median	71.4	84.1	99.6	86.8	104.1	95.0	103.4	123.2	92.8	97.1	115.1	134.6

Source: Hypothetical.

Table 17.22 *Seasonal index calculation based on the mean ratio for wine sales*

Month (1)	Mean of ratios to 12-month moving average (2)	Seasonal index (%) Col. (2) × 1.007641
January	71.5	72.0
February	83.4	84.0
March	98.6	99.4
April	87.6	88.3
May	101.2	102.0
June	94.0	94.7
July	101.1	101.9
August	120.3	121.2
September	94.9	95.6
October	97.4	98.1
November	113.6	114.5
December	127.3	128.3
Total	1190.9	1200.0
Average	99.24	100.0

Source: Hypothetical.

Note: Adjustment factor $\frac{1200}{1190.9} = 100.7641$.

Business mathematics and statistics

The same principle is illustrated for the median when applying the adjustment factor of .99404. In this case (Table 17.23), since the median total was greater than 1200, each month's index has to be reduced; hence, the adjustment factor was less than 1.

Table 17.23 *Seasonal index calculation based on the median ratio for wine sales*

Month (1)	Median of ratios to 12-month moving average (2)	Seasonal index (%) Col. (2) × .99404
January	71.4	71.0
February	84.1	83.6
March	99.6	99.0
April	86.8	86.3
May	104.1	103.5
June	95.0	94.4
July	103.4	102.8
August	123.2	122.5
September	92.8	92.2
October	97.1	96.5
November	115.1	114.4
December	134.6	133.8
Total	1207.2	1200.0
Average	100.6	100.0

Source: Hypothetical.

Note: Adjustment factor $\dfrac{1200}{1207.2} = .99404$.

The seasonal indexes determined by the simple average method (Table 17.16) and the moving average method are similar, with a distinct periodic pattern; January, for example, is always the month of lowest sales and December the highest.

In review, the ratio-to-moving average is the preferred method of determining seasonal indexes and, through its application, the seasonal and irregular effects are smoothed and the long-term change becomes obvious (see Figure 17.19). When cyclical and secular variations are in balance with little or no extremes, the simple average method may be useful, the results being similar to those of the ratio-to-moving average method.

Finally, as you will have noticed, the calculations are many if the seasonal index is determined by using a hand-held calculator, as in the previous examples. However, since most businesses have access to computers, it is a relatively simple operation if the computer does all the calculating of seasonal data. But now, at least, you will understand the basis of and logic behind seasonal index construction and application.

Figure 17.19 *Sales by Besoffen Valley winemakers, 1993–97: 12-month moving average method*

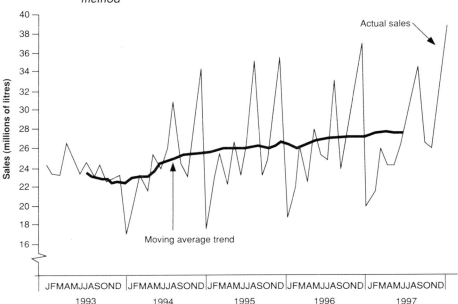

Source: Hypothetical.

Use of the seasonal index

As described earlier, the seasonal index is invaluable for forecasting and deseasonalising purposes.

The seasonal index as a forecasting tool
The seasonal index assists management in forward planning and the control of elements affected by seasonal variation, such as financial requirements, inventory needs, production runs and staffing levels. The use of a seasonal index can also improve forecasting reliability, as the index is normally updated (as long as there have been no major or sudden changes in the seasonal pattern). For example, if we assume that on a quarterly basis the expected number of overseas visitors to Australia, based on past trend patterns, is 800 000; then to determine the expected number in the June quarter we must adjust this expected trend value of 800 000 by the corresponding seasonal index for the June quarter, which according to Table 17.19 is 86.5. Therefore, the estimated trend value must be multiplied by the seasonal index—that is, 800 000 × 86.5, or 692 000, visitors are expected to arrive in Australia during the June quarter.

The seasonal index as a deseasonalising tool
The use of the seasonal index to remove the effects of seasonal variation from a time series is widespread. For example, government unemployment statistics, international trade statistics, and production and retail statistics are all seasonally adjusted.

Business mathematics and statistics

Seasonally adjusted time series or deseasonalised data (as it is often called) illustrate what result would have occurred if only the trend, cyclical and irregular movements had prevailed. In other words, by deseasonalising data you eliminate the impact that seasonal fluctuations make on them. It is a fairly simple calculation; you divide the original data by the appropriate seasonal index. The basic formula is:

$$\text{Period's seasonally adjusted data} = \frac{\text{Period's actual value}}{\text{Seasonal index for period}} \times 100$$

To illustrate the deseasonalisation of data, let us examine the original data on the total number of visitors to Australia (Table 17.13) and the corresponding seasonal index calculated by the ratio-to-moving average method (Table 17.19).

Table 17.24 *Total number of visitors ('000s) to Australia, 1991–95: seasonally adjusted*

Year and quarter (1)		Actual data (2)	Seasonal index data (3)	Seasonally adjusted data (2) ÷ (3) × 100
1991	March	559	104.9	533
	June	502	86.5	580
	September	606	94.5	641
	December	707	114.1	620
1992	March	652	104.9	622
	June	556	86.5	643
	September	608	94.5	643
	December	787	114.1	690
1993	March	746	104.9	711
	June	644	86.5	745
	September	708	94.5	749
	December	898	114.1	787
1994	March	862	104.9	822
	June	703	86.5	813
	September	800	94.5	847
	December	997	114.1	874
1995	March	936	104.9	892
	June	800	86.5	925
	September	903	94.5	956
	December	1086	114.1	952

Source: Australian Bureau of Statistics, *Australian Economic Indicators* (Cat. No. 1350.0).

According to the formula, the actual number of visitors for each quarter must be divided by the corresponding seasonal index for that same quarter. In other words, the seasonally adjusted result for the December quarter 1993 is 787 ('000) versus the actual number of visitors of 898 ('000)—that is:

Chapter 17 – Time series

$$\text{December quarter 1993} = \frac{898}{114.1} \times 100$$
$$= 787$$

In other words, the December quarter's results are normally 114.1% of the quarterly result for all quarters and thus allowance is made for this fact by dividing by this factor. Likewise, all the actual data in Table 17.13 are adjusted the same way.

Interestingly, throughout the 1991–95 period the actual number of visitors to Australia during the June quarter represented the lowest quarterly result each year and yet, when seasonally adjusted, it hit the lowest rung only in 1994.

If we examine other data, we observe that the December quarter's actual results were always far higher than any other quarter and yet, in seasonally adjusted terms, only in three of the five years were the number of visitors during the December quarter the highest for that year.

It should also be noted that as a result of seasonally adjusting the data there is a dramatically reduced range of results on a per annum basis. For example, in 1995 the range of the actual number of visitors was 286 (1086 – 800), but on a seasonally adjusted basis it was reduced to 64 (956 – 892).

Problem: Deseasonalise the sales of wine by Besoffen Valley winemakers (Table 17.20) by the seasonal index as determined by the ratio to moving-average method (Table 17.23).

Solution: See Table 17.25.

Table 17.25 *Sales by Besoffen Valley winemakers (millions of litres), 1993–97: seasonally adjusted results*

Month	Year				
	1993	1994	1995	1996	1997
January	34.1	23.8	24.8	25.9	28.2
February	28.0	23.9	26.4	26.1	26.0
March	23.6	23.7	26.0	26.8	26.5
April	30.6	25.1	25.7	26.0	28.4
May	23.9	24.7	26.0	27.1	23.8
June	24.8	25.3	24.6	27.0	28.1
July	24.0	24.8	27.0	24.2	29.9
August	18.9	25.3	28.9	27.3	28.2
September	26.4	26.6	25.2	26.1	29.0
October	23.4	23.9	26.0	29.4	27.2
November	20.0	26.0	26.2	29.3	28.1
December	17.3	25.7	26.7	28.0	29.1

Source: Hypothetical.

Several interesting observations can be made by examining Table 17.20 and the seasonally adjusted data (Table 17.25). In 1994, September's wine sales were the largest of the year in seasonally adjusted terms, but in actual sales September was the sixth best month of the year. In 1995, July's actual result was 7.9 (million litres) less than December's, yet on a seasonally adjusted basis July's result was 0.3 (million litres) higher than that of December (i.e. 27.0 versus 26.7). Similarly, in 1996 actual sales in December were 9.0 (million litres) greater than in October, but in seasonally adjusted terms October was 1.4 (million litres) greater.

Once again, the importance of seasonally adjusting data is emphasised, for otherwise management may be quite pleased with an actual sales result, which in seasonally adjusted terms demands attention. On the other hand, it may consider that it has a major problem on its hands because of a poor January result, for example, although in seasonally adjusted terms the results are in fact promising.

Finally, let's apply this seasonal index to forecast a result. If, for example, you expected wine sales in 1999 to reach 420 (million litres), and you had to target or plan for monthly sales, you could apply the calculated seasonal indexes (as per Table 17.23). For instance, projected sales in April would be 30.2 (million litres) since 420 (million litres) was forecast for the year, which averages out to 35 (million litres) per month. Since the seasonal index for April is 86.3, this indicates that April sales are normally 86.3% of the average monthly sales, hence 86.3% of 35 equals 30.2. A similar procedure would be followed to determine other monthly projections.

In the light of the above commentary, the importance and mechanics of the seasonal index should be thoroughly understood. By applying a seasonal index to actual data, management may well find that the index results, which give an improved indication of performance, differ considerably from actual results.

17.3 Competency check*

1. Measuring seasonal variation can be helpful to most organisations for a number of reasons. Briefly discuss two such reasons, relating each to your own workplace.

2. Briefly describe three types of businesses/industries which were not discussed in the text but which are affected by seasonal variation.

3. Jericho Master Builders is determined to develop a seasonal index to improve its planning, inventory control and appropriate staffing levels throughout the year. Below are the collected data per quarter.

   (a) Calculate the seasonal index by the ratio-to-moving average method (use the arithmetic mean for averaging).
   (b) If management believe that in 2001 the company will build 200 dwellings, forecast the number to be built each quarter of that year.

*Optional

Chapter 17 – Time series

Table 17.26 *Jericho Master Builders' dwelling completions, 1994–97*

Year	Quarter			
	March	June	September	December
1994	16	25	32	17
1995	18	29	34	20
1996	17	26	28	16
1997	19	28	30	18

Source: Hypothetical.

4. Use the simple average method and the ratio-to-moving average method to calculate the seasonal index per quarter (to one decimal place) for the retail sales listed in Table 17.27.

Table 17.27 *Quarterly retail sales ($ million), 1992–97*

Year	Quarter			
	March	June	September	December
1992	9.4	9.4	9.4	11.4
1993	9.7	10.1	10.3	12.3
1994	10.4	10.2	10.8	13.1
1995	11.2	11.9	12.2	14.8
1996	12.3	13.2	12.4	15.9
1997	13.3	14.0	16.7	19.9

Source: Hypothetical.

5. (a) Deseasonalise the data in question 4 for all quarters in 1995 using the seasonal index as determined by the ratio-to-moving average method.
   (b) If the number of staff employed is a function of retail sales and if during the December quarter approximately 200 people are employed, how many should be employed during the March quarter?

17.4 Summary

It would be difficult to find a successful business not reliant on statistics of past and current trends which assist it in more accurately defining and forecasting future needs and expectations. A time-series analysis—that is, the evaluation of business activity as affected by secular trends and cyclical, seasonal and irregular variations—is invaluable to business managers, enabling them to improve decision-making when dealing with uncertainty.

Secular trends—that is, the direction of change in business activity over a lengthy period (e.g. 10 to 20 years)—can be measured by any of five methods, the first three listed below resulting in straight-line trends and the fourth and fifth resulting in a non-linear trend.

1. *Freehand graphic method.* This is the simplest and fastest method and entails plotting data onto a graph, then carefully examining the direction of the trend. A straight line is then drawn—one which the analyst feels best fits the line's direction; hence, this method is subjective rather than objective and, as such, is frequently not satisfactory.
2. *Semi-average method.* This is the easiest method of finding a straight-line trend without being subjective. The data are divided into two equal groups, the arithmetic mean of each group is calculated and, finally, each mean centred in the middle period of each respective group is plotted. The straight line drawn through the two points is the secular trend line.
3. *Least-squares method.* This is the most reliable method of evaluating secular trend, when the sum of the squared deviations of individual values around the trend line is minimised.
4. *Moving average method.* When the data indicate that perhaps a straight line is inappropriate to describe the data's direction, the moving average method can be used. This technique consists in successively averaging the data in the cycle selected. The cycle represents the number of periods to be averaged, and the moving averages are calculated by adding the values in the cycle and then dividing this sum by the number of periods in the cycle.
5. *Exponential smoothing method.* This is another method used to define data which do not approximate a linear trend. The time series is smoothed by replacing each observation by the weighted sum of all past observations, the weights declining exponentially as the observations recede in time. It is sometimes preferred to the ratio-to-moving average method, since weights for exponential smoothing are allocated according to the importance of the time-frame. In other words, if the more recent data are deemed more relevant than older data, then α (alpha) will be closer to one.

Analysis of the secular trend provides information to management on the long-term trend of the business activity in question, permits comparison of data over different periods as well as comparison of different groups of data over the same period, and assists in forecasting future activity.

If a periodic, repetitive pattern of activity recurs at least once over a period of a year, it is referred to as a seasonal variation. A seasonal index is constructed on the basis of either the simple average method or the ratio-to-moving average method, the latter being used to a far greater extent by government and business. The calculated seasonal index defines each period's performance (weekly, monthly or quarterly) in terms of the year's result as a whole. It provides management with data which should assist in current and future planning and forecasting, as well as permitting the data to be seasonally adjusted.

The calculation of the secular trend and seasonal variation was most tedious and time-consuming, since a hand-held calculator was used. However, the use of a computer with an appropriate spreadsheet should enable you to collect similar calculations in a fraction of the time and with little effort.

Finally, we must remember that when using the various techniques explained, we are working under the assumption that past patterns will persist and that the collected data were accurate and reliable over the period being studied. Also, you will find that forecasts using time-series analysis may well be incorrect. (Just look at economic forecasts on GDP, unemployment, price level changes and the like.) However, this does not imply that the techniques are of no use; at the very least, a statistically based forecast is favoured to a hunch or one based on little but intuition.

17.5 Multiple-choice questions

1. Cyclical variations can be best described as:

   (a) periodic movements in a time series usually seen in the form of the expansion and contraction of business activity;
   (b) periodic movements which repeat themselves year after year with predictable intensities;
   (c) a relatively persistent long-term movement;
   (d) unpredictable, random patterns in a time series.

2. The semi-average method of defining a secular trend can be thought of as being all of the following except:

   (a) objective;
   (b) simple to calculate;
   (c) not affected by extreme values;
   (d) only appropriate when fitting a straight-line trend.

3. The most widely used statistical method for determining the straight-line secular trend is the:

   (a) freehand graphic method;
   (b) semi-average method;
   (c) least-squares method;
   (d) exponential smoothing method.

4. The moving average method should be used for all of the following reasons except:

   (a) to present a straight-line secular trend line;
   (b) to eliminate short-term variations;
   (c) to describe a smooth curve;
   (d) to eliminate the impact of cyclical variations on the secular trend.

5. Examining Table 17.11, what is the smoothed series result for 1998 if actual sales of Green Thumbs in that year reaches $50 (million)?

   (a) $47.92 (million)   (b) $53.76 (million)
   (c) $47.20 (million)   (d) $50.00 (million)

17.6 Exercises

1. Select an industry you are familiar with and briefly examine how it is influenced by each of the four components of a time series.

2. Nominate three businesses, not covered in the text, with sales influenced by cyclical variations.

3. Identify two products and/or services for which a secular trend line, if drawn, would illustrate a gradual reduction in sales. To what do you attribute this reduction?

4. The freehand graphic method is the simplest to use in deriving a straight-line secular trend. However, because of the subjectiveness of its presentation, it is not highly favoured. Under what circumstances would its use generally be acceptable?

5. Of the time-series components, which is the most predictable? Briefly discuss, using an appropriate example.

6. There are two ways of smoothing a time series. Briefly describe each of the two available methods.

7. Under what circumstances is the smoothing of a time series appropriate?

8. Captain Hook, the owner and operator of Tinkerbell Cruise Line, has listed the number of passengers he has carried over each of the last eight years (Table 17.28).

    (a) Plot the data on a graph.
    (b) Draw a trend line using the freehand graphic method.
    (c) Use this trend line (drawn in (b)) to forecast passenger numbers for the year 2000.

Table 17.28 *Number of passengers on Tinkerbell Cruise Line, 1990–97*

Year	Number of passengers ('000)
1990	50
1991	58
1992	64
1993	76
1994	84
1995	87
1996	95
1997	104

Source: Hypothetical.

9. (a) For Tinkerbell Cruise Line, calculate the secular trend equation with the least-squares method.
   (b) How many passengers can Captain Hook expect in the year 2000?

Chapter 17 – Time series

10. Calculate the three-year moving average for Tinkerbell Cruise Line. Do you think the moving average more accurately describes the data than the straight-line description found in question 9?

11. Table 17.29 shows the unit turnover of carpets over the 1987–97 period sold by the Flying Carpet Factory.

    (a) Plot the data on a graph.
    (b) Construct a trend line using the semi-average method.
    (c) Use this trend line (drawn in (b)) to estimate the 2002 sales results.

Table 17.29 *Unit turnover for the Flying Carpet Factory, 30 June 1987–97*

Year (30 June)	Units sold ('00)
1987	4.0
1988	4.2
1989	4.5
1990	4.7
1991	5.0
1992	5.8
1993	6.4
1994	6.4
1995	6.5
1996	7.0
1997	7.3

Source: Hypothetical.

12. A directive has just been sent from Head Office that all estimates are to be based on the least-squares method.

    (a) What is the linear equation that describes the trend in the number of carpets sold?
    (b) Is the line drawn from (a) a good fit? Briefly discuss.
    (c) Estimate the number of carpets that will be sold in 2003.
    (d) Do you believe this estimate to be reasonably reliable? Briefly explain.

13. Compare the forecast results for 2002 using the semi-average method and the least-squares method.

14. Calculate the five-year moving averages for the Flying Carpet Factory and comment on the results.

15. (a) Calculate the exponentially smoothed values for the Flying Carpet Factory if $\alpha = 0.2$.
    (b) Draw a line based on (a) and forecast the sales for 1998.

Business mathematics and statistics

16. Table 17.30 lists the sales turnover for Florenti Furnishings for 1983–97
    (a) Plot the data.
    (b) Sketch a trend line using the freehand graphic method.
    (c) Forecast the sales turnover for 1999.

    Table 17.30 *Sales turnover for Florenti Furnishings, 30 June 1983–97*

Year (30 June)	Sales turnover ($ million)
1983	11
1984	17
1985	21
1986	20
1987	19
1988	21
1989	24
1990	29
1991	36
1992	42
1993	47
1994	52
1995	51
1996	50
1997	51

    Source: Hypothetical.

17. On second thought, your employer feels the freehand graphic method is too subjective to rely on for your forecast.
    (a) Calculate the trend line (to the nearest million) by referring to Table 17.30 and using the semi-average method;
    (b) Forecast the sales results for 1999.

18. Your employer is a bit worried about the use of the freehand and semi-average methods and, instead, believes the least-squares method is the most accurate means of describing secular trend.
    (a) Find the equation of the trend line in Table 17.30.
    (b) Is this trend line a good fit? Briefly discuss.
    (c) Using the equation derived in (b), forecast the sales result for 1999.

19. Let's assume that a three-year cycle is appropriate for defining the sales turnover trend of Table 17.30.
    (a) Calculate a three-year moving average.
    (b) Briefly explain the fit you would expect if a five-year rather than a three-year moving average was calculated.
    (c) Calculate the exponentially smoothed values in Table 17.30 and draw the trend line the graph. (Let $\alpha = 0.3$.)

Chapter 17 – Time series

20. (a) Determine the five-year moving average for the data presented in Table 17.31. Graph your results and determine how appropriate a fit it is to the original data.
    (b) Use the least-squares method to calculate the equation of the secular trend.
    (c) Use the equation calculated in (b) to estimate the number of motor vehicles likely to be sold in 2008.
    (d) What considerations should be taken into account before acting on the estimate made in (c)?

Table 17.31 *Motor vehicle sales, 1988–97*

Year	Number of motor vehicles sold	Year	Number of motor vehicles sold
1988	413	1993	357
1989	459	1994	368
1990	365	1995	405
1991	398	1996	390
1992	412	1997	322

Source: Hypothetical.

21.* Captain Hook now believes that it may prove useful to construct a seasonal index using the ratio-to-moving average method. Accordingly, Table 17.32 contains the quarterly results for each of the eight years listed in Table 17.28.

   (a) Calculate the seasonal index using a four-quarter centred moving average (use the arithmetic mean for the average).
   (b) Deseasonalise the 1996 results for each quarter.
   (c) In question 9(b) you calculated the number of passengers expected in 2000. Forecast the number of passengers that the captain can expect for each quarter of 2000.
   (d) Liquor is served on board, with approximately 2000 litres consumed per 10 000 passengers. How many litres should the captain have ordered for each quarter of 1997?

Table 17.32 *Number of passengers ('000s) on Tinkerbell Cruise Line, 1990–97*

Year	Quarter			
	March	June	September	December
1990	14	9	10	17
1991	15	12	11	20
1992	16	14	12	22
1993	19	16	12	29
1994	22	15	13	34
1995	21	17	14	35
1996	23	20	12	40
1997	24	21	15	44

Source: Hypothetical.

*Optional

Business mathematics and statistics

22.*(a) Calculate the seasonal index per quarter (to one decimal place) for the quarterly sales of About-Face Industries listed in Table 17.33 by using the ratio-to-moving average method (use the arithmetic mean for the average).
(b) Deseasonalise each quarter of 1994 and analyse the results.
(c) If About-Face Industries expect sales of $5.8 billion in 2001, what are the forecast sales (to the nearest $10 million) for each quarter of that year?

Table 17.33 *Quarterly sales of About-Face Industries ($ millions), 1991–97*

Year	Quarter			
	March	June	September	December
1991	432	412	441	502
1992	512	510	564	618
1993	602	600	620	670
1994	674	614	669	760
1995	742	704	750	800
1996	781	784	809	850
1997	860	842	889	955

Source: Hypothetical.

23.*(a) Determine the seasonal index (to one decimal place) for the data in Table 17.34 through the use of the ratio-to-moving average method.
(b) Deseasonalise the data for 1997 and briefly comment on the result.

Table 17.34 *Quarterly sales of groceries ($ millions), 1991–97*

Year	Quarter			
	March	June	September	December
1991	1.4	1.4	1.5	1.7
1992	1.6	1.6	1.7	1.9
1993	1.7	2.0	2.0	2.3
1994	2.2	2.3	2.4	2.6
1995	2.4	2.5	2.5	2.8
1996	2.6	2.7	2.8	3.1
1997	2.9	3.0	3.1	3.4

Source: Hypothetical.

*Optional

APPENDIX A
Amount of compound interest

Business mathematics and statistics

Amount at compound interest $(1 + i)^n$

n \ i	$\frac{5}{12}$ % .00416666	$\frac{1}{2}$ % .00500000	$\frac{2}{3}$ % .00666666	$\frac{3}{4}$ % .00750000	1% .01000000	$1\frac{1}{4}$ % .01250000
1	1.00416667	1.00500000	1.00666667	1.00750000	1.01000000	1.01250000
2	1.00835069	1.01002500	1.01337778	1.01505625	1.02010000	1.02515625
3	1.01255216	1.01507513	1.02013363	1.02266917	1.03030100	1.03797070
4	1.01677112	1.02015050	1.02693452	1.03033919	1.04060401	1.05094534
5	1.02100767	1.02525125	1.03378075	1.03806673	1.05101005	1.06408215
6	1.02526187	1.03037751	1.04067262	1.04585224	1.06152015	1.07738318
7	1.02953379	1.03552940	1.04761044	1.05369613	1.07213535	1.09085047
8	1.03382352	1.04070704	1.05459451	1.06159885	1.08285671	1.10448610
9	1.03813111	1.04591058	1.06162514	1.06956084	1.09368527	1.11829218
10	1.04245666	1.05114013	1.06870264	1.07758255	1.10462213	1.13227083
11	1.04680023	1.05639583	1.07582732	1.08566441	1.11566835	1.14642422
12	1.05116190	1.06167781	1.08299951	1.09380690	1.12682503	1.16075452
13	1.05554174	1.06698620	1.09021950	1.10201045	1.13809328	1.17526395
14	1.05993983	1.07232113	1.09748763	1.11027553	1.14947421	1.18995475
15	1.06435625	1.07768274	1.10480422	1.11860259	1.16096896	1.20482918
16	1.06879106	1.08307115	1.11216958	1.12699211	1.17257864	1.21988955
17	1.07324436	1.08848651	1.11958404	1.13544455	1.18430443	1.23513817
18	1.07771621	1.09392894	1.12704794	1.14396039	1.19614748	1.25057739
19	1.08220670	1.09939858	1.13456159	1.15254009	1.20810895	1.26620961
20	1.08671589	1.10489558	1.14212533	1.16118414	1.22019004	1.28203723
21	1.09124387	1.11042006	1.14973950	1.16989302	1.23239194	1.29806270
22	1.09579072	1.11597216	1.15740443	1.17866722	1.24471586	1.31428848
23	1.10035652	1.12155202	1.16512046	1.18750723	1.25716302	1.33071709
24	1.10494134	1.12715978	1.17288793	1.19641353	1.26973465	1.34735105
25	1.10954526	1.13279558	1.18070718	1.20538663	1.28243200	1.36419294
26	1.11416836	1.13845955	1.18857873	1.21442703	1.29525631	1.38124535
27	1.11881073	1.14415185	1.19650242	1.22353523	1.30820888	1.39851092
28	1.12347244	1.14987261	1.20447911	1.23271175	1.32129097	1.41599230
29	1.12815358	1.15562197	1.21250897	1.24195709	1.33450388	1.43369221
30	1.13285422	1.16140008	1.22059236	1.25127176	1.34784892	1.45161336
31	1.13757444	1.16720708	1.22872964	1.26065630	1.36132740	1.46975853
32	1.14231434	1.17304312	1.23692117	1.27011122	1.37494068	1.48813051
33	1.14707398	1.17890833	1.24516731	1.27963706	1.38869009	1.50673214
34	1.15185346	1.18480288	1.25346843	1.28923434	1.40257699	1.52556629
35	1.15665284	1.19072689	1.26182489	1.29890359	1.41660276	1.54463587
36	1.16147223	1.19668052	1.27023705	1.30864537	1.43076878	1.56394382
37	1.16631170	1.20266393	1.27870530	1.31846021	1.44507647	1.58349312
38	1.17117133	1.20867725	1.28723000	1.32834866	1.45952724	1.60328678
39	1.17605121	1.21472063	1.29581153	1.33831128	1.47412251	1.62332787
40	1.18095142	1.22079424	1.30445028	1.34834861	1.48886373	1.64361946
41	1.18587206	1.22689821	1.31314661	1.35846123	1.50375237	1.66416471
42	1.19081319	1.23303270	1.32190092	1.36864969	1.51878989	1.68496677
43	1.19577491	1.23919786	1.33071360	1.37891459	1.53397779	1.70602885
44	1.20075731	1.24539385	1.33958502	1.38925642	1.54931757	1.72734521
45	1.20576046	1.25162082	1.34851559	1.39967584	1.56481075	1.74894614
46	1.21078446	1.25787892	1.35750569	1.41017341	1.58045885	1.77080797
47	1.21582940	1.26416832	1.36655573	1.42074971	1.59626344	1.79294306
48	1.22089536	1.27048916	1.37566610	1.43140533	1.61222608	1.81535485
49	1.22598242	1.27684161	1.38483721	1.44214087	1.62834834	1.83804679
50	1.23109068	1.28322581	1.39406946	1.45295693	1.64463182	1.86102237
51	1.23622022	1.28964194	1.40336325	1.46385411	1.66107814	1.88428515
52	1.24137114	1.29609015	1.41271901	1.47483301	1.67768892	1.90783872
53	1.24654352	1.30257060	1.42213713	1.48589426	1.69446581	1.93168670
54	1.25173745	1.30908346	1.43161805	1.49703847	1.71141047	1.95583279
55	1.25695302	1.31562887	1.44116217	1.50826626	1.72852457	1.98028070
56	1.26219033	1.32220702	1.45076992	1.51957825	1.74580982	2.00503420
57	1.26744946	1.32881805	1.46044172	1.53097509	1.76326792	2.03009713
58	1.27273050	1.33546214	1.47017799	1.54245740	1.78090060	2.05547335
59	1.27803354	1.34213946	1.47997918	1.55402383	1.79870960	2.08116676
60	1.28335868	1.3485015	1.48984571	1.56568103	1.81669670	2.10718135

Appendix A

Amount at compound interest $(1 + i)^n$—(continued)

n \ i	$1\frac{1}{2}\%$ .01500000	$1\frac{3}{4}\%$ .01750000	2% .02000000	$2\frac{1}{4}\%$ .02250000	$2\frac{1}{2}\%$ .02500000	$2\frac{3}{4}\%$ .02750000
1	1.01500000	1.01750000	1.02000000	1.02250000	1.02500000	1.02750000
2	1.03022500	1.03530625	1.04040000	1.04550625	1.05062500	1.05575625
3	1.04567838	1.05342411	1.06120800	1.06903014	1.07689062	1.08478955
4	1.06136355	1.07185903	1.08243216	1.09308332	1.10381289	1.11462126
5	1.07728400	1.09061656	1.10408080	1.11767769	1.13140821	1.14527334
6	1.09344326	1.10970235	1.12616242	1.14282544	1.15969342	1.17676836
7	1.10984491	1.12912215	1.14868567	1.16853901	1.18868575	1.20912949
8	1.12649259	1.14888178	1.17165938	1.19483114	1.21840290	1.24238055
9	1.14338998	1.16898721	1.19509257	1.22171484	1.24886297	1.27654602
10	1.16054083	1.18944449	1.21899442	1.24920343	1.28008454	1.31165103
11	1.17794894	1.21025977	1.24337431	1.27731050	1.31208666	1.34772144
12	1.19561817	1.23143931	1.26824179	1.30604999	1.34488882	1.38478378
13	1.21355244	1.25298950	1.29360663	1.33543611	1.37851104	1.42286533
14	1.23175573	1.27491682	1.31947876	1.36548343	1.41297382	1.46199413
15	1.25023207	1.29722786	1.34586834	1.39620680	1.44829817	1.50219896
16	1.26898555	1.31992935	1.37278571	1.42762146	1.48450562	1.54350944
17	1.28802033	1.34302811	1.40024142	1.45974294	1.52161826	1.58595595
18	1.30734064	1.36653111	1.42824625	1.49258716	1.55965872	1.62956973
19	1.32695075	1.39044540	1.45681117	1.52617037	1.59865019	1.67438290
20	1.34685501	1.41477820	1.48594740	1.56050920	1.63861644	1.72042843
21	1.36705783	1.43953681	1.51566634	1.59562066	1.67958185	1.76774021
22	1.38756370	1.46472871	1.54597967	1.63152212	1.72157140	1.81635307
23	1.40837715	1.49036146	1.57689926	1.66823137	1.76461068	1.86630278
24	1.42950281	1.51644279	1.60843725	1.70576626	1.80872595	1.91762610
25	1.45094535	1.54298054	1.64060599	1.74414632	1.85394410	1.97036082
26	1.47270953	1.56998269	1.67341811	1.78338962	1.90029270	2.02454575
27	1.49480018	1.59745739	1.70688648	1.82351588	1.94780002	2.08022075
28	1.51722218	1.62541290	1.74102421	1.86454499	1.99649502	2.13742682
29	1.53998051	1.65385762	1.77584469	1.90649725	2.04640739	2.19620606
30	1.56308022	1.68280013	1.81136158	1.94939344	2.09756758	2.25660173
31	1.58652642	1.71224913	1.84758882	1.99325479	2.15000677	2.31865828
32	1.61032432	1.74221349	1.88454059	2.03810303	2.20375594	2.38242138
33	1.63447918	1.77270223	1.92223140	2.08396034	2.25885086	2.44793797
34	1.65899637	1.80372452	1.96067603	2.13084945	2.31532213	2.51525626
35	1.68388132	1.83528970	1.99988955	2.17879356	2.37320519	2.58442581
36	1.70913954	1.86740727	2.03988734	2.22781642	2.43253532	2.65549752
37	1.73477663	1.90008689	2.08068509	2.27794229	2.49334870	2.72852370
38	1.76079828	1.93333841	2.12229879	2.32919599	2.55568242	2.80355810
39	1.78721025	1.96717184	2.16474477	2.38160290	2.61957448	2.88065595
40	1.81401841	2.00159734	2.20803966	2.43518897	2.68506384	2.95987399
41	1.84122868	2.03662530	2.25220046	2.48998072	2.75219043	3.04127052
42	1.86884712	2.07226624	2.29724447	2.54600528	2.82099520	3.12490546
43	1.89687982	2.10853090	2.34318936	2.60329040	2.89152008	3.21084036
44	1.92533302	2.14543019	2.39005314	2.66186444	2.96380808	3.29913847
45	1.95421301	2.18297522	2.43785421	2.72175639	3.03790328	3.38986478
46	1.98352621	2.22117728	2.48661129	2.78299590	3.11385086	3.48308606
47	2.01327910	2.26004789	2.53634352	2.84561331	3.19169713	3.57887093
48	2.04347829	2.29959872	2.58707039	2.90963961	3.27148956	3.67728988
49	2.07413046	2.33984170	2.63881179	2.97510650	3.35327680	3.77841535
50	2.10524242	2.38078893	2.69158803	3.04204640	3.43710872	3.88232177
51	2.13682106	2.42245274	2.74541979	3.11049244	3.52303644	3.98908562
52	2.16887337	2.46484566	2.80032819	3.18047852	3.61111235	4.09878547
53	2.20140647	2.50798046	2.85633475	3.25203929	3.70139016	4.21150208
54	2.23442757	2.55187012	2.91346144	3.32521017	3.79392491	4.32731838
55	2.26794398	2.59652785	2.97173067	3.40002740	3.88877303	4.44631964
56	2.30196314	2.64196708	3.03116529	3.47652802	3.98599236	4.56859343
57	2.33649259	2.68820151	3.09178859	3.55474990	4.08564217	4.69422975
58	2.37153998	2.73524503	3.15362436	3.63473177	4.18778322	4.82332107
59	2.40711308	2.78311182	3.21669685	3.71651324	4.29247780	4.95596239
60	2.44321978	2.83181628	3.28103079	3.80013479	4.39978975	5.09225136

Amount at compound interest $(1 + i)^n$—(continued)

n \ i	3% .03000000	$3\frac{1}{2}$% .03500000	4% .04000000	$4\frac{1}{2}$% .04500000	5% .05000000	6% .06000000
1	1.03000000	1.03500000	1.04000000	1.04500000	1.05000000	1.06000000
2	1.06090000	1.07122500	1.08160000	1.09202500	1.10250000	1.12360000
3	1.09272700	1.10871787	1.12486400	1.14116613	1.15762500	1.19101600
4	1.12550881	1.14752300	1.16985856	1.19251860	1.21550625	1.26247696
5	1.15927407	1.18768631	1.21665290	1.24618194	1.27628156	1.33822558
6	1.19405230	1.22925533	1.26531902	1.30226012	1.34009564	1.41851911
7	1.22987387	1.27227926	1.31593178	1.36086183	1.40710042	1.50363026
8	1.26677008	1.31680904	1.36856905	1.42210061	1.47745544	1.59384807
9	1.30477318	1.36289735	1.42331181	1.48609514	1.55132822	1.68947896
10	1.34391638	1.41059876	1.48024428	1.55296942	1.62889463	1.79084770
11	1.38423387	1.45996972	1.53945406	1.62285305	1.71033936	1.89829856
12	1.42576089	1.51106866	1.60103222	1.69588143	1.79585633	2.01219647
13	1.46853371	1.56395606	1.66507351	1.77219610	1.88564914	2.13292826
14	1.51258972	1.61869452	1.73167645	1.85194492	1.97993160	2.26090396
15	1.55796742	1.67534883	1.80094351	1.93528244	2.07892818	2.39655819
16	1.60470644	1.73398604	1.87298125	2.02237015	2.18287459	2.54035168
17	1.65284763	1.79467555	1.94790050	2.11337681	2.29201832	2.69277279
18	1.70243306	1.85748920	2.02581652	2.20847877	2.40661923	2.85433915
19	1.75350605	1.92250132	2.10684918	2.30786031	2.52695020	3.02559950
20	1.80611123	1.98978886	2.19112314	2.41171402	2.65329771	3.20713547
21	1.86029457	2.05943147	2.27876807	2.52024116	2.78596259	3.39956360
22	1.91610341	2.13151158	2.36991879	2.63365201	2.92526072	3.60353742
23	1.97358651	2.20611448	2.46471554	2.75216635	3.07152376	3.81974966
24	2.03279411	2.28332849	2.56330416	2.87601383	3.22509994	4.04893464
25	2.09377793	2.36324498	2.66583633	3.00543446	3.38635494	4.29187072
26	2.15659127	2.44595856	2.77246978	3.14067901	4.55567269	4.54938296
27	2.22128901	2.53156711	2.88336858	3.28200956	3.73345632	4.82234594
28	2.28792768	2.62017196	2.99870332	3.42969999	3.92012914	5.11168670
29	2.35656551	2.71187798	3.11865145	3.58403649	4.11613560	5.41838790
30	2.42726247	2.80679370	3.24339751	3.74531813	4.32194238	5.74349117
31	2.50008035	2.90503148	3.37313341	3.91385745	4.53803949	6.08810064
32	2.57508276	3.00670759	3.50805875	4.08998104	4.76494147	6.45338668
33	2.65233524	3.11194235	3.64838110	4.27403018	5.00318854	6.84058988
34	2.73190530	3.22086033	3.79431634	4.46636154	5.25334797	7.25102528
35	2.81386245	3.33359045	3.94608899	4.66734781	5.51601537	7.68608679
36	2.89827833	3.45026611	4.10393255	4.87737846	5.79181614	8.14725200
37	2.98522668	3.57102543	4.26808986	5.09686049	6.08140694	8.63608712
38	3.07478348	3.69601132	4.43881345	5.32621921	6.38547729	9.15425235
39	3.16702698	3.82537171	4.61636599	5.56589908	6.70475115	9.70350749
40	3.26203779	3.95925972	4.80102063	5.81636454	7.03998871	10.28571794
41	3.35989893	4.09783381	4.99306145	6.07810094	7.39198815	10.90286101
42	3.46069589	4.24125799	5.19278391	6.35161548	7.76158756	11.55703267
43	3.56451677	4.38970202	5.40049527	6.63743818	8.14966693	12.25045463
44	3.67145227	4.54334160	5.61651508	6.93612290	8.55715028	12.98548191
45	3.78159584	4.70235855	5.84117568	7.24824843	8.98500779	13.76461083
46	3.89504372	4.86694110	6.07482271	7.57441961	9.43425818	14.59048748
47	4.01189503	5.03728404	6.31781562	7.91526849	9.90597109	15.46591673
48	4.13225188	5.21358898	6.57052824	8.27145557	10.40126965	16.39387173
49	4.25621944	5.39606459	6.83334937	8.64367107	10.92133313	17.37750403
50	4.38390602	5.58492686	7.10668335	9.03263627	11.46739979	18.42015427
51	4.51542320	5.78039930	7.39095068	9.43910490	12.04076978	19.52536353
52	4.65088590	5.98271327	7.68658871	9.86386463	12.64280826	20.69688534
53	4.79041247	6.19210824	7.99405226	10.30773853	13.27494868	21.93869846
54	4.93412485	6.40883202	8.31381435	10.77158677	13.93869611	23.25502037
55	5.08214859	6.63314114	8.64636692	11.25630817	14.63563092	24.65032159
56	5.23461305	6.86530108	8.99222160	11.76284204	15.36741246	26.12934089
57	5.39165144	7.10558662	9.35191046	12.29216993	16.13578309	27.69710134
58	5.55340098	7.35428215	9.72598688	12.84531758	16.94257221	29.35892742
59	5.72000301	7.61168203	10.11502635	13.42335687	17.78870085	31.12046307
60	5.89160310	7.87809090	10.51962741	14.02740793	18.67918589	32.98769085

Appendix A

Amount at compound interest $(1 + i)^n$—*(continued)*

n \ i	7% .07000000	8% .08000000	9% .09000000	10% .10000000	11% .11000000	12% .12000000
1	1.07000000	1.08000000	1.09000000	1.10000000	1.11000000	1.12000000
2	1.14490000	1.16640000	1.18810000	1.21000000	1.23210000	1.25440000
3	1.22504300	1.25971200	1.29502900	1.33100000	1.36763100	1.40492800
4	1.31079601	1.36048896	1.41158161	1.46410000	1.51807041	1.57351936
5	1.40255173	1.46932808	1.53862395	1.61051000	1.68505816	1.76234168
6	1.50073035	1.58687432	1.67710011	1.77156100	1.87041455	1.97382269
7	1.60578148	1.71382427	1.82803912	1.94871710	2.07616015	2.21068141
8	1.71818618	1.85093021	1.99256264	2.14358881	2.30453777	2.47596318
9	1.83845921	1.99900463	2.17189328	2.35794769	2.55803692	2.77307876
10	1.96715136	2.15892500	2.36736367	2.59374246	2.83942099	3.10584821
11	2.10485195	2.33163900	2.58042641	2.85311671	3.15175729	3.47854999
12	2.25219159	2.51817012	2.81266478	3.13842838	3.49845060	3.89597599
13	2.40984500	2.71962373	3.06580461	3.45227121	3.88328016	4.36349311
14	2.57853415	2.93719362	3.34172703	3.79749834	4.31044098	4.88711229
15	2.75903154	3.17216911	3.64248246	4.17724817	4.78458949	5.47356576
16	2.95216375	3.42594264	3.97030588	4.59497299	5.31089433	6.13039365
17	3.15881521	3.70001805	4.32763341	5.05447028	5.89509271	6.86604089
18	3.37993228	3.99601950	4.71712042	5.55991731	6.54355291	7.68996580
19	3.61652754	4.31570106	5.14166125	6.11590904	7.26334373	8.61276169
20	3.86968446	4.66095714	5.60441077	6.72749995	8.06231154	9.64629309
21	4.14056237	5.03383372	6.10880774	7.40024994	8.94916581	10.80384826
22	4.43040174	5.43654041	6.65860043	8.14027494	9.93357404	12.10031006
23	4.74052986	5.87146365	7.25787447	8.95430243	11.02626719	13.55234726
24	5.07236695	6.34118074	7.91108317	9.84973268	12.23915658	15.17862893
25	5.42743264	6.84847520	8.62308066	10.83470594	13.58546380	17.00006441
26	5.80735292	7.39635321	9.39915792	11.91817654	15.07986482	19.04007214
27	6.21386763	7.98806147	10.24508213	13.10999419	16.73864995	21.32488079
28	6.64883836	8.62710639	11.16713952	14.42099361	18.57990145	23.88386649
29	7.11425705	9.31727490	12.17218208	15.86309297	20.62369061	26.74993047
30	7.61225504	10.06265689	13.26767847	17.44940227	22.89229657	29.95992212
31	8.14511290	10.86766944	14.46176953	19.19434250	25.41044919	33.55511278
32	8.71527080	11.73708300	15.76332879	21.11377675	28.20559861	37.58172631
33	9.32533975	12.67604964	17.18202838	23.22515442	31.30821445	42.09153347
34	9.97811354	13.69013361	18.72841093	25.54766986	34.75211804	47.14251748
35	10.67658148	14.78534429	20.41396792	28.10243695	38.57485103	52.79961958
36	11.42394219	15.96817184	22.25122503	30.91268053	42.81808464	59.13557393
37	12.22361814	17.24562558	24.25383528	34.00394859	47.52807395	66.23184280
38	13.07927141	18.62527563	26.43668046	37.40434342	52.75616209	74.17966394
39	13.99482041	20.11529768	28.81598170	41.14477779	58.55933991	83.08122361
40	14.97445784	21.72452150	31.40942005	45.25925557	65.00086731	93.05097044
41	16.02266989	23.46248322	34.23626786	49.78518112	72.15096271	104.21708689
42	17.14425678	25.33948187	37.31753197	54.76369924	80.08756861	116.72313732
43	18.34435475	27.36664042	40.67610984	60.24006916	88.89720115	130.72991380
44	19.62845959	29.55597166	44.33695973	66.26407608	98.67589328	146.41750346
45	21.00245176	31.92044939	48.32728610	72.89048369	109.53024154	163.98760387
46	22.47262338	34.47408534	52.67674185	80.17953205	121.57856811	183.66611634
47	24.04570702	37.23201217	57.41764862	88.19748526	134.95221060	205.70605030
48	25.72890651	40.21057314	62.58523700	97.01723378	149.79695377	230.39077633
49	27.52992997	43.42741899	68.21790833	106.71895716	166.27461868	258.03766949
50	29.45702506	46.90161251	74.35752008	117.39085288	184.56482674	289.00218983
51	31.51901682	50.65374151	81.04969688	129.12993817	204.86695768	323.68245261
52	33.72534799	54.70604084	88.34416960	142.04293198	227.40232303	362.52434692
53	36.08612235	59.08252410	96.29514487	156.24722518	252.41657856	406.02726855
54	38.61215092	63.80912603	104.96170790	171.87194770	280.18240220	454.75054078
55	41.31500148	68.91385611	114.40826162	189.05914247	311.00246644	509.32060567
56	44.20705159	74.42696460	124.70500516	207.96505672	345.21273775	570.43907835
57	47.30154520	80.38112177	135.92845563	228.76156239	383.18613890	638.89176776
58	50.61265336	86.81161151	148.16201663	251.63771863	425.33661418	715.55877989
59	54.15553910	93.75654043	161.49659813	276.80149049	472.12364174	801.42583347
60	57.94642683	101.25706367	176.03129196	304.48163954	524.05724234	897.59693349

APPENDIX B
Present value at compound interest

Present value at compound interest $(1 + i)^{-n}$

n \ i	$\frac{5}{12}\%$ .00416666	$\frac{1}{2}\%$ .00500000	$\frac{2}{3}\%$ .00666666	$\frac{3}{4}\%$ .00750000	1% .01000000	$1\frac{1}{4}\%$ .01250000
1	.99585062	.99502488	.99337748	.99255583	.99009901	.98765432
2	.99171846	.99007450	.98679882	.98516708	.98029605	.97546106
3	.98760345	.98514876	.98026373	.97783333	.97059015	.96341833
4	.98350551	.98024752	.97377192	.97055417	.96098034	.95152428
5	.97942457	.97537067	.96732310	.96332920	.95146569	.93977706
6	.97536057	.97051808	.96091699	.95615802	.94204524	.92817488
7	.97131343	.96568963	.95455330	.94904022	.93271805	.91671593
8	.96728308	.96088520	.94823175	.94197540	.92348322	.90539845
9	.96326946	.95610468	.94195207	.93496318	.91433982	.89422069
10	.95927249	.95134794	.93571398	.92800315	.90528695	.88318093
11	.95529211	.94661487	.92951720	.92109494	.89632372	.87227746
12	.95132824	.94190534	.92336145	.91423815	.88744923	.86150860
13	.94738082	.93721924	.91724648	.90743241	.87866260	.85087269
14	.94344978	.93255646	.91117200	.90067733	.86996297	.84036809
15	.93953505	.92791688	.90513775	.89397254	.86134947	.82999318
16	.93563657	.92330037	.89914346	.88731766	.85282126	.81974635
17	.93175426	.91870684	.89318886	.88071231	.84437749	.80962602
18	.92788806	.91413616	.88727371	.87415614	.83601731	.79963064
19	.92403790	.90958822	.88139772	.86764878	.82773992	.78975866
20	.92020372	.90506290	.87556065	.86118985	.81954447	.78000855
21	.91638544	.90056010	.86976224	.85477901	.81143017	.77037881
22	.91258301	.89607971	.86400222	.84841589	.80339621	.76086796
23	.90879636	.89162160	.85828035	.84210014	.79544179	.75147453
24	.90502542	.88718567	.85259638	.83583140	.78756613	.74219707
25	.90127013	.88277181	.84695004	.82960933	.77976844	.73303414
26	.89753042	.87837991	.84134110	.82343358	.77204796	.72398434
27	.89380623	.87400986	.83576931	.81730380	.76440392	.71504626
28	.89009749	.86966155	.83023441	.81121966	.75683557	.70621853
29	.88640414	.86533488	.82473617	.80518080	.74934215	.69749978
30	.88272611	.86102973	.81927434	.79918690	.74192292	.68888867
31	.87906335	.85674600	.81384868	.79323762	.73457715	.68038387
32	.87541578	.85248358	.80845896	.78733262	.72730411	.67198407
33	.87178335	.84824237	.80310492	.78147158	.72010307	.66368797
34	.86816599	.84402226	.79778635	.77565418	.71297334	.65549429
35	.86456365	.83982314	.79250299	.76988008	.70591420	.64740177
36	.86097624	.83564492	.78725463	.76414896	.69892495	.63940916
37	.85740373	.83148748	.78204102	.75846051	.69200490	.63151522
38	.85384604	.82735073	.77686194	.75281440	.68515337	.62371873
39	.85030311	.82323455	.77171716	.74721032	.67836967	.61601850
40	.84677488	.81913886	.76660645	.74164796	.67165314	.60841334
41	.84326129	.81506354	.76152959	.73612701	.66500311	.60090206
42	.83976228	.81100850	.75648635	.73064716	.65841892	.59348352
43	.83627779	.80697363	.75147650	.72520809	.65189992	.58615656
44	.83280776	.80295884	.74649984	.71980952	.64544546	.57892006
45	.82935212	.79896402	.74155613	.71445114	.63905492	.57177290
46	.82591083	.79498907	.73664516	.70913264	.63272764	.56471397
47	.82248381	.79103390	.73176672	.70385374	.62646301	.55774219
48	.81907102	.78709841	.72692058	.69861414	.62026041	.55085649
49	.81567238	.78318250	.72210654	.69341353	.61411921	.54405579
50	.81228785	.77928607	.71732437	.68825165	.60803882	.53733905
51	.80891736	.77540902	.71257388	.68312819	.60201864	.53070524
52	.80556086	.77155127	.70785485	.67804286	.59605806	.52415332
53	.80221828	.76771270	.70316707	.67299540	.59015649	.51768229
54	.79888957	.76389324	.69851033	.66798551	.58431336	.51129115
55	.79557468	.76009277	.69388444	.66301291	.57852808	.50497892
56	.79227354	.75631122	.68928918	.65807733	.57280008	.49874461
57	.78898610	.75254847	.68472435	.65317849	.56712879	.49258727
58	.78571230	.74880445	.68018975	.64831612	.56151365	.48650594
59	.78245208	.74507906	.67568518	.64348995	.55593411	.48049970
60	.77920539	.74137220	.67121044	.63869970	.55044962	.47456760

Appendix B

Present value at compound interest $(1 + i)^{-n}$—(continued)

n \ i	$1\frac{1}{2}\%$ .01500000	$1\frac{3}{4}\%$ .01750000	2% .02000000	$2\frac{1}{4}\%$ .02250000	$2\frac{1}{2}\%$ .02500000	$2\frac{3}{4}\%$ .02750000
1	.98522167	.98280098	.98039216	.97799511	.97560976	.97323601
2	.97066175	.96589777	.96116878	.95647444	.95181440	.94718833
3	.95631699	.94928528	.94232233	.93542732	.92859941	.92183779
4	.94218423	.93295851	.92384543	.91484335	.90595064	.89716573
5	.92826033	.91691254	.90573081	.89471232	.88385429	.87315400
6	.91454219	.90114254	.88797138	.87502427	.86229687	.84978491
7	.90102679	.88564378	.87056018	.85576946	.84126524	.82704128
8	.88771112	.87041157	.85349037	.83693835	.82074657	.80490635
9	.87459224	.85544135	.83675527	.81852161	.80072836	.78336385
10	.86166723	.84072860	.82034830	.80051013	.78119840	.76239791
11	.84893323	.82626889	.80426304	.78289499	.76214478	.74199310
12	.83638742	.81205788	.78849318	.76566748	.74355589	.72213440
13	.82402702	.79809128	.77303253	.74881905	.72542038	.70280720
14	.81184928	.78436490	.75787502	.73234137	.70772720	.68399728
15	.79985150	.77087459	.74301473	.71622628	.69046556	.66569078
16	.78803104	.75761631	.72844581	.70046580	.67362493	.64787424
17	.77638526	.74458605	.71416256	.68505212	.65719506	.63053454
18	.76491159	.73177990	.70015937	.66997763	.64116591	.61365892
19	.75360747	.71919401	.68643076	.65523484	.62552772	.59723496
20	.74247042	.70682458	.67297133	.64081647	.61027094	.58125057
21	.73149795	.69466789	.65977582	.62671538	.59538629	.56569398
22	.72068763	.68272028	.64683904	.61292457	.58086467	.55055375
23	.71003708	.67097817	.63415592	.59943724	.56669724	.53581874
24	.69954392	.65943800	.62172149	.58624668	.55287535	.52147809
25	.68920583	.64809632	.60953087	.57334639	.53939059	.50752126
26	.67902052	.63694970	.59757928	.56072997	.52623472	.49393796
27	.66898574	.62599479	.58586204	.54839117	.51339973	.48071821
28	.65909925	.61522829	.57437455	.53632388	.50087778	.46785227
29	.64935887	.60464697	.56311231	.52452213	.48866125	.45533068
30	.63976243	.59424764	.55207089	.51298008	.47674269	.44314421
31	.63030781	.58402716	.54124597	.50169201	.46511481	.43128391
32	.62099292	.57398247	.53063330	.49065233	.45377055	.41974103
33	.61181568	.56411053	.52022873	.47985558	.44270298	.40850708
34	.60277407	.55440839	.51002817	.46929641	.43190534	.39757380
35	.59386608	.54487311	.50002761	.45896960	.42137107	.38693314
36	.58508974	.53550183	.49022315	.44887002	.41109372	.37657727
37	.57644309	.52629172	.48061093	.43899268	.40106705	.36649856
38	.56792423	.51724002	.47118719	.42933270	.39128492	.35668959
39	.55953126	.50834400	.46194822	.41988528	.38174139	.34714316
40	.55126232	.49960098	.45289042	.41064575	.37243062	.33785222
41	.54311559	.49100834	.44401021	.40160954	.36334695	.32880995
42	.53508925	.48256348	.43530413	.39277216	.35448483	.32000968
43	.52718153	.47426386	.42676875	.38412925	.34583886	.31144495
44	.51939067	.46610699	.41840074	.37567653	.33740376	.30310944
45	.51171494	.45809040	.41019680	.36740981	.32917440	.29499702
46	.50415265	.45021170	.40215373	.35932500	.32114576	.28710172
47	.49670212	.44246850	.39426836	.35141809	.31331294	.27941773
48	.48936170	.43485848	.38653761	.34368518	.30567116	.27193940
49	.48212975	.42737934	.37895844	.33612242	.29821576	.26466122
50	.47500468	.42002883	.37152788	.32872608	.29094221	.25757783
51	.46798491	.41280475	.36424302	.32149250	.28384606	.25068402
52	.46106887	.40570492	.35710100	.31441810	.27692298	.24397471
53	.45425505	.39872719	.35009902	.30749936	.27016876	.23744497
54	.44754192	.39186947	.34323433	.30073287	.26357928	.23109000
55	.44092800	.38512970	.33650425	.29411528	.25715052	.22490511
56	.43441182	.37850585	.32990613	.28764330	.25087855	.21888575
57	.42799194	.37199592	.32343738	.28131374	.24475956	.21302749
58	.42166694	.36559796	.31709547	.27512347	.23878982	.20732603
59	.41543541	.35931003	.31087791	.26906940	.23296568	.20177716
60	.40929597	.35313025	.30478227	.26314856	.22728359	.19637679

Present value at compound interest $(1 + i)^{-n}$—*(continued)*

n \ i	3% .03000000	$3\frac{1}{2}$% .03500000	4% .04000000	$4\frac{1}{2}$% .04500000	5% .05000000	6% .06000000
1	.97087379	.96618357	.96153846	.95693780	.95238095	.94339623
2	.94259591	.93351070	.92455621	.91572995	.90702948	.88999644
3	.91514166	.90194271	.88899636	.87629660	.86383760	.83961928
4	.88848705	.87144223	.85480419	.83856134	.82270247	.79209366
5	.86260878	.84197317	.82192711	.80245105	.78352617	.74725817
6	.83748426	.81350064	.79031453	.76789574	.74621540	.70496054
7	.81309151	.78599096	.75991781	.73482846	.71068133	.66505711
8	.78940923	.75941156	.73069021	.70318513	.67683936	.62741237
9	.76641673	.73373097	.70258674	.67290443	.64460892	.59189846
10	.74409391	.70891881	.67556417	.64392768	.61391325	.55839478
11	.72242128	.68494571	.64958093	.61619874	.58467929	.52678753
12	.70137988	.66178330	.62459705	.58966386	.55683742	.49696936
13	.68095134	.63940415	.60057409	.56427164	.53032135	.46883902
14	.66111781	.61778179	.57747508	.53997286	.50506795	.44230096
15	.64186195	.59689062	.55526450	.51672044	.48101710	.41726506
16	.62316694	.57670591	.53390818	.49446932	.45811152	.39364628
17	.60501645	.55720378	.51337325	.47317639	.43629669	.37136442
18	.58739461	.53836114	.49362812	.45280037	.41552065	.35034379
19	.57028603	.52015569	.47464242	.43330179	.39573396	.33051301
20	.55367575	.50256588	.45638695	.41464286	.37688948	.31180473
21	.53754928	.48557090	.43883360	.39678743	.35894236	.29415540
22	.52189250	.46915063	.42195539	.37970089	.34184987	.27750510
23	.50669175	.45328563	.40572633	.36335013	.32557131	.26179726
24	.49193374	.43795713	.39012147	.34770347	.31006791	.24697855
25	.47760557	.42314699	.37511680	.33273060	.29530277	.23299863
26	.46369473	.40883767	.36068923	.31840248	.28124073	.21981003
27	.45018906	.39501224	.34681657	.30469137	.26784832	.20736795
28	.43707675	.38165434	.33347747	.29157069	.25509364	.19563014
29	.42434636	.36874815	.32065141	.27901502	.24294632	.18455674
30	.41198676	.35627841	.30831867	.26700002	.23137745	.17411013
31	.39998715	.34423035	.29646026	.25550241	.22035947	.16425484
32	.38833703	.33258971	.28505794	.24449991	.20986617	.15495740
33	.37702625	.32134271	.27409417	.23397121	.19987254	.14618622
34	.36604490	.31047605	.26355209	.22389589	.19035480	.13791153
35	.35538340	.29997686	.25341547	.21425444	.18129029	.13010522
36	.34503243	.28983272	.24366872	.20502817	.17265741	.12274077
37	.33498294	.28003161	.23429685	.19619921	.16443563	.11579318
38	.32522615	.27056194	.22528543	.18775044	.15660536	.10923885
39	.31575355	.26141250	.21662061	.17966549	.14914797	.10305552
40	.30655684	.25257247	.20828904	.17192870	.14204568	.09722219
41	.29762800	.24403137	.20027793	.16452507	.13528160	.09171905
42	.28895922	.23577910	.19257493	.15744026	.12883962	.08652740
43	.28054294	.22780594	.18516820	.15066054	.12270440	.08162962
44	.27237178	.22010231	.17804635	.14417276	.11686133	.07700908
45	.26443862	.21265924	.17119841	.13796437	.11129651	.07265007
46	.25673653	.20546787	.16461834	.13202332	.10599668	.06853781
47	.24925876	.19851968	.15828256	.12633810	.10094921	.06465831
48	.24199880	.19180645	.15219476	.12089771	.09614211	.06099840
49	.23495029	.18532024	.14634112	.11569158	.09156391	.05754566
50	.22810708	.17905337	.14071262	.11070965	.08720373	.05428836
51	.22146318	.17299843	.13530059	.10594225	.08305117	.05121544
52	.21501280	.16714824	.13009672	.10138014	.07909635	.04831645
53	.20875029	.16149589	.12509300	.09701449	.07532986	.04558156
54	.20267019	.15603467	.12028173	.09283683	.07174272	.04300147
55	.19676717	.15075814	.11565551	.08883907	.06832640	.04056742
56	.19103609	.14566004	.11120722	.08501347	.06507276	.03827115
57	.18547193	.14073433	.10693002	.08135260	.06197406	.03610486
58	.18006984	.13597520	.10281733	.07784938	.05902291	.03406119
59	.17482508	.13137701	.09886282	.07449701	.05621236	.03213320
60	.16973309	.12693431	.09506040	.07128901	.05353552	.03031434

Appendix B

Amount at compound interest $(1 + i)^{-n}$—*(continued)*

n \ i	7% .07000000	8% .08000000	9% .09000000	10% .10000000	11% .11000000	12% .12000000
1	.93457944	.92592593	.91743119	.90909091	.90090090	.89285714
2	.87343873	.85733882	.84167999	.82644628	.81162243	.79719388
3	.81629788	.79383224	.77218348	.75131480	.73119138	.71178025
4	.76289521	.73502985	.70842521	.68301346	.65873097	.63551808
5	.71298618	.68058320	.64993139	.62092132	.59345133	.56742686
6	.66634222	.63016963	.59626733	.56447393	.53464084	.50663112
7	.62274974	.58349040	.54703424	.51315812	.48165841	.45234922
8	.58200910	.54026888	.50186628	.46650738	.43392650	.40388323
9	.54393374	.50024897	.46042778	.42409762	.39092477	.36061002
10	.50834929	.46319349	.42241081	.38554329	.35218448	.32197324
11	.47509280	.42888286	.38753285	.35049390	.31728331	.28747610
12	.44401196	.39711376	.35553473	.31863082	.28584082	.25667509
13	.41496445	.36769792	.32617865	.28966438	.25751426	.22917419
14	.38781724	.34046104	.29924647	.26333125	.23199482	.20461981
15	.36244602	.31524170	.27453804	.23939205	.20900435	.18269626
16	.33873460	.29189047	.25186976	.21762914	.18829220	.16312166
17	.31657439	.27026895	.23107318	.19784467	.16963262	.14564434
18	.29586392	.25024903	.21199374	.17985879	.15282218	.13003959
19	.27650833	.23171206	.19448967	.16350799	.13767764	.11610678
20	.25841900	.21454821	.17843089	.14864363	.12403391	.10366677
21	.24151309	.19865575	.16369806	.13513057	.11174226	.09255961
22	.22571317	.18394051	.15018171	.12284597	.10066870	.08264251
23	.21094688	.17031528	.13778139	.11167816	.09069252	.07378796
24	.19714662	.15769934	.12640494	.10152560	.08170498	.06588210
25	.18424918	.14601790	.11596784	.09229600	.07360809	.05882331
26	.17219549	.13520176	.10639251	.08390545	.06631359	.05252081
27	.16093037	.12518682	.09760781	.07627768	.05974197	.04689358
28	.15040221	.11591372	.08954845	.06934335	.05382160	.04186927
29	.14056282	.10732752	.08215454	.06303941	.04848793	.03738327
30	.13136712	.09937733	.07537114	.05730855	.04368282	.03337792
31	.12277301	.09201605	.06914783	.05209868	.03935389	.02980172
32	.11474113	.08520005	.06343838	.04736244	.03545395	.02660868
33	.10723470	.07888893	.05820035	.04305676	.03194050	.02375775
34	.10021934	.07304531	.05339481	.03914251	.02877522	.02121227
35	.09366294	.06763454	.04898607	.03558410	.02592363	.01893953
36	.08753546	.06262458	.04494135	.03234918	.02335462	.01691029
37	.08180884	.05798572	.04123059	.02940835	.02104020	.01509848
38	.07645686	.05369048	.03782623	.02673486	.01895513	.01348078
39	.07145501	.04971341	.03470296	.02430442	.01707670	.01203641
40	.06678038	.04603093	.03183758	.02209493	.01538441	.01074680
41	.06241157	.04262123	.02920879	.02008630	.01385983	.00959536
42	.05832857	.03946411	.02679706	.10826027	.01248633	.00856728
43	.05451268	.03654084	.02458446	.01660025	.01124895	.00764936
44	.05094643	.03383411	.02255455	.01509113	.01013419	.00682978
45	.04761349	.03132788	.02069224	.01371921	.00912990	.00609802
46	.04449859	.02900730	.01898371	.01247201	.00822513	.00544466
47	.04158747	.02685861	.01741625	.01133819	.00741003	.00486131
48	.03886679	.02486908	.01597821	.01030745	.00667570	.00434045
49	.03632410	.02302693	.091465891	.00937041	.00601415	.00387540
50	.03394776	.02132123	.01344854	.00851855	.00541815	.00346018
51	.03172688	.01974188	.01233811	.00774414	.00488122	.00308945
52	.02965129	.0 1 827952	.01131397	.00704013	.00439749	.00275844
53	.02771148	.01692548	.01038474	.00640011	.00396170	.00246289
54	.02589858	.01567174	.00952728	.00581829	.00356910	.00219901
55	.02420428	.01451087	.00874063	.00528935	.00321541	.00196340
56	.02262083	.01343599	.00801892	.00480850	.00289676	.00175304
57	.02114096	.01244073	.00735681	.00437136	.00260970	.00156521
58	.01975791	.01151920	.00674937	.00397397	.00235108	.00139751
59	.01846533	.01066592	.00619208	.00361270	.00211809	.00124778
60	.01725732	.00987585	.00568081	.00328427	.00190819	.00111409

APPENDIX C
Future value (amount) of an annuity

Business mathematics and statistics

Future value (amount) of an annuity $s_{\overline{n}|i} = \dfrac{(1+i)^n - 1}{i}$

i \ n	$\frac{5}{12}\%$ .00416666	$\frac{11}{24}\%$ .00458333	$\frac{1}{2}\%$ .00500000	$\frac{13}{24}\%$ .00541666	$\frac{7}{12}\%$ .00583333	$\frac{5}{8}\%$ .00625000
1	1.00000000	1.00000000	1.00000000	1.00000000	1.00000000	1.00000000
2	2.00416667	2.00458333	2.00500000	2.00541667	2.00583333	2.00625000
3	3.01251736	3.01377101	3.01502500	3.01627934	3.01753403	3.01878906
4	4.02506952	4.02758412	4.03010013	4.03261752	4.03513631	4.03765649
5	5.04184064	5.04604388	5.05025063	5.05446086	5.05867460	5.06289185
6	6.06284831	6.06917159	6.07550188	6.08183919	6.08818354	6.09453492
7	7.08811018	7.09698862	7.10587939	7.11478249	7.12369794	7.13262576
8	8.11764397	8.12951649	8.14140879	8.15332090	8.16525285	8.17720468
9	9.15146749	9.16677677	9.18211583	9.19748472	9.21288349	9.22831220
10	10.18959860	10.20879116	10.22802641	10.24730443	10.26662531	10.28598916
11	11.23205526	11.25558146	11.27916654	11.30281066	11.32651396	11.35027659
12	12.27885549	12.30716954	12.33556237	12.36403422	12.39258529	12.42121582
13	13.33001739	13.36357740	13.39724018	13.43100607	13.46487537	13.49884842
14	14.38555913	14.42482713	14.46422639	14.50375735	14.54342048	14.58321622
15	15.44549896	15.49094092	15.53654752	15.58231937	15.62825710	15.67436132
16	16.50985520	16.56194107	16.61423026	16.66672360	16.71942193	16.77232608
17	17.57864627	17.63784996	17.69730141	17.75700169	17.81695189	17.87715312
18	18.65189063	18.71869011	18.78578791	18.85318544	18.92088411	18.98888532
19	19.72960684	19.80448410	19.87971685	19.95530687	20.03125593	20.10756586
20	20.81181353	20.89525466	20.97911544	21.06339811	21.14810493	21.23323814
21	21.89852942	21.99102457	22.08401101	22.17749152	22.27146887	22.36594588
22	22.98977330	23.09181677	23.19443107	23.29761960	23.40138577	23.50573304
23	24.08556402	24.19765426	24.31040322	24.42381504	24.53789386	24.65264387
24	25.18592053	25.30856018	25.43195524	25.55611070	25.68103157	25.80672290
25	26.29086187	26.42455775	26.55911502	26.69453963	26.83083759	26.96801492
26	27.40040713	27.54567030	27.69191059	27.83913506	27.98735081	28.13656501
27	28.51457549	28.67192129	28.83037015	28.98993037	29.15061036	29.31241854
28	29.63338622	29.80333426	29.97452200	30.14695916	30.32065558	30.49562116
29	30.75685866	30.93993288	31.12439461	31.31025519	31.49752607	31.68621879
30	31.88501224	32.08174090	32.28001658	32.47985241	32.68126164	32.88425766
31	33.01786646	33.22878222	33.44141666	33.65578494	33.87190233	34.08978427
32	34.15544090	34.38108080	34.60862375	34.83808711	35.06948843	35.30284542
33	35.29775524	35.53866076	35.78166686	36.02679341	36.27406045	36.52348820
34	36.44482922	36.70154628	36.96057520	37.22193854	37.48565913	37.75176000
35	37.59668268	37.86976171	38.14537807	38.42355738	38.70432548	38.98770850
36	38.75333552	39.04333145	39.33610496	39.63168498	39.93010071	40.23138168
37	39.91480775	40.22228005	40.53278549	40.84635661	41.16302630	41.48282782
38	41.08111945	41.40663217	41.73544942	42.06760771	42.40314395	42.74209549
39	42.25229078	42.59641256	42.94412666	43.29547391	43.65049562	44.00923359
40	43.42834199	43.79164612	44.15884730	44.52999106	44.90512352	45.28429130
41	44.60929342	44.99235783	45.37964153	45.77119518	46.16707007	46.56731812
42	45.79516547	46.19857281	46.60653974	47.01912249	47.43637798	47.85836386
43	46.98597866	47.41031626	47.83957244	48.27380940	48.71309018	49.15747863
44	48.18175357	48.62761355	49.07877030	49.53529254	49.99724988	50.46471287
45	49.38251088	49.85049011	50.32416415	50.80360871	51.28890050	51.78011733
46	50.58827134	51.07897152	51.57578497	52.07879492	52.58808575	53.10374306
47	51.79905581	52.31308347	52.83366390	53.36088839	53.89484959	54.43564146
48	53.01488521	53.55285177	54.09783222	54.64992654	55.20923621	55.77586421
49	54.23578056	54.79830234	55.36832138	55.94594697	56.53129009	57.12446337
50	55.46176298	56.04946123	56.64516299	57.24898752	57.86105595	58.48149126
51	56.69285366	57.30635459	57.92838880	58.55908620	59.19857877	59.84700058
52	57.92907388	58.56900872	59.21803075	59.87628125	60.54390331	61.22104434
53	59.17044502	59.83745001	60.51412090	61.20061111	61.89707659	62.60367586
54	60.41698854	61.11170499	61.81669150	62.53211442	63.25814287	63.99494884
55	61.66872600	62.39180030	63.12577496	63.87083004	64.62714870	65.39471727
56	62.92567902	63.67776272	64.44140384	65.21679703	66.00414040	66.80363550
57	64.18786935	64.96961913	65.76361086	66.57005469	67.38916455	68.22115822
58	65.45531881	66.26739655	67.09242891	67.93064248	68.78226801	69.64754046
59	66.72804930	67.57112212	68.42789105	69.29860013	70.18349791	71.08283759

Appendix C

Future value (amount) of an annuity $s_{\overline{n}|i} = \dfrac{(1+i)^n - 1}{i}$ —*(continued)*

i n	$\frac{5}{12}\%$ .00416666	$\frac{11}{24}\%$ .00458333	$\frac{1}{2}\%$ .00500000	$\frac{13}{24}\%$ .00541666	$\frac{7}{12}\%$ .00583333	$\frac{5}{8}\%$ .00625000
60	68.00608284	68.88082310	69.77003051	70.67396755	71.59290165	72.52710532
61	69.28944152	70.19652687	71.11888066	72.05678487	73.01052691	73.98039973
62	70.57814753	71.51826095	72.47447507	73.44709245	74.43642165	75.44277723
63	71.87222314	72.84605298	73.83684744	74.84493087	75.87063411	76.91429459
64	73.17169074	74.17993073	75.20603168	76.25034091	77.31321281	78.39500893
65	74.47657278	75.51992207	76.58206184	77.66336359	78.76420655	79.88497774
66	75.78689183	76.86605505	77.96497215	79.08404015	80.22366442	81.38425885
67	77.10267055	78.21835780	79.35479701	80.51241203	81.69163580	82.89291046
68	78.42393168	79.57685861	80.75157099	81.94852093	83.16817034	84.41099115
69	79.75069806	80.94158588	82.15532885	83.39240875	84.65331800	85.93855985
70	81.08299264	82.31256815	83.56610549	84.84411763	86.14712902	87.47567585
71	82.42083844	83.68983408	84.98393602	86.30368994	87.64965394	89.02239882
72	83.76425860	85.07341249	86.40885570	87.77116826	89.16094359	90.57878882
73	85.11327634	86.46333230	87.84089998	89.24659542	90.68104909	92.14490625
74	86.46791499	87.85962257	89.28010448	90.73001448	92.21002188	93.72081191
75	87.82819797	89.26231251	90.72650500	92.22146872	93.74791367	95.30656698
76	89.19414880	90.67143144	92.18013752	93.72100168	95.29477650	96.90223303
77	90.56579108	92.08700883	93.64103821	95.22865710	96.85066270	98.50787198
78	91.94314855	93.50907429	95.10924340	96.74447900	98.41562490	100.12354618
79	93.32624500	94.93765755	96.58478962	98.26851159	99.98971604	101.74931835
80	94.71510435	96.37278848	98.06771357	99.80079936	101.57298939	103.38525159
81	96.10975062	97.81449709	99.55805214	101.34138702	103.16549849	105.03140941
82	97.51020792	99.26281354	101.05584240	102.89031954	104.76729723	106.68785572
83	98.91650045	100.71776810	102.56112161	104.44764210	106.37843980	108.35465482
84	100.32865253	102.17939120	104.07392722	106.01340016	107.99898070	110.03187141
85	101.74668859	103.64771341	105.59429685	107.58763941	109.62897475	111.71957061
86	103.17063312	105.12276543	107.12226834	109.17040579	111.26847710	113.41781792
87	104.60051076	106.60457811	108.65787968	110.76174549	112.91754322	115.12667928
88	106.03634622	108.09318242	110.20116908	112.36170495	114.57622889	116.84622103
89	107.47816433	109.58860951	111.75217492	113.97033085	116.24459022	118.57650991
90	108.92599002	111.09089064	113.31093580	115.58767014	117.92268367	120.31761310
91	110.37984831	112.60005722	114.87749048	117.21377002	119.61056599	122.06959818
92	111.83976434	114.11614081	116.45187793	118.84867794	121.30829429	123.83253317
93	113.30576336	115.63917313	118.03413732	120.49244167	123.01592601	125.60648650
94	114.77787071	117.16918600	119.62430800	122.14510901	124.73351891	127.39152704
95	116.25611184	118.70621144	121.22242954	123.80672835	126.46113110	129.18772408
96	117.74051230	120.25028158	122.82854169	125.47734812	128.19882103	130.99514736
97	119.23109777	121.80142870	124.44268440	127.15701709	129.94664749	132.81386703
98	120.72789401	123.35968525	126.06489782	128.84578427	131.70466960	134.64395370
99	122.23092690	124.92508380	127.69522231	130.54369893	133.47294684	136.48547841
100	123.74022243	126.49765711	129.33369842	132.25081064	135.25153903	138.33851265
101	125.25580669	128.07743803	130.98036692	133.96716919	137.04050634	140.20312836
102	126.77770589	129.66445962	132.63526875	135.69282469	138.83990929	142.07939791
103	128.30594633	131.25875506	134.29844509	137.42782750	140.64980877	143.96739414
104	129.84055444	132.86035769	135.96993732	139.17222823	142.47026598	145.86719036
105	131.38155675	134.46930100	137.64978701	140.92607780	144.30134253	147.77886030
106	132.92897990	136.08561863	139.33803594	142.68942738	146.14310037	149.70247817
107	134.48285065	137.70934438	141.03472612	144.46232845	147.99560178	151.63811866
108	136.04319586	139.34051221	142.73989975	146.24483273	149.85890946	153.58585690
109	137.61004251	140.97915622	144.45359925	148.03699224	151.73308643	155.54576851
110	139.18341769	142.62531069	146.17586725	149.83885928	153.61819610	157.51792956
111	140.76334859	144.27901003	147.90674658	151.65048644	155.51430225	159.50241662
112	142.34986255	145.94028883	149.64628032	153.47192657	157.42146901	161.49930673
113	143.94298697	147.60918182	151.39451172	155.30323284	159.33976091	163.50867739
114	145.54274942	149.28572390	153.15148428	157.14445868	161.26924285	165.53060663
115	147.14917754	150.96995013	154.91724170	158.99565783	163.20998010	167.56517292
116	148.76229911	152.66189574	156.69182791	160.85688431	165.16203832	169.61245525
117	150.38214203	154.36159940	158.47528704	162.72819244	167.12548354	171.67253310
118	152.00873429	156.06908674	160.26766348	164.60963681	169.10038220	173.74548643
119	153.64210401	157.78440339	162.06900180	166.50127235	171.08680109	175.83139572
120	155.28227945	159.50758191	163.87934681	168.40315424	173.08480743	177.93034194

Business mathematics and statistics

Future value (amount) of an annuity $s_{\overline{n}|i} = \dfrac{(1+i)^n - 1}{i}$ —(continued)

i n	$\tfrac{2}{3}\%$ .00666666	$\tfrac{17}{24}\%$ .00708333	$\tfrac{3}{4}\%$ .00750000	$\tfrac{19}{24}\%$ .00791666	$\tfrac{5}{6}\%$ .00833333	$\tfrac{7}{8}\%$ .00875000
1	1.00000000	1.00000000	1.00000000	1.00000000	1.00000000	1.00000000
2	2.00666667	2.00708333	2.00750000	2.00791667	2.00833333	2.00875000
3	3.02004444	3.02130017	3.02255625	3.02381267	3.02506944	3.02632656
4	4.04017807	4.04270105	4.04522542	4.04775119	4.05027836	4.05280692
5	5.06711259	5.07133685	5.07556461	5.07979589	5.08403068	5.08826898
6	6.10089335	6.10725882	6.11363135	6.12001094	6.12639760	6.13279133
7	7.14156597	7.15051857	7.15948358	7.16846102	7.17745091	7.18645326
8	8.18917641	8.20116807	8.21317971	8.22521134	8.23726300	8.24933472
9	9.24377092	9.25925968	9.27477856	9.29032760	9.30590686	9.32151640
10	10.30539606	10.32484610	10.34433940	10.36387602	10.38345608	10.40307967
11	11.37409870	11.39798043	11.42192194	11.44592338	11.46998489	11.49410662
12	12.44992602	12.47871613	12.50758636	12.53653694	12.56556809	12.59468005
13	13.53292553	13.56710703	13.60139325	13.63578452	13.67028116	13.70488350
14	14.62314503	14.66320737	14.70340370	14.74373448	14.78420017	14.82480123
15	15.72063266	15.76707176	15.81367923	15.86045571	15.90740184	15.95451824
16	16.82543688	16.87875518	16.93228183	16.98601765	17.03996352	17.09412028
17	17.93760646	17.99831303	18.05927394	18.12049029	18.18196322	18.24369383
18	19.05719051	19.12580108	19.19471849	19.26394418	19.33347958	19.40332615
19	20.18423844	20.26127551	20.33867888	20.41645040	20.49459191	20.57310526
20	21.31880003	21.40479288	21.49121897	21.57808063	21.66538017	21.75311993
21	22.46092536	22.55641016	22.65240312	22.74890710	22.84592501	22.94345973
22	23.61066487	23.71618473	23.82229614	23.92900262	24.03630772	24.14421500
23	24.76806930	24.88417437	25.00096336	25.11844056	25.23661028	25.35547688
24	25.93318976	26.06043727	26.18847059	26.31729488	26.44691537	26.57733730
25	27.10607769	27.24503204	27.38488412	27.52564013	27.66730633	27.80988900
26	28.28678488	28.43801768	28.59027075	28.74355145	28.89786721	29.05322553
27	29.47536344	29.63945364	29.80469778	29.97110456	30.13868277	30.30744126
28	30.67186587	30.84939977	31.02823301	31.20837581	31.38983846	31.57263137
29	31.87634497	32.06791635	32.26094476	32.45544211	32.65142045	32.84889189
30	33.08885394	33.29506409	33.50290184	33.71238103	33.92351562	34.13631970
31	34.30944630	34.53090413	34.75417361	34.97927071	35.20621158	35.43501249
32	35.53817594	35.77549804	36.01482991	36.25618994	36.49959668	36.74506885
33	36.77509711	37.02890781	37.28494113	37.54321811	37.80375999	38.06658820
34	38.02026443	38.29119591	38.56457819	38.84043525	39.11879132	39.39967085
35	39.27373286	39.56242521	39.85381253	40.14792203	40.44478125	40.74441797
36	40.53555774	40.84265906	41.15271612	41.46575905	41.78182109	42.10093163
37	41.80579479	42.13196123	42.46136149	42.79403035	43.13000293	43.46931478
38	43.08450009	43.43039595	43.77982170	44.13281642	44.48941962	44.84967128
39	44.37173009	44.73802792	45.10817037	45.48220122	45.86016479	46.24210591
40	45.66754163	46.05492229	46.44648164	46.84226864	47.24233283	47.64672433
41	46.97199191	47.38114466	47.79483026	48.21310327	48.63601893	49.06363317
42	48.28513852	48.71676110	49.15329148	49.59479034	50.04131909	50.49293996
43	49.60703944	50.06183815	50.52194117	50.98741576	51.45833008	51.93475319
44	50.93775304	51.41644284	51.90085573	52.39106614	52.88714950	53.38918228
45	52.27733806	52.78064264	53.29011215	53.80582874	54.32787575	54.85633762
46	53.62585365	54.15450553	54.68978799	55.23179155	55.78060805	56.33633058
47	54.98335934	55.53809994	56.09996140	56.66904324	57.24544645	57.82927347
48	56.34991507	56.93149482	57.52071111	58.11767316	58.72249183	59.33527961
49	57.72558117	58.33475957	58.95211644	59.57777141	60.21184593	60.85446331
50	59.11041837	59.74796412	60.39425732	61.04942877	61.71361131	62.38693986
51	60.50448783	61.17117887	61.84721424	62.53273674	63.22789141	63.93282559
52	61.90785108	62.60447472	63.31106835	64.02778758	64.75479050	65.49223781
53	63.32057009	64.04792308	64.78590136	65.53467423	66.29441376	67.06529489
54	64.74270722	65.50159587	66.27179562	67.05349040	67.84686721	68.65211622
55	66.17432527	66.96556551	67.76883409	68.58433053	69.41225777	70.25282224
56	67.61548744	68.43990493	69.27710035	70.12728981	70.99069325	71.86753443
57	69.06625736	69.92468759	70.79667860	71.68246419	72.58228236	73.49637536
58	70.52669907	71.41998746	72.32765369	73.24995037	74.18713471	75.13946864
59	71.99687706	72.92587904	73.87011109	74.82984581	75.80536083	76.79693900

Appendix C

Future value (amount) of an annuity $s_{\overline{n}|i} = \dfrac{(1+i)^n - 1}{i}$ —*(continued)*

n \ i	$\tfrac{2}{3}\%$ .00666666	$\tfrac{17}{24}\%$ .00708333	$\tfrac{3}{4}\%$ .00750000	$\tfrac{19}{24}\%$ .00791666	$\tfrac{5}{6}\%$ .00833333	$\tfrac{7}{8}\%$ .00875000
60	73.47685625	74.44243735	75.42413693	76.42224875	77.43707217	78.46891221
61	74.96670195	75.96973794	76.98981795	78.02725822	79.08238111	80.15551519
62	76.46647997	77.50785592	78.56724159	79.64497402	80.74140095	81.85687595
63	77.97625650	79.05687091	80.15649590	81.27549673	82.41424596	83.57312362
64	79.49609821	80.61685708	81.75766962	82.91892774	84.10103134	85.30438845
65	81.02607220	82.18789315	83.37085214	84.57536926	85.80187327	87.05080185
66	82.56624601	83.77005739	84.99613353	86.24492426	87.51688888	88.81249636
67	84.11668765	85.36342863	86.63360453	87.92769658	89.24619629	90.58960571
68	85.67746557	86.96808625	88.28335657	89.62379084	90.98991459	92.38226476
69	87.24864867	88.58411019	89.94548174	91.33331252	92.74816388	94.19060957
70	88.83030633	90.21158097	91.62007285	93.05636791	94.52106524	96.01477741
71	90.42250837	91.85057967	93.30722340	94.79306416	96.30874079	97.85490671
72	92.02532510	93.50118795	95.00702758	96.54350925	98.11131363	99.71113714
73	93.63882726	95.16348803	96.71958028	98.30781203	99.92890791	101.58360959
74	95.26308611	96.83756273	98.44497714	100.08608221	101.76164881	103.47246618
75	96.89817335	98.52349547	100.18331446	101.87843036	103.60966255	105.37785025
76	98.54416118	100.22137023	101.93468932	103.68496793	105.47307640	107.29990644
77	100.20112225	101.93127160	103.69919949	105.50580726	107.35201870	109.23878063
78	101.86912973	103.65328478	105.47694349	107.34106157	109.24661886	111.19461996
79	103.54825726	105.38749554	107.26802056	109.19084497	111.15700735	113.16757288
80	105.23857898	107.13399030	109.07253072	111.05527250	113.08331575	115.15778914
81	106.94016950	108.89285607	110.89057470	112.93446007	115.02567671	117.16541980
82	108.65310397	110.66418046	112.72225401	114.82852455	116.98422402	119.19061722
83	110.37745799	112.44805174	114.56767091	116.73758370	118.95909255	121.23353512
84	112.11330771	114.24455878	116.42692845	118.66175624	120.95041832	123.29432855
85	113.86072977	116.05379107	118.30013041	120.60116181	122.95833847	125.37315393
86	115.61980130	117.87583875	120.18738139	122.55592101	124.98299129	127.47016903
87	117.39059997	119.71079261	122.08878675	124.52615538	127.02451622	129.58553301
88	119.17320397	121.55874406	124.00445265	126.51198744	129.08305386	131.71940642
89	120.96769200	123.41978516	125.93448604	128.51354068	131.15874597	133.87195123
90	122.77414328	125.29400864	127.87899469	130.53093954	133.25173552	136.04333080
91	124.59263757	127.18150787	129.83808715	132.56430948	135.36216665	138.23370994
92	126.42325515	129.08237688	131.81187280	134.61377693	137.49018471	140.44325491
93	128.26607685	130.99671039	133.80046185	136.67946933	139.63593625	142.67213339
94	130.12118403	132.92409831	135.80396515	138.76151513	141.79956905	144.92051455
95	131.98865859	134.86615303	137.82249505	140.86004379	143.98123212	147.18856906
96	133.86858298	136.82145495	139.85616377	142.97518580	146.18107572	149.47646903
97	135.76104020	138.79060692	141.90508499	145.10707269	148.39925136	151.78438814
98	137.66611380	140.77370705	143.96937313	147.25583702	150.63591178	154.11250153
99	139.58388790	142.77085414	146.04914343	149.42161239	152.89121105	156.46098592
100	141.51444715	144.78214769	148.14451201	151.60453349	155.16530447	158.83001955
101	143.45787680	146.80768791	150.25559585	153.80473605	157.45834868	161.21978222
102	145.41426264	148.84757569	152.38251281	156.02235687	159.77050158	163.63045532
103	147.38369106	150.90191269	154.52538166	158.25753387	162.10192243	166.06222180
104	149.36624900	152.97080124	156.68432202	160.51040601	164.45277178	168.51526624
105	151.36202399	155.05434441	158.85945444	162.78111339	166.82321155	170.98977482
106	153.37110415	157.15264602	161.05090035	165.06979720	169.21340498	173.48593535
107	155.39357818	159.26581059	163.25878210	167.37659977	171.62351669	176.00393728
108	157.42953537	161.39394342	165.48322296	169.70166451	174.05371266	178.54397174
109	159.47906560	163.53715052	167.72434714	172.04513602	176.50416026	181.10623149
110	161.54225937	165.69553867	169.98227974	174.40716002	178.97502827	183.69091101
111	163.61920777	167.86921540	172.25714684	176.78783337	181.46646684	186.29820648
112	165.71000249	170.05828901	174.54907544	179.18745411	183.97870756	188.92831579
113	167.81473584	172.26286856	176.85819351	181.60602146	186.51186346	191.58143855
114	169.93350074	174.48306358	179.18462996	184.04373579	189.06612898	194.25777614
115	172.06639075	176.71898558	181.52851468	186.50074870	191.64168006	196.95753168
116	174.21350002	178.97074506	183.88997854	188.97721296	194.23869406	199.68091009
117	176.37492335	181.23845450	186.26915338	191.47328256	196.85734984	202.42811805
118	178.55075618	183.52222689	188.66617203	193.98911272	199.49782776	205.19936408
119	180.74109455	185.82217600	191.08116832	196.52485986	202.16030966	207.99485852
120	182.94603518	188.13841641	193.51427708	199.08068167	204.84497890	210.81481353

Business mathematics and statistics

Future value (amount) of an annuity $s_{\overline{n}|i} = \dfrac{(1+i)^n - 1}{i}$ —(continued)

n \ i	$\tfrac{11}{12}\%$ .00916666	$\tfrac{23}{24}\%$ .00958333	1% .01000000	$1\tfrac{1}{4}\%$ .01250000	$1\tfrac{3}{8}\%$ .01375000	$1\tfrac{1}{2}\%$ .01500000
1	1.00000000	1.00000000	1.00000000	1.00000000	1.00000000	1.00000000
2	2.00916667	2.00958333	2.01000000	2.01250000	2.01375000	2.01500000
3	3.02758403	3.02884184	3.03010000	3.03765625	3.04143906	3.04522500
4	4.05533688	4.05786824	4.06040100	4.07562695	4.08325885	4.09090337
5	5.09251080	5.09675615	5.10100501	5.12657229	5.13940366	5.15226693
6	6.13919215	6.14560006	6.15201506	6.19065444	6.21007046	6.22955093
7	7.19546808	7.20449539	7.21353521	7.26803762	7.29545893	7.32299419
8	8.26142654	8.27353847	8.28567056	8.35888809	8.39577149	8.43283911
9	9.33715628	9.35282655	9.36852727	9.46337420	9.51121335	9.55933169
10	10.42274688	10.44245780	10.46221254	10.58166637	10.64199253	10.70272167
11	11.51828873	11.54253136	11.56683467	11.71393720	11.78831993	11.86326249
12	12.62387304	12.65314728	12.68250301	12.86036142	12.95040933	13.04121143
13	13.73959188	13.77440661	13.80932804	14.02111594	14.12847745	14.23682960
14	14.86553813	14.90641134	14.94742132	15.19637988	15.32274402	15.45038205
15	16.00180557	16.04926445	16.09689554	16.38633463	16.53343175	16.68213778
16	17.14848879	17.20306990	17.25786449	17.59116382	17.76076644	17.93236984
17	18.30568327	18.36793266	18.43044314	18.81105336	19.00497697	19.20135539
18	19.47348536	19.54395868	19.61474757	20.04619153	20.26629541	20.48937572
19	20.65199231	20.73125495	20.81089504	21.29676893	21.54495697	21.79671636
20	21.84130224	21.92992947	22.01900399	22.56297854	22.84120013	23.12366710
21	23.04151418	23.14009130	23.23919403	23.84501577	24.15526663	24.47052211
22	24.25272806	24.36185051	24.47158598	25.14307847	25.48740155	25.83757994
23	25.47504473	25.59531824	25.71630183	26.45736695	26.83785332	27.22514364
24	26.70856598	26.84060671	26.97346485	27.78808403	28.20687380	28.63352080
25	27.95339450	28.09782919	28.24319950	29.13543508	29.59471832	30.06302361
26	29.20963395	29.36710005	29.52563150	30.49962802	31.00164569	31.51396896
27	30.47738892	30.64853476	30.82088781	31.88087337	32.42791832	32.98667850
28	31.75676499	31.94224988	32.12909969	33.27938429	33.87380220	34.48147867
29	33.04786867	33.24836311	33.45038766	34.69537659	35.33956698	35.99870085
30	34.35080746	34.56699326	34.78489153	36.12906880	36.82548602	37.53868137
31	35.66568987	35.89826028	36.13274045	37.58068216	38.33183646	39.10176159
32	36.99262536	37.24228527	37.49406785	39.05044069	39.85889921	40.68828801
33	38.33172442	38.59919051	38.86900853	40.53857120	41.40695907	42.29861233
34	39.68309856	39.96909941	40.25769862	42.04530334	42.97630476	43.93309152
35	41.04686030	41.35213662	41.66027560	43.57086963	44.56722895	45.59208789
36	42.42312319	42.74842793	43.07687836	45.11550550	46.18002835	47.27596921
37	43.81200182	44.15810036	44.50764714	46.67944932	47.81500374	48.98510874
38	45.21361183	45.58128216	45.95272361	48.26294243	49.47246004	50.71988538
39	46.62806994	47.01810278	47.41225085	49.86622921	51.15270636	52.48068366
40	48.05549391	48.46869293	48.88637336	51.48955708	52.85605608	54.26789391
41	49.49600261	49.93318457	50.37523709	53.13317654	54.58282685	56.08191232
42	50.94971597	51.41171092	51.87898946	54.79734125	56.33334072	57.92314100
43	52.41675503	52.90440648	53.39777936	56.48230801	58.10792415	59.79198812
44	53.89724195	54.41140705	54.93175715	58.18833687	59.90690811	61.68886794
45	55.39130000	55.93284970	56.48107472	59.91569108	61.73062809	63.61420096
46	56.89905358	57.46887258	58.04585547	61.66463721	63.57942423	65.56841398
47	58.42062824	59.01961620	59.62634432	63.43544518	65.45364131	67.55194018
48	59.95615067	60.58522086	61.22260777	65.22838824	67.35362888	69.56521929
49	61.50574872	62.16582923	62.83483385	67.04374310	69.27974128	71.60869758
50	63.06955141	63.76158509	64.46318218	68.88178989	71.23233772	73.68282804
51	64.64768897	65.37263361	66.10781401	70.74281226	73.21178237	75.78807046
52	66.24029278	66.99912135	67.76889215	72.62709741	75.21844437	77.92489152
53	67.84749547	68.64119627	69.44658107	74.53493613	77.25269798	80.09376489
54	69.46943084	70.29900773	71.14104688	76.46662283	79.31492258	82.29517136
55	71.10623396	71.97270655	72.85245735	78.42245562	81.40550277	84.52959893
56	72.75804110	73.66244499	74.58098192	80.40273631	83.52482843	86.79754292
57	74.42498981	75.36837676	76.32679174	82.40777052	85.67329482	89.09950606
58	76.10721889	77.09065703	78.09005703	84.43786765	87.85130262	91.43599865
59	77.80486839	78.82944250	79.87096025	86.49334099	90.05925804	93.80753863

Appendix C

Future value (amount) of an annuity $s_{\overline{n}|i} = \dfrac{(1+i)^n - 1}{i}$ —(continued)

n	i → $\frac{11}{12}\%$ .00916666	$\frac{23}{24}\%$ .00958333	1% .01000000	$1\frac{1}{4}\%$ .01250000	$1\frac{3}{8}\%$ .01375000	$1\frac{1}{2}\%$ .01500000
60	79.51807969	80.58489132	81.66966986	88.57450776	92.29757283	96.21465171
61	81.24699542	82.35716320	83.48636655	90.68168910	94.56666446	98.65787149
62	82.99175954	84.14641934	85.32123022	92.81521022	96.86695610	101.13773956
63	84.75251734	85.95282253	87.17444252	94.97540034	99.19887674	103.65480565
64	86.52941541	87.77653708	89.04618695	97.16259285	101.56286130	106.20962774
65	88.32260172	89.61772889	90.93664882	99.37712526	103.95935064	108.80277215
66	90.13222557	91.47656546	92.84601532	101.61933933	106.38879171	111.43481374
67	91.95843764	93.35321588	94.77447546	103.88958107	108.85163760	114.10633594
68	93.80138998	95.24785086	96.72222021	106.18820083	111.34834761	116.81793098
69	95.66123606	97.16064277	98.68944242	108.51555334	113.87938739	119.57019995
70	97.53813072	99.09176559	100.67633684	110.87199776	116.44522897	122.36375295
71	99.43223025	101.04139501	102.68310021	113.25789773	119.04635087	125.19920924
72	101.34369236	103.00970838	104.70993121	115.67362145	121.68323819	128.07719738
73	103.27267621	104.99688476	106.75703052	118.11954172	124.35638272	130.99835534
74	105.21934241	107.00310490	108.82460083	120.59603599	127.06628298	133.96333067
75	107.18385305	109.02855132	110.91284684	123.10348644	129.81344437	136.97278063
76	109.16637170	111.07340827	113.02197530	125.64228002	132.59837923	140.02737234
77	111.16706344	113.13786177	115.15219506	128.21280852	135.42160695	143.12778292
78	113.18609486	115.22209961	117.30371701	130.81546863	138.28365404	146.27469967
79	115.22363406	117.32631140	119.47675418	133.45066199	141.18505429	149.46882016
80	117.27985070	119.45068855	121.67152172	136.11879526	144.12634878	152.71085247
81	119.35491600	121.59542432	123.88823694	138.82028020	147.10808608	156.00151525
82	121.44900273	123.76071380	126.12711931	141.55553370	150.13082226	159.34153798
83	123.56228526	125.94675397	128.38839050	144.32497787	153.19512107	162.73166105
84	125.69493954	128.15374370	130.67227440	147.12904010	156.30155398	166.17263597
85	127.84714315	130.38188374	132.97899715	149.96815310	159.45070035	169.66522551
86	130.01907530	132.63137679	135.30878712	152.84275501	162.64314748	173.21020389
87	132.21091682	134.90242749	137.66187499	155.75328945	165.87949076	176.80835695
88	134.42285022	137.19524242	140.03849374	158.70020557	169.16033375	180.46044823
89	136.65505968	139.51003016	142.43887868	161.68395814	172.48628834	184.16738950
90	138.90773107	141.84700128	144.86326746	164.70500762	175.85797481	187.92990030
91	141.18105193	144.20636838	147.31190014	167.76382021	179.27602196	191.74884889
92	143.47521158	146.58834607	149.78501914	170.86086796	182.74106726	195.62508162
93	145.79040102	148.99315106	152.28286933	173.99662801	186.25375694	199.55945784
94	148.12681302	151.42100209	154.80569803	177.17158667	189.81474610	203.55284971
95	150.48464214	153.87212002	157.35375501	180.38623151	193.42469886	207.60614246
96	152.86408470	156.34672784	159.92729256	183.64105940	197.08428847	211.72023459
97	155.26533881	158.84505065	162.52656548	186.93657264	200.79419743	215.89603811
98	157.68860441	161.36731572	165.15183114	190.27327980	204.55511765	220.13447868
99	160.13408329	163.91375249	167.80334945	193.65169580	208.36775051	224.43649586
100	162.60197905	166.48459262	170.48138294	197.07234200	212.23280708	228.80304330
101	165.09249719	169.08006997	173.18619677	200.53574627	216.15100818	233.23508895
102	167.60584508	171.70042064	175.91805874	204.04244310	220.12308454	237.73361529
103	170.14223200	174.34588300	178.67723933	207.59297364	224.14977696	242.29961951
104	172.70186912	177.01669772	181.46401172	211.18788581	228.23183639	246.93411381
105	175.28496959	179.71310773	184.27865184	214.82773438	232.37002414	251.63812551
106	177.89174848	182.43535835	187.12143836	218.51308106	236.56511197	256.41269740
107	180.52242284	185.18369720	189.99265274	222.24449459	240.81788226	261.25888786
108	183.17721171	187.95837430	192.89257927	226.02255076	245.12912814	266.17777118
109	185.85633615	190.75964205	195.82150506	229.84783264	249.49965365	271.17043774
110	188.56001924	193.58775529	198.77972011	233.72093055	253.93027389	276.23799431
111	191.28848608	196.44297128	201.76751731	237.64244218	258.42181516	281.38156422
112	194.04196387	199.32554975	204.78519248	241.61297271	262.97511512	286.60228769
113	196.82068187	202.23575294	207.83304441	245.63313487	267.59102295	291.90132200
114	199.62487145	205.17384557	210.91137485	249.70354905	272.27039951	297.27984183
115	202.45476611	208.14009492	214.02048860	253.82484342	277.01411751	302.73903946
116	205.31060146	211.13477083	217.16069349	257.99765396	281.82306162	308.28012505
117	208.19261531	214.15814572	220.33230042	262.22262463	286.69812872	313.90432693
118	211.10104762	217.21049462	223.53562343	266.50040744	291.64022799	319.61289183
119	214.03614055	220.29209519	226.77097966	270.83166253	296.65028113	325.40708521
120	216.99813851	223.40322777	230.03868946	275.21705832	301.72922249	331.28819149

Business mathematics and statistics

Future value (amount) of an annuity $s_{\overline{n}|i} = \dfrac{(1+i)^n - 1}{i}$ —(continued)

i \ n	$1\frac{1}{8}\%$ .01625000	$1\frac{3}{4}\%$ .01750000	$1\frac{7}{8}\%$ .01875000	2% .02000000	$2\frac{1}{8}\%$ .02125000	$2\frac{1}{4}\%$ .02250000
1	1.00000000	1.00000000	1.00000000	1.00000000	1.00000000	1.00000000
2	2.01625000	2.01750000	2.01875000	2.02000000	2.02125000	2.02250000
3	3.04901406	3.05280625	3.05660156	3.06040000	3.06420156	3.06800625
4	4.09856054	4.10623036	4.11391284	4.12160800	4.12931585	4.13703639
5	5.16516215	5.17808939	5.19104871	5.20404016	5.21706381	5.23011971
6	6.24909603	6.26870596	6.28838087	6.30812096	6.32792641	6.34779740
7	7.35064385	7.37840831	7.40628801	7.43428338	7.46239485	7.49062284
8	8.47009181	8.50753045	8.54515591	8.58296905	8.62097074	8.65916186
9	9.60773080	9.65641224	9.70537759	9.75462843	9.80416637	9.85399300
10	10.76385643	10.82539945	10.88735342	10.94972100	11.01250490	11.07570784
11	11.93876909	12.01484394	12.09149129	12.16871542	12.24652063	12.32491127
12	13.13277409	13.22510371	13.31820675	13.41208973	13.50675920	13.60222177
13	14.34618167	14.45654303	14.56792313	14.68033152	14.79377783	14.90827176
14	15.57930712	15.70953253	15.84107169	15.97393815	16.10814561	16.24370788
15	16.83247086	16.98444935	17.13809178	17.29341692	17.45044370	17.60919130
16	18.10599851	18.28167721	18.45943100	18.63928525	18.82126563	19.00539811
17	19.40022099	19.60160656	19.80554534	20.01207096	20.22121753	20.43301957
18	20.71547458	20.94463468	21.17689931	21.41231238	21.65091840	21.89276251
19	22.05210104	22.31116578	22.57396617	22.84055863	23.11100041	23.38534966
20	23.41044768	23.70161119	23.99722804	24.29736980	24.60210917	24.91152003
21	24.79086746	25.11638938	25.44717606	25.78331719	26.12490399	26.47202923
22	26.19371905	26.55592620	26.92431062	27.29898354	27.68005820	28.06764989
23	27.61936699	28.02065490	28.42914144	28.84496321	29.26825944	29.69917201
24	29.06818170	29.51101624	29.96218784	30.42186247	30.89020995	31.36740338
25	30.54053966	31.02745915	31.52397886	32.03029972	32.54662691	33.07316996
26	32.03682343	32.57043969	33.11505347	33.67090572	34.23824274	34.81731628
27	33.55742181	34.14042238	34.73596072	35.34432383	35.96580539	36.60070590
28	35.10272991	35.73787977	36.38725998	37.05121031	37.73007876	38.42422178
29	36.67314927	37.36329267	38.06952111	38.79223451	39.53184293	40.28876677
30	38.26908795	39.01715029	39.78332463	40.56807921	41.37189459	42.19526402
31	39.89096063	40.69995042	41.52926197	42.37944079	43.25104735	44.14465746
32	41.53918874	42.41219955	43.30793563	44.22702961	45.17013211	46.13791226
33	43.21420055	44.15441305	45.11995942	46.11157020	47.12999742	48.17601528
34	44.91643131	45.92711527	46.96595866	48.03380160	49.13150986	50.25997563
35	46.64632332	47.73083979	48.84657038	49.99447763	51.17555445	52.39082508
36	48.40432608	49.56612949	50.76244358	51.99436719	53.26303498	54.56961864
37	50.19089637	51.43353675	52.71423940	54.03425453	55.39487447	56.79743506
38	52.00649844	53.33362365	54.70263138	56.11493962	57.57201556	59.07537735
39	53.85160404	55.26696206	56.72830572	58.23723841	59.79542089	61.40457334
40	55.72669261	57.23413390	58.79196146	60.40198318	62.06607358	63.78617624
41	57.63225136	59.23573124	60.89431073	62.61002284	64.38497764	66.22136521
42	59.56877544	61.27235654	63.03607906	64.86222330	66.75315842	68.71134592
43	61.53676805	63.34462278	65.21800554	67.15946777	69.17166304	71.25735121
44	63.53674053	65.45315367	67.44084315	69.50265712	71.64156087	73.86064161
45	65.56921256	67.59858386	69.70535895	71.89271027	74.16394404	76.52250605
46	67.63471226	69.78155908	72.01233443	74.33056447	76.73992785	79.24426243
47	69.73377634	72.00273636	74.36256571	76.81717576	79.37065132	82.02725834
48	71.86695020	74.26278425	76.75686381	79.35351927	82.05727766	84.87287165
49	74.03478814	76.56238298	79.19605501	81.94058966	84.80099481	87.78251126
50	76.23785345	78.90222468	81.68098104	84.57940145	87.60301595	90.75761776
51	78.47671857	81.28301361	84.21249943	87.27098948	90.46458004	93.79966416
52	80.75196525	83.70546635	86.79148380	90.01640927	93.38695237	96.91015661
53	83.06418468	86.17031201	89.41882412	92.81673746	96.37142510	100.09063513
54	85.41397768	88.67829247	92.09542707	95.67307221	99.41931789	103.34267442
55	87.80195482	91.23016259	94.82221633	98.58653365	102.53197839	106.66788460
56	90.22873659	93.82669043	97.60013289	101.55826432	105.71078293	110.06791200
57	92.69495356	96.46865752	100.43013538	104.58942961	108.95713707	113.54444002
58	95.20124655	99.15685902	103.31320042	107.68121820	112.27247623	117.09918992
59	97.74826681	101.89210405	106.25032292	110.83484257	115.65826635	120.73392169

Appendix C

Future value (amount) of an annuity $s_{\overline{n}|i} = \dfrac{(1+i)^n - 1}{i}$ —(continued)

n \ i	$1\tfrac{1}{8}\%$ .01625000	$1\tfrac{3}{4}\%$ .01750000	$1\tfrac{7}{8}\%$ .01875000	2% .02000000	$2\tfrac{1}{8}\%$ .02125000	$2\tfrac{1}{4}\%$ .02250000
60	100.33667614	104.67521588	109.24251648	114.05153942	119.11600451	124.45043493
61	102.96714713	107.50703215	112.29081366	117.33257021	122.64721961	128.25056972
62	105.64036327	110.38840522	115.39626642	120.67922161	126.25347303	132.13620754
63	108.35701918	113.32020231	118.55994641	124.09280604	129.93635933	136.10927221
64	111.11782074	116.30330585	121.78294541	127.57466216	133.69750696	140.17173083
65	113.92348532	119.33861370	125.06637564	131.12615541	137.53857899	144.32559477
66	116.77474196	122.42703944	128.41137018	134.74867852	141.46127379	148.57292066
67	119.67233152	125.56951263	131.81908337	138.44365209	145.46732586	152.91581137
68	122.61700690	128.76697910	135.29069118	142.21252513	149.55850653	157.35641713
69	125.60953327	132.02040124	138.82739164	146.05677563	153.73662480	161.89693651
70	128.65068818	135.33075826	142.43040524	149.97791118	158.00352807	166.53961758
71	131.74126186	138.69904653	146.10097533	153.97746937	162.36110305	171.28675898
72	134.88205737	142.12627984	149.84036862	158.05701875	166.81127649	176.14071106
73	138.07389080	145.61348974	153.64987553	162.21815913	171.35601611	181.10387705
74	141.31759153	149.16172581	157.53081070	166.46252231	175.99733145	186.17871429
75	144.61400239	152.77205601	161.48451340	170.79177276	180.73727475	191.36773536
76	147.96397993	156.44556699	165.51234803	175.20760821	185.57794183	196.67350941
77	151.36839460	160.18336441	169.61570455	179.71176638	190.52147310	202.09866337
78	154.82813102	163.98657329	173.79599901	184.30599558	195.57005440	207.64588329
79	158.34408814	167.85633832	178.05467399	188.99211549	200.72591806	213.31791567
80	161.91717958	171.79382424	182.39319913	193.77195780	205.99134382	219.11756877
81	165.54833374	175.80021617	186.81307162	198.64739696	211.36865987	225.04771407
82	169.23849417	179.87671995	191.31581671	203.62034490	216.86024390	231.11128763
83	172.98861970	184.02456255	195.90298827	208.69275180	222.46852408	237.31129160
84	176.79968477	188.24499239	200.57616930	213.86660683	228.19598021	243.65079567
85	180.67267965	192.53927976	205.33697248	219.14393897	234.04514479	250.13293857
86	184.60861069	196.90871716	210.18704071	224.52681775	240.01860412	256.76092969
87	188.60850061	201.35461971	215.12804772	230.01735411	246.11899946	263.53805060
88	192.67338875	205.87832555	220.16169862	235.61770119	252.34902820	270.46765674
89	196.80433132	210.48119625	225.28973047	241.33005521	258.71144505	277.55317902
90	201.00240170	215.16461718	230.51391291	247.15665632	265.20906325	284.79812555
91	205.26869073	219.92999798	235.83604878	253.09978944	271.84475585	292.20608337
92	209.60430695	224.77877295	241.25797469	259.16178523	278.62145691	299.78072025
93	214.01037694	229.71240148	246.78156172	265.34502094	285.54216287	307.52578645
94	218.48804557	234.73236850	252.40871600	271.65192135	292.60993383	315.44511665
95	223.03847631	239.84018495	258.14137943	278.08495978	299.82789492	323.54263177
96	227.66285155	245.03738819	263.98153029	284.64665898	307.19923769	331.82234099
97	232.36237288	250.32554248	269.93118398	291.33959216	314.72722149	340.28834366
98	237.13826144	255.70623947	275.99239368	298.16638400	322.41517495	348.94483139
99	241.99175819	261.18109866	282.16725107	305.12971168	330.26649742	357.79609010
100	246.92412426	266.75176789	288.45788702	312.23230591	338.28466049	366.84650213
101	251.93664128	272.41992383	294.86647241	319.47695203	346.47320952	376.10054842
102	257.03061170	278.18727250	301.39521476	326.86649107	354.83576522	385.56281076
103	262.20735914	284.05554976	308.04637911	334.40382089	363.37602524	395.23797401
104	267.46822873	290.02652188	314.82224872	342.09189731	372.09776577	405.13082842
105	272.81458744	296.10198602	321.72516589	349.93373526	381.00484329	415.24627206
106	278.24782449	302.28377077	328.75751275	357.93240094	390.10119621	425.58931318
107	283.76935164	308.57373676	335.92171611	366.09105816	399.39084663	436.16507273
108	289.38060360	314.97377716	343.22024829	374.41287932	408.87790212	446.97878687
109	295.08303841	321.48581826	350.65562794	382.90113691	418.56655754	458.03580957
110	300.87813779	328.11182007	358.23042097	391.55915965	428.46109689	469.34161529
111	306.76740752	334.85377693	365.94724136	400.39034284	438.56589520	480.90180163
112	312.75237790	341.71371802	373.80875214	409.39814970	448.88542047	492.72209217
113	318.83460404	348.69370809	381.81766624	418.58611269	459.42423566	504.80833924
114	325.01566635	355.79584798	389.97674748	427.95783495	470.18700067	517.16652687
115	331.29717093	363.02227532	398.28881150	437.51699165	481.17847443	529.80277373
116	337.68074996	370.35516514	406.75672671	447.26733148	492.40351701	542.72333614
117	344.16806215	377.85673053	415.38341534	457.21267811	503.86709175	555.93461120
118	350.76079316	385.46922331	424.17185437	467.35693167	515.57426745	569.44313995
119	357.46065604	393.21493472	433.12507664	477.70407030	527.53022063	583.25561060
120	364.26939171	401.09619608	442.24617183	488.25815171	539.74023782	597.37886184

639

Business mathematics and statistics

Future value (amount) of an annuity $s_{\overline{n}|i} = \dfrac{(1+i)^n - 1}{i}$ —(continued)

n \ i	$2\tfrac{3}{8}\%$ .02375000	$2\tfrac{1}{2}\%$ .02500000	$2\tfrac{5}{8}\%$ .02625000	$2\tfrac{3}{4}\%$ .02750000	$2\tfrac{7}{8}\%$ .02875000	3% .03000000
1	1.00000000	1.00000000	1.00000000	1.00000000	1.00000000	1.00000000
2	2.02375000	2.02500000	2.02625000	2.02750000	2.02875000	2.03000000
3	3.07181406	3.07562500	3.07943906	3.08325625	3.08707656	3.09090000
4	4.14476965	4.15251562	4.16027434	4.16804580	4.17583001	4.18362700
5	5.24320793	5.25632852	5.26948154	5.28266706	5.29588513	5.30913581
6	6.36773411	6.38773673	6.40780543	6.42794040	6.44814182	6.46840988
7	7.51896780	7.54743015	7.57601032	7.60470876	7.63352590	7.66246218
8	8.69754328	8.73611590	8.77488059	8.81383825	8.85298977	8.89233605
9	9.90410994	9.95451880	10.00522121	10.05621880	10.10751323	10.15910613
10	11.13933255	11.20338177	11.26785827	11.33276482	11.39810423	11.46387931
11	12.40389170	12.48346631	12.56363954	12.64441585	12.72579973	12.80779569
12	13.69848412	13.79555297	13.89343508	13.99213729	14.09166647	14.19202956
13	15.02382312	15.14044179	15.25813775	15.37692107	15.49680188	15.61779045
14	16.38063892	16.51895284	16.65866387	16.79978639	16.94233494	17.08632416
15	17.76967910	17.93192666	18.09595580	18.26178052	18.42942707	18.59891389
16	19.19170897	19.38022483	19.57097258	19.76397948	19.95927309	20.15688130
17	20.64751206	20.86473045	21.08471061	21.30748892	21.53310220	21.76158774
18	22.13789047	22.38634871	22.63818427	22.89344487	23.15217888	23.41443537
19	23.66366537	23.94600743	24.23243660	24.52301460	24.81780403	25.11686844
20	25.22567743	25.54465761	25.86853807	26.19739750	26.53131589	26.87037449
21	26.82478726	27.18327045	27.54758719	27.91782593	28.29409122	28.67648572
22	28.46187596	28.86285590	29.27071135	29.68556615	30.10754635	30.53678030
23	30.13784552	30.58442730	31.03906753	31.51091921	31.97313830	32.45288370
24	31.85361935	32.34903798	32.85384305	33.36822199	33.89236603	34.42647022
25	33.61014281	34.15776393	34.71625643	35.28584810	35.86677155	36.45926432
26	35.40838370	36.01170803	36.62755816	37.25620892	37.89794124	38.55304225
27	37.24933281	37.91200073	38.58903156	39.28075467	39.98750705	40.70963352
28	39.13400446	39.85980075	40.60199364	41.36097542	42.13714787	42.93092252
29	41.06343707	41.85629577	42.66779597	43.49840224	44.34859088	45.21885020
30	43.03869370	43.90270316	44.78782562	45.69460831	46.62361286	47.57541571
31	45.06086268	46.00027074	46.96350604	47.95121003	48.96404173	50.00267818
32	47.13105817	48.15027751	49.19629807	50.26986831	51.37175793	52.50275852
33	49.25042080	50.35403445	51.48770090	52.65228696	53.84869597	55.07784128
34	51.42011829	52.61288531	53.83925305	55.10022765	56.39684598	57.73017652
35	53.64134610	54.92820744	56.25253344	57.61548391	59.01825530	60.46208181
36	55.91532807	57.30141263	58.72916244	60.19990972	61.71503014	63.27594427
37	58.24331711	59.73394794	61.27080296	62.85540724	64.48933726	66.17422259
38	60.62659589	62.22729664	63.87916153	65.58393094	67.34340571	69.15944927
39	63.06647755	64.78297906	66.55598952	68.38748904	70.27952862	72.23423275
40	65.56430639	67.40255354	69.30308425	71.26814499	73.30006507	75.40125973
41	68.12145866	70.08761737	72.12229021	74.22801898	76.40744194	78.66329753
42	70.73934331	72.83980781	75.01550033	77.26928950	79.60415590	82.02319645
43	73.41940271	75.66080300	77.98465721	80.39419496	82.89277538	85.48389234
44	76.16311352	78.55232308	81.03175646	83.60503532	86.27594267	89.04840911
45	78.97198747	81.51613116	84.15883802	86.90417379	89.75637602	92.71986139
46	81.84757217	84.55403443	87.36800752	90.29403857	93.33687183	96.50145723
47	84.79145201	87.66788530	90.66141771	93.77712463	97.02030690	100.39650095
48	87.80524900	90.85958243	94.04127993	97.35599556	100.80964072	104.40839598
49	90.89062366	94.13107199	97.50986353	101.03328544	104.70791789	108.54064785
50	94.04927597	97.48434879	101.06949745	104.81170079	108.71827053	112.79686729
51	97.28294628	100.92145751	104.72257175	108.69402256	112.84392081	117.18077331
52	100.59341625	104.44449395	108.47153926	112.68310818	117.08818353	121.69619651
53	103.98250989	108.05560629	112.31891717	116.78189365	121.45446881	126.34708240
54	107.45209450	111.75699645	116.26728874	120.99339573	125.94628479	131.13749488
55	111.00408174	115.55092136	120.31930507	125.32071411	130.56724047	136.07161972
56	114.64042868	119.43969440	124.47768683	129.76703375	135.32104864	141.15376831
57	118.36313886	123.42568676	128.74522611	134.33562718	140.21152879	146.38838136
58	122.17426341	127.51132893	133.12478830	139.02985692	145.24261024	151.78003280
59	126.07590217	131.69911215	137.61931399	143.85317799	150.41833528	157.33343379

Appendix C

Future value (amount) of an annuity $s_{\overline{n}|i} = \dfrac{(1+i)^n - 1}{i}$ —*(continued)*

i \ n	$2\frac{3}{8}\%$ .02375000	$2\frac{1}{2}\%$ .02500000	$2\frac{5}{8}\%$ .02625000	$2\frac{3}{4}\%$ .02750000	$2\frac{7}{8}\%$ .02875000	3% .03000000
60	130.07020485	135.99158995	142.23182098	148.80914038	155.74286242	163.05343680
61	134.15937221	140.39137970	146.96540628	153.90139174	161.22046972	168.94503991
62	138.34565730	144.90116419	151.82324820	159.13368002	166.85555822	175.01339110
63	142.63136666	149.52369330	156.80860846	164.50985622	172.65265552	181.26379284
64	147.01886162	154.26178563	161.92483443	170.03387726	178.61641937	187.70170662
65	151.51055958	159.11833027	167.17536134	175.70980889	184.75164142	194.33275782
66	156.10893537	164.09628853	172.56371457	181.54182863	191.06325111	201.16274055
67	160.81652259	169.19869574	178.09351208	187.53422892	197.55631958	208.19762277
68	165.63591500	174.42866314	183.76846677	193.69142022	204.23606377	215.44355145
69	170.56976798	179.78937971	189.59238902	200.01793427	211.10785061	222.90685800
70	175.62079997	185.28411421	195.56918924	206.51842746	218.17720131	230.59406374
71	180.79179397	190.91621706	201.70288045	213.19768422	225.44979585	238.51188565
72	186.08559908	196.68912249	207.99758107	220.06062054	232.93147748	246.66724222
73	191.50513205	202.60635055	214.45751757	227.11228760	240.62825746	255.06725949
74	197.05337894	208.67150931	221.08702740	234.35787551	248.54631986	263.71927727
75	202.73339669	214.88829705	227.89056187	241.80271709	256.69202655	272.63085559
76	208.54831486	221.26050447	234.87268912	249.45229181	265.07192232	281.80978126
77	214.50133734	227.79201709	242.03809721	257.31222983	273.69274008	291.26407469
78	220.59574410	234.48681751	249.39159726	265.38831615	282.56140636	301.00199693
79	226.83489302	241.34898795	256.93812669	273.68649485	291.68504679	311.03205684
80	233.22222173	248.38271265	264.68275252	282.21287345	301.07099189	321.36301855
81	239.76124950	255.59228047	272.63067477	290.97372747	310.72678291	332.00390910
82	246.45557918	262.98208748	280.78722998	299.97550098	320.66017792	342.96402638
83	253.30889918	270.55663966	289.15789477	309.22483137	330.87915803	354.25294717
84	260.32498554	278.32055566	297.74828951	318.72851423	341.39193382	365.88053558
85	267.50770394	286.27856955	306.56418211	328.49354837	352.20695192	377.85695165
86	274.86101191	294.43553379	315.61149189	338.52712095	363.33290179	390.19266020
87	282.38896094	302.79642213	324.89629355	348.83661678	374.77872272	402.89844001
88	290.09569877	311.36633268	334.42482126	359.42962374	386.55361099	415.98539321
89	297.98547161	320.15049100	344.20347282	370.31393839	398.66702731	429.46495500
90	306.06262656	329.15425328	354.23881398	381.49757170	411.12870434	443.34890365
91	314.33161394	338.38310961	364.53758284	392.98875492	423.94865459	457.64937076
92	322.79698978	347.84268735	375.10669439	404.79594568	437.13717841	472.37885189
93	331.46341828	357.53875453	385.95324512	416.92783418	450.70487229	487.55021744
94	340.33567447	367.47722339	397.08451781	429.39334962	464.66263737	503.17672397
95	349.41864674	377.66415398	408.50798640	442.20166674	479.02168820	519.27202568
96	358.71733960	388.10575783	420.23132104	455.36221257	493.79356173	535.85018646
97	368.23687641	398.80840177	432.26239322	468.88467342	508.99012663	552.92569205
98	377.98250223	409.77861182	444.60928104	482.77900194	524.62359277	570.51346281
99	387.95958665	421.02307711	457.28027467	497.05542449	540.70652106	588.62886669
100	398.17362684	432.54865404	470.28388188	511.72444867	557.25183355	607.28773270
101	408.63025047	444.36237039	483.62883378	526.79687100	574.27282376	626.50636468
102	419.33521892	456.47142965	497.32409066	542.28378496	591.78316744	646.30155562
103	430.29443037	468.88321539	511.37884804	558.19658904	609.79693351	666.69060228
104	441.51392309	481.60529578	525.80254280	574.54699524	628.32859535	687.69132035
105	452.99987877	494.64542817	540.60485955	591.34703761	647.39304246	709.32205996
106	464.75862589	508.01156388	555.79573712	608.60908114	667.00559243	731.60172176
107	476.79664325	521.71185298	571.38537522	626.34583088	687.18200321	754.54977342
108	489.12056353	535.75464930	587.38424131	644.57034123	707.93848581	778.18626662
109	501.73717691	550.14851553	603.80307765	663.29602561	729.29171727	802.53185462
110	514.65343486	564.90222842	620.65290844	682.53666631	751.24885415	827.60781026
111	527.87645394	580.02478413	637.94504728	702.30642464	773.85754620	853.43604456
112	541.41351972	595.52540373	655.69110478	722.61985131	797.10595066	880.03912590
113	555.27209082	611.41353883	673.90299628	743.49189723	821.02274674	907.44029968
114	569.45980297	627.69887730	692.59294993	764.93792440	845.62715071	935.66350867
115	583.98447329	644.39134923	711.77351486	786.97371732	870.93893129	964.73341393
116	598.85410454	661.50113296	731.45756963	809.61549455	896.97842556	994.67541635
117	614.07688952	679.03866129	751.65833083	832.87992065	923.76655530	1025.51567884
118	629.66121564	697.01462782	772.38936202	856.78411846	951.32484376	1057.28114920
119	645.61566952	715.43999351	793.66458277	881.34568172	979.67543302	1089.99958368
120	661.94904167	734.32599335	815.49827807	906.58268197	1008.84110172	1123.69957119

Business mathematics and statistics

Future value (amount) of an annuity $s_{\overline{n}|i} = \dfrac{(1+i)^n - 1}{i}$ —*(continued)*

n \ i	$3\tfrac{1}{4}\%$ .03250000	$3\tfrac{1}{2}\%$ .03500000	$3\tfrac{3}{4}\%$ .03750000	4% .04000000	$4\tfrac{1}{4}\%$ .04250000	$4\tfrac{1}{2}\%$ .04500000
1	1.00000000	1.00000000	1.00000000	1.00000000	1.00000000	1.00000000
2	2.03250000	2.03500000	2.03750000	2.04000000	2.04250000	2.04500000
3	3.09855625	3.10622500	3.11390625	3.12160000	3.12930625	3.13702500
4	4.19925933	4.21494287	4.23067773	4.24646400	4.26230177	4.27819113
5	5.33573526	5.36246588	5.38932815	5.41632256	5.44344959	5.47070973
6	6.50914665	6.55015218	6.59142796	6.63297546	6.67479620	6.71689166
7	7.72069392	7.77940751	7.83860650	7.89829448	7.95847504	8.01915179
8	8.97161647	9.05168677	9.13255425	9.21422626	9.29671023	9.38001362
9	10.26319401	10.36849581	10.47502503	10.58279531	10.69182041	10.80211423
10	11.59674781	11.73139316	11.86783847	12.00610712	12.14622278	12.28820937
11	12.97364212	13.14199192	13.31288241	13.48635141	13.66243725	13.84117879
12	14.39528548	14.60196164	14.81211550	15.02580546	15.24309083	15.46403184
13	15.86313226	16.11303030	16.36756983	16.62683768	16.89092219	17.15991327
14	17.37868406	17.67698636	17.98135370	18.29191119	18.60878638	18.93210937
15	18.94349129	19.29568088	19.65565447	20.02358764	20.39965980	20.78405429
16	20.55915476	20.97102971	21.39274151	21.82453114	22.26664534	22.71933673
17	22.22732729	22.70501575	23.19496932	23.69751239	24.21297777	24.74170689
18	23.94971543	24.49969130	25.06478067	25.64541288	26.24202933	26.85508370
19	25.72808118	26.35718050	27.00470994	27.67122940	28.35731557	29.06356246
20	27.56424382	28.27968181	29.01738656	29.77807858	30.56250149	31.37142277
21	29.46008174	30.26947068	31.10553856	31.96920172	32.86140780	33.78313680
22	31.41753440	32.32890215	33.27199626	34.24796979	35.25801763	36.30337795
23	33.43860426	34.46041373	35.51969612	36.61788858	37.75648338	38.93702996
24	35.52535890	36.66652821	37.85168472	39.08260412	40.36113392	41.68919631
25	37.67993307	38.94985669	40.27112290	41.64590829	43.07648211	44.56521015
26	39.90453089	41.31310168	42.78129001	44.31174462	45.90723260	47.57064460
27	42.20142815	43.75906024	45.38558838	47.08421440	48.85828999	50.71132361
28	44.57297456	46.29062734	48.08754794	49.96758298	51.93476732	53.99333317
29	47.02159623	48.91079930	50.89083099	52.96628630	55.14199493	57.42303316
30	49.54979811	51.62267728	53.79923715	56.08493775	58.48552971	61.00706966
31	52.16016655	54.42947098	56.81670855	59.32833526	61.97116472	64.75238779
32	54.85537196	57.33450247	59.94733512	62.70146867	65.60493922	68.66624524
33	57.63817155	60.34121005	63.19536019	66.20952742	69.39314914	72.75622628
34	60.51141213	63.45315240	66.56518619	69.85790851	73.34235798	77.03025646
35	63.47803302	66.67401274	70.06138067	73.65222486	77.45940819	81.49661800
36	66.54106909	70.00760318	73.68868245	77.59831385	81.75143304	86.16396581
37	69.70365384	73.45786930	77.45200804	81.70224640	86.22586895	91.04134427
38	72.96902259	77.02889472	81.35645834	85.97033626	90.89046838	96.13820476
39	76.34051582	80.72490604	85.40732553	90.40914971	95.75331328	101.46442398
40	79.82158259	84.55027775	89.61010024	95.02551570	100.82282910	107.03032306
41	83.41578402	88.50953747	93.97047900	99.82653633	106.10779933	112.84668760
42	87.12679700	92.60737128	98.49437196	104.81959778	111.61738080	118.92478854
43	90.95841791	96.84862928	103.18791091	110.01238169	117.36111949	125.27640402
44	94.91456649	101.23833130	108.05745757	115.41287696	123.34896707	131.91384220
45	98.99928990	105.78167290	113.10961223	121.02939204	129.59129817	138.84996510
46	103.21676682	110.48403145	118.35122269	126.87056772	136.09892834	146.09821353
47	107.57131174	115.35097255	123.78939354	132.94539043	142.88313279	153.67263314
48	112.06737937	120.38825659	129.43149579	139.26320604	149.95566594	161.58790163
49	116.70956920	125.60184557	135.28517689	145.83373429	157.32878174	169.85935720
50	121.50263020	130.99791016	141.35837102	152.66708366	165.01525496	178.50302828
51	126.45146568	136.58283702	147.65930993	159.77376700	173.02840330	187.53566455
52	131.56113832	142.36323631	154.19653405	167.16471768	181.38211044	196.97476946
53	136.83687531	148.34594958	160.97890408	174.85130639	190.09085013	206.83863408
54	142.28407376	154.53805782	168.01561298	182.84535865	199.16971127	217.14637262
55	147.90830616	160.94688984	175.31619847	191.15917299	208.63442399	227.91795938
56	153.71532611	167.58003099	182.89055591	199.80553991	218.50138701	239.17426756
57	159.71107421	174.44533207	190.74895176	208.79776151	228.78769596	250.93710960
58	165.90168412	181.55091869	198.90203745	218.14967197	239.51117304	263.22927953
59	172.29348885	188.90520085	207.36086386	227.87565885	250.69039789	276.07459711
60	178.89302724	196.51688288	216.13689625	237.99068520	262.34473980	289.49795398

Appendix C

Future value (amount) of an annuity $s_{\overline{n}|i} = \dfrac{(1+i)^n - 1}{i}$ —(continued)

n \ i	$4\frac{3}{4}\%$ .04750000	5% .05000000	$5\frac{1}{4}\%$ .05250000	$5\frac{1}{2}\%$ .05500000	$5\frac{3}{4}\%$ .05750000	6% .06000000
1	1.00000000	1.00000000	1.00000000	1.00000000	1.00000000	1.00000000
2	2.04750000	2.05000000	2.05250000	2.05500000	2.05750000	2.06000000
3	3.14475625	3.15250000	3.16025625	3.16802500	3.17580625	3.18360000
4	4.29413217	4.31012500	4.32616970	4.34226638	4.35841511	4.37461600
5	5.49810345	5.52563125	5.55329361	5.58109103	5.60902398	5.63709296
6	6.75926336	6.80191281	6.84484153	6.88805103	6.93154286	6.97531854
7	8.08032837	8.14200845	8.20419571	8.26689384	8.33010657	8.39383765
8	9.46414397	9.54910888	9.63491598	9.72157300	9.80908770	9.89746791
9	10.91369081	11.02656432	11.14074907	11.25625951	11.37311024	11.49131598
10	12.43209112	12.57789254	12.72563840	12.87535379	13.02706408	13.18079494
11	14.02261545	14.20678716	14.39373441	14.58349825	14.77612027	14.97164264
12	15.68868969	15.91712652	16.14940547	16.38559065	16.62574718	16.86994120
13	17.43390245	17.71298285	17.99724926	18.28679814	18.58172764	18.88213767
14	19.26201281	19.59863199	19.94210484	20.29257203	20.65017698	21.01506593
15	21.17695842	21.57856359	21.98906535	22.40866350	22.83756216	23.27596988
16	23.18286395	23.65749177	24.14349128	24.64113999	25.15072198	25.67252808
17	25.28404998	25.84036636	26.41102457	26.99640269	27.59688850	28.21287976
18	27.48504236	28.13238467	28.79760336	29.48120483	30.18370959	30.90565255
19	29.79058187	30.53900391	31.30947754	32.10267110	32.91927289	33.75999170
20	32.20563451	33.06595410	33.95322511	34.86831801	35.81213108	36.78559120
21	34.73540215	35.71925181	36.73576943	37.78607550	38.87132862	39.99272668
22	37.38533375	38.50521440	39.66439732	40.86430965	42.10643001	43.39229028
23	40.16113710	41.43047512	42.74677818	44.11184669	45.52754974	46.99582769
24	43.06879111	44.50199887	45.99098403	47.53799825	49.14538385	50.81557735
25	46.11455869	47.72709882	49.40551070	51.15258816	52.97124342	54.86451200
26	49.30500023	51.11345376	52.99930001	54.96598051	57.01708991	59.15638272
27	52.64698774	54.66912645	56.78176326	58.98910943	61.29557258	63.70576568
28	56.14771966	58.40258277	60.76280583	63.23351045	65.82006801	68.52811162
29	59.81473634	62.32271191	64.95285313	67.71135353	70.60472192	73.63979832
30	63.65593632	66.43884750	69.36287792	72.43547797	75.66449343	79.05818622
31	67.67959329	70.76078988	74.00442901	77.41942926	81.01520180	84.80167739
32	71.89437398	75.29882937	78.88966154	82.67749787	86.67357590	90.88977803
33	76.30935674	80.06377084	84.03136877	88.22476025	92.65730652	97.34316471
34	80.93405119	85.06695938	89.44301563	94.07712207	98.98510164	104.18375460
35	85.77841862	90.32030735	95.13877395	100.25136378	105.67674499	111.43477987
36	90.85289350	95.83632272	101.13355958	106.76518879	112.75315782	119.12086666
37	96.16840594	101.62813886	107.44307146	113.63727417	120.23646440	127.26811866
38	101.73640522	107.70954580	114.08383271	120.88732425	128.15006110	135.90420578
39	107.56888447	114.09502309	121.07323393	128.53612708	136.51868962	145.05845813
40	113.67840648	120.79977424	128.42957871	136.60561407	145.36851427	154.76196562
41	120.07813079	127.83976295	136.17213159	145.11892285	154.72720384	165.04768356
42	126.78184201	135.23175110	144.32116850	154.10046360	164.62401806	175.95054457
43	133.80397950	142.99333866	152.89802985	163.57598910	175.08989910	187.50757724
44	141.15966853	151.14300559	161.92517642	173.57266850	186.15756830	199.75803188
45	148.86475278	159.70015587	171.42624818	184.11916527	197.86162847	212.74351379
46	156.93582854	168.68516366	181.42612621	195.24571936	210.23867211	226.50812462
47	165.39028039	178.11942185	191.95097833	206.98423292	223.32739576	241.09861210
48	174.24631871	188.02539294	203.02842522	219.36836679	237.16872101	256.56452882
49	183.52301885	198.42666259	214.68741754	232.43362696	251.80592247	272.95840055
50	193.24036225	209.34799572	226.95850696	246.21747645	267.28476301	290.33590458
51	203.41927945	220.81539550	239.87382858	260.75943765	283.65363689	308.75605886
52	214.08169523	232.85616528	253.46720458	276.10120672	300.96372101	328.28142239
53	225.25057575	245.49897354	267.77423282	292.28677309	319.26913497	348.97830773
54	236.94997810	258.77392222	282.83238004	309.36254561	338.62711023	370.91700620
55	249.20510206	272.71261833	298.68107999	327.37748562	359.09816906	394.17202657
56	262.04234441	287.34824924	315.36183669	346.38324733	380.74631379	418.82234816
57	275.48935577	302.71566171	332.91833312	366.43432593	403.63922683	444.95168905
58	289.57510017	318.85144479	351.39654561	387.58821386	427.84848237	472.64879040
59	304.32991742	335.79401703	370.84486425	409.90556562	453.44977011	502.00771782
60	319.78558850	353.58371788	391.31421963	433.45037173	480.52313189	533.12818089

Business mathematics and statistics

Future value (amount) of an annuity $s_{\overline{n}|i} = \dfrac{(1+i)^n - 1}{i}$ —(continued)

n \ i	$6\frac{1}{2}\%$ .06500000	7% .07000000	$7\frac{1}{2}\%$ .07500000	8% .08000000	$8\frac{1}{2}\%$ .08500000	9% .09000000
1	1.00000000	1.00000000	1.00000000	1.00000000	1.00000000	1.00000000
2	2.06500000	2.07000000	2.07500000	2.08000000	2.08500000	2.09000000
3	3.19922500	3.21490000	3.23062500	3.24640000	3.26222500	3.27810000
4	4.40717462	4.43994300	4.47292188	4.50611200	4.53951413	4.57312900
5	5.69364098	5.75073901	5.80839102	5.86660096	5.92537283	5.98471061
6	7.06372764	7.15329074	7.24402034	7.33592907	7.42902952	7.52333456
7	8.52286994	8.65402109	8.78732187	8.92280336	9.06049702	9.20043468
8	10.07685648	10.25980257	10.44637101	10.63662763	10.83063927	11.02847380
9	11.73185215	11.97798875	12.22984883	12.48755784	12.75124361	13.02103644
10	13.49442254	13.81644796	14.14708750	14.48656247	14.83509932	15.19292972
11	15.37156001	15.78359932	16.20811906	16.64548746	17.09608276	17.56029339
12	17.37071141	17.88845127	18.42372799	18.97712646	19.54924979	20.14071980
13	19.49980765	20.14064286	20.80550759	21.49529658	22.21093603	22.95338458
14	21.76729515	22.55048786	23.36592066	24.21492030	25.09886559	26.01918919
15	24.18216933	25.12902201	26.11834047	27.15211393	28.23226916	29.36091622
16	26.75401034	27.88805355	29.07724206	30.32428304	31.63201204	33.00339868
17	29.49302101	30.84021730	32.25803521	33.75022569	35.32073306	36.97370456
18	32.41006738	33.99903251	35.67738785	37.45024374	39.32299538	41.30133797
19	35.51672176	37.37896479	39.35319194	41.44626324	43.66544998	46.01845839
20	38.82530867	40.99549232	43.30468134	45.76196430	48.37701323	51.16011964
21	42.34895373	44.86517678	47.55253244	50.42292144	53.48905936	56.76453041
22	46.10163573	49.00573916	52.11897237	55.45675516	59.03562940	62.87333815
23	50.09824205	53.43614090	57.02789530	60.89329557	65.05365790	69.53193858
24	54.35462778	58.17667076	62.30498744	66.76475922	71.58321882	76.78981305
25	58.88767859	63.24903772	67.97786150	73.10593995	78.66779242	84.70089623
26	63.71537769	68.67647036	74.07620112	79.95441515	86.35455478	93.32397689
27	68.85687725	74.48382328	80.63191620	87.35076836	94.69469193	102.72313481
28	74.33257427	80.69769091	87.67930991	95.33882983	103.74374075	112.96821694
29	80.16419159	87.34652927	95.25525816	103.96593622	113.56195871	124.13535646
30	86.37486405	94.46078632	103.39940252	113.28321111	124.21472520	136.30753855
31	92.98923021	102.07304137	112.15435771	123.34586800	135.77297684	149.57521702
32	100.03353017	110.21815426	121.56593454	134.21353744	148.31367987	164.03698655
33	107.53570963	118.93342506	131.68337963	145.95062044	161.92034266	179.80031534
34	115.52553076	128.25876481	142.55963310	158.62667007	176.68357179	196.98234372
35	124.03469026	138.23687835	154.25160558	172.31680368	192.70167539	215.71075465
36	133.09694513	148.91345984	166.82047600	187.10214797	210.08131780	236.12472257
37	142.74824656	160.33740202	180.33201170	203.07031981	228.93822981	258.37594760
38	153.02688259	172.56102017	194.85691258	220.31594540	249.39797935	282.62978288
39	163.97362996	185.64029158	210.47118102	238.94122103	271.59680759	309.06646334
40	175.63191590	199.63511199	227.25651960	259.05651871	295.68253624	337.88244504
41	188.04799044	214.60956983	245.30075857	280.78104021	321.81555182	369.29186510
42	201.27110981	230.63223972	264.69831546	304.24352342	350.16987372	403.52813296
43	215.35373195	247.77649650	285.55068912	329.58300530	380.93431299	440.84566492
44	230.35172453	266.12085125	307.96699080	356.94964572	414.31372959	481.52177477
45	246.32458662	285.74931084	332.06451511	386.50561738	450.53039661	525.85873450
46	263.33568475	306.75176260	357.96935375	418.42606677	489.82548032	574.18602060
47	281.45250426	329.22438598	385.81705528	452.90015211	532.46064615	626.86276245
48	300.74691704	353.27009300	415.75333442	490.13216428	578.71980107	684.28041107
49	321.29546665	378.99899951	447.93483451	530.34273742	628.91098416	746.86564807
50	343.17967198	406.52892947	482.52994709	573.77015642	683.36841782	815.08355640
51	366.48635066	435.98595454	519.71969313	620.67176893	742.45473333	889.44107647
52	391.30796345	467.50497135	559.69867011	671.32551044	806.56338566	970.49077336
53	417.74298108	501.23031925	602.67607037	726.03155128	876.12127345	1058.83494296
54	445.89627485	537.31644170	648.87677565	785.11407538	951.59158169	1155.13008783
55	475.87953271	575.92859262	698.54253382	848.92320141	1033.47686613	1260.09179573
56	507.81170234	617.24359410	751.93322386	917.83705752	1122.32239975	1374.50005734
57	541.81946299	661.45064569	809.32821564	992.26402213	1218.71980373	1499.20506251
58	578.03772808	708.75219089	871.02783182	1072.64514390	1323.31098705	1635.13351813
59	616.61018041	759.36484425	931.35491920	1159.45675541	1430.79242095	1783.29553476
60	657.68984213	813.52038335	1008.65653814	1253.21329584	1559.91977673	1944.79213289

Appendix C

Future value (amount) of an annuity $s_{\overline{n}|i} = \dfrac{(1+i)^n - 1}{i}$ —*(continued)*

n \ i	$9\frac{1}{2}\%$ .09500000	10% .10000000	$10\frac{1}{2}\%$ .10500000	11% .11000000	$11\frac{1}{2}\%$ .11500000	12% .12000000
1	1.00000000	1.00000000	1.00000000	1.00000000	1.00000000	1.00000000
2	2.09500000	2.10000000	2.10500000	2.11000000	2.11500000	2.12000000
3	3.29402500	3.31000000	3.32602500	3.34210000	3.35822500	3.37440000
4	4.60695737	4.64100000	4.67525762	4.70973100	4.74442088	4.77932800
5	6.04461833	6.10510000	6.16615968	6.22780141	6.29002928	6.35284736
6	7.61885707	7.71561000	7.81360644	7.91258957	8.01338264	8.11518904
7	9.34264849	9.48717100	9.63403512	9.78327412	9.93492165	10.08901173
8	11.23020009	11.43588810	11.64560881	11.85943427	12.07743764	12.29969314
9	13.29706910	13.57947691	13.86839773	14.16397204	14.46634296	14.77565631
10	15.56029067	15.93742460	16.32457949	16.72200896	17.12997240	17.54873507
11	18.03851828	18.53116706	19.03866034	19.56142995	20.09991923	20.65458328
12	20.75217752	21.38428377	22.03771967	22.71318724	23.41140994	24.13313327
13	23.72363438	24.52271214	25.35168024	26.21163784	27.10372209	28.02910926
14	26.97737965	27.97498336	29.01360666	30.09491800	31.22065013	32.39260238
15	30.54023072	31.77248169	33.06003536	34.40535898	35.81102489	37.27971466
16	34.44155263	35.94972986	37.53133908	39.18994847	40.92929275	42.75328042
17	38.71350013	40.54470285	42.47212968	44.50084281	46.63616142	48.88367407
18	43.39128265	45.59917313	47.93170330	50.39593551	52.99931998	55.74971496
19	48.51345450	51.15909045	53.96453214	56.93948842	60.09424178	63.43968075
20	54.12223267	57.27499949	60.63080802	64.20283215	68.00507958	72.05244244
21	60.26384478	64.00249944	67.99702806	72.26514368	76.82566374	81.69873554
22	66.98891003	71.40274939	76.13673236	81.21430949	86.66061507	92.50258380
23	74.35285649	79.54302433	85.13108926	91.14788353	97.62658580	104.60289386
24	82.41637785	88.49732676	95.06953833	102.17415072	109.85364317	118.15524112
25	91.24593375	98.34705943	106.05218826	114.41330730	123.48681213	133.33387006
26	100.91429745	109.18176538	118.18766803	127.99877110	138.68779553	150.33393446
27	111.50115571	121.09994192	131.59737317	143.07863592	155.63689201	169.37400660
28	123.09376551	134.20993611	146.41509736	159.81728587	174.53513459	190.69888739
29	135.78767323	148.63092972	162.78868258	178.39718732	195.60667507	214.58275388
30	149.68750218	164.49402269	180.88149425	199.02087793	219.10144270	241.33268434
31	164.90781489	181.94342496	200.87405114	221.91317450	245.29810861	271.29260646
32	181.57405731	201.13776745	222.96582651	247.32362369	274.50739111	304.84771924
33	199.82359275	222.25154420	247.37723830	275.52922230	307.07574108	342.42944555
34	219.80683406	245.47669862	274.35184832	306.83743675	343.38945131	384.52097901
35	241.68848330	271.02436848	304.15879239	341.58955480	383.87923821	431.66349649
36	265.64888921	299.12680533	337.09546560	380.16440582	429.02535060	484.46311607
37	291.88553369	330.03948586	373.49048948	422.98249046	479.36326592	543.59869000
38	320.61465939	364.04343445	413.70699088	470.51056441	535.49004150	609.83053280
39	352.07305203	401.44777789	458.14622492	523.26672650	598.07139627	684.01019674
40	386.51999197	442.59255568	507.25157854	581.82606641	667.84960685	767.09142034
41	424.23939121	487.85181125	561.51299428	646.82693372	745.65231163	860.14239079
42	465.54213337	537.63699237	621.47185868	718.97789643	832.40232747	964.35947768
43	510.76863604	592.40069161	687.72640384	799.06546504	929.12859513	1081.08261500
44	560.29165647	652.64076077	760.93767625	887.96266619	1036.97838357	1221.81252880
45	614.51936383	718.90483685	841.83613225	986.63855947	1157.23089768	1358.23003226
46	673.89870340	791.79532054	931.22892614	1096.16880101	1291.31245091	1522.21763613
47	738.91908022	871.97485259	1030.00796339	1217.74736912	1440.81338277	1705.88375246
48	810.11639284	960.17233785	1139.15879954	1352.69957973	1607.50692179	1911.58980276
49	888.07745016	1057.18957163	1259.77047349	1502.49653350	1793.37021779	2141.98057909
50	973.44480793	1163.90852880	1393.04637321	1668.77115218	2000.60779284	2400.01824858
51	1066.92206468	1281.29938168	1540.31624240	1853.33597892	2231.67768902	2689.02043841
52	1169.27966082	1410.42931984	1703.04944785	2058.20293661	2489.32062325	3012.70289102
53	1281.36122860	1552.47225183	1882.86963987	2285.60525963	2776.59249493	3375.22723795
54	1404.09054532	1708.71947701	2081.57095206	2538.02183819	3096.90063184	3781.25450650
55	1538.47914713	1880.59142471	2301.13590203	2818.20424039	3454.04420451	4236.00504728
56	1685.63466610	2069.65056718	2543.75517174	3129.20670684	3852.25928802	4745.32565295
57	1846.76995938	2277.61562390	2811.84946477	3474.41944459	4296.26910615	5315.76473131
58	2023.21310552	2506.37718629	3108.09365857	3857.60558349	4791.34005335	5954.65649906
59	2216.41835055	2758.01490492	3435.44349272	4282.94219768	5343.34415949	6670.21527895
60	2427.97809385	3034.81639541	3797.16505946	4755.06583942	5958.82873783	7471.64111242

APPENDIX D
Present value of an annuity

Present value of an annuity $a_{\overline{n}|i} = \dfrac{1-(1+i)^{-n}}{i}$ —(continued)

n \ i	$\frac{5}{12}\%$ .00416666	$\frac{11}{24}\%$ .00458333	$\frac{1}{2}\%$ .00500000	$\frac{13}{24}\%$ .00541666	$\frac{7}{12}\%$ .00583333	$\frac{5}{8}\%$ .00625000
1	.99585062	.99543758	.99502488	.99461252	.99420050	.99378882
2	1.98756908	1.98633355	1.98509938	1.98386657	1.98263513	1.98140504
3	2.97517253	2.97270863	2.97024814	2.96779104	2.96533732	2.96288699
4	3.95867804	3.95458346	3.95049566	3.94641462	3.94234034	3.93827279
5	4.93810261	4.93197856	4.92586633	4.91976589	4.91367722	4.90760029
6	5.91346318	5.90491437	5.89638441	5.88787325	5.87938083	5.87090712
7	6.88477661	6.87341123	6.86207404	6.85076494	6.83948384	6.82823068
8	7.85205970	7.83748941	7.82295924	7.80846906	7.79401874	7.77960813
9	8.81532916	8.79716905	8.77906392	8.76101357	8.74301780	8.72507640
10	9.77460165	9.75247023	9.73041186	9.70842626	9.68651314	9.66467220
11	10.72989376	10.70341292	10.67702673	10.65073478	10.62453667	10.59843200
12	11.68122200	11.65001701	11.61893207	11.58796663	11.55712014	11.52639205
13	12.62860283	12.59230229	12.55615131	12.52014916	12.48429509	12.44858837
14	13.57205261	13.53028846	13.48870777	13.44730956	13.40609288	13.36505676
15	14.51158766	14.46399515	14.41662465	14.36947491	14.32254470	14.27583281
16	15.44722422	15.39344188	15.33992502	15.28667210	15.23368156	15.18095186
17	16.37897848	16.31864807	16.25863186	16.19892791	16.13953427	16.08044905
18	17.30686654	17.23963309	17.17276802	17.10626895	17.04013350	16.97435931
19	18.23090443	18.15641618	18.08235624	18.00872171	17.93550969	17.86271733
20	19.15110815	19.06901652	18.98741915	18.90631251	18.82569315	18.74555759
21	20.06749359	19.97745320	19.88797925	19.79906756	19.71071398	19.62291438
22	20.98007661	20.88174520	20.78405896	20.68701291	20.59060213	20.49482174
23	21.88887297	21.78191144	21.67568055	21.57017447	21.46538738	21.36131353
24	22.79389839	22.67797074	22.56286622	22.44857800	22.33509930	22.22242338
25	23.69516853	23.56994184	23.44563803	23.32224915	23.19976732	23.07818473
26	24.59269895	24.45784339	24.32401794	24.19121341	24.05942070	23.92863079
27	25.48650517	25.34169396	25.19802780	25.05549614	24.91408852	24.77379457
28	26.37660266	26.22151203	26.06768936	25.91512256	25.76379968	25.61370889
29	27.26300680	27.09731600	26.93302423	26.77011776	26.60858295	26.44840635
30	28.14573291	27.96912418	27.79405397	27.62050668	27.44846689	27.27791935
31	29.02479626	28.83695480	28.65079997	28.46631414	28.28347993	28.10228010
32	29.90021205	29.70082601	29.50328355	29.30756483	29.11365030	28.92152060
33	30.77199540	30.56075588	30.35152592	30.14428330	29.93900610	29.73567265
34	31.64016139	31.41676239	31.19554818	30.97649396	30.75957524	30.54476585
35	32.50472504	32.26886343	32.03537132	31.80422109	31.57538549	31.34883761
36	33.36570128	33.11707683	32.87101624	32.62748886	32.38646445	32.14791315
37	34.22310501	33.96142032	33.70250372	33.44632129	33.19283955	32.94202550
38	35.07695105	34.80191156	34.52985445	34.26074227	33.99453808	33.73120546
39	35.92725416	35.63856812	35.35308900	35.07077557	34.79158716	34.51548369
40	36.77402904	36.47140750	36.17222786	35.87644482	35.58401374	35.29489062
41	37.61729033	37.30044712	36.98729141	36.67777355	36.37184465	36.06945652
42	38.45705261	38.12570431	37.79829991	37.47478513	37.15510653	36.83921145
43	39.29333040	38.94719633	38.60527354	38.26750282	37.93382588	37.60418529
44	40.12613816	39.76494025	39.40823238	39.05594976	38.70802904	38.36440774
45	40.95549028	40.57895348	40.20719640	39.84014896	39.47774221	39.11990831
46	41.78140111	41.38925274	41.00218547	40.62012329	40.24299143	39.87071634
47	42.60388492	42.19585507	41.79321937	41.39589552	41.00380258	40.61686096
48	43.42295594	42.99877734	42.58031778	42.16748829	41.76020141	41.35837114
49	49.23862832	43.79803634	43.36350028	42.93492412	42.51221349	42.09527566
50	45.05091617	44.59364878	44.14278635	43.69822540	43.25986428	42.82760314
51	45.85983353	45.38563131	44.91819537	44.45741441	44.00317907	43.55538201
52	46.66539439	46.17400047	45.68974664	45.21251329	44.74218301	44.27864050
53	47.46761267	46.95877276	46.45745934	45.96354409	45.47690108	44.99740671
54	48.26650224	47.73996459	47.22135258	46.71052873	46.20735816	45.71170853
55	49.06207692	48.51759229	47.98144535	47.45348900	46.93357895	46.42157370
56	49.85435046	49.29167213	48.73775657	48.19244658	47.65558802	47.12702976
57	50.64333656	50.06222029	49.49030505	48.92742304	48.37340980	47.82810411
58	51.42904885	50.82925288	50.23910950	49.65843982	49.08706856	48.52482396
59	52.21150093	51.59278594	50.98418855	50.38551826	49.79658846	49.21721636

Appendix D

Present value of an annuity $a_{\overline{n}|i} = \dfrac{1-(1+i)^{-n}}{i}$ —*(continued)*

n \ i	$\tfrac{5}{12}\%$ .00416666	$\tfrac{11}{24}\%$ .00458333	$\tfrac{1}{2}\%$ .00500000	$\tfrac{13}{24}\%$ .00541666	$\tfrac{7}{12}\%$ .00583333	$\tfrac{5}{8}\%$ .00625000
60	52.99070632	52.35283545	51.72556075	51.10867958	50.50199350	49.90530818
61	53.76667850	53.10941728	52.46324453	51.82794488	51.20330754	50.58912614
62	54.53943087	53.86254727	53.19725824	52.54333515	51.90055431	51.36869679
63	55.30897680	54.61224117	53.92762014	53.25487126	52.59375739	51.94404650
64	56.07532959	55.35851464	54.65434839	53.96257399	53.28294024	52.61520149
65	56.83850250	56.10138330	55.37746109	54.66646398	53.96812617	53.28218781
66	57.59850871	56.84086268	56.09697621	55.36656177	54.64933836	53.94503137
67	58.35536137	57.57696825	56.81291165	56.06288779	55.32659986	54.60375788
68	59.10907357	58.30971538	57.52528522	56.75546237	55.99993358	55.25839293
69	59.85965832	59.03911942	58.23411465	57.44430571	56.66936230	55.90896191
70	60.60712862	59.76519561	58.93941756	58.12943792	57.33490867	56.55549010
71	61.35149738	60.48795913	59.64121151	58.81087900	57.99659520	57.19800259
72	62.09277748	61.20742510	60.33951394	59.48864882	58.65444427	57.83652431
73	62.83098172	61.92360856	61.03434222	60.16276716	59.30847815	58.47108006
74	63.56612287	62.63652449	61.72571366	60.83325370	59.95871896	59.10169447
75	64.29821365	63.34618779	62.41364543	61.50012801	60.60518869	59.72839202
76	65.02726670	64.05261331	63.09815466	62.16340954	61.24790922	60.35119704
77	65.75329464	64.75581582	63.77925836	62.82311765	61.88690229	60.97013370
78	66.47631002	65.45581003	64.45697350	63.47927160	62.52218952	61.58522604
79	67.19632533	66.15261056	65.13131691	64.13189053	63.15379239	62.19649793
80	67.91335303	66.84623200	65.80230538	64.78099348	63.78173229	62.80397309
81	68.62740550	67.53668884	66.46995561	65.42659940	64.40603044	63.40767512
82	69.33849511	68.22399553	67.13428419	66.06872713	65.02670798	64.00762745
83	70.04663413	68.90816643	67.79530765	66.70739540	65.64378590	64.60385337
84	70.75183482	69.58921586	68.45304244	67.34262286	66.25728507	65.19637602
85	71.45410936	70.26715805	69.10750491	67.97442804	66.86722625	65.78521840
86	72.15346991	70.94200719	69.75871135	68.60282938	67.47363007	66.37040338
87	72.84992854	71.61377737	70.40667796	69.22784522	68.07651706	66.95195367
88	73.54349730	72.28248266	71.05142086	69.84949380	68.67590759	67.52989185
89	74.23418818	72.94813703	71.69295608	70.46779325	69.27182197	68.10420435
90	74.92201313	73.61075441	72.33129958	71.08276162	69.86428033	68.67502146
91	75.60698403	74.27034864	72.96646725	71.69441687	70.45330273	69.24225735
92	76.28911272	74.92693353	73.59847487	73.30277682	71.03890910	69.80597004
93	76.96841101	75.58052280	74.22733818	72.90785925	71.62111923	70.36618141
94	77.64489063	76.23113012	74.85307282	73.50968181	72.19995264	70.92291320
95	78.31856329	76.87876910	75.47569434	74.10826206	72.77542950	71.47618703
96	78.98944062	77.52345327	76.09521825	74.70361746	73.34756869	72.02602438
97	79.65753422	78.16519612	76.71165995	75.29576540	73.91638975	72.57244659
98	80.32285566	78.80401107	77.32503478	75.88472315	74.48191193	73.11547487
99	80.98541642	79.43991148	77.93535799	76.47050790	75.04415436	73.65513030
100	81.64522797	80.07291064	78.54264477	77.05313674	75.60313606	74.19143384
101	82.30230172	80.70302179	79.14691021	77.63262668	76.15887596	74.72440630
102	82.95664901	81.33025810	79.74816937	78.20899463	76.71139283	75.25406838
103	83.60828117	81.95463270	80.34643718	78.78225740	77.26070538	75.78044062
104	84.25720947	82.57615864	80.94172854	79.35243173	77.80683219	76.30354348
105	84.90344511	83.19484892	81.53405825	79.91953425	78.34979174	76.82339724
106	85.54699928	83.81071647	82.12344104	80.48358152	78.88960240	77.34002210
107	86.18788310	84.42377417	82.70989158	81.04458999	79.42628241	77.85343812
108	86.82610765	85.03403484	83.29342446	81.60257603	79.95984996	78.36366521
109	87.46168397	85.64151125	83.87405419	82.15755594	80.49032307	78.87072319
110	88.09462304	86.24621609	84.45179522	82.70954590	81.01771971	79.37463174
111	88.72493581	86.84816202	85.02666191	83.25856202	81.54205770	79.87541043
112	89.35263317	87.44736161	85.59866856	83.80462033	82.06335480	80.37307868
113	89.97772598	88.04382740	86.16782942	84.34778675	82.58162863	80.86765583
114	90.60022504	88.63757186	86.73415862	84.88792715	83.09689674	81.35916108
115	91.22014112	89.22860741	87.29767027	85.42520728	83.60917654	81.84761349
116	91.83748493	89.81694641	87.85837838	85.95959281	84.11848537	82.33303204
117	92.45226715	90.40260115	88.41629690	86.49109936	84.62484047	82.81543557
118	93.06449841	90.98558389	88.97143970	87.01974242	85.12825896	83.29484280
119	93.67418929	91.56590682	89.52382059	87.54553743	85.62875787	83.77127235
120	94.28135033	92.14358207	90.07345333	88.06849972	86.12635414	84.24474271

Business mathematics and statistics

Present value of an annuity $a_{\overline{n}|i} = \dfrac{1-(1+i)^{-n}}{i}$ —(continued)

n \ i	$\frac{2}{3}\%$ .00666666	$\frac{17}{24}\%$ .00708333	$\frac{3}{4}\%$ .00750000	$\frac{19}{24}\%$ .00791666	$\frac{5}{6}\%$ .00833333	$\frac{7}{8}\%$ .00875000
1	.99337748	.99296649	.99255583	.99214551	.99173554	.99132590
2	1.98017631	1.97894893	1.97772291	1.97649824	1.97527491	1.97405294
3	2.96044004	2.95799646	2.95555624	2.95311938	2.95068586	2.94825570
4	3.93421196	3.93015784	3.92611041	3.92206966	3.91803557	3.91400813
5	4.90153506	4.89548151	4.88943961	4.88340933	4.87739065	4.87138352
6	5.86245205	5.85401557	5.84559763	5.83719818	5.82881717	5.82045454
7	6.81700535	6.80580776	6.79463785	6.78349551	6.77238066	6.76129323
8	7.76523710	7.75090552	7.73661325	7.72236016	7.70814611	7.69397098
9	8.70718917	8.68935591	8.67157642	8.65385051	8.63617796	8.61855859
10	9.64290315	9.62120570	9.59957958	9.57802448	9.55654013	9.53512624
11	10.57242035	10.54650132	10.52067452	10.49493954	10.46929600	10.44374348
12	11.49578180	11.46528886	11.43491267	11.40465271	11.37450843	11.34447929
13	12.41302828	12.37761409	12.34234508	12.30722055	12.27223976	12.23740202
14	13.32420028	13.28352247	13.24302242	13.20269918	13.16255183	13.12257945
15	14.22933802	14.18305914	14.13699495	14.09114428	14.04550595	14.00007876
16	15.12848148	15.07626890	15.02431261	14.97261111	14.92116292	14.86996656
17	16.02167035	15.96319626	15.90502492	15.84715447	15.78958306	15.73230885
18	16.90894405	16.84388540	16.77918107	16.71482875	16.65082618	16.58717111
19	17.79034177	17.71838021	17.64682984	17.57568788	17.50495158	17.43461820
20	18.66590242	18.58672425	18.50801969	18.42978542	18.35201810	18.27471445
21	19.53566466	19.44896078	19.36279870	19.27717445	19.19208406	19.10752361
22	20.39966688	20.30513275	20.21121459	20.11790768	20.02520734	19.93310891
23	21.25794723	21.15528283	21.05331473	20.95203739	20.85144529	20.75153300
24	22.11054361	21.99945337	21.88914614	21.77961543	21.67085483	21.56285799
25	22.95749365	22.83768643	22.71875547	22.60069328	22.48349240	22.36714547
26	23.79883475	23.67002376	23.54218905	23.41532198	23.28941395	23.16445647
27	24.63460406	24.49650683	24.35949286	24.22355219	24.08867499	23.95485152
28	25.46483847	25.31717683	25.17071251	25.02543417	24.88133057	24.73839060
29	26.28957464	26.13207464	25.97589331	25.82101778	25.66743527	25.51513319
30	27.10884898	26.94124085	26.77508021	26.61035249	26.44704325	26.28513823
31	27.92269766	27.74471578	27.56831783	27.39348738	27.22020818	27.04846417
32	28.73115662	28.54253946	28.35565045	28.17047115	27.98698332	27.80516894
33	29.53426154	29.33475163	29.13712203	28.94135211	28.74742147	28.55530998
34	30.33204789	30.12139177	29.91277621	29.70617820	29.50157501	29.29994422
35	31.12455088	30.90249907	30.68265629	30.46499697	30.24949588	30.03612809
36	31.91180551	31.67811244	31.44680525	31.21785562	30.99123559	30.76691757
37	32.69384653	32.44827053	32.20526576	31.96480094	31.72684521	31.49136810
38	33.47070848	33.21301169	32.95808016	32.70587940	32.45637541	32.20953467
39	34.24242564	33.97237404	33.70529048	33.44113706	33.17987644	32.92147179
40	35.00903209	34.72639541	34.44693844	34.17061966	33.89739813	33.62723350
41	35.77056168	35.47511336	35.18306545	34.89437254	34.60898988	34.32687335
42	36.52704803	36.21856519	35.91371260	35.61244072	35.31470070	35.02044446
43	37.27852453	36.95678794	36.63892070	36.32486884	36.01457921	35.70799947
44	38.02502437	37.68981839	37.35873022	37.03170121	36.70867360	36.38959055
45	38.76658050	38.41769306	38.07318136	37.73298177	37.39703167	37.06526944
46	39.50322566	39.14044822	38.78231401	38.42875413	38.07970083	37.73508743
47	40.23499238	39.85811987	39.48616775	39.11906156	38.75672809	38.39909535
48	40.96191296	40.57074377	40.18478189	39.80394698	39.42816009	39.05734359
49	41.68401949	41.27835542	40.87819542	40.48345298	40.09404307	39.70988212
50	42.40134387	41.98099008	41.56644707	41.15672180	40.75442288	40.35676047
51	43.11391775	42.67868274	42.24957525	41.82649538	41.40934500	40.99802772
52	43.82177260	43.37146817	42.92761812	42.49011530	42.05885455	41.63373256
53	44.52493967	44.05938089	43.60061351	43.14852283	42.70299624	42.26392324
54	45.22345000	44.74245517	44.26859902	43.80175891	43.34181446	42.88864757
55	45.91733444	45.42072503	44.93161193	44.44986415	43.97535318	43.50795298
56	46.60662361	46.09422428	45.58968926	45.09287886	44.60365605	44.12188647
57	47.29134796	46.76298646	46.24286776	45.73084302	45.22676633	44.73049465
58	47.97153771	47.42704489	46.89118388	46.36379630	45.84472694	45.33382369
59	48.64722289	48.08643266	47.53467382	46.99177805	46.45758043	45.93191939

Appendix D

Present value of an annuity $a_{\overline{n}|i} = \dfrac{1-(1+i)^{-n}}{i}$ —*(continued)*

n \ i	$\frac{2}{3}\%$ .00666666	$\frac{17}{24}\%$ .00708333	$\frac{3}{4}\%$ .00750000	$\frac{19}{24}\%$ .00791666	$\frac{5}{6}\%$ .00833333	$\frac{7}{8}\%$ .00875000
60	49.31843334	48.74118261	48.17337352	47.61482734	47.06536902	46.52482716
61	49.98519868	49.39132738	48.80731863	48.23298289	47.66813457	47.11259198
62	50.64754836	50.03689934	49.43654455	48.84628315	48.26591858	47.69525846
63	51.30551161	50.67793067	50.06108640	49.45476625	48.85876223	48.27287084
64	51.95911749	51.31445329	50.68097906	50.05847003	49.44670634	48.84547296
65	52.60839486	51.94649892	51.29625713	50.65743202	50.02979141	49.41310826
66	53.25337238	52.57409905	51.90695497	51.25168948	50.60805760	49.97581984
67	53.89407852	53.19728495	52.51310667	51.84127935	51.18154473	50.53365039
68	54.53054158	53.81608766	53.11474607	52.42623830	51.75029229	51.08664227
69	55.16278965	54.43053802	53.71190677	53.00660269	52.31433946	51.63483745
70	55.79085064	55.04066663	54.30462210	53.58240863	52.87372509	52.17827752
71	56.41475230	55.64650389	54.89292516	54.15369190	53.42848769	52.71700374
72	57.03452215	56.24807999	55.47684880	54.72048804	53.97866548	53.25105699
73	57.65018756	56.84542490	56.05642461	55.28283228	54.52429634	53.78047781
74	58.26177573	57.43856837	56.63168795	55.84075960	55.06541786	54.30530638
75	58.86931363	58.02753997	57.20266794	56.39430469	55.60206730	54.82558253
76	59.47282811	58.61236902	57.76939746	56.94350196	56.13428162	55.34134575
77	60.07234581	59.19308467	58.33190815	57.48838558	56.66209747	55.85263520
78	60.66789319	59.76971585	58.89023141	58.02898941	57.18555121	56.35948966
79	61.25949654	60.34229129	59.44439842	58.56534708	57.70467889	56.86194762
80	61.84718200	60.91083951	59.99444012	59.09749194	58.21951625	57.36004721
81	62.43097549	61.47538884	60.54038722	59.62545707	58.73009876	57.85382623
82	63.01090281	62.03596740	61.08227019	60.14927530	39.23646158	58.34332216
83	63.58698954	62.59260313	61.62011930	60.66897922	59.73863959	58.82857215
84	64.15926114	63.14532375	62.15396456	61.18460113	60.23666736	59.30961304
85	64.72774285	63.69415681	62.68383579	61.69617309	60.73057920	59.78648133
86	65.29245979	64.23912964	63.20976257	62.20372692	61.22040912	60.25921321
87	65.85343687	64.78026940	63.73177427	62.70729417	61.70619087	60.72784457
88	66.41069888	65.31760304	64.24990002	63.20690617	62.18795788	61.19241097
89	66.96427041	65.85115734	64.76416875	63.70259396	62.66574336	61.65294768
90	67.51417591	66.38095889	65.27460918	64.19438839	63.13958019	62.10948965
91	68.06043964	66.90703406	65.78124981	64.68232002	63.60950101	62.56207152
92	68.60308574	67.42940908	66.28411892	65.16641920	64.07553819	63.01072765
93	69.14213815	67.94810997	66.78324458	65.64671603	64.53772383	63.45549210
94	69.67762068	68.46316257	67.27865467	66.12324038	64.99608975	63.89639861
95	70.20955696	68.97459254	67.77037685	66.59602188	65.45066752	64.33348066
96	70.73797049	69.48242536	68.25843856	67.06508991	65.90148845	64.76677141
97	71.26288460	69.98668633	68.74286705	67.53047366	66.34858358	65.19630375
98	71.78432245	70.48740057	69.22368938	67.99220206	66.79198372	65.62211028
99	72.30230707	70.98459304	69.70093239	68.45030383	67.23171939	66.04422333
100	72.81686133	71.47828850	70.17462272	68.90480743	67.66782088	66.46267492
101	73.32800794	71.96851154	70.64478682	69.35574115	68.10031823	66.87749683
102	73.83576948	72.45528659	71.11145094	69.80313301	68.52924122	67.28872052
103	74.34016835	72.93863791	71.57464113	70.24701084	68.95461939	67.69637722
104	74.84122684	73.41858956	72.03438325	70.68740224	69.37648204	68.10049786
105	75.33896706	73.89516548	72.49070298	71.12433459	69.79485823	68.50111312
106	75.83341099	74.36838938	72.94362579	71.55783507	70.20977675	68.89825341
107	76.32458045	74.83828487	73.39317696	71.98793062	70.62126620	69.29194885
108	76.81249714	75.30487533	73.83938160	72.41464799	71.02935491	69.68222935
109	77.29718259	75.76818403	74.28226461	72.83801371	71.43407099	70.06912451
110	77.77865820	76.22823404	74.72185073	73.25805412	71.83544230	70.45266370
111	78.25694523	76.68504828	75.15816450	73.67479532	72.23349650	70.83287604
112	78.73206480	77.13864951	75.59123027	74.08826324	72.62826099	71.20979037
113	79.20403788	77.58906033	76.02107223	74.49848357	73.01976296	71.58343531
114	79.67288531	78.03630319	76.44771437	74.90548184	73.40802938	71.95383922
115	80.13862779	78.48040035	76.87118052	75.30928335	73.79308699	72.32103020
116	80.60128589	78.92137495	77.29149431	75.70991320	74.17496231	72.68503614
117	81.06088002	79.35924956	77.70867922	76.10739632	74.55368163	73.04588465
118	81.51743048	79.79403819	78.12275853	76.50175740	74.92927103	73.40360312
119	81.97095743	80.22577230	78.53375536	76.89302099	75.30175640	73.75821871
120	82.92148089	80.65446981	78.94169267	77.28121140	75.67116337	74.10975832

Business mathematics and statistics

Present value of an annuity $a_{\overline{n}|i} = \dfrac{1-(1+i)^{-n}}{i}$ —(continued)

i / n	$\frac{11}{12}\%$ .00916666	$\frac{23}{24}\%$ .00958333	1% .01000000	$1\frac{1}{4}\%$ .01250000	$1\frac{3}{8}\%$ .01375000	$1\frac{1}{2}\%$ .01500000
1	.99091660	.99050764	.99009901	.98765432	.98643650	.98522167
2	1.97283230	1.97161301	1.97039506	1.96311538	1.95949346	1.95588342
3	2.94582887	2.94340538	2.94098521	2.92653371	2.91935237	2.91220042
4	3.90998732	3.90197313	3.90196555	3.87805798	3.86619222	3.85438465
5	4.86538793	4.85940385	4.85343124	4.81783504	4.80018962	4.78264497
6	5.81211025	5.80378425	5.79547647	5.74600992	5.72151874	5.69718717
7	6.75023312	6.73920024	6.72819453	6.66272585	6.63035140	6.59821396
8	7.67983463	7.66573693	7.65167775	7.56812429	7.52685712	7.48592508
9	8.60099220	8.58347860	8.56601758	8.46234498	8.41120308	8.36051732
10	9.51378253	9.49250872	9.47130453	9.34552591	9.28355421	9.22218455
11	10.41828162	10.39291000	10.36762825	10.21780337	10.14407320	10.07111779
12	11.31456477	11.28476434	11.25507747	11.07931197	10.99292054	10.90750521
13	12.20270663	12.16815288	12.14474007	11.93018466	11.83025454	11.73153222
14	13.08278113	13.04315597	13.00370304	12.77055275	12.65623136	12.54338150
15	13.95486157	13.90985321	13.86505252	13.60054592	13.47100504	13.34323301
16	14.81902055	14.76832344	14.71787378	14.42029227	14.27472754	14.13126405
17	15.67533002	15.61864476	15.56225127	15.22991829	15.06754874	14.90764931
18	16.52386129	16.46089452	16.39826858	16.02954893	15.84961651	15.67256089
19	17.36488502	17.29514934	17.22600850	16.81930759	16.62107671	16.42616837
20	18.19787120	18.12148511	18.04555297	17.59931613	17.38207320	17.16863879
21	19.02348921	18.93969699	18.85698313	18.36969495	18.13274792	17.90013673
22	19.84160781	19.75069946	19.66037934	19.13056291	18.87324086	18.62082437
23	20.65229510	20.55372625	20.45582113	19.88203744	19.60369012	19.33086145
24	21.45561860	21.34913042	21.24338726	20.62423451	20.32423193	20.03040537
25	22.25164518	22.13698432	22.02315570	21.35726865	21.03500067	20.71961120
26	23.04044114	22.91735962	22.79520366	22.08125299	21.73612890	21.39863172
27	23.82207214	23.69032732	23.55960759	22.79629925	22.42774737	22.06761746
28	24.59660328	24.45595772	24.31644316	23.50251778	23.10998508	22.72671671
29	25.36409904	25.21432048	35.06578530	24.20001756	23.78296925	23.37607558
30	26.12462333	25.96548459	25.80770822	24.88890623	24.44682540	24.01583801
31	26.87823946	26.70951837	26.54228537	25.56299010	25.10167734	24.64614825
32	27.62501020	27.44648951	27.26958947	26.24127418	25.74764719	25.26713874
33	28.36499773	28.17646506	27.98969255	26.90496215	26.38485543	25.87895442
34	29.09826364	28.89951141	28.70266589	27.56045644	27.01342089	26.48172849
35	29.82486901	29.61569433	29.40858009	28.20785822	27.63346080	27.07559458
36	30.54487433	30.32507899	30.10750504	28.84726737	28.24509080	27.66068431
37	31.25833955	31.02772992	30.79950994	29.47878259	28.84842496	28.23712740
38	31.96532408	31.72371102	31.48466330	30.10250133	29.44357579	28.80505163
39	32.66588678	32.41308562	32.16303298	30.71851983	30.03065430	29.36458288
40	33.36008599	33.09591642	32.83468611	31.32693316	30.60976996	29.91584520
41	34.04797952	33.77226554	33.49968922	31.92783522	31.18103079	30.45896079
42	34.72962462	34.44219451	34.15810814	32.52131874	31.74454332	30.99405004
43	35.40507807	35.10576427	34.81000806	33.10747530	32.30041264	31.52123157
44	36.07439611	35.76303518	35.45545352	33.68639536	32.84874243	32.04062223
45	36.73763446	36.41406704	36.09450844	34.25816825	33.38963495	32.55233718
46	37.39484835	37.05891906	36.72723608	34.82288222	33.92319108	33.05648983
47	38.04609250	37.69764992	37.35369909	35.38062442	34.44951031	33.55319195
48	38.69142114	38.33031771	37.97395949	35.93148091	34.96869081	34.04255365
49	39.33088800	38.95697998	38.58807871	36.47553670	35.48082941	34.52468339
50	39.86454633	39.57769375	39.19611753	37.01287575	35.98602161	34.99968807
51	40.59244888	40.19251548	39.79813617	37.54358099	36.48436164	35.46768298
52	41.21464794	40.80150109	40.39419423	38.06773431	36.97594243	35.92874185
53	41.83119532	41.40470599	40.98435072	38.58541660	37.46085566	36.38299690
54	42.44214234	42.00218505	41.56866408	39.09670776	37.93919178	36.83053882
55	43.04753990	42.59399262	42.14719216	39.60168667	38.41103998	37.27146681
56	43.64743838	43.18018254	42.71999224	40.10043128	38.87648496	37.70587863
57	44.24188774	43.76080813	43.28712102	40.59301855	39.33562344	38.13387058
58	44.83093748	44.33592221	43.84863468	41.07952449	39.78853114	38.55553751
59	45.41463664	44.90557709	44.40458879	41.56002419	40.23529582	38.97097292

Present value of an annuity $a_{\overline{n}|i} = \dfrac{1-(1+i)^{-n}}{i}$ —(continued)

n \ i	$1\frac{11}{12}\%$ .00916666	$1\frac{23}{24}\%$ .00958333	1% .01000000	$1\frac{1}{4}\%$ .01250000	$1\frac{3}{8}\%$ .01375000	$1\frac{1}{2}\%$ .01500000
60	45.99303383	45.46982461	44.95503841	42.03459179	40.67600081	39.38026889
61	46.56617721	46.02871608	45.50003803	42.50330054	41.11072829	39.78351614
62	47.13411449	46.58230235	46.03964161	42.96622275	41.53955935	40.18080408
63	47.69689297	47.13063377	46.57390258	43.42342988	41.96257396	40.57222077
64	48.25455951	47.67376024	47.10287385	43.87499247	42.37985101	40.95785298
65	48.80716054	48.21173115	47.62660777	44.32098022	42.79146832	41.33778618
66	49.35474207	48.74459544	48.14515621	44.76146195	43.19750266	41.71210461
67	49.89734970	49.27240159	48.65857050	45.19650563	43.59802975	42.08089125
68	50.43502860	49.79519762	49.16690149	45.62617840	43.99312429	42.44422783
69	50.96782355	50.31303107	49.67019949	46.05054656	44.38285997	42.80219490
70	51.49577891	50.82594906	50.16851435	46.46967562	44.76730946	43.15487183
71	52.01893864	51.33399824	50.66189539	46.88363024	45.14654448	43.50233678
72	52.53734630	51.83722484	51.15039148	47.29247431	45.52063573	43.84466677
73	53.05104506	52.33567462	51.63405097	47.69627093	45.88965300	44.18193771
74	53.56007768	52.82939294	52.11292175	48.09508240	46.25366511	44.51422434
75	54.06448655	53.31842470	52.58705124	48.48897027	46.61273994	44.84160034
76	54.56431367	53.80281440	53.05648638	48.87799533	46.96694445	45.16413826
77	55.05960067	54.28260609	53.52127364	49.26221761	47.31634471	45.48190962
78	55.55038877	54.75784342	53.98145905	49.64169640	47.66100588	45.79498485
79	56.03671885	55.22856963	54.43708817	50.01649027	48.00099224	46.10343335
80	56.51863139	55.69482753	54.88820611	50.38665706	48.33636719	46.40732349
81	56.99616653	56.15665955	55.33485753	50.75225389	48.66719328	46.70672265
82	57.46936403	56.61410768	55.77708666	51.11333717	48.99353221	47.00169720
83	57.93826328	57.06721355	56.21493729	51.46996264	49.31544484	47.29231251
84	58.40290334	57.51601837	56.64845276	51.82218532	49.63299122	47.57863301
85	58.86332288	57.96056298	57.07767600	52.17005958	49.94623055	47.86072218
86	59.31956024	58.40088780	57.50264951	52.51363909	50.25522125	48.13864254
87	59.77165342	58.83703291	57.92341535	52.85297688	50.56002096	48.41245571
88	60.21964005	59.26903796	58.34001520	53.18812531	50.86068653	48.68222237
89	60.66355744	59.69694226	58.75249030	53.51913611	51.15727401	48.94800234
90	61.10344255	60.12078474	59.16088148	53.84606036	51.44983873	49.20985452
91	61.53933201	60.54060395	59.56522919	54.16894850	51.73843524	49.46783696
92	61.97126211	60.95643809	59.96557346	54.48785037	52.02311738	49.72200686
93	62.39926881	61.36832497	60.36195392	54.80281518	52.30393823	49.97242055
94	62.82338775	61.77630208	60.75440982	55.11389154	52.58095016	50.21913355
95	63.24365426	62.18040652	61.14298002	55.42112744	52.85420484	50.46220054
96	63.66010331	62.58067505	61.52770299	55.72457031	53.12375324	50.70167541
97	64.07276959	62.97714408	61.90861682	56.02426698	53.38964561	50.93761124
98	64.48168745	63.36984969	62.28575923	56.32026368	53.65193155	51.17006034
99	64.88689095	63.75882759	62.65916755	56.61260610	53.91065998	51.39907422
100	65.28841383	64.14411317	63.02887877	56.90133936	54.16587914	51.62470367
101	65.68628951	64.52574149	63.39492947	57.18650801	54.41763664	51.84699869
102	66.08055112	64.90374724	63.75735591	57.46815606	54.66597942	52.06600856
103	66.47123150	65.27816483	64.11619397	57.74632697	54.91095380	52.28178183
104	66.85836317	65.64902831	64.47147918	58.02106368	55.15260548	52.49436634
105	67.24197837	66.01637141	64.82324671	58.29240857	55.39097951	52.70380920
106	67.62210903	66.38022757	65.17153140	58.56040353	55.62612035	52.91015685
107	67.99878682	66.74062986	65.51636772	58.82508990	55.85807187	53.11345502
108	68.37204309	67.09761109	65.85778983	59.08650855	56.08687730	53.31374879
109	68.74190893	67.45120372	66.19583151	59.34469980	56.31257934	53.51108255
110	69.10841512	67.80143992	66.53052625	59.59970350	56.53522006	53.70550005
111	69.47159219	68.14835155	66.86190718	59.85155902	56.75484100	53.89704438
112	69.83147038	68.49197017	67.19000710	60.10030520	56.97148311	54.08575801
113	70.18807965	68.83232704	67.51485852	60.34598045	57.18518679	54.27168277
114	70.54144970	69.16945311	67.83649358	60.58862266	57.39599190	54.45485987
115	70.89160994	69.50337906	68.15494414	60.82826930	57.60393775	54.63532993
116	71.23858954	69.83413527	68.47024172	61.06495733	57.80906314	54.81313293
117	71.58241738	70.16175181	68.78241755	61.29872329	58.01140630	54.98830831
118	71.92312209	70.48625850	69.09150252	61.52960325	58.21100498	55.16089488
119	72.26073205	70.80768486	69.39752725	61.75763284	58.40789641	55.33093092
120	72.59527536	71.12606011	69.70052203	61.98284725	58.60211729	55.49845411

Business mathematics and statistics

Present value of an annuity $a_{\overline{n}|i} = \dfrac{1-(1+i)^{-n}}{i}$ —(continued)

n \ i	$1\frac{1}{8}\%$ .01625000	$1\frac{3}{4}\%$ .01750000	$1\frac{7}{8}\%$ .01875000	2% .02000000	$2\frac{1}{8}\%$ .02125000	$2\frac{1}{4}\%$ .02250000
1	.98400984	.98280098	.98159509	.98039216	.97919217	.97799511
2	1.95228521	1.94869875	1.94512402	1.94156094	1.93800947	1.93446955
3	2.90507769	2.89798403	2.89091928	2.88388327	2.87687585	2.86989687
4	3.84263488	3.83094254	3.81930727	3.80772870	3.79620647	3.78474021
5	4.76520037	4.74785508	4.73060836	4.71345951	4.69640780	4.67945253
6	5.67301389	5.64899762	5.62523704	5.60143089	5.57787790	5.55447680
7	6.56631134	6.53464139	6.50320200	6.47199107	6.44100651	6.41024626
8	7.44532481	7.40505297	7.36510626	7.32548144	7.28617528	7.24718461
9	8.31028271	8.26049432	8.21114725	8.16223671	8.11375793	8.06570622
10	9.16140980	9.10122291	9.04161693	8.98258501	8.92412037	8.86621635
11	9.99892724	9.92749181	9.85680190	9.78684805	9.71762092	9.64911134
12	10.82305263	10.73954969	10.65698346	10.57534122	10.49461045	10.41477882
13	11.63400013	11.53764097	11.44243775	11.34837375	11.25543251	11.16359787
14	12.43198045	12.32200587	12.21343583	12.10624877	12.00042351	11.89593924
15	13.21720093	13.09288046	12.97024376	12.84926350	12.72991286	12.61216551
16	13.98986562	13.85049677	13.71312271	13.57770931	13.44422312	13.31263131
17	14.75017527	14.59508282	14.44232904	14.29187188	14.14367013	13.99768343
18	15.49832745	15.32686272	15.15811439	14.99203125	14.82856316	14.66766106
19	16.23451655	16.04605673	15.86072578	15.67846201	15.49920506	15.32289590
20	16.95893388	16.75288130	16.55040568	16.35143334	16.15589234	15.96371237
21	17.67176765	17.44754919	17.22739208	17.01120916	16.79891539	16.59042775
22	18.37320310	18.13026948	17.89191860	17.65804820	17.42855852	17.20335232
23	19.06342249	18.80124764	18.54421458	18.29220412	18.04510015	17.80278955
24	19.74260515	19.46068565	19.18450511	18.91392560	18.64881287	18.38903624
25	20.41092758	20.10878196	19.81301115	19.52345647	19.23996364	18.96238263
26	21.06856342	20.74753166	20.42994960	20.12103576	19.81881385	19.52311260
27	21.71568357	21.37172644	21.03553334	20.70689780	20.38561944	20.07150376
28	22.35245615	21.98695474	21.62997138	21.28127236	20.94063103	20.60782764
29	22.97904665	22.59160171	22.21346884	21.84438466	21.48409403	21.13234977
30	23.59561786	23.18584934	22.78622708	22.39645555	22.01624874	21.64532985
31	24.20232999	23.76987650	23.34844376	22.93770152	22.53733047	22.14702186
32	24.79934071	24.34385897	23.90031290	23.46833482	23.04756962	22.63767419
33	25.38680512	24.90796951	24.44202493	23.98856355	23.54719179	23.11752977
34	25.96487589	25.46237789	24.97376680	24.49859172	24.03641791	23.58682618
35	26.53370321	26.00725100	25.49572201	24.99861933	24.51546429	24.04579577
36	27.09343490	26.54275283	26.00807069	25.48884248	24.98454276	24.49466579
37	27.64421638	27.06904455	26.51098963	25.96945341	25.44386072	24.93365848
38	28.18619078	27.58628457	27.00465240	26.44064060	25.89362127	25.36299118
39	28.71949892	28.09462857	27.48922935	26.90258883	26.33402327	25.78287646
40	29.24427938	28.59422955	27.96488770	27.35547924	26.76526147	26.19352221
41	29.76066852	29.08523789	28.43179161	27.79948945	27.18752653	26.59513174
42	30.26880051	29.56780136	28.89010220	28.23479358	27.60100517	26.98790390
43	30.76880739	30.04206522	29.33997762	28.66156233	28.00588021	27.37203316
44	31.26081908	30.50817221	29.78157312	29.07996307	28.40233069	27.74770969
45	31.74496342	30.96626261	30.21504110	29.49015987	28.79053188	28.11511950
46	32.22136622	31.41647431	30.64053114	29.89231360	29.17065546	28.47444450
47	32.69015127	31.85894281	31.05819008	30.28658196	29.54286948	28.82586259
48	33.15144036	32.29380129	31.46816204	30.67311957	29.90733854	29.16954777
49	33.60535337	32.72118063	31.87058850	31.05207801	30.26422378	29.50567019
50	34.05200823	33.14120946	32.26560835	31.42360589	30.61368302	29.83439627
51	34.49152102	33.55401421	32.65335789	31.78784892	30.95587076	30.15588877
52	34.92400592	33.95971913	33.03397093	32.14494992	31.29093832	30.47030687
53	35.34957532	34.35844632	33.40757883	32.49504894	31.61903385	30.77780623
54	35.76833980	34.75031579	33.77431051	32.83828327	31.94030243	31.07853910
55	36.18040817	35.13544550	34.13429252	33.17478752	32.25488610	31.37265438
56	36.58588750	35.51395135	34.48764910	33.50469365	32.56292396	31.66029768
57	36.98488314	35.88594727	34.83450219	33.82813103	32.86455223	31.94161142
58	37.37749879	36.25154523	35.17497147	34.14522650	33.15990426	32.21673489
59	37.76383645	36.61085526	35.50917445	34.45610441	33.44911066	32.48580429

Appendix D

Present value of an annuity $a_{\overline{n}|i} = \dfrac{1-(1+i)^{-n}}{i}$ —(continued)

n \ i	$1\frac{1}{8}\%$ .01625000	$1\frac{3}{4}\%$ .01750000	$1\frac{7}{8}\%$ .01875000	2% .02000000	$2\frac{1}{8}\%$ .02125000	$2\frac{1}{4}\%$ .02250000
60	38.14399650	36.96398552	35.83722645	34.76088668	33.73229930	32.74895285
61	38.51807774	37.31104228	36.15924069	35.05969282	34.00959540	33.00631086
62	38.88617736	37.65213000	36.47532828	35.35264002	34.28112157	33.25800573
63	39.24839100	37.98735135	36.78559832	35.63984316	34.54699786	33.50416208
64	39.60481280	38.31680723	37.09015786	35.92141486	34.80734185	33.74490179
65	39.95553535	38.64059678	37.38911201	36.19746555	35.06226864	33.98034405
66	40.30064979	38.95881748	37.68256393	36.46810348	35.31189096	34.21060543
67	40.64024579	39.27156509	37.97061490	36.73343478	35.55631917	34.43579993
68	40.97441161	39.57893375	38.25336432	36.99356315	35.79566137	34.65603905
69	41.30323405	39.88101597	38.53090976	37.24859168	36.03002337	34.87143183
70	41.62679858	40.17790267	38.80334701	37.49861929	36.25950881	35.08208492
71	41.94518925	40.46968321	39.07077007	37.74374441	36.48421915	35.28810261
72	42.25848881	40.75644542	38.33327123	37.98406314	36.70425376	35.48958691
73	42.56677865	41.03827560	39.59094109	38.21966975	36.91970993	35.68663756
74	42.87013890	41.31525857	39.84386855	38.45065662	37.13068291	35.87935214
75	43.16864836	41.58747771	40.09214091	38.67711433	37.33726601	36.06782605
76	43.46238461	41.85501495	40.33584384	38.89913170	37.53955056	36.25215262
77	43.75142397	42.11795081	40.57506144	39.11679578	37.73762601	36.43242310
78	44.03584155	42.37636443	40.80987626	39.33019194	37.93157993	36.60872675
79	44.31571124	42.63033359	41.04036933	39.53940386	38.12149810	36.78115085
80	44.59110577	42.87993474	41.26662020	39.74451359	38.30746448	36.94978079
81	44.86209670	43.12524298	41.48870695	39.94560136	38.48956130	37.11470004
82	45.12875444	43.36633217	41.70670621	40.14274663	38.66786908	37.27599026
83	45.39114828	43.60327486	41.92069321	40.33602611	38.84246667	37.43373130
84	45.64934640	43.83614237	42.13074180	40.52551579	39.01343125	37.58800127
85	45.90341589	44.06500479	42.33692447	40.71128999	39.18083844	37.73887655
86	46.15342277	44.28993099	42.53931236	40.89342156	39.34476224	37.88643183
87	46.39943200	44.51098869	42.73797532	41.07198192	39.50527514	38.03074018
88	46.64150750	44.72824441	42.93298191	41.24704110	39.66244812	38.17187304
89	46.87971218	44.94176355	43.12439942	41.41866774	39.81635067	38.30990028
90	47.11410793	45.15161037	43.31229391	41.58692916	39.96705084	38.44489025
91	47.34475565	45.35784803	43.49673022	41.75189133	40.11461526	38.57690978
92	47.57171528	45.56053860	43.67777199	41.91361895	40.25910919	38.70602423
93	47.79504578	45.75974310	43.85548171	42.07217545	40.40059652	38.83229754
94	48.01480520	45.95552147	44.02992070	42.22762299	40.53913980	38.95579221
95	48.23105062	46.14793265	44.20114915	42.38002254	40.67480029	39.07656940
96	48.44383825	46.33703455	44.36922616	42.52943386	40.80763749	39.19468890
97	48.65322337	46.52288408	44.53420973	42.67591555	40.93771161	39.31020920
98	48.85926039	46.70553718	44.69615679	42.81952505	41.06507869	39.42318748
99	49.06200285	46.88504882	44.85512323	42.96031867	41.18979553	39.53367968
100	49.26150341	47.06147304	45.01116391	43.09835164	41.31191729	39.64174052
101	49.45781394	47.23486294	45.16433267	43.23367808	41.43149796	39.74742349
102	49.65098543	47.40527071	45.31468237	43.36635106	41.54859041	39.85078092
103	49.84106807	47.57274762	45.46226491	43.49642261	41.66324643	39.95186398
104	50.02811126	47.73734410	45.60713120	43.62394373	41.77551670	40.05072272
105	50.21216360	47.89910968	45.74933124	43.74896444	41.88545087	40.14740609
106	50.39327292	48.05809305	45.88891410	43.87153377	41.99309754	40.24196194
107	50.57148627	48.21434207	46.02592795	43.99169977	42.09850433	40.33443711
108	50.74684995	48.36790375	46.16042007	44.10950958	42.20171782	40.42487737
109	50.91940955	48.51882432	46.29243688	44.22500939	42.30278367	40.51332750
110	51.08920989	48.66714921	46.42202393	44.33824450	42.40174656	40.59983129
111	51.25629509	48.81292306	46.54922595	44.44925932	42.49865024	40.68443158
112	51.42070858	48.95618974	46.67408682	44.55809737	42.59353756	40.76717025
113	51.58249307	49.09699237	46.79664964	44.66480134	42.68645049	40.84808827
114	51.74169059	49.23537334	46.91695670	44.76941308	42.77743010	40.92722569
115	51.89834253	49.37137429	47.03504952	44.87197361	42.86651662	41.00462170
116	52.05248957	49.50503616	47.15096885	44.97252314	42.95374945	41.08031462
117	52.20417178	49.63639917	47.26475470	45.07110112	43.03916715	41.15434193
118	52.35342857	49.76550287	47.37644633	45.16774620	43.12280749	41.22674027
119	52.50029871	49.89238611	47.48608229	45.26249627	43.20470745	41.29754550
120	52.64482038	50.01708709	47.59370041	45.35538850	43.28490326	41.36679266

Business mathematics and statistics

Present value of an annuity $a_{\overline{n}|i} = \dfrac{1-(1+i)^{-n}}{i}$ —(continued)

i / n	$2\frac{3}{8}\%$ .02375000	$2\frac{1}{2}\%$ .02500000	$2\frac{5}{8}\%$ .02625000	$2\frac{3}{4}\%$ .02750000	$2\frac{7}{8}\%$ .02875000	3% .03000000
1	.97680098	.97560976	.97442144	.97323601	.97205346	.97087379
2	1.93094113	1.92742415	1.92391857	1.92042434	1.91694140	1.91346970
3	2.86294615	2.85602356	2.84912894	2.84226213	2.83542299	2.82861135
4	3.77332958	3.76197421	3.75067375	3.73942787	3.72823620	3.71709840
5	4.66259299	4.64582850	4.62915835	4.61258186	4.59609837	4.57970719
6	5.53122637	5.50812536	5.48517257	5.46236678	5.43970680	5.41719144
7	6.37970829	6.34939060	6.31929117	6.28940806	6.25973929	6.23028296
8	7.20850627	7.17013717	7.13207423	7.09431441	7.05685472	7.01969219
9	8.01807694	7.97086553	7.92406746	7.87767826	7.83169353	7.78610892
10	8.80886637	8.75206393	8.69580264	8.64007616	8.58487828	8.53020284
11	9.58131025	9.51420871	9.44779794	9.38206926	9.31701412	9.25262411
12	10.33583419	10.25776460	10.18055828	10.10420366	10.02868931	9.95400399
13	11.07285391	10.98318497	10.89457567	10.80701086	10.72047563	10.63495533
14	11.79277549	11.69091217	11.59032952	11.49100814	11.39292893	11.29607314
15	12.49599559	12.38137773	12.26828699	12.15669892	12.04658948	11.93793509
16	13.18290168	13.05500266	12.92890328	12.80457315	12.68148248	12.56110203
17	13.85387221	13.71219772	13.57262195	13.43510769	13.29961845	13.16611847
18	14.50927689	14.35336363	14.19987523	14.04876661	13.89999363	13.75351308
19	15.14947681	14.97889134	14.81108427	14.64600157	14.48359041	14.32379911
20	15.77482473	15.58916229	15.40665946	15.22725213	15.05087768	14.87747486
21	16.38566518	16.18454857	15.98700069	15.79294612	15.60231123	15.41502414
22	16.98233473	16.76541324	16.55249762	16.34349987	16.13833412	15.93691664
23	17.56516213	17.33211048	17.10352996	16.87931861	16.65937703	16.44360839
24	18.13446850	17.88498583	17.64046769	17.40079670	17.16585860	16.93554212
25	18.69056752	18.42437642	18.16367131	17.90831795	17.65818576	17.41314769
26	19.23376559	18.95061114	18.67349215	18.40225592	18.13675408	17.87684242
27	19.76436199	19.46401087	19.17027249	18.88297413	18.60194807	18.32703147
28	20.28264908	19.96488866	19.65434591	19.35082640	19.05414150	18.76410823
29	20.78891241	20.45354991	20.12603743	19.80615708	19.49369769	19.18845459
30	21.28343092	20.93029259	20.58566376	20.24930130	19.92096981	19.60044135
31	21.76647709	21.39540741	21.03353350	20.68058520	20.33630115	20.00042849
32	22.23831706	21.84917796	21.46994738	21.10032623	20.74002542	20.38876553
33	22.69921080	22.29188094	21.89519842	21.50883332	21.13246700	20.76579178
34	23.14941226	22.72378628	22.30957216	21.90640712	21.51394119	21.13183668
35	23.58916949	23.14515734	22.71334680	22.29334026	21.88475450	21.48722007
36	24.01872477	23.55625107	23.10679347	22.66991753	22.24520486	21.83225250
37	24.43831480	23.95731812	23.49017634	23.03641609	22.59558188	22.16723544
38	24.84817074	24.34860304	23.86375283	23.39310568	22.93616707	22.49246159
39	25.24851843	24.73034443	24.22777377	23.74024884	23.26723409	22.80821513
40	25.63957844	25.10277505	24.58248358	24.07810106	23.58904894	23.11477197
41	26.02156624	25.46612200	24.92812042	24.40691101	23.90187017	23.41239997
42	26.39469230	25.82060683	25.26491636	24.72692005	24.20594913	23.70135920
43	26.75916220	26.16644569	25.59309755	25.03836563	24.50153014	23.98190213
44	27.11517675	26.50384945	25.91288434	25.34147507	24.78885068	24.25427392
45	27.46293211	26.83302386	26.22449144	25.63647209	25.06814161	24.51871254
46	27.80261989	27.15416962	26.52812807	25.92357381	25.33922733	24.77544907
47	28.13442724	27.46748255	26.82399812	26.20299154	25.60352596	25.02470783
48	28.45853699	27.77315371	27.11230024	26.47493094	25.86004953	25.26670664
49	28.77512771	28.07136947	27.39322801	26.73959215	26.10940416	25.50165693
50	29.08437383	28.36231168	27.66697004	26.99716998	26.35179019	25.72976401
51	29.38644574	28.64615774	27.93371015	27.24785400	26.58740238	25.95122719
52	29.68150988	28.92308072	28.19362743	27.49182871	26.81643001	26.16623999
53	29.96972882	29.19324948	28.44689640	27.72927368	27.03905712	26.37499028
54	30.25126137	29.45682876	28.69368711	27.96036368	27.25546257	26.57766047
55	30.52626263	29.71397928	28.92416528	28.18526879	27.46582024	26.77442764
56	30.79488413	29.96485784	29.16849235	28.40415454	27.67029914	26.96546373
57	31.05727388	30.20961740	29.39682568	28.61718203	27.86906356	27.15093566
58	31.31357644	30.44840722	29.61931857	28.82450806	28.06227321	27.33100549
59	31.56393303	30.68137290	29.83612041	29.02628522	28.25008331	27.50583058

Appendix D

Present value of an annuity $a_{\overline{n}|i} = \dfrac{1-(1+i)^{-n}}{i}$ —(continued)

n \ i	$2\frac{3}{8}\%$ .02375000	$2\frac{1}{2}\%$ .02500000	$2\frac{5}{8}\%$ .02625000	$2\frac{3}{4}\%$ .02750000	$2\frac{7}{8}\%$ .02875000	3% .03000000
60	31.80848159	30.90865649	30.04737676	29.22266201	28.43264477	27.67556367
61	32.04735686	31.13039657	30.25322949	29.41378298	28.61010428	27.84035307
62	32.28069046	31.34672836	30.45381680	29.59978879	28.78260440	28.00034279
63	32.50861095	31.55778377	30.64927337	29.78081634	28.95028374	28.15567261
64	32.73124391	31.76369148	30.83973045	29.95699887	29.11327703	28.30647826
65	32.94871200	31.96457705	31.02531591	30.12846605	29.27171522	28.45289152
66	33.16113504	32.16056298	31.20615436	30.29534409	29.42572560	28.59504031
67	33.36863008	32.35176876	31.38236722	30.45775581	29.57543194	28.73304884
68	33.57131143	32.53831099	31.55407280	30.61582074	29.72095449	28.86703771
69	33.76929078	32.72030340	31.72138641	30.76965522	29.86241020	28.99712399
70	33.96267719	32.89785698	31.88442038	30.91937247	29.99991271	29.12342135
71	34.15157723	33.07107998	32.04328417	31.06508270	30.13357250	29.24604015
72	34.33609498	33.24007803	32.19808445	31.20689314	30.26349696	29.36508752
73	34.51633209	33.40495417	32.34892516	31.34490816	30.38979049	29.48066750
74	34.69238788	33.56580895	32.49590759	31.47922936	30.51255454	29.59288107
75	34.86435935	33.72274044	32.63913042	31.60995558	30.63188777	29.70182628
76	35.03234124	33.87584433	32.77868981	31.73718304	30.74788605	29.80759833
77	35.19642612	34.02521398	32.91467947	31.86100540	30.86064257	29.91028964
78	35.35670439	34.17094047	33.04719072	31.98151377	30.97024794	30.00998994
79	35.51326436	34.31311265	33.17631251	32.09879685	31.07679023	30.10678635
80	35.66619230	34.45181722	33.30213156	32.21294098	31.18035502	30.20076345
81	35.81557245	34.58713875	33.42473234	32.32403015	31.28102553	30.29200335
82	35.96148713	34.71915976	33.54419716	32.43214613	31.37888266	30.38058577
83	36.10401673	34.84796074	33.66060625	32.53736850	31.47400501	30.46658813
84	36.24323979	34.97362023	33.77403776	32.63977469	31.56646903	30.55008556
85	36.37923300	35.09621486	33.88456785	32.73944009	31.65634900	30.63115103
86	36.51207131	35.21581938	33.99227074	32.83643804	31.74371713	30.70985537
87	36.64182790	35.33250671	34.09721875	32.93083994	31.82864362	30.78626735
88	36.76857426	35.44634801	34.19948234	33.02271527	31.91119672	30.86045374
89	36.89238023	35.55741269	34.29913017	33.11213165	31.99144274	30.93247936
90	37.01331402	35.66576848	34.39622916	33.19915489	32.06944616	31.00240714
91	37.13144227	35.77148144	34.49084449	33.28384905	32.14526966	31.07029820
92	37.24683005	35.87461604	34.58303970	33.36627644	32.21897415	31.13621184
93	37.35954096	35.97523516	34.67287668	33.44649776	32.29061886	31.20020567
94	37.46963708	36.07340016	34.76041577	33.52457202	32.36026135	31.26233560
95	37.57717907	36.16917089	34.84571573	33.60055671	32.42795757	31.32265592
96	37.68222620	36.26260574	34.92883384	33.67450775	32.49376191	31.38121934
97	37.78483634	36.35376170	35.00982591	33.74647956	32.55772725	31.43807703
98	37.88506602	36.44269434	35.08874632	33.81652512	32.61990499	31.49327867
99	37.98297047	36.52945790	35.16564806	33.88469598	32.68034506	31.54687250
100	38.07860363	36.61410526	35.24058276	33.95104232	32.73909605	31.59890534
101	38.17201820	36.69668806	35.31360074	34.01561296	32.79620515	31.64942266
102	38.26326564	36.77725665	35.38475103	34.07845544	32.85171825	31.69846860
103	38.35239623	36.85586014	35.45408139	34.13961600	32.90567996	31.74608602
104	38.43945908	36.93254648	35.52163838	34.19913966	32.95813361	31.79231652
105	38.52450215	37.00736242	35.58746737	34.25707023	33.00912138	31.83720051
106	38.60757231	37.08035358	35.65161254	34.31345034	33.05868420	31.88077719
107	38.68871532	37.15156447	35.71411697	34.36832150	33.10686192	31.92308465
108	38.76797589	37.22103851	35.77502262	34.42172409	33.15369324	31.96415986
109	38.84539770	37.28881806	35.83437040	34.47369741	33.19921579	32.00403870
110	38.92102339	37.35494444	35.89220015	34.52427972	33.24346614	32.04275602
111	38.99489465	37.41945799	35.94855069	34.57350824	33.28647984	32.08034565
112	39.06705216	37.48239804	36.00345987	34.62141921	33.32829146	32.11684043
113	39.13753568	37.54380297	36.05696455	34.66804790	33.36893459	32.15227227
114	39.20638406	37.60371021	36.10910066	34.71342861	33.40844189	32.18667210
115	39.27363523	37.66215631	36.15990320	34.75759475	33.44684509	32.22007000
116	39.33932623	37.71917688	36.20940628	34.80057884	33.48417506	32.25249515
117	39.40349326	37.77480672	36.25764315	34.84241249	33.52046178	32.28397587
118	39.46617169	37.82907972	36.30464619	34.88312561	33.55573442	32.31453968
119	39.52739603	37.88202900	36.35044696	34.92275086	33.59002131	32.34421328
120	39.58720003	37.93368683	36.39507621	34.96131471	33.62334999	32.37302261

Business mathematics and statistics

Present value of an annuity $a_{\overline{n}|i} = \dfrac{1-(1+i)^{-n}}{i}$ —(continued)

n \ i	$3\frac{1}{4}\%$ .03250000	$3\frac{1}{2}\%$ .03500000	$3\frac{3}{4}\%$ .03750000	4% .04000000	$4\frac{1}{4}\%$ .04250000	$4\frac{1}{2}\%$ .04500000
1	.96852300	.96618357	.96385542	.96153846	.95923261	.95693780
2	1.90655981	1.89969428	1.89287270	1.88609467	1.87935982	1.87266775
3	2.81507003	2.80163698	2.78831103	2.77509103	2.76197585	2.74896435
4	3.69498308	3.67307921	3.65138413	3.62989522	3.60860993	3.58752570
5	4.54719911	4.51505238	4.48326181	4.45182233	4.42072895	4.38997674
6	5.37258994	5.32855302	5.28507162	5.24213686	5.19974000	5.15787248
7	6.17199994	6.11454398	6.05790036	6.00205467	5.94699280	5.89270094
8	6.94624692	6.87395554	6.80279553	6.73274487	6.66378206	6.59588607
9	7.69612292	7.60768651	7.52076677	7.43533161	7.35134970	7.26879050
10	8.42239508	8.31660532	8.21278725	8.11089578	8.01088700	7.91271818
11	9.12580637	9.00155104	8.87979494	8.76047671	8.64353669	8.52891692
12	9.80707639	9.66333433	9.52269392	9.38507376	9.25039491	9.11858078
13	10.46690207	10.30273849	10.14235558	9.98564785	9.83251310	9.68285242
14	11.10595842	10.92052028	10.73961984	10.56312293	10.39089986	10.22282528
15	11.72489920	11.51741090	11.31529623	11.11838743	10.92652265	10.73954573
16	12.32435758	12.09411681	11.87016504	11.65229561	11.44030949	11.23401505
17	12.90494681	12.65132059	12.40497835	12.16566885	11.93315059	11.70719143
18	13.46726083	13.18968173	12.92046106	12.65929697	12.40589985	12.15999180
19	14.01187490	13.70983742	13.41731187	13.13393940	12.85937636	12.59329359
20	14.53934615	14.21240330	13.89620421	13.59032634	13.29436581	13.00793645
21	15.05021419	14.69797420	14.35778719	14.02915995	13.71162188	13.40472388
22	15.54500163	15.16712484	14.80268645	14.45111533	14.11186751	13.78442476
23	16.02421466	15.62041047	15.23150501	14.85684167	14.49579617	14.14777489
24	16.48834349	16.05836760	15.64482411	15.24696314	14.86407307	14.49547837
25	16.93786295	16.48151459	16.04320396	15.62207994	15.21733627	14.82820896
26	17.37323288	16.89035226	16.42718454	15.98276918	15.55619787	15.14661145
27	17.79489867	17.28536451	16.79728630	16.32958575	15.88124496	15.45130282
28	18.20329169	17.66701885	17.15401089	16.66306322	16.19304072	15.74287351
29	18.59882973	18.03576700	17.49784183	16.98371463	16.49212539	16.02188853
30	18.98191741	18.39204541	17.82924513	17.29203330	16.77901717	16.28888854
31	19.35294664	18.73627576	18.14867001	17.58849356	17.05421311	16.54439095
32	19.71229699	19.06886547	18.45654941	17.87355150	17.31819003	16.78889086
33	20.06033607	19.39020818	18.75330063	18.14764567	17.57140531	17.02286207
34	20.39741992	19.70068423	19.03932591	18.41119776	17.81429766	17.24675796
35	20.72389339	20.00066110	19.31501293	18.66461323	18.04728792	17.46101240
36	21.04009045	20.29049381	19.58073535	18.90828195	18.27077978	17.66604058
37	21.34633457	20.57052542	19.83685335	19.14257880	18.48516046	17.86223979
38	21.64293905	20.84108736	20.08371407	18.36786423	18.69080140	18.04999023
39	21.93020732	21.10249987	20.32165212	19.58448484	18.88805890	18.22965572
40	22.20843324	21.35507234	20.55098999	19.79277388	19.07727472	18.40158442
41	22.47790144	21.59910371	20.77203855	19.99305181	19.25877671	18.56610949
42	22.73888759	21.83488281	20.98509739	20.18562674	19.43287934	18:72354975
43	22.99165869	22.06268870	21.19045532	20.37079494	19.59988426	18.87421029
44	23.23647330	22.28279102	21.38839067	20.54884129	19.76008082	19.01838305
45	23.47358189	22.49545026	21.57917173	20.72003970	19.91374659	19.15634742
46	23.70322701	22.70091813	21.76305709	20.88465356	20.06114781	19.28837074
47	23.92564360	22.89943780	21.94029599	21.04293612	20.20253987	19.41470884
48	24.14105917	23.09124425	22.11112866	21.19513088	20.33816774	19.53560654
49	24.34969412	23.27656450	22.27578666	21.34147200	20.46826642	19.65129813
50	24.55176185	23.45561787	22.43449317	21.48218462	20.59306131	19.76200778
51	24.74746911	23.62861630	22.58746330	21.61748521	20.71276864	19.86795003
52	24.93701609	23.79576454	22.73490438	21.74758193	20.82759582	19.96933017
53	25.12059669	23.95726043	22.87701627	21.87267493	20.93774179	20.06634466
54	25.29839873	24.11329510	23.01399159	21.99295667	21.04339740	20.15918149
55	25.47060410	24.26405323	23.14601599	22.10861218	21.14474571	20.24802057
56	25.63738896	24.40971327	23.27326842	22.21981940	21.24196231	20.33303404
57	25.79892393	24.55044760	23.39592137	22.32674943	21.33521565	20.41438664
58	25.95537427	24.68642281	23.51414108	22.42956676	21.42466729	20.49223602
59	26.10690002	24.81779981	23.62808779	22.52842957	21.51047222	20.56673303
60	26.25365619	24.94473412	23.73791594	22.62348997	21.59277911	20.63802204

Present value of an annuity $a_{\overline{n}|i} = \dfrac{1-(1+i)^{-n}}{i}$ —(continued)

n \ i	$4\tfrac{3}{4}\%$ .04750000	5% .05000000	$5\tfrac{1}{4}\%$ .05250000	$5\tfrac{1}{2}\%$ .05550000	$5\tfrac{3}{4}\%$ .05750000	6% .06000000
1	.95465394	.95238095	.95011876	.94786730	.94562648	.94339623
2	1.86601808	1.85941043	1.85284443	1.84631971	1.83983591	1.83339267
3	2.73605545	2.72324803	2.71054103	2.69793338	2.68542403	2.67301195
4	3.56664004	3.54595050	3.52545466	3.50515012	3.48503454	3.46510561
5	4.35956090	4.32947667	4.29971939	4.27028448	4.24116742	4.21236379
6	5.11652592	5.07569207	5.03536284	4.99553031	4.95618668	4.91732433
7	5.83916556	5.78637340	5.73431149	5.68296712	5.63232783	5.58238144
8	6.52903633	6.46321276	6.39839571	6.33456599	6.27170481	6.20979381
9	7.18762418	7.10782168	7.02935460	6.95219525	6.87631660	6.80169227
10	7.81634767	7.72173493	7.62884047	7.53762583	7.44805352	7.36008705
11	8.41656102	8.30641422	8.19842325	8.09253633	7.98870310	7.88687458
12	8.98955706	8.86325164	8.73959454	8.61851785	8.49995565	8.38384394
13	9.53656998	9.39357299	9.25377153	9.11707853	8.98340960	8.85268296
14	10.05877803	9.89864094	9.74230074	9.58964790	9.44057645	9.29498393
15	10.55730599	10.37965804	10.20646151	10.03758094	9.87288553	9.71224899
16	11.03322768	10.83776956	10.64746937	10.46216203	10.28168845	10.10589527
17	11.48756819	11.27406625	11.06647921	10.86460856	10.66826331	10.47725969
18	11.92130615	11.68958690	11.46458833	11.24607447	11.03381873	10.82760348
19	12.33537580	12.08532086	11.84283926	11.60765352	11.37949762	11.15811649
20	12.73066902	12.46221034	12.20222258	11.95038248	11.70638072	11.46992122
21	13.10803725	12.82115271	12.54367941	12.27524406	12.01549005	11.76407662
22	13.46829332	13.16300258	12.86810395	12.58316973	12.30779201	12.04158172
23	13.81221319	13.48857388	13.17634528	12.87504239	12.58420048	12.30337898
24	14.14053765	13.79864179	13.46921216	13.15169895	12.84557965	12.55035753
25	14.45397389	14.09394457	13.74746999	13.41393266	13.09274671	12.78335616
26	14.75319703	14.37518530	14.01184797	13.66249541	13.32647443	13.00316619
27	15.03885158	14.64303362	14.26303845	13.89809991	13.54749355	13.21053414
28	15.31155282	14.89812726	14.50169924	14.12142172	13.75649509	13.40616428
29	15.57188814	15.14107358	14.72845533	14.33310116	13.95413247	13.59072102
30	15.82041827	15.37245103	14.94390055	14.53374517	14.14102361	13.76483115
31	16.05767854	15.59281050	15.14859910	14.72392907	14.31775282	13.92908599
32	16.28417999	15.80267667	15.34308703	14.90419817	14.48487265	14.08404339
33	16.50041049	16.00254921	15.52787367	15.07506936	14.64290558	14.23022961
34	16.70683579	16.19290401	15.70344291	15.23703257	14.79234570	14.36814114
35	16.90390052	16.37419429	15.87025455	15.39054273	14.93366024	14.49824636
36	17.09202913	16.54685171	16.02874541	15.53606843	15.06729100	14.62098713
37	17.27162686	16.71128734	16.17933056	15.67399851	15.19365579	14.73678031
38	17.44308053	16.86789271	16.32240433	15.80473793	15.31314969	14.84601916
39	17.60675946	17.01704067	16.45834141	15.92866154	15.42614628	14.94907468
40	17.76301619	17.15908635	16.58749778	16.04612469	15.53299884	15.04629687
41	17.91218729	17.29436796	16.71021166	16.15746416	15.63404146	15.13801592
42	18.05459407	17.42320758	16.82680443	16.26299920	15.72959003	15.22454332
43	18.19054327	17.54591198	16.93788141	16.36303242	15.81994329	15.30617294
44	18.32032770	17.66277331	17.04283269	16.45784116	15.90538373	15.38318202
45	18.44422692	17.77406982	17.14283391	16.54772572	15.98617847	15.45583209
46	18.56250780	17.88006650	17.23784695	16.63291537	16.06258011	15.52436990
47	18.67542511	17.98101571	17.32812061	16.71362594	16.13482753	15.58902821
48	18.78322206	18.07715782	17.41389132	16.79020271	16.20314660	15.65002661
49	18.88613085	18.16872173	17.49538368	16.86275139	16.26775092	15.70757227
50	18.98437312	18.25592546	17.57281109	16.93151790	16.32884248	15.76186064
51	19.07816050	18.22897663	17.64637634	16.99669943	16.38661227	15.81307607
52	19.16769499	18.41807298	17.71627205	17.05848287	16.44124092	15.86139252
53	19.25316944	18.49340284	17.78268129	17.11704538	16.49289921	15.90697408
54	19.33476796	18.56514556	17.84577794	17.17255486	16.54174867	15.94997554
55	19.41266631	18.63347196	17.90572726	17.22517048	16.58794200	15.99054297
56	19.48703228	18.69854473	17.96268624	17.27504311	16.63162364	16.02881412
57	19.55802604	18.76051879	18.01680402	17.32231575	16.67293016	16.06491898
58	19.62580052	18.81954170	18.06822235	17.36712393	16.71199069	16.09898017
59	19.69050169	18.87575400	18.11707587	17.40959614	16.74892737	16.13111337
60	19.75226891	18.92928953	18.16349251	17.44985416	16.78385567	16.16142771

Business mathematics and statistics

Present value of an annuity $a_{\overline{n}|i} = \dfrac{1-(1+i)^{-n}}{i}$ —(continued)

n \ i	$6\tfrac{1}{2}\%$ .06500000	7% .07000000	$7\tfrac{1}{2}\%$ .07500000	8% .08000000	$8\tfrac{1}{2}\%$ .08500000	9% .09000000
1	.93896714	.93457944	.93023256	.92592593	.92165899	.91743119
2	1.82062642	1.80801817	1.79556517	1.78326475	1.77111427	1.75911119
3	2.64847551	2.62431604	2.60052574	2.57709699	2.55402237	2.53129467
4	3.42579860	3.38721126	3.34932627	3.31212684	3.27559666	3.23971988
5	4.15567944	4.10019744	4.04588490	3.99271004	3.94064208	3.88965126
6	4.84101356	4.76653966	4.69384642	4.62287966	4.55358717	4.48591859
7	5.48451977	5.38928940	5.29660132	5.20637006	5.11851352	5.03295284
8	6.08875096	5.97129851	5.85730355	5.74663894	5.63918297	5.53481911
9	6.65610419	6.51523225	6.37888703	6.24688791	6.11906264	5.99524689
10	7.18883022	7.02358154	6.86408096	6.71008140	6.56134806	6.41765770
11	7.68904246	7.49867434	7.31542415	7.13896426	6.96898439	6.80519055
12	8.15872532	7.94268630	7.73527827	7.53607802	7.34468607	7.16072528
13	8.59974208	8.35765074	8.12584026	7.90377594	7.69095490	7.48690392
14	9.01384233	8.74546799	8.48915373	8.24423698	8.01009668	7.78615039
15	9.40266885	9.10791401	8.82711975	8.55947869	8.30423658	8.06068843
16	9.76776418	9.44664860	9.14150674	8.85136916	8.57533325	8.31255819
17	10.11057670	9.76322299	9.43395976	9.12163811	8.82519194	8.54363137
18	10.43246638	10.05908691	9.70600908	9.37188714	9.05547644	8.75562511
19	10.73471022	10.33559524	9.95907821	9.60359920	9.26772022	8.95011478
20	11.01850725	10.59401425	10.19449136	9.81814741	9.46333661	9.12854567
21	11.28498333	10.83552733	10.41348033	10.01680316	9.64362821	9.29224373
22	11.53519562	11.06124050	10.61719101	10.20074366	9.80979559	9.44242544
23	11.77013673	11.27218738	10.80668931	10.37105895	9.96294524	9.58020683
24	11.99073871	11.46933400	10.98296680	10.52875828	10.10409700	9.70661177
25	12.19787673	11.65358318	11.14694586	10.67477619	10.23419078	9.82257960
26	12.39237251	11.82577867	11.29944852	10.80997795	10.35409288	9.92897211
27	12.57499766	11.98670904	11.44138095	10.93516477	10.46460174	10.02657992
28	12.74647668	12.13711125	11.57337763	11.05107849	10.56645321	10.11612837
29	12.90748984	12.27767407	11.69616524	11.15840601	10.66032554	10.19828291
30	13.05867591	12.40904118	11.81038627	11.25778334	10.74684382	10.27365404
31	13.20063465	12.53181419	11.91663839	11.34979939	10.82658416	10.34280187
32	13.33392925	12.64655532	12.01547757	11.43499944	10.90007757	10.40624025
33	13.45908850	12.75379002	12.10742099	11.51388837	10.96781343	10.46444060
34	13.57660892	12.85400936	12.19294976	11.58693367	11.03024279	10.51783541
35	13.68695673	12.94767230	12.27251141	11.65456822	11.08778717	10.56682148
36	13.79056970	13.03520776	12.34652224	11.71719279	11.14081233	10.61176282
37	13.88785887	13.11701660	12.41536952	11.77517851	11.18968878	10.65299342
38	13.97921021	13.19347345	12.47941351	11.82886999	11.23473620	10.69081965
39	14.06498611	13.26492846	12.53898931	11.87858240	11.27625457	10.72552261
40	14.14552687	13.33170884	12.59440866	11.92461333	11.31452034	10.75736020
41	14.22115199	13.39412041	12.64596155	11.96723457	11.34978833	10.78656899
42	14.29216149	13.45244898	12.69391772	12.00669867	11.38229339	10.81336604
43	14.35883708	13.50696167	12.73852811	12.04323951	11.41225197	10.83795050
44	14.42144327	13.55790810	12.78002615	12.07707362	11.43986357	10.86050504
45	14.48022842	13.60552159	12.81862898	12.10840150	11.46531205	10.88119729
46	14.53542575	13.65002018	12.85453858	12.13740880	11.48876686	10.90018100
47	14.58725422	13.69160764	12.88794247	12.16426741	11.51038420	10.91759725
48	14.63591946	13.73047443	12.91901662	12.18913649	11.53030802	10.93357546
49	14.68161451	13.76679853	12.94792244	12.21216341	11.54867099	10.94823436
50	14.72452067	13.80074629	12.97481157	12.23348464	11.56559538	10.96168290
51	14.76480814	13.83247317	12.99982472	12.25322652	11.58119390	10.97402101
52	14.80263675	13.86212446	13.02309276	12.27150604	11.59557041	10.98534038
53	14.83815658	13.88983594	13.04473745	12.28843152	11.60882066	10.99572512
54	14.87150852	13.91573453	13.06487205	12.30410326	11.62103287	11.00525240
55	14.90282490	13.93993881	13.08360190	12.31861413	11.63228835	11.01399303
56	14.93222996	13.96255964	13.10102503	12.33205012	11.64266208	11.02201195
57	14.95984033	13.98370059	13.11723258	12.34449085	11.65222311	11.02936876
58	14.98576557	14.00345850	13.13230938	12.35601005	11.66103513	11.03611813
59	15.01010852	14.02192383	13.14633431	12.36667597	11.66915680	11.04231021
60	15.03296574	14.03918115	13.15938075	12.37655182	11.67664221	11.04799102

Appendix D

Present value of an annuity $a_{\overline{n}|i} = \dfrac{1-(1+i)^{-n}}{i}$ —*(continued)*

i \ n	$9\frac{1}{2}\%$ .09500000	10% .10000000	$10\frac{1}{2}\%$ .10500000	11% .11000000	$11\frac{1}{2}\%$ .11500000	12% .12000000
1	.91324201	.90909091	.90497738	.90090090	.89686099	.89285714
2	1.74725298	1.73553719	1.72396143	1.71252333	1.70122062	1.69005102
3	2.50890683	2.48685199	2.46512346	2.44371472	2.42261939	2.40183127
4	3.20448112	3.16986545	3.13585834	3.10244569	3.06961380	3.03734935
5	3.83970879	3.79078677	3.74285822	3.69589702	3.64987785	3.60477620
6	4.41982538	4.35526070	4.29217939	4.23053785	4.17029403	4.11140732
7	4.94961222	4.86841882	4.78930261	4.71219626	4.63703501	4.56375654
8	5.43343581	5.33492620	5.23918789	5.14612276	5.05563678	4.96763977
9	5.87528385	5.75902382	5.64632388	5.53704753	5.43106437	5.32824979
10	6.27879803	6.14456711	6.01477274	5.88923201	5.76777074	5.65022303
11	6.64730414	6.49506101	6.34821062	6.20651533	6.06974954	5.93769913
12	6.98383940	6.81369182	6.64996437	6.49235615	6.34058255	6.19437423
13	7.29117753	7.10335620	6.92304468	6.74987040	6.58348211	6.42354842
14	7.57185163	7.36668746	7.17017618	6.98186523	6.80132924	6.62816823
15	7.82817500	7.60607951	7.39382459	7.19086958	6.99670784	6.81086449
16	8.06226028	7.82370864	7.59622135	7.37916178	7.17193528	6.97398615
17	8.27603678	8.02155331	7.77938584	7.54879440	7.32908994	7.11963049
18	8.47126647	8.20141210	7.94514556	7.70161657	7.47003582	7.24967008
19	8.64955842	8.36492009	8.09515435	7.83929421	7.59644468	7.36577686
20	8.81238212	8.51356372	8.23090891	7.96332812	7.70981586	7.46944362
21	8.96107956	8.64869429	8.35376372	8.07507038	7.81149404	7.56200324
22	9.09687631	8.77154026	8.46494455	8.17573908	7.90268524	7.64464575
23	9.22089161	8.88321842	8.56556067	8.26643160	7.98447107	7.71843370
24	9.33414759	8.98474402	8.65661599	8.34813658	8.05782159	7.78431581
25	9.43757770	9.07704002	8.73901900	8.42174466	8.12360680	7.84313911
26	9.53203443	9.16094547	8.81359186	8.48805826	8.18260700	7.89565992
27	9.61829629	9.23722316	8.88107860	8.54780023	8.23552197	7.94255350
28	9.69707423	9.30656651	8.94215258	8.60162133	8.28297935	7.98442277
29	9.76901756	9.36960591	8.99742315	8.65010976	8.32554202	8.02180604
30	9.83471924	9.42691447	9.04744176	8.69379257	8.36371481	8.05518397
31	9.89472076	9.47901315	9.09270748	8.73314646	8.39795050	8.08498569
32	9.94951668	9.52637559	9.13367193	8.76860042	8.42865516	8.11159436
33	9.99955861	9.56943236	9.17074383	8.80054092	8.45619297	8.13535211
34	10.04525901	9.60857487	9.20429305	8.82931614	8.48089056	8.15656438
35	10.08699453	9.64415897	9.23465435	8.85523977	8.50304086	8.17550391
36	10.12510916	9.67650816	9.26213063	8.87859438	8.52290660	8.19241421
37	10.15991704	9.70591651	9.28699605	8.89963458	8.54072341	8.20751269
38	10.19170506	9.73265137	9.30949868	8.91858971	8.55670261	8.22099347
39	10.22073521	9.75695579	9.32986306	8.93566641	8.57103373	8.23302988
40	10.24724677	9.77905072	9.34829237	8.95105082	8.58388675	8.24377668
41	10.27145824	9.79913702	9.36497047	8.96491065	8.59541413	8.25337204
42	10.29356917	9.81739729	9.38006717	8.97739968	8.60575258	8.26193932
43	10.31376180	9.83399753	9.39372287	8.98864593	8.61502474	8.26958868
44	10.33220255	9.84908867	9.40608404	8.99878011	8.62334057	8.27641846
45	10.34904343	9.86280788	9.41727063	9.00791001	8.63079872	8.28251648
46	10.36442322	9.87527989	9.42739423	9.01613615	8.63748764	8.28796115
47	10.37846870	9.88661808	9.43655587	9.02354518	8.64348667	8.29282245
48	10.39129561	9.89692553	9.44484694	9.03022088	8.64886697	8.29716290
49	10.40300969	9.90629594	9.45235017	9.03623503	8.65369235	8.30103831
50	10.41370748	9.91481449	9.45914043	9.04165318	8.65802004	8.30449849
51	10.42347715	9.92255862	9.46528545	9.04653439	8.66190139	8.30758794
52	10.43239922	9.92959875	9.47084656	9.05093189	8.66538241	8.31034637
53	10.44054724	9.93599886	9.47587924	9.05489359	8.66850440	8.31280926
54	10.44798834	9.94181715	9.48043370	9.05846270	8.67130440	8.31500827
55	10.45478388	9.94710650	9.48455539	9.06167810	8.67381560	8.31697167
56	10.46098984	9.95191500	9.48828542	9.06457487	8.67606780	8.31872470
57	10.46665739	9.95628636	9.49166101	9.06718457	8.67808772	8.32028991
58	10.47183323	9.96026033	9.49471585	9.06953565	8.67989939	8.32168742
59	10.47656003	9.96387303	9.49748041	9.07165373	8.68152403	8.32293520
60	10.48087674	9.96715730	9.49998227	9.07356192	8.68298120	8.32404928

APPENDIX E
Formulae used in the text

Business mathematics and statistics

The mathematical and statistical formulae used in the text are listed below, together with the corresponding chapter citation.

FORMULA   CHAPTER CITATION

1. $a^n \times a^m = a^{n+m}$ (1.4)

2. $a^n \div a^m = a^{n-m}$ (1.4)

3. $a^{-n} = \dfrac{1}{a^n}$ (1.4)

4. $\sqrt[n]{a} = a^{\frac{1}{n}}$ (1.4)

5. $\left(\sqrt[n]{a}\right)^m = a^{\frac{m}{n}}$ (1.4)

6. $a^0 = 1$ (1.4)

7. *If commission only:*

    Commission (C) = Sales (S) × Commission rate (R) (2.3)

8. *If retainer plus commission:*

    Earnings (E) = Sales (S) × Commission rate (R) + Retainer (2.4)

9. *Discount rate* $(R) = \dfrac{\text{Discount } (D)}{\text{List Price } (LP)}$ (2.4)

10. *Discount* (D) = List price (LP) × Discount rate (R) (2.4)

11. *Combined discount rate* $(CDR) = 1 - (100\% - 1^{st} \text{ discount})$ (2.4)
    $\times (100\% - 2^{nd} \text{ discount}) \ldots (100\% - n^{th} \text{ discount})$

12. *Amount of company tax* = Tax (Y) × Company tax rate (R) (2.5)

13. *Amount of sales tax* = Wholesale price (WP) × Sales tax rate (R) (2.5)

14. *Selling (or list) price* (SP) − Cost (C) = Mark-up (M) (2.6)

15. *Selling Price* (SP) = Cost (C) + Mark-up (M) (2.6)

16. *Mark-up rate on cost* $(R) = \dfrac{\text{Selling Price } (SP)}{\text{Cost}} - 1$ (2.6)

17. *Mark-up rate selling price* $(R) = \dfrac{\text{Mark-up } (M)}{\text{Selling Price } (SP)}$ (2.6)

Appendix E

18. *Mark-up rate (R)* = $\dfrac{\text{Selling price }(SP) - \text{Cost }(C)}{\text{Cost }(C)}$ (2.6)

19. *Simple Interest Amount (I)* = Principal (P) × Rate (R) × Time (T) (3.2)

20. *Maturity Value (S)* = Principal (P) + Principal (P) × Rate (R) × Time (T) (3.2)

21. *Principal Amount (P)* = $\dfrac{\text{Maturity Value }(S)}{(1 + \text{Rate }(R) \times \text{Time }(T))}$ (3.2)

22. *Interest Accumulated (T)* = Maturity Value (S) − Principal (P) (3.2)

23. *Effective rate of interest (ER)*

$$\dfrac{2(\text{Number of repayments }(N)) \times \text{Rate of interest }(R)}{\text{Number of repayments }(N) + 1}$$ (3.3)

24. *Compound interest formula* for the accumulated value (S) is:

$$S = P(1 + i)^n$$ (4.1)

25. *Present value (P) at compound interest* of S is:

$$P = S(1 + i)^{-n}$$ (4.2)

26. *Compound interest rate (i) per period* is:

$$i = \sqrt[n]{\dfrac{S}{P}} - 1$$ (4.2)

27. *Future value of an annuity:*

$$S = P\dfrac{(1+i)^n - 1}{i}$$ (5.2)

28. *Periodic payment amount of an annuity:*

$$P = \dfrac{S}{\dfrac{(1+i)^n - 1}{i}}$$ (5.3)

29. *Present value of an annuity:*

$$PV = P\dfrac{(1 - (1+i))^{-n}}{i}$$ (5.4)

30. Using *straight-line depreciation method:*

$$\text{Annual depreciation amount} = \dfrac{\text{Acquisition cost - Residual value}}{\text{Estimated life}}$$ (6.2)

$$\text{Annual depreciation rate} = \frac{100}{\text{Estimated life}} \quad (6.2)$$

$$\text{Annual depreciation rate} = \frac{\text{Annual depreciation}}{\text{Original cost}} \times 100\% \quad (6.2)$$

31. To calculate *book value using reducing balance method*:

32. Book value $(A) =$
    Purchase price $(P) (1 - \text{interest rate } (i)/\text{period})^{\text{Number of periods }(n)}$ \quad (6.3)

33. To solve for *interest rate (i) using the reducing balance method*:

$$i = 1 - \sqrt[n]{\frac{A}{P}} \quad (6.4)$$

34. To solve for *depreciation expense using units-of-production method*:

$$\text{Depreciation expense per unit} = \frac{\text{Acquisition cost} - \text{Residual value}}{\text{Estimated units of use}} \quad (6.5)$$

35. A *linear equation* may be written as:

$$y = mx + b \quad \text{or} \quad y = a + bx \quad (7.1)$$

36. In *break-even analysis*:

$$\text{Total costs } (TC) = \text{Fixed cost } (FC) + \text{Variable cost } (VC) \quad (7.3)$$

$$\text{Total revenue } (TR) = \text{Quarterly sold } (Q) \times \text{Price } (P) \quad (7.3)$$

$$\text{Total profit } (TP) = \text{Total revenue } (TR) - \text{Total cost } (TC) \quad (7.3)$$

$$\text{Break-even point: Total revenue } (TR) = \text{Total cost } (TC) \quad (7.3)$$

37. To calculate the *mean of a sample of ungrouped data*:

$$\text{Mean } (\bar{x}) = \frac{\text{Sum of values } (\Sigma x)}{\text{Number of values } (n)} \quad (10.1)$$

38. To calculate the *mean of a population of ungrouped data*:

$$\text{Population mean } (\mu) = \frac{\text{Sum of values } (\Sigma x)}{\text{Number of values in the population } (N)} \quad (10.1)$$

$$\text{Weighted mean average} = \frac{\Sigma(\text{Weight } (W)) \times \text{Each value inset } (x)}{\Sigma(\text{Weight } (W))} \quad (10.1)$$

Appendix E

39. *The median for grouped data:*

$$Md = L + \frac{\frac{n}{2} - C}{f}(i) \qquad (10.2)$$

where: L = red lower limit of median class
C = cumulative frequency of class preceding median class
f = frequency of median class
n = total frequency in distribution
i = class interval size

40. *The mode for grouped data:*

$$Mo = L + \frac{d_1}{d_1 + d_2}(i) \qquad (10.3)$$

where: L = red lower limit of modal class
d_1 = frequency of modal class – frequency of previous class
d_2 = frequency of modal class – frequency of next class
i = class interval size

41. *Interquartile range* = $Q_3 - Q_1$ \qquad (11.2)

42. *Quartile deviation (QD)* = $\dfrac{Q_3 - Q_1}{2}$ \qquad (11.2)

43. *First quartile for grouped data* = $L + \dfrac{\left(\frac{n}{4} - C\right)}{f}(i)$ \qquad (11.2)

44. *Third quartile for grouped data* = $L + \dfrac{\left(\frac{3n}{4} - C\right)}{f}(i)$ \qquad (11.2)

45. *Mean deviation for ungrouped data (MD)* = $\dfrac{\Sigma \mid x - \bar{x} \mid}{n}$ \qquad (11.3)

46. *Mean deviation for grouped data (MD)* = $\dfrac{\Sigma f \mid x - \bar{x} \mid}{n}$ \qquad (11.3)

47. *Standard deviation for ungrouped data for a population* (σ) =

$$\sqrt{\frac{\Sigma(x - \mu)^2}{N}} \qquad \text{where } x = \text{value of each observation} \qquad (11.4)$$

48. *Standard deviation for grouped data for a population:*

$$\sigma = \sqrt{\frac{\Sigma f(m - \mu)^2}{N}} \qquad \text{where } m = \text{midpoint of each class} \qquad (11.4)$$

Business mathematics and statistics

49. **Short-hand method** to determine *standard deviation for grouped data for a population*:

$$\sigma = (i)\sqrt{\frac{\Sigma fd^2}{N} - \left(\frac{\Sigma fd}{N}\right)^2}$$

where d = deviation of mid point of class interval from arbitrary origin in terms of class intervals   (11.4)

50. *Coefficient of variation for a population* $(V) = \dfrac{\text{Standard deviation }(\sigma)}{\text{Mean }(\mu)} \times 100$ (11.5)

51. *Coefficient of variation for a sample* $(V) = \dfrac{\text{Standard deviation }(s)}{\text{Mean }(\bar{x})} \times 100$ (11.5)

52. $\text{Skewness} = \dfrac{3(\text{Mean} - \text{Median})}{\text{Standard deviation}} = \dfrac{3(\bar{x} - Md)}{s}$ (11.A)

53. *Standard score* (z score) =

$$\frac{\text{Selected value }(x) - \text{Mean of all values in the population }(\mu)}{\text{Standard deviation of }x\text{ in the population }(\sigma)} \quad (14.1)$$

54. Standard error of the mean $(\sigma_{\bar{x}}) = \dfrac{\sigma}{\sqrt{n}}$ (14.2)

where n = sample size
σ = standard deviation of population

$$z = \frac{\bar{x} - \mu}{\sigma_{\bar{x}}} \quad (14.2)$$

55. *Coefficient of correlation* $(r) = \dfrac{\Sigma(X - \bar{X})(Y - \bar{Y})}{\sqrt{\Sigma(X - \bar{X})^2 \Sigma(Y - \bar{Y})^2}}$ (15.2)

56. **Shorthand method** to determine *coefficient of correlation*:

$$r = \frac{\Sigma XY - \frac{1}{n}\Sigma X \Sigma Y}{\sqrt{(\Sigma X^2 - \frac{1}{n}(\Sigma X)^2)(\Sigma Y^2 - \frac{1}{n}(\Sigma Y)^2)}} \quad (15.2)$$

57. *Coefficient of determination* $= r^2$ (15.2)

58. *Rank correlation* $r_s = 1 - \dfrac{6\Sigma D^2}{n(n^2 - 1)}$ (15.3)

where D = difference between the rankings for the Z variables for each item in the series
n = number of paired observations

Appendix E

59. *Least-squares regression line* $Y = a + bX$ (15.4)

60. *Regression coefficients a and b are determined by calculating:*

$$a = \overline{Y} - b\overline{X} \quad \text{or} \quad \frac{\Sigma Y - b\Sigma X}{n}$$

$$b = \frac{n\Sigma XY - \Sigma X \Sigma Y}{n\Sigma X^2 - (\Sigma X)^2} \tag{15.5}$$

61. *Standard error of the estimate:*

$$Se = \sqrt{\frac{\Sigma(Y_a - Y_c)^2}{n}} \tag{15.5}$$

where Y_a = actual value of Y
Y_c = calculated value of Y
n = number of values

62. *Simple price index* $= \dfrac{P_g}{P_b} \times 100$ (16.1)

where P_g = price in given period
P_b = price in base period

63. *Simple quantity index* $= \dfrac{Q_g}{Q_b} \times 100$ (16.1)

where Q_g = quantity in given period
Q_b = quantity in base period

64. *Simple value index* $= \dfrac{V_g}{V_b} \times 100$ (16.1)

where V_g = value in given period
V_b = value in base period

65. *Simple aggregate price index* $= \left(\dfrac{\Sigma P_g}{\Sigma P_b}\right) \times 100$ (16.2)

66. *Average of relative price index* $= \dfrac{\Sigma\left(\dfrac{P_g}{P_b}\right) \times 100}{n}$ (16.2)

Business mathematics and statistics

67. The Laspeyres index $= \dfrac{\Sigma(P_g Q_b)}{\Sigma(P_b Q_b)} \times 100$ (16.3)

68. The Paasche index $= \dfrac{\Sigma(P_g Q_g)}{\Sigma(P_b Q_g)} \times 100$ (16.3)

69. Weighted average of price relative $= \dfrac{\Sigma\left(\dfrac{P_g}{P_b} \times w \times 100\right)}{\Sigma w}$ (16.3)

70. Weighted average of quantity relative $= \dfrac{\Sigma\left(\dfrac{Q_g}{Q_b} \times w \times 100\right)}{\Sigma w}$ (16.3)

71. Weighted value index $= \dfrac{\Sigma(P_g \times Q_g)}{\Sigma(P_b \times Q_b)} \times 100$ (16.3)

72. Straight-line secular trend line using least squares method $= Y_c = a + bX$ (17.1)

    where Y_c = calculated trend
           X = time period
           a = value of Y and X

73. The formula for finding a in $Y_c = a + bX$ is: $a = \dfrac{\Sigma Y}{n}$ (17.2)

    where Y = dependent variable
           n = number of periods in data

74. The formula for finding b in $Y_c = a + bX$ is: $b = \dfrac{\Sigma XY}{\Sigma X^2}$ (17.2)

    where X = time deviation from middle period selected as origin

75. *Exponential smoothing:* $\Sigma_t = (1 - \alpha)_{\Sigma t-1} + \alpha Y_t$ (17.2)

    where Σ_{t-1} = value of the exponentially smoothed series already calculated in time period $t - 1$
           Y_t = observed value of the time series period t

Appendix E

a = subjectivity assigned weight or smoothing coefficient
Σ_t = value of the exponentially smoothed series being calculated in time period t

76. Period's *seasonally adjusted data* = $\dfrac{\text{Period's actual value}}{\text{Seasonal index for period}} \times 100$ (17.3)

APPENDIX F
*Working with your calculator**

---

* This appendix was written by Peter Wheatley.

Business mathematics and statistics

As you have studied this book, you have been shown how to do many different questions. Whilst it is important that you have an understanding of the method of solving different types of business mathematics problems, many of the questions can be done more quickly using a calculator—*once* the underlying principles have been understood !

There is a wide range of scientific calculators on the market today but it is true to say that the CASIO and SHARP brands are the most popular. That is not to say other brands are not as good but rather the 'popularity' of these two brands is a result of their being the most commonly used in high school—after which they are put in a drawer and forgotten about!

All scientific calculators come supplied with an instruction booklet which can be confusing to read because of the seemingly haphazard use of up to five different languages: English, French, German, Spanish and Japanese, usually - all mixed together in the one booklet!

In order to help you practically, a number of questions given in different chapters of this book will be worked through with the calculator steps set out for you to follow. However, before we start, please note that the instructions below refer specifically to the CASIO fx-100s and the SHARP EL-531GH (which has Direct Algebraic Logic or DAL). The instructions carry over to other models and will at least set you in the right direction if you are using another model or brand of calculator.

First of all on both calculators you will find the following buttons:

CASIO	SHARP	Meaning
$\sqrt{}$	$\sqrt{}$	Square root
x^2	x^2	Square
$a\,b/c$	$a\,b/c$	Mixed number
EXP	Exp	Scientific Notation
x^y	y^x	Power
MODE	MODE	Gives access to different 'modes' of calculation
SHIFT above ◯	2ndF	Gives access to functions above keys
$+/-$	$+/-$	sign change (or plus/minus)

For the CASIO, the ON button is actually labelled AC (for All Clear) with the word 'ON' above and to the right of the AC button. On the SHARP, ON/C (C for clear) is in the top right hand corner. Conversely, the CASIO has the OFF button clearly labelled in the top right hand corner whilst to switch off the SHARP you press 2ndF followed by the ON/C button (OFF is in yellow above the button).

Appendix F

Both calculators switch themselves off after a number of minutes (about 7 minutes for the CASIO and 10 minutes for the SHARP, though this may vary). However, a great feature on both is the fact that you can bring back the last answer! On the CASIO, switch on and then press SHIFT |Ans| and hey presto!

For the SHARP, (2ndF) then the (=) button and the last answer is brought back.

Before starting on exercises from earlier chapters, here are some introductory examples. The first four can be done without a calculator (just to give you a feeling for what's going on):

1. $25 - 5 \times 3 + 18 \div 6$ (Answer = 13)

   This calculation is done identically on a CASIO and a SHARP. The steps are as follows:

   (2) (5) (−) (5) (×) (3) (+) (1) (8) (÷) (6) (=)

   N.B. In mathematics the convention is that division and multiplication are done before subtraction and addition, working from left to right; i.e. we actually have $25 - 15 + 3 = 10 + 3$ or 13.

2. $\dfrac{14 - 6 - 5}{11 - 8 + 2}$ (Answer = 3)

   Here the value of the numerator (15) is divided by the value of the denominator (5). If you simply press $14 + 6 - 5 \div 11 - 8 + 2 =$ on a calculator you obtain 13.545454 . . . In a question like this, you can think 'It's all of the top divided by all of the bottom' where the two words 'all of ' imply brackets. On the CASIO, brackets sit above the (9) button whilst on the SHARP they are to the right of (9).

   There are in fact two *main* ways of answering a question like this on a calculator, with steps as follows:

CASIO	SHARP
Method 1	Method 1
[(---)] 14 + 6 − 5 (---)] ÷ [(---)] 11 − 8 + 2 (---)] =	(14 + 6 − 5) ÷ (11 − 8 + 2 } =
Method 2 11 − 8 + 2 = SHIFT MR 14 + 6 − 5 = ÷ MR = MR stands for Memory Read (or Recall) whilst SHIFT MR gives Min or 'Memory in'	Method 2 11 − 8 + 2 = (STO) 14 + 6 − 5 = ÷ RCL = (STO) stands for Store whilst RCL stands for Recall

   Note in the above that the button outline ◯ was not put around all the buttons that had to be pressed—I hope you'll agree that it is reasonably obvious which buttons need to be pressed.

Business mathematics and statistics

3. $2\frac{3}{4} + 5\frac{1}{3} \times 4\frac{1}{8} - 1\frac{4}{5} \div \frac{3}{25}$ (Answer = $9\frac{3}{4}$)

   This is done identically on both brands. The steps are laid out as follows:

   2 [a b/c] 3 a b/c 4 + 5 a b/c 1 a b/c 3 × 4 a b/c 1 a b/c 8 − 1 a b/c 4 a b/c 5 ÷ 3 a b/c 25

   Notice that the calculators show the answer as either 9⌐3⌐4 or as 9⌐3⌐4 for $9\frac{3}{4}$. This way of displaying a mixed number by a calculator is quite common but not universal.

4. $7 \times 3^2 \div 2^2$ (Answer = 15.75 = $15\frac{3}{4}$)

   Again this is done identically on both brands. The steps are as follows:

   $7 \times 3 \,[x^2] \div 2 \,[x^2] =$

5. $(2.34 \times 10^6)^2 \times (1.6 \times 10^{-7})^3 \div (-4.5 \times 10^5)^{-4}$

   This is a humdinger of a question but here goes! The steps vary slightly on the two calculators, but on both you start by pressing '2' then '.' then '3' and so on:

CASIO	SHARP
2 . 3 4 [EXP] 6 [x²] × 1 . 6 [EXP] 7 [+/−] [xʸ] 3 ÷ 4 . 5 [+/−] EXP 5 [xʸ] 4 [+/−] =	2 . 3 4 [Exp] 6 [x²] × 1 . 6 [Exp] 7 [+/−] [xʸ] 3 ÷ 4 . 5 [+/−] [Exp] 5 [xʸ] 4 [+/−] =

   (Answer = $9.19690537 \times 10^{14}$)

   Notice that the display on the CASIO is easier to interpret than on the SHARP where you have to read 9.19690537 ^{14} as meaning $9.19690537 \times 10^{14}$.

   By the way, if an **M** appears in the display, some value has been put in memory. To clear memory on the CASIO, press 0 (zero) then SHIFT MR (to access Min). On the SHARP, press 0 then **STO** then =.

   Now let's look at how we could use the calculator to answer four questions taken from various exercises within the various chapters of this book.

   Firstly we'll attempt question 4 from 1.7 Multiple Choice on page 19 of Chapter 1. The question says:

   Determine the value of $\left(\sqrt{16}\right)^3 \left(\sqrt[3]{16}\right)^{1/2}$

   On both calculators the steps are as shown in the table:

Appendix F

CASIO	SHARP
[(---) √ 1 6 (---)] x^y 3 × [(--- 3 SHIFT x^y 6 4 ---)] x^y 1 (a b/c) 2 =	((√ 1 6 ( )) y^x 3 × (3 (2ndF) y^x 6 4) y^x 1 a b/c 2 =

In both cases the correct answer is 128.

Now let's look at the following equation which occurs in Chapter 5 on Annuities. The expression is:

$$PV = \$2000 \, \frac{1 - (1 + 0.025)^{-12}}{0.025}$$

The steps to calculate the right hand side vary slightly on both calculators. The steps are as shown in the table:

CASIO	SHARP
2 0 0 0 × [(---) 1 − [(--- 1 + • 0 2 5 ---)] x^y 1 2 (+/−) ---)] ÷ • 0 2 5 =	2 0 0 0 × (1 − (1 + • 0 2 5) y^x 1 2 (+/−)) ÷ • 0 2 5 =

In both cases the correct answer is $20 515.53

Next we will try question 6 (b) from Competency Check 10.1 on page 290. In this question we are asked to calculate the mean for a grouped frequency distribution table. The steps are somewhat different on the two calculators so please be careful ! Before we can use either calculator, we need the class centre for each class. These are shown below with the corresponding frequencies:

Class centre	Frequency
61.0	22
64.0	25
67.0	49
70.0	64
73.0	58
76.0	27
79.0	11

CASIO
After switching on the machine, press the MODE button once—if you press it too many times, keep pressing the MODE button until the display cycles back to

COMP	SD	LR	CMPLX
1	2	3	4

COMP is short for COMPutation, SD stands for Statistical Display, LR stands for Linear Regression whilst CMPLX is short for CoMPLeX. At the moment, we are only interested in SD. Press the ②button so that the calculator is switched to Statistical Display. Notice the small SD in the display? If you see DEG or something else above the SD, that's OK—it won't affect our calculations! First of all with any new statistical calculation, you have to clear statistical memory. To do this press SHIFT then (AC). The steps carry on from here as:

61 × 22 (M+)
64 × 25 (M+)
67 × 49 (M+)
70 × 64 (M+)
73 × 58 (M+)
76 × 27 (M+)
79 × 11 (M+)
SHIFT 1

By pressing the MODE button 3 times, you will arrive at

FIX	SCI	NORM
1	1	3

Now you can FIX your answer at 2 decimal places by pressing the 1 button to select FIX followed by pressing the 2 button to give 2 decimal places. The other two possibilities give SCIentific notation and NORMal view. Note that if you press SHIFT followed by 3 you obtain the *sample* standard deviation whereas SHIFT followed by 2 gives the *population* standard deviation. To remove SD, go to

COMP	SD	LR	CMPLX
1	2	3	4

and choose 1.

SHARP
With the calculator turned on, press the MODE button once—if you keep on pressing it, it will cycle back and forth between normal display and ? 0–I. When this display appears, press 1. With any new statistical calculation, you have to clear statistical memory. To do this press 2ndF then → to access **CA** (Clear All). The steps carry on from here as:

61 × 22 **M+**
64 × 25 **M+**
67 × 49 **M+**
70 × 64 **M+**
73 × 58 **M+**
76 × 27 **M+**
79 × 11 **M+**

Notice that the display gives a running total of the frequencies. To display the mean ($\bar{x}$), press 2ndF then (. Note that 2ndF followed by × gives the sample standard deviation (σx) whilst 2ndF followed by ÷ gives the population standard deviation (σx). To fix your answer for the mean at 2 decimal places, press 2ndF followed by (•) to access **FSE** (Fix Scientific Engineering). Now press 2ndF followed by (+/−) then 2 to TAB to 2 decimal places. To clear the Fix, keep on pressing 2ndF then (•) until the display returns to 'normal'. If you press MODE then 0, you will remove STAT from the display.

Appendix F

The value of the mean is 69.765625 or 69.77 correct to 2 decimal places..

Finally, let's do question 3 in Competency Check 15.4. On the CASIO, you enter the data and the calculator supplies you with 'r' (the correlation coefficient), 'A' (the y-intercept) and 'B' (the gradient) and even makes predictions based on the linear regression equation $Y = A + BX$. However, on the SHARP, you will have to use the formula as given in the text.

CASIO	SHARP

Switch on the machine and press the MODE button once—if you press it too many times, keep pressing the MODE button until the display cycles back to
    COMP   SD   LR   CMPLX
     1   2   3   4
 At the moment, we are interested in LR. Press the ③ button so that the calculator is switched to Linear Regression. Notice the small LR in the display? First of all you have to clear statistical memory. To do this press **SHIFT** then AC. Now look under the $[(---)$ button for $\lfloor x_D, y_D \rfloor$ in blue. From now on when you are told to press 'THAT BUTTON', you press the $[(---)$ button to access $\lfloor x_D, y_D \rfloor$! The steps carry on from here as:
     15 'THAT BUTTON' 64 **M+**
     13 'THAT BUTTON' 64 **M+**
     17 'THAT BUTTON' 73 **M+**
     19 'THAT BUTTON' 79 **M+**
     16 'THAT BUTTON' 70 **M+**
       SHIFT 7 (to access A)
       SHIFT 8 (to access B)
 To predict the attendance when the temperature is 14°C, press 1 and 4 then $(---)]$ to access $\hat{y}$ in blue. Similarly, press 2 followed by 1 then $(---)]$ to access $\hat{y}$ (y 'hat' is a symbol for 'the predicted value of y').
     By pressing the MODE button 3 times, you will arrive at
      FIX   SCI   NORM
       1   1   3
 Now you can FIX your answer at 2 decimal places by pressing the 1 button to select FIX followed by pressing the 2 button to give 2 decimal places. To remove LR, go to
	Unfortunately, the SHARP EL-531GH does not have the capacity to do Linear Regression calculations. However, other (more expensive) SHARP models are able to do LR questions. A trip to the calculator section of your newsagent or a large department store will unearth such models!

Business mathematics and statistics

CASIO (continued)
COMP   SD   LR   CMPLX 1   2   3   4 and choose 1. To clear FIX, go back to FIX   SCI   NORM 1   1   3 press 3 then press 1.

The answers are: A = 26.8
                    B = 2.7
                    When temperature = 14°C, the attendance is predicted to be 64 600.
                    When temperature = 21°C, the attendance is predicted to be 83 500.

ANSWERS

Answers to Competency checks and Exercises that require discussion and/or graphic representation have, in general, not been included here. Also, your answers for questions in Part 2 of the book (Business Statistics) may differ slightly from those below, due to rounding off in the computations involved in determining the final answer.

1.1 Competency check

1. (a) $\frac{1}{2}$ (b) $\frac{2}{7}$ (c) $\frac{1}{2}$ (d) $\frac{6}{7}$ (e) $\frac{3}{8}$
3. (a) $\frac{5}{21}$ (b) $\frac{7}{30}$ (c) $2\frac{29}{40}$ (d) $-\frac{13}{15}$
5. (a) $\frac{9}{14}$ (b) $2\frac{2}{5}$ (c) $\frac{176}{285}$ (d) $\frac{20}{33}$

1.2 Competency check

1. (a) 0.33   (b) 0.015   (c) 0.38
3. (a) 0.387   (b) 1.542   (c) 14.2456
5. (a) 1.360791 (b) 0.499849   (c) 0.0987

1.3 Competency check

1. 6
3. −69
5. 8.6

1.4 Competency check

3. 1
5. (a) $7^2 = 49$ (b) $2^{-1} = \frac{1}{2}$ (c) $2^4 = 16$ (d) $4^{-5} = \frac{1}{1024}$
   (e) $8^{-2} = \frac{1}{64}$

1.5 Competency check

3. (a) 8.75×10^{-3} (b) 9.6×10^{-4} (c) 6×10^{-3}
   (d) 9.87×10^{-5}
5. (a) 0.003 27   (b) 0.000 003 672   (c) 0.000 082 070 4

1.7 Multiple-choice questions

1. (a)   2. (c)   3. (d)   4. (b)   5. (c)

1.8 Exercises

1. (a) $\frac{13}{8}$ (b) $\frac{78}{11}$ (c) $\frac{26}{3}$ (d) $\frac{176}{15}$ (e) $\frac{65}{7}$
3. (a) $\frac{6}{7}$ (b) $-\frac{3}{8}$ (c) $\frac{19}{28}$ (d) $\frac{17}{24}$ (e) $\frac{3}{40}$
   (f) $\frac{10}{21}$ (g) $\frac{14}{27}$
5. (a) 0.429 (b) 0.222 (c) 0.333 (d) 0.113 (e) 0.529
   (f) 0.143
7. (a) 12.352 (b) 0.2415 (c) 1.1523 (d) 101

Business mathematics and statistics

9. (a) 54 (b) 43.90 (c) −10
11. (a) $2^4 = 16$ (b) $3^2 = 9$
13. (a) $24\sqrt[4]{8}$ (b) $10\sqrt[4]{3}$
15. $6\frac{1}{3}$ hours
17. $132 000
19. (a) 1×10^0 (b) 4.4×10^1 (c) 1.2349×10^2

2.1 Competency check

3. (a) 0.15 (b) 0.01 (c) 0.17 (d) 0.245 (e) 5.682
   (f) 0.84̇3
5. (a) $\frac{3}{20}$ (b) $2\frac{2}{20}$ (c) $\frac{1}{200}$ (d) $\frac{13}{200}$ (e) $1\frac{17}{20}$

2.2 Competency check

1. 27%
3. 289 805 US tourists graced Australia's shores in 1994.
5. $15 million

2.3 Competency check

1. $3675
3. $418.65
5. $661

2.4 Competency check

1. (a) 15% (b) $408
3. (a) $1554
5. (a) $141 (b) $96

2.5 Competency check

1. Personal income tax is progressive, that is, the percentage tax paid increases as income grows whilst the company tax rate is fixed irrespective of income.
3. $34.34
5. $4265

2.6 Competency check

1. $47.61
3. (a) $200 (b) $120
5. 46.4%

2.8 Multiple-choice questions

1. (a) 2. (b) 3. (d) 4. (c) 5. (a)

2.9 Exercises

1. (a) 280 (b) 6.885 (c) 0.4 (d) 0.04 (e) 20.7 (f) 46.06

3. (a) 50% (b) 2% (c) 200% (d) 50% (e) 400%
5. USSR = 56%; Germany = 32%; Great Britain = 23%; France = 20%; Australia = 10%
7. Men = 26%; Women = 74% of part-time labour force.
9. Grooms = 16.2%; Brides = 19%
11. $1.5625 million
13. $126.66 billion
15. $131 billion
17. $300 million
19. 8.5%
21. $36 250
23. Straight commission = $43 750; Retainer plus commission = $40 980
25. (a) $46 274   (b) $40 000
27. 30%
29. $600
31. $9180
33. (a) No tax payable
    (b) $1860 tax payable
    (c) $16 452 tax payable
35. $21
37. 22%
39. $201.40
41. $392
43. 12.5%
45. (a) $375   (b) $150
47. (a) $28.74   (b) $41.67   (c) 31.03%
49. (a) $3348.21 (b) 10.7%

3.2 Competency check
3. $8500
5. 7.5 years

3.3 Competency check
3. Effective rate = 19.5%
5. (a) Simple interest = $1275   (b) Effective rate = 16.3%

3.5 Multiple-choice questions
1. (b)   2. (c)   3. (b)   4. (c)   5. (a)

3.6 Exercises
9. (a) 5%   (b) 7.5%   (c) 11%   (d) 10.25%
11. (a) $12 500   (b) $6250   (c) $14 000   (d) $5000
13. (a) $115   (b) 22.5%
15. Monique: $10 560   Dana: $15 840   Robyn: $18 480

Business mathematics and statistics

17. $24 375
19. (a) $13 162.80 (b) 15% (c) 29.5%

4.1 Competency check
3. (a) 24   (b) 42   (c) 6   (d) 71
5. (a) Accumulated Value = $1308.65; Compound Interest = $308.65
   (b) Accumulated Value = $30 460.07; Compound Interest = $5460.07
   (c) Accumulated Value = $160.36; Compound Interest = $110.36
   (d) Accumulated Value = $348.77; Compound Interest = $73.77

4.2 Competency check
3. (a) $3378   (b) $1,184   (c) $96   (d) $50 073
5. (a) 10.2% p.a. (i.e. 0.00848 × 12)   (b) 10.0% p.a. (i.e. 0.02506 × 4)
   (c) 7.8% p.a. (i.e. 0.0389 × 2)

4.4 Multiple-choice questions
1. (c)   2. (a)   3. (c)   4. (d)   5. (b)   6. (a)

4.5 Exercises
5. $8954
7. 48.1 cents
9. (a) Simple Interest = $6,000
   (b) Compound Interest = $6386.16
   Therefore option (b)—compound interest—is more favourable.
11. $5584
13. (a) 13.5% p.a.   (b) 14.4% p.a.   (c) 6.7% p.a.
15. (a) $10 095   (b) $10 068
17. $4225
19. $2651

5.2 Competency check
3. $9411.43
5. $17 839.84

5.3 Competency check
3. $1287.60
5. $299.72

5.4 Competency check
3. $22 158
5. $4098

5.5 Competency check
3. $68 808
5. $1076.19

5.7 Multiple-choice questions
1. (a) 2. (a) 3. (b) 4. (b) 5. (c)

5.8 Exercises
1. (a) $97 855 (b) $25 462 (c) $22 865 (d) $65 030 (e) $8615
3. $17 443
5. $84 678
7. $7539.05
9. $30 927
11. $10 185.22
13. $76.27
15. $13 680 + $8000 = $21 680
17. (a) $408.87 (b) $15 983.40
19. (a) $842.58 (b) $53 721

6.2 Competency check
3. (a) $2100 (b) $14 100
5. (a) $3600 (b) 8.33%
   (c)

Year	Original cost ($)	Opening WDV ($)	Depreciation ($)	Closing WDV ($)
1	48 000		3600	44 400
2		44 400	3600	40 800
3		40 800	3600	37 200

(d) At end of 3rd year, book value is $37 200

6.3 Competency check
5. (a) 22.5% (b) 15% (c) 30%
7. $A = P(1-i)^n$

$$\frac{A}{(1-i)^n} = P$$

6.4 Competency check
3. Reducing balance rate = 30% and depreciation amount = $480
5. Reducing balance rate = 3.75%

Year	Original cost ($)	Opening WDV ($)	Depreciation ($)	Closing WDV ($)
1	5 000 000		187 500	4 812 500
2		4 812 500	180 469	4 632 031
3		4 632 031	173 701	4 458 330

Business mathematics and statistics

6.5 Competency check
3. (a) (i) $600   (ii) $32   (iii) $270
   (b) (i) $434   (ii) $25   (iii) $173
5. $7731

6.6 Competency check
5. (a) 0.7 cents/kilometre   (b) $142

6.8 Multiple-choice questions
1. (c)   2. (d)   3. (a)   4. (a)   5. (a)

6.9 Exercises
3. $5800
5. (a) $42 500   (b) $10 875   (c) $6900   (d) $82 750
7. $100
9. $685
11. (a) $51 000   (b) $5100   (c) $4550
13. $13 050
15. (a) $37 239   (b) $3067   (c) $972   (d) $15 754
17. 15%
19. (a) 30%   (b) $159
21. (a)

Year	Original cost ($)	Opening WDV ($)	Depreciation ($)	Closing WDV ($)
1	145 000		29 000	116 000
2		116 000	29 000	87 000
3		87 000	29 000	58 000
4		58 000	29 000	29 000
5		29 000	29 000	0

(b)

Year	Original cost ($)	Opening WDV ($)	Depreciation at 30% ($)	Closing WDV ($)
1	145 000		43 500	101 500
2		101 500	30 450	71 050
3		71 050	21 315	49 735
4		49 735	14 921	34 814
5		34 814	10 444	24 370

23. $7392

7.1 Competency check
3. The (point of) origin.

5. The graph of $3y = 2x$ or $y = 2x/3$

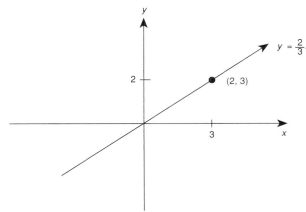

7.2 Competency check
1. No. The lines may be parallel and so there won't be a point of intersection.
3. $x = 5; y = 1$
5. racquets = \$75; shoes = \$55

7.3 Competency check
3. 106 000 units

7.5 Multiple-choice questions
1. (c) 2. (a) 3. (c) 4. (a) 5. (d)

7.6 Exercises
5. Slope = 4; y–intercept = 2
7. Slope = -1; y–intercept = 3
13. $x = 3; y = -1$
15. $x = -2; y = -10$
19. 120 000 units

8.5 Multiple-choice questions
1. (c) 2. (b) 3. (d) 4. (a) 5. (d)

9.5 Multiple-choice questions
1. (c) 2. (c) 3. (a) 4. (c) 5. (a)

10.1 Competency check
1. The mean.
3. The mean is affected by extreme values which pull the answer away from the "centre" of the distribution.
5. 67.4%

Business mathematics and statistics

10.2 Competency check
1. Arrange the data into ascending order.
3. The major drawback of the median is that it ignores values above and below the middle value. Consequently, the mean is the preferred measure of central tendency provided there are no extreme values.
5. (a) $63 250 (b) 70 cents (c) $145

10.3 Competency check
1. The value(s) which occurs most often is the mode.
3. (a) there is no modal class (b) 16 and 19 (c) 28
5. (a) $62 830 (b) 70.6 cents (c) $146

10.6 Multiple-choice questions
1. (a) 2. (a) 3. (b) 4. (b) 5. (a)

10.7 Exercises
3. The mean is preferred as long as there are no extreme values in the data.
5. $600 daily expenditure by delegates compared with $79 for the average overseas tourist.
7. Mean = 293 500; Median = 292 000; Mode = 292 000
9. 77%
11. $\overline{X}$ = 46.4; Md = 42; Mo = 40, 41, 42
13. (a) Wines/spirits—positively skewed; sporting goods—negatively skewed; graphics—positively skewed; beef—symmetrical.
15. $\overline{X}$ = $5250; Md = $5270; Mo = $5360
17. X = $79.65; Md = $80.00; Mo = $81.38
19. (a) $\overline{X}$ = 12.4; Md = 11.5; Mo = 4, 6, 7, 11, 13, 14, 16
    (b) Number of days lost for 30 workers.

Days lost	Frequency
0–4	5
5–9	7
10–14	8
15–19	6
20–24	2
25–29	1
30+	1

(c) $\overline{X}$ = 12; Md = 11.9; Mo = 11.7

11.1 Competency check
1. The range is dependent on the largest and smallest values which may not be typical of the set of values being studied.
3. (a) 13 (b) 36
5. 167(000) government employees.

11.2 Competency check
3. (a) (i) $1 325 000   (ii) $Q_1 = \$982\,500$; $Q_3 = \$1\,850\,000$
   (iii) $867 500
5. (a) (i) $62   (ii) $31

11.3 Competency check
3. 48(000) government employees
5. 3.5 cents

11.4 Competency check
3. 56(000) government employees
5. 4.5 cents

11.5 Competency check
3. Collingwood = 14.3%; Hawthorne = 13.9% with Hawthorne having the smaller relative dispersion.
5. Salaries of executives = 24.9%; price of petrol = 6.65%

11.6 Multiple-choice questions
1. (a)   2. (b)   3. (b)   4. (b)   5. (d)

11.8 Exercises
3. (a) 50   (b) Interquartile range = 5; Q.D. = 2.5   (c) 8   (d) 13
5. (a) 22 500   (b) 7700   (c) 40%
   (d) There is a high degree of variation in the number of people migrating from NSW to other states.
7. $S = 8.46$
9. (a) 85   (b) 104
11. Highest risk is R-U-Nuts Pty. Ltd ; lowest risk is I-B-4-U Pty Ltd.
13. $\sigma = \$39$; $V = 27\%$
15. (a) (i) $\mu = \$44$; Median = $45   (ii) $12   (iii) 16
    (iv) 36%
17. (a) $\mu = 63\%$; Median = 63%; (b) Mean deviation = 7%; $\sigma = 10\%$
    (c) 15%
19. (a) 7   (b) 25%   (c) 17.5 to 38.5 years

12.4 Multiple-choice questions
1. (d)   2. (e)   3. (c)   4. (b)   5. (d)

12.5 Exercises
11. Use a systematic or random sample.
15. Use a systematic or random sample.
17. Cluster or multistage sampling.

13.1 Competency check

1. (a) At least one error in an account ledger.
   (b) There are 6 cars in a garage and no more than 2 have flat batteries.
   (c) Jim passes an exam.
   (d) A committee of 5 people is chosen, of which less than 2 are women.
3. Elements of each set are referred to by the initial letter of the surname.
   (a) {R, F, G, T}; Prob = 4/10 = 2/5
   (b) {R, J, H, D, S, T, W}; Prob = 7/10
   (c) {H, T, W}; Prob = 3/10
   (d) {R, H, G, T, W}; Prob = 5/10 = 1/2
   (e) {H, G, T}; Prob = 3/10
   (f) {B, F, D, W}; Prob = 4/10 = 2/5

13.2 Competency check

1. (a) Dependent (b) Dependent (c) Independent (d) Independent (e) Dependent
3. $\left(\dfrac{15}{27}\right)^2 = 0.039$; $2 \times \dfrac{15 \times 12}{27 \times 27} = 0.494$; $\left(\dfrac{12}{27}\right)^2 = 0.198$
5. $P[A \cup B] = P[A] + P[B] - P[A \cap B]$ which means $P[A \cap B]$ must be nonzero, so there must be an overlap of the sets, hence not mutually exclusive.

13.3 Competency check

1. $3/7 = 0.429$
3. 6×10^{-8}
5. (a) 1/8 (b) 7/8

13.4 Competency check

1. $\dfrac{0.5 \times 0.8}{0.5 \times 0.8 + 0.5 \times 0.6} = 0.571$

13.6 Multiple-choice questions

1. (c) 2. (d) 3. (b) 4. (c) 5. (d)
6. (c) 7. (c) 8. (b) 9. (b) 10. (c)

13.7 Exercises

1. $24/60 = 2/5$
3. (a) $\dfrac{18}{87}$ (b) $\dfrac{36}{87}$ (c) $\dfrac{3}{87}$ (d) $\dfrac{79}{87}$ (e) $\dfrac{9}{87}$
5. (a) $(0.65)^3 = 0.275$ (b) $(0.35)^3 = 0.043$
   (c) $(0.65)^2 (0.35) = 0.148$ (d) $3(0.65)^2 (0.35) = 0.444$
7. (a) $\dfrac{20}{120}$ (b) $\dfrac{10}{120}$ (c) $\dfrac{50}{120}$ (d) $\dfrac{70}{120}$ (e) $\dfrac{90}{120}$
9. (a) $\dfrac{75}{256}$ (b) (i) $\dfrac{5}{27}$ (ii) $\dfrac{107}{216}$
11. (a) 0.9 (b) 0.1
13. (a) 0.74 (b) 0.265 (c) 0.75

15. (a) $\frac{165}{400} = 0.413$ (b) $\frac{27}{400} = 0.068$ (c) 0.205
17. (a) 0.149 (b) 0.851
19. $\frac{1}{25} = 0.04$

14.1 Competency check

1. (a) 0.45994 or 45.994% (b) 0.05301 or 5.301% (c) 0.75619 or 75.619%
3. (a) 14% (b) 4% (c) 25%
5. (a) 6.7% (b) 98.6% (c) 0.6%

14.4 Multiple-choice questions

1. (d)   2. (a)   3. (a)   4. (a)   5. (b)

14.5 Exercises

1. The mean and the standard deviation.
3. 93 bulls-eyes
5. (a) 2% (b) 9%
7. (a) $190 000 (b) $392 000
9. (a) 3.3% (b) 1.5% (c) 25.9% (d) 36.6%
11. (a) 1.5% (b) 12.4% (c) 34.2%
13. (a) 13.1% (b) 4.6% (c) 31.0% (d) 8.9%
15. (a) 0.1% (b) 0.2% (c) 7.6% (d) 40% (e) 32.2%
17. (i) 0.3% (ii) 8.5% (iii) 66.8%
19. (a) 50% (b) 0.13% (c) 44% (d) 21%
21. 94%

15.1 Competency check

3. (a) inverse (b) direct (c) no relationship (d) direct
   (e) direct (f) inverse (g) inverse

15.2 Competency check

3. $r = 0.85$ $r^2 = 0.72$
5. $r = 0.98$ $r^2 = 0.96$

15.3 Competency check

3. A rank correlation of -0.8 suggests that the top graduates' performance in the workforce was very nearly in reverse order to their academic prowess; i.e. once they started work, the top students did not perform as well as the average students.
5. $r_s = -0.10$. A rank correlation of this size suggests that height has little bearing on average points scored per game.

15.4 Competency check

3. (a) $Y = 26.8 + 2.7X$

(b) On average, each 1°C increase in temperature increases attendance by 2700.
(c) (i) 64 600   (ii) 83 500

5. $Y = 82.47 - 0.21X$ (a) 73   (b) 79   (c) 51

15.5 Competency check

3. The limits are 42 ± 12; i.e. between 30 and 54.
5. (a) 5.43% or 5 % to nearest %.

15.6 Multiple-choice questions

1. (a)   2. (c)   3. (c)   4. (a)   5. (a)

15.7 Exercises

11. $Y = -1.537 + 0.014X$ (c) (i) 22 hours   (ii) 7 hours
13. $r = 0.181$; $r^2 = 0.033$. There does not appear to be any significant correlation between number of beehives and number of marriages. Only 3.3% (= 0.033) of the variation in number of beehives can be explained by knowing the number of marriages - to say nothing of the number of "stings"!
15. (b) $Y = 33.186 + 1.439X$ (c) 63%
17. (a) $r = -0.958$; $r^2 = 0.918$ (b) $Y = 1488.675 - 34.570X$ (c) 797
19. (a) $Y = 91.572 + 6.253X$ (b) $13 200   (c) $192 000   (d) $r^2 = 0.77$
21. (b) $Y = -1.99 + 1.01X$ (c) 79%   (d) $r = 0.90$; $r^2 = 0.81$

16.1 Competency check

3. (a) 1996 = 92; 1997 = 86.7;   (b) 1995 = 115.4; 1996 = 106.2
5. (a) 1994/95 = 90.5; 1995/96 = 66.1
   (b) 1993/94 = 151.2; 1994/95 = 136.8

16.2 Competency check

5. (a) 104.0
   (b) Each article used different units of measurement.

16.3 Competency check

5. (a) 120.6   (b) 105.35   (c) 126.42

16.4 Competency check

3. 1994/95 = 1 048 288 ; 1995/96 = 1 065 712
5.

Year	Other than House building	House building
1991/92	100.0	100.0
1992/93	101.9	100.6
1993/94	106.8	100.8
1994/95	110.0	102.3
1995/96	110.2	104.5

The Price Index for House Building's materials rose more than twice the index for Other than House Building.

16.6 Multiple-choice questions

1. (c)   2. (e)   3. (b)   4. (c)   5. (b)

16.7 Exercises

3. It means that inflation during the 1995/96 financial year was 4.2%.
5. You would not use the 1994/95 result as it was a record year for dwelling commencements whereas a base period should be one experiencing a more typical number of dwelling commencements.
7. 1993/94 = 116.9 ; 1994/95 = 138.4
   Travel credit was 16.9% higher in 1993/94 compared with the 1992/93 base period and 38.4% higher in 1994/95 relative to 1992/93.
9. 1995 = 115.2; 1996 = 130.9 ; 1997 = 155.4
11. In money terms, AWE rose $79 or 12.6% (79/625 × 100). However, in real terms, AWE grew by $16 ($593 − $577) or 2.8%. In other words, inflation eroded almost 80% of the money increase.
13. 125.4
15. (a) 122.7   (b) 120.9

17.

Year	GDP ($ billion)	CPI (1989/90 = 100)	CPI (1990/91 = 100)	Real GDP (1990/91 = 100)
1990/91	379.0	107.3	100.0	379.0
1991/92	387.2	108.4	101.0	383.4
1992/93	405.8	110.3	102.8	394.7
1993/94	430.4	113.9	106.2	405.3
1994/95	455.6	118.7	110.6	411.9

In money terms, the GDP rose by $76.6 (billion) or 20.2% (76.6/379 × 100) but in real terms (based on 1990/91) the GDP increased by $32.9 (billion) ($411.9 − $379.0) or 8.7%. Thus inflation eroded almost 60% of the money gain.

19. (a) 117.34   (b) 117.56

17.1 Competency check

5. (a) Secular
   (b) Seasonal
   (c) Irregular
   (d) Cyclical

17.2 Competency check

1. The common assumption for all three is that when the long-term trend is plotted, a straight line will result.

5. (a)

Year (30 June)	Sales ($ million)	Smoothed series (using alpha = 0.2)
1989	1	1.0
1990	3	2.60 (0.2 × 1.00 + 0.8 × 3)
1991	6	5.32 (0.2 × 2.60 + 0.8 × 6)
1992	7	6.66 (0.2 × 5.32 + 0.8 × 7)
1993	10	9.33 (0.2 × 6.66 + 0.8 × 10)
1994	14	13.07 (0.2 × 9.33 + 0.8 × 14)
1995	15	14.61 (0.2 × 13.07 + 0.8 × 15)
1996	15	14.92 (0.2 × 14.61 + 0.8 × 15)
1997	16	15.78 (0.2 × 14.92 + 0.8 × 16)

(b)

Year	Sales ($ million)	Smoothed series (using alpha = 0.6)
1989	1	1.0
1990	3	1.8 (0.6 × 1.00 + 0.4 × 3)
1991	6	3.48 (0.6 × 1.80 + 0.4 × 6)
1992	7	4.89 (0.6 × 3.48 + 0.4 × 7)
1993	10	6.93 (0.6 × 4.89 + 0.4 × 10)
1994	14	9.76 (0.6 × 6.93 + 0.4 × 14)
1995	15	11.86 (0.6 × 9.76 + 0.4 × 15)
1996	15	13.12 (0.6 × 11.86 + 0.4 × 15)
1997	16	14.27 (0.6 × 13.12 + 0.4 × 16)

An alpha level of 0.2 places small weight on past values and a proportionally greater emphasis on present values. The opposite is true for an α level of 0.6.

17.3 Competency check

3. (a) March 75.78   (b) 2001 March 38
    June 116.59   June 58
    Sept 133.34   Sept 67
    Dec 74.29   Dec 37
    ―――――   ―――
    400.00   200

5. (a)

Year	March	June	September	December
1995	11.9	12.5	12.8	12.9

(b) 165 people

17.4 Multiple-choice questions

1. (a)   2. (c)   3. (c)   4. (a)   5. (a)

17.6 Exercises

5. Seasonal variations are the most predictable and should be explained with an appropriate example.

9. (a) The equation will vary depending on which year is chosen as the base year and it is chosen.

   (b) However, the prediction for year 2000 will still be 127 000 passengers.

15.

Year	Units sold	Smoothed series (using alpha = 0.2)
1987	4.0	4.0
1988	4.2	4.16 ($0.2 \times 4 + 0.8 \times 4.2$)
1989	4.5	4.43 ($0.2 \times 4.16 + 0.8 \times 4.5$)
1990	4.7	4.65 ($0.2 \times 4.43 + 0.8 \times 4.7$)
1991	5.0	4.93 ($0.2 \times 4.65 + 0.8 \times 5.0$)
1992	5.8	5.63 ($0.2 \times 4.93 + 0.8 \times 5.8$)
1993	6.4	6.25 ($0.2 \times 5.63 + 0.8 \times 6.4$)
1994	6.4	6.37 ($0.2 \times 6.25 + 0.8 \times 6.4$)
1995	6.5	6.47 ($0.2 \times 6.37 + 0.8 \times 6.5$)
1996	7.0	6.89 ($0.2 \times 6.47 + 0.8 \times 7.0$)
1997	7.3	7.22 ($0.2 \times 6.89 + 0.8 \times 7.3$)

19. Florentine Furnishings

(a)

Year	Sales turnover	3-year moving total	3-year moving average
1983	11		
1984	17	49	16.3
1985	21	58	19.3
1986	20	60	20.0
1987	19	60	20.0
1988	21	64	21.3
1989	24	74	24.7
1990	29	89	29.7
1991	36	107	35.7
1992	42	125	41.7
1993	47	141	47.0
1994	52	150	50.0
1995	51	153	51.0
1996	50	152	50.7
1997	51		

Business mathematics and statistics

(c)

Year	Units sold	Smoothed series (using alpha = 0.3)
1983	11	11.0
1984	17	15.20 (0.3 × 11 + 0.7 × 17)
1985	21	19.26 (0.3 × 15.2 + 0.7 × 21)
1986	20	19.78 (0.3 × 19.26 + 0.7 × 21)
1987	19	19.23 (0.3 × 19.78 + 0.7 × 19)
1988	21	20.47 (0.3 × 19.23 + 0.7 × 21)
1989	24	22.94 (0.3 × 20.47 + 0.7 × 24)
1990	29	27.18 (0.3 × 22.94 + 0.7 × 29)
1991	36	33.35 (0.3 × 27.18 + 0.7 × 36)
1992	42	39.40 (0.3 × 33.35 + 0.7 × 42)
1993	47	44.72 (0.3 × 39.4 + 0.7 × 47)
1994	52	49.82 (0.3 × 44.72 + 0.7 × 52)
1995	51	50.65 (0.3 × 49.82 + 0.7 × 51)
1996	50	50.20 (0.3 × 50.65 + 0.7 × 50)
1997	51	50.76 (0.3 × 50.20 + 0.7 × 51)

21. (a) March   104.57
         June   83.18
         Sept.   66.43
         Dec.   145.82
                   400.00

(b) 1991 March = 22.0   June = 24.0   Sept. = 18.1   Dec. = 27.4
(c) March = 33 200   June = 26 400   Sept. = 21 100   Dec. = 46 300
(d) March = 4800 Litres   June = 4200 L   Sept. = 3000 L   Dec. = 8800 L

GLOSSARY

Abscissa The *x*-coordinate of a point on a graph and is the distance of the point from the *y*-axis.

Accumulated annuities The future value of the sum of total periodic deposits and accrued interest over the life of the deposits, given the amount of periodic payment, the compound interest rate, and the term of the account. The formula is:

$$S = p\frac{(1 + i)^n - 1}{i}$$

where S = the future value (i.e. the total deposit and accrued interest), p = the periodic payment, i = the interest rate/period, and n = the number of periods.

Accumulated depreciation See ***Total depreciation***

Alternative hypothesis (H_1) The hypothesis assumed true if the null hypothesis is rejected.

Amortisation Equal periodic payments with each such payment covering interest on the balance of the principal resulting in a decline of the principal, so that by the end of the term of the loan, after the last payment, there is zero current account balance. The formula is:

$$PV = \frac{1 - (1 + i)^n}{i}$$

where p = the periodic payment, PV = the present value, i = the interest rate/period, and n = the number of periods.

Annuities Periodic payments (monthly, quarterly, semi-annually or annually) of equal amounts of money over a period of time.

Arithmetic mean The most commonly used measure of central tendency. It is calculated by summing the values of all items and dividing by the number of items and is affected by extreme values. The formula for ungrouped data is:

$$\bar{x} = \frac{\Sigma x}{n}$$

where $\bar{x}$ = the mean, Σ = summation, x = the set of values, and n = the total number of values in the set. The formula for grouped data is:

$$\bar{x} = \frac{\Sigma fm}{n}$$

where $\bar{x}$ = the mean, Σ = summation, f = the frequency of the individual class, m = the midpoint of the individual class, and n = the total number of frequencies.

Arithmetic progression A sequence of terms, listed one after the other, with each term generated from the one before by adding a constant amount, referred to as the common difference.

Asset Any resource that has value to the owner (e.g. bank deposits, cash accounts, receivables and property inventories).

Average A single value indicating a typical representative figure of the data examined.

Balance sheet A statement of the financial position of a business at the end of a period, divided into a liability group and an asset group.

Bar chart A visual presentation of data with rectangular bars of equal width but varying height or length representing the magnitude of the specific data; the common types of bar charts include simple, component and multiple.

Base period The period to which all other periods are compared. The base is given a value of 100 and the period selected as the base should be recent and fairly normal.

Basic requirements of probability
- The value of a probability ranges from zero to unity:

$0 \leq P[E_i] \leq 1$ for each and every outcome, i; an impossible event has a probability of zero; a certain event has a probability of unity.

- The sum of the probability for all the outcomes in an experiment is unity:

$$\Sigma P[E_i] = 1.$$

Bayes' Theorem A useful extension of conditional probability. It provides a way of revising our prior probabilities in the light of new information in order to provide a posterior probability:

$$P[E_k \mid A] = \frac{P[E_k]P[A \mid E_k]}{P[E_1]P[A \mid E_1] + P[E_2]P[A \mid E_2] + \mathrm{K} + P[E_n]P[A \mid E_n]}$$

Book value (Closing written-down value) The current value of an asset derived by subtracting the total depreciation up to that point in time from the original cost of the asset.

Break-even analysis Determination that production and sales level where the firm's total revenue from sales is equal to the firm's total costs incurred in realising that revenue.

Brokerage A charge in the form of a fee made by the broker (e,g stockbroker) to a client for the provision of a service such as acting as the intermediary in the purchase of shares for the client.

Cartesian coordinates Each point on a graph can be represented by an ordered pair of values (x,y) referred to as the Cartesian (after the French mathematician Descartes) coordinates of the point.

Cash discount A reduction in the price of a commodity or service in return for payment in the form of cash or a cheque.

Central Limit Theorem As the size of the sample (i.e. the number of observations in each sample) gets to a certain point (e.g. 30), the sampling distribution of the mean can be approximated by the normal distribution. This is true irrespective of the shape of the distribution of the population's individual values.

Chain discount More than one discount on an item or service, often applied to clear old stock and/or for cash-flow purposes.

Classical (theoretical) method The outcomes of the experiment are assumed to be equally likely:

$$P[E] = \frac{n(E)}{n(S)}$$

Class size In a frequency distribution, the difference between the real upper limit and the real lower limit of a particular class.

Closing written down value See **Book value**

Cluster sampling Data obtained when a population is divided into clusters or groups exhibiting a particular characteristic or description (e.g. single-parent families) and members of each cluster observed.

Coefficient of correlation Used to examine the relationship between two variables. The formula is:

$$r = \frac{\Sigma(X - \overline{X})(Y - \overline{Y})}{\sqrt{\Sigma(X - \overline{X})^2 \Sigma(Y - \overline{Y})^2}}$$

where r = the coefficient of correlation, Σ = the symbol representing summation, $X - \overline{X}$ = the deviation of the X value from its mean, and $Y - \overline{Y}$ = the deviation of the Y value from its mean.

Coefficient of determination The square of the coefficient of correlation (i.e. r^2). It indicates the degree (defined as a percentage) to which the dependent variable's changes are a result of variations in the independent variable.

Coefficient of variation The calculated measure of dispersion (e.g. standard deviation) in relation to the mean, expressed as a percentage. The formula is:

$$V = \frac{s}{\overline{x}} \times 100$$

where V = the coefficient of variation, s = the standard deviation, and $\overline{x}$ = the mean.

Commission Income received as a result of selling a product or service, normally based on a percentage of the sales value.

Complement of an event The event that does not happen — all the outcomes that are not in M — written M^c or $\tilde{M}$ or $\overline{M}$ (referred to as 'not M').

Glossary

Component bar chart A type of bar chart where each bar is divided into sections representing components making up the total. For example, a component bar chart of the major components of Federal taxation revenue would include income tax, company tax, sales tax and excise tax.

Compound interest Interest received from a previous period/s, which itself earns interest and continues to do so. The formula is:

$$S = P(1 + i)^n$$

where S = the sum of the principal and the compounded interest, P = the principal, r = the interest rate/period, and n = the number of periods.

Conditional probability The conditional probability of A is based on the occurrence of B is written:

$$p[B \mid A] = \frac{P[A \cap B]}{P[A]} = \frac{P[B]P[A \mid B]}{P[A]}$$

Confidence coefficient $(1 - \alpha)$ The probability that a confidence interval will contain the estimated parameter.

Confidence interval See ***Interval estimate***

Confidence level See ***Confidence coefficient***

Consumer Price Index (CPI) A weighted average price relative of retail prices commonly used in Australia as a measure of the inflation rate.

Coordinates Ordered pairs of values in the form (x, y).

Correlation analysis A measure of the degree of closeness in the relationship between two or more variables.

Critical value A (tabular) value of a statistic with which the calculated statistic value is compared to see if the null hypothesis is accepted or rejected. It is written with an asterisk (*), e.g. t*.

Cyclical variation A time series component indicating the expansion and contraction of business activity, describing the tendency for a particular firm, industry or economy to recover from a recession, become prosperous, contract again, and so on.

Decision tree A tree diagram used in the context of decision-making.

Degrees of freedom These indicate how many values are free to vary in a random sample of given size. The number of degrees of freedom lost is equal to the number of population parameters estimated; e.g. if s is used to estimate s because s is unknown, then one degree of freedom is lost.

Depreciation For taxation purposes, the deduction allowed for wear and tear of plant and equipment used in generating income; it may be straight-line (prime-cost) or diminishing-value (reducing-balance).

Descriptive statistics Methods involving the collection, presentation and characterisation of data to describe or summarise its features accurately.

Discount A reduction in price given to enhance the likelihood of selling an item (the reduction is the discount).

Discounted bill of exchange A note that has had an amount (the discount) deducted from its face value at the outset of the loan period. The discount rate is expressed as a percentage rate per annum to the face value of the note.

Dispersion The degree to which the original values of a set of data are clustered around the measure of central tendency (i.e. the average).

Effective rate The rate of interest actually paid in a year when a flat rate is cited. The formula is:

$$ER = \frac{2N \times R}{N + 1}$$

where ER = the effective rate, N = the number of repayment periods, and R = the flat rate of interest.

Efficiency ratio Measurement of the company's ability to control, for example, inventory, debtor turnover, and average collection period per debtor.

Elimination method One means of solving a simultaneous equation.

Event A collection of outcomes from a specified sample space.

Experiment A process which produces a well-defined outcome.

Exponent A quantity which when multiplied by itself one or more times has a superscript number at the upper right-hand side of the quantity (e.g. $2^3 = 8$).

Exponential equation An equation of the form $a^x = b$ where a and b are constant and x is a variable.

Exponential function This is a function in the form of $y = a^x$ where a is a constant, b is the horizontal variable and y is the vertical variable.

Exponential smoothing A technique used to smooth a time series and determine long-term movements in the data.

External data Data obtained from outside the business organisation (e.g. ABS statistics, journals, magazines and newsletters).

Face value The full amount of a security, loan or investment before interest is added on or discounted.

Fixed cost Costs inucrred which remain constant irrespective of the level of output (e.g., leasing charges, insurance premiums)

Frequency distribution A tool for grouping data items into classes and then recording the number of observations in each class.

Frequency polygon A line graph of a frequency distribution; normally drawn on a histogram.

Future value of annuities See **Accumulated annuities**

Geometric mean Used to average ratios, rates, and growth progression. It is defined as the nth root of the product of n values. The formula is:

$$Gm = \sqrt[n]{\text{Product of } n \text{ values}}$$

Geometric progression Sequence of terms listed one after the other where each is generated from the one before by multiplying by a constant ratio, referred to as the common ratio.

Grouped data Data which has been presented in the form of a frequency distribution as the frequency of observations in each class interval is given.

Harmonic mean Used to determine average rates and is found by determining the arithmetic mean of the reciprocal of the values contained in the data.

Histogram A frequency distribution displayed in the form of a bar chart (but without any space between respective bars) with the width of each bar equal to the size of the class interval and the height representing the corresponding frequency.

Improper fraction A fraction where the numerator is larger than the denominator, that is, the fraction is larger than 1, e.g., 12/11. 3/2

Independent events Events which have no influence on the probability of each other's occurrence.

Index number A relative comparison between two figures with one of these figures used as the base period.

Inferential statistics Methods of estimating the characteristics of a population from sample results.

Integer A whole number such as 1, 2, 3,. . .7, 8. . .

Internal data Data obtained from within the business organisation concerned (e.g. sales, profit, and inventory results).

Interval estimate The two numbers (called 'upper and lower limits') believed to contain the value of a population parameter.

Interval estimator A rule that tells us how to calculate two numbers (for a population parameter) from sample data.

Irregular movement A time series component representing leftover or residual variations after secular trend, cyclical, and seasonal factors have been examined. It may, for example, be the result of freak forces of nature or erratic business activity (e.g. cyclones, strikes, or abrupt changes in the value of the Australian dollar).

Joint probability One that relates to more than one category of classification—determined from the intersection of two (or more) of the criteria in question.

Judgment (Non-random) sample A sample which is selected on a subjective rather than on a objective, statistical basis.

Laspeyres method A weighted index where the weights are derived from the quantities of each item consumed or produced during the base period.

Least-squares method Determination of the regression line of the observed data that minimises the sum of the squares of the errors from the line; referred to as line of best fit since it minimises the error between the estimated points on the line and the actual observed points that were used to draw it.

Liability Obligation of a company incurred in the operating of the business (e.g. payments to suppliers, accounts payable, and long-term debt).

Linear (Line) graph The graphic representation of the relationship between two variables which can be described by a straight line.

Liquidity ratio Evaluation of the current monetary commitments of a firm versus its access to assets available, which may be necessary for immediate conversion into a means of payment.

Marginal probabilities Can be expressed as the sum of joint probabilities and are the values in the margins of a contingency table.

Mean See ***Arithmetic mean.***

Mean deviation The mean average of the deviations of individual values in a series of data from the mean, with all deviations assigned an absolute (positive) value. The formula is:

$$MD = \frac{\Sigma f \mid x - \bar{x} \mid}{n}$$

where MD = the mean deviation, Σ = represents summation, f = the frequency of the individual class, x = the midpoint of each class, $\bar{x}$ = the arithmetic mean, and n = the total number of frequencies.

Median Considers the middle value of an array of items, ignoring other values. It is not affected by extreme values as only the middle value is examined. The formula is:

$$Md = L + \frac{\left(\frac{n}{2} - C\right)}{f}(i)$$

where Md = the median, L = the real lower limit of the median class, n = the total number of frequencies, C = the cumulative frequency of the class prior to the median class, i = the size of the class interval, and f = the frequency of the median class.

Mixed number A combination of a whole number and a proper fraction (e.g, 4/3 = 1 1/3; 7/5 = 1 2/5

Mode The most frequently occurring value. It ignores other values outside the modal range, thus is not affected by extreme values unless the mode itself is located at an extreme. The formula for grouped data is:

$$Mo = L + \frac{d_1}{d_1 + d_2}(i)$$

where Mo = the mode, L = the real lower limit of the modal class, d_1 = the frequency of the modal class minus the frequency of the previous class, d_2 = the frequency of the modal class minus the frequency of the subsequent class, and i = the size of the class interval in the modal class.

Monomial An algebraic expression containing only one term (e.g. ab, ab^2).

Multi-stage sampling Data selected by stages with sampling units at each stage sub-sampled from the larger units chosen at the previous stage. For example, the division of geographical regions into areas or towns with each such area subsequently subdivided into smaller zones. A random or systematic sample is then selected from these smaller zones.

Mutually exclusive or disjoint events Cannot occur together—the occurrence of one automatically precludes the occurrence of the others.

Negative correlation When the dependent and independent variables move in opposite directions (e.g. theatre attendance and video rentals).

Nominal interest rate The current rate of interest.

Non-exclusive or joint events Have an overlap or common region.

Non-linear graph Data which when illustrated in the form of a graphic representation is not portrayed by a straight line.

Non-random sample See *Judgment sample*

Normal distribution When the determination of central tendency of the data results in the mean, median, and mode being of equal value. When graphed it is a symmetrical, bell-shaped curve.

Null hypothesis (H_0) A statement to be proved or disproved which says that there is no significant difference between a sample and a population or between two samples, and that any difference is due to random error.

Ogive A graph of cumulative frequencies. It can be used to determine the median value of data but lacks precision.

Open-ended data Data for which the upper or lower limits of a frequency distribution are undefined (e.g. under 20, 65+).

Ordinate The *y*-coordinate in the ordered pair of values (*x*,*y*) is called the ordinate of the point and is its distance from the *x*-axis.

Paasche method A weighted index calculated by using as the weights the quantities of each item consumed or produced during the current period.

Parameter A characteristic of a population, be it a population of widgets, cars, people, etc.

Personal income tax A Federal Government imposed tax that all income earners must pay (if their income exceeds a threshold of $5,400); it is a progressive tax, that is, as the income earned increases, the percentage tax paid also rises.

Pictogram A picture or pictorial presentation of data.

Pie chart A visual display of proportions in the form of the segments of a pie or circle.

Point estimate The single number used to represent the population parameter.

Point estimator A rule that tells us how to calculate a single number from sample data, to obtain a population parameter.

Polynomial An algebraic expression containing two or more terms (e.g. $a + b + cd + d^2$).

Pooled variance estimate An estimate of *s* based on the standard deviations of two (or more) samples:

$$s_p^2 = \frac{(n_1 - 1)s_1^2 + (n_2 - 1)s_2^2}{n_1 + n_2 - 2}$$

Positive correlation When the dependent and independent variables move in the same direction (e.g. obesity and incidence of heart disease).

Posterior probabilities Revised probabilities based on later information.

Present value of an annuity Sum of the present values of the payments (each discounted by the rate of interest for the time it is paid to the present). It is used for calculating the amortisation of, for example, home-loan repayments. The formula is:

$$PV = \frac{1 - (1 + i)^n}{i}$$

where *PV* = the present value, *p* = the periodic payment, *i* = the interest rate/period, and *n* = the number of periods.

Primary data Collection and presentation of numerical information by the source (e.g. the results of a questionnaire prepared and administered by you), rather than using the results of someone else (secondary data).

Principal The actual amount borrowed or lent.

Prior probabilities Initial estimates of the probabilities of events.

Price index Summarises the change in the price of a item(s), usually over time.

Probability The study of assigning numerical values to questions of chance.

Product moment formula The formula used to determine the coefficient of correlation.

Profitability ratio The relationship of profit (gross and/or net) to another financial variable (e.g. total assets or sales).

Proper fraction A fraction in which the numerator is less than the denominator, that is, the fraction is less than 1, e.g., 1/2, 5/8.

Proportion A statement equating two ratios. If can also express the comparison of a part to a whole, normally in fractional terms.

Quadratic equation The product of two terms with the highest power of any of the resulting terms being 2; normally written:

$$ax^2 + bx + c = 0$$

Quantity index Comparison between quantity of products over time in index form.

Quartile deviation The examination of the values of the first and third quartiles of data which have been arranged in the form of an array. It includes the middle 50% of values, which although an improvement on the range (which observes only two values and is affected by extremes) still ignores the remaining 50% of values. The formula is:

$$QD = \frac{Q_3 - Q_1}{2}$$

where QD = the quartile deviation, Q_3 = the third quartile (i.e. the value above which are 25% of the items), and Q_1 = the first quartile (i.e. the value below which are 25% of the items).

Quota sample Each interviewer is given an assigned number of individuals from whom to collect information within prescribed parameters (e.g. mothers with young children or people over 50).

Random sample A sample selected such that each member of the population has an equal chance of selection (e.g. a lottery).

Range The simplest and easiest method of determining dispersion, based on the difference between the highest and lowest values of the data. As its result depends solely on two values, it is considered only a rough approximation of the degree of dispersion.

Rank correlation (Spearman rank correlation)) A measure of the degree of association between two sets of rankings.

Ratio The description of the relationship between two or more numbers. It can be written as a fraction, decimal, or percentage and can be used to express rates (e.g. kilometres/hour).

Real interest rate The rate of interest expressed in inflation-adjusted terms (e.g., If the nominal interest rate on a bank account is 4% and inflation is 3%, then the real rate of interest is 1%).

Reducing balance (diminishing value) method of depreciation A fixed rate of depreciation which results in the greatest amount of depreciation at the outset of the article's use, with the amount becoming less with each successive period.

Relative frequency (empirical) method Probabilities are assigned on experimentation or historical data.

Relative index The expression of a commodity's price or quantity as a percentage of the base period result. It is used when comparing different units of production (e.g. tonnes and litres).

Ratio scale See *Semi-logarithmic scale*

Regression analysis Describes the pattern of the relationship between variables.

Rules of probability

- The subtraction rule:
  $P[\overline{M}] + P[M] = 1$ or $P[\overline{M}] = 1 - P[M]$
- The rule of addition for mutually exclusive events:
  $P[A \text{ or } B] = P[A \cup B] = P[A] + P[B]$
- The rule of addition for non-exclusive events: $P[A \text{ or } B] = P[A \cup B] = P[A] + P[B] - P[A \cap B]$
- The rule of multiplication for two independent events:
  $P[A \text{ and } B] = P[A]P[B]P[B]$
- Multiplication rule for dependent events:
  $P[A \text{ and } B] = P[A \cap B] = P[A]P[B \mid A]$
  $P[A \text{ and } B] = P[A \cap B] = P[B]P[A \mid B]$.

Sales tax A tax applied on the wholesale price of a manufactured good.

Sample A portion of a population which is surveyed; the larger the sample in relation to the population, the greater the reliability of results.

Sample point or ***element*** The individual outcome of an experiment.

Sample space The complete statement of all possible experimental outcomes—denoted by the symbol S.

Sample statistics Value calculated from the sample of the population.

Sampling error The difference between the sample and the population that is due to the particular statistic being measured.

Sampling with replacement The sampled item is returned to the original set; the sample space thus remains constant.

Sampling without replacement The sampled item is not returned to the original set; the sample space is thus reduced by one element.

Scatter diagram A technique used to determine if a correlation exists between two variables, found by plotting the points on a graph and subjectively judging the approximation of these points on a straight line.

Seasonal variation A time series component of repeating periodic movements recurring one or more times a year due to, for example, climatic conditions and traditional holidays.

Secondary data Collection and presentation of numerical information by another party (e.g. the use of ABS publications).

Secular trend A time series component describing a relatively long-term movement, indicating the general direction of a change of values. Normally, consideration of a period of 10 to 20 years is necessary to allow persistent patterns to emerge.

Semi-logarithmic (ratio) scale A scale which allows the rate of change over time to be illustrated, both between variables expressed in the same unit as well as those expressed in different units.

Simple aggregate (or composite) index An index number indicating the movement of several prices and/or quantities at a point in time. It is found by totalling the prices or quantities of the commodities in the given year and expressing this sum as a percentage of the base year result.

Simple index number Ratio or percentage change of price, quantity, or value of a commodity in a given period versus the price, quantity, or value of the same commodity in another period, referred to as the *base period*.

Simple interest The interest amount received per annum is calculated on the principal only (i.e. interest does not earn interest). The formula is:

$$I = P \times R \times T$$

where I = the interest earned, P = the principal, R = the rate of interest, and T = the time over which the principal has been deposited.

Simultaneous equations Two or more equations which describe the same set of conditions.

Sinking fund The determination of the periodic payment, given the interest rate, term and future value of an annuity. The formula is:

$$S = p \frac{(1 + i)^n - 1}{i}$$

where p = the periodic payment, S = the future value, i = the interest rate/period, and n = the number of periods.

Spurious correlation A high correlation but one with no direct relationship between the variables. Instead, a coincidental appearance of a relationship exists (e.g. tobacco consumption and life expectancy have been increasing for years; however, this does not mean that increased consumption of tobacco leads to prolonged life).

Stamp duty A state duty imposed on certain legal instruments and on several commercial and financial transactions. The duty varies from state to state.

Standard deviation A measure of dispersion which takes into account all values within the data and measures the deviation of the individual values from the mean. Instead of taking the absolute deviation of all values (as does the mean deviation) the deviations are squared and, subsequently, the square root is determined. It is the most widely used measure of dispersion and the greater the standard deviation, the less reliable is the mean as a measure of central tendency. The formula for ungrouped data is:

$$s = \sqrt{\frac{\Sigma(m - \mu)^2}{N}}$$

where s = the standard deviation, Σ = the symbol representing summation, m = the midpoint of the individual class, μ = arithmetic mean, and n = the total number of frequencies. The formula for grouped data is:

$$s = \sqrt{\frac{\Sigma f(m - \mu)^2}{N}}$$

where f = the frequency of the individual class.

Standard error of estimate The degree of variation of the actual data from the computed regression line.

Standard error of the mean The standard deviation of $\bar{x}$—the average absolute distance between the sample mean and the population mean. The average distance should decline as the sample size increases.

Statistical inference A conclusion derived about a population based on sample results.

Statistical method According to the order of applications in a statistical study, statistical methods may be divided into a number of basic steps including the collection, organisation, presentation, analysis and interpretation of the data.

Statistics As a subject, statistics embraces the collection, presentation, analysis and interpretation of information and is used to assist decision-making. A statistic is a summary measure or characteristic of a sample taken from a population.

Straight-line (prime-cost) method of depreciation A constant depreciated amount deducted every year for the expected life of the article.

Stratified sampling The division of the population into groups exhibiting common characteristics rather than selecting a random or systematic sample of the entire population.

Subjective method Probabilities are assigned on the basis of (personal) judgment.

Sum-of-the-years digit method A depreciation method with greater depreciation computed during the earlier years of an asset's life and less in later years; rarely used in Australia as the diminishing-value method is the preferred option by the Australian Taxation Department.

Systematic sampling A sample obtained by initially selecting a starting-point at random, with each successive number selected systematically (e.g. selecting every fifth item).

t distribution A family of distribution curves (very similar to the normal curve) which allows us to make estimations based on small samples. It must be used when the sample size is small ($n \leq 30$) and the population standard deviation is unknown. The underlying distribution should also be normal or close to normal.

Table A list of related pairs of items; the most detailed form of visual presentation.

Time series A set of statistical observations, arranged in chronological sequence, normally at equal time intervals (e.g. yearly sales results, GDP, unemployment, CPI, strikes and exports).

Total depreciation The cumulative sum of all depreciation of an asset since its purchase.

Total costs The total cost facing a firm in the short run is the total fixed costs and total variable costs.

Trade discount Price reduction given by one business to another for the purchase of materials used in their particular profession.

Tree diagram A useful visual approach for systematically listing the events and sample space of an experiment.

Turning point The point on the graph which has reached a peak or trough.

Type I error The error of rejecting the null hypothesis when it is true.

Type II error The error of accepting the null hypothesis when it is false.

Ungrouped (raw) data Data presented in its original state.

Variable cost Costs which vary with the level of production (e.g. wages, fuel).

Venn diagram A diagrammatic method used to clarify the relationships between sets or events.

Weighted aggregate (or composite) index An index number which is used to attribute the appropriate importance (or weight) to each of the items being examined (e.g. the CPI).

Weighted mean average The allocation of weights to items in relation to their degree of importance to the whole.

Weighted price index An index indicating the change in the cost of purchasing a constant basket of goods in a given period versus the base period.

Yield The actual return on an investment expressed as a percentage.

INDEX

abscissa 167
Access Economics 205, 210
accumulated depreciation 142
accumulated value 103
acid test ratios 80
A.C. Nielsen 211
acquisition cost 143
All Ordinaries Index 551–2
amortisation 132–5
annuities 117–18
   amortisation 132–5
   deferred 118
   future value 118–24
      table 631–45
   ordinary 117
   present value 118, 128–31
      table 647–61
   sinking fund 125–8
annuity due 117
approximation 22–4
area-probability sampling *see* multi-stage sampling
areas under the normal curve 431–46, 461
arithmetic mean *see* mean
asset turnover 77
assets 141–2
Australian Bureau of Statistics (ABS) 199, 204, 205, 210
   definition of employment 213
   multi-stage sampling 378
   Population Census 205, 206
   sample statistics 371
   stratification 376–7
Australian Stock Exchange, indexes 551
average of relative index 523–6, 532–7
averages 207–8, 279–80

balance of payments 89
balance sheet 75
bar charts 243–7, 248
   component 246–7, 248, 249
   multiple 246, 247
   simple 243–4
base period 515, 542

selection of 209, 542–3
   shifting of 543–5
Bayes' Theorem 418–21
bias 212–13, 542
BIS-Shrapnel 205
book value 142–3
break-even analysis 183–7
brokerage 36–7
budget deficit/surplus 88
Budget Papers 205
business activity indexes 514

calculator, working with 673–80
Cartesian coordinates 168
cash discounts 41–2
cause and effect relationships 220–1
census 205, 206, 373
Central Limit Theorem 447–55
central tendency, measurement of 279–80
chain discounts 43–5
charts
   bar 243–7, 248, 249
   pie 240–3, 244
circle charts *see* pie charts
class interval (size) 252
closing written-down value *see* book value
cluster sampling 377–8
coefficient of correlation 473–5
coefficient of determination 473–5
coefficient of skewness 366–7
coefficient of variation 352–4
commission 32, 34–6
   graduated 33–4
   retainer plus 32–3
company tax 46–7
complement of an event 395
complex fraction 4
component bar charts 246–7, 248, 249
composite (aggregate) indexes 515
   weighted *see* weighted composite (aggregate) indexes
compound interest 103–4
   calculation 104–6

..., 534, 565

correlation 473–5
  coefficient of 473–5
  negative 475
  positive 475
  spurious 479
correlation analysis 201, 464–5, 473–8
cumulative frequency curves 259–61
current working capital ratios 80
cyclical variation 562, 563–4, 581

data 198
  analysis 200
  bimodal/multimodal 300
  collection 200, 203–6
  continuous 199
  deseasonalised 608
  discrete 199
  external 204–5
  internal 204–5
  interpretation 201
  non-comparable 219–20
  primary 200, 205, 206
  secondary 200, 205–6
  visual presentation of 200, 227–71
debtor turnover 79
decimals 8–10
  rounding off 23–4
deflation 545–6
denominator 3
dependent variables 465, 469, 496
depreciation 142–3
  reducing balance (diminishing value) method 143, 148–52, 153–6
  straight-line (prime-cost) method 143, 144–7
  sum-of-the-digits method 143
  units-of-production method 143, 157–9
deregulation, of financial system 88

descriptive statistics 199, 200
  see also visual presentation of data
determination, coefficient of 473–5
deviation
  mean *see* mean (average) deviation
  quartile *see* quartile deviation
  standard *see* standard deviation
diminishing value depreciation method 143, 148–52, 153–6
discounts 38–40
  cash 41–2
  chain 43–5
  trade 42
dispersion 320–1
  coefficient of skewness 366–7
  coefficient of variation 352–4
  mean deviation 330–6
  quartile deviation 323–9
  range 321–2
  standard deviation 338–50
distortion
  graphic 213–17
  statistical 218–19
distribution
  frequency 247–8, 250–61
  normal 305, 429–31

earnings per share 76
Economic Roundup (Treasury) 205
effective rate of interest 95–7
efficiency ratios 78–9
element *see* sample point
errors
  non-sampling 210, 383
  sampling 210, 378, 382, 542
estimated life 143
estimating line *see* regression line
event 394
  complement of 395
events
  independent 402–3
  mutually exclusive (disjoint) 400
  non-exclusive (joint) 400–2
experiment 392
exponential smoothing 590–2
exponents 12
  fractional 14–15
  negative 13–14
  positive 12–13

financial ratios 75
    see also percentages
fixed costs 183
flat rate of interest 95–7
forecasting 607, 610
foreign exchange rate 88–9
fractional exponents 14–15
fractions 3–7
freehand graphic method 568–70, 581
frequency distribution 247–8, 250–61
frequency polygons 247, 254, 257, 258
future value 103–4
    of an ordinary annuity 118–24
        table 121–4, 631–45

Gaussian distribution 430
geometric mean 317–18
graphic distortion 213–17
graphs
    non-linear 188–91
    of linear equations 167–75
    straight-line 167–87, 261–5
    to solve simultaneous equations
        176–82
    turning point 188
'gun levy' 47

Hering illusion 214
histograms 247, 254, 255, 256

ideograms *see* pictograms
illusions, perceptual 213–15
improper fractions 4
independent variables 465, 496
index number construction 542
    selection of base period 542–3
    shifting of base period 543–5
index number (power) *see* exponents
index numbers 201, 513–15
    simple 515
        price 515–18
        quantity 518–19
        value 519–21
    simple aggregate
        average of relative price/quantity
            523–6
        price 522–3
    uses 545–6
        limitation in use of 542

weighted composite (aggregate) indexes
    average of price relative 532–7
    average of quantity relative 539–40
    price 527–31
    quantity 537–9
    value 540–1
inferential statistics 199–200, 372
inflation 87–8, 534, 536, 537
Information Age 201
integers 3
interest rates
    compound 103–8
    simple 87–95
International Monetary Fund 205

Jastrow illusion 214
judgment samples 373, 379
just-in-time system 79

kanban system 79

Laspeyres method 527–8, 530–1, 537–8
least-squares method
    regression analysis 468, 486–8, 489,
        491–4
    secular trend analysis 572–80, 581
line graphs 167, 261–5
    break-even analysis 183–7
    linear equations 167–75
    non-linear 188–91
    simultaneous equations 176–82
linear equations 167–75
linked relative calculation 545
liquidity ratios 79

mark-up 53
maturity value 93
mean 280, 305
    advantages 289
    arithmetic 572
    disadvantages 290
    geometric 317–18
    of grouped data 285–9
    of ungrouped data 280–3
    standard error 447
    weighted 283–4
mean (average) deviation 330, 336
    for grouped data 332–5
    for ungrouped data 331–2

measure of central tendency 280
median 280, 292, 297, 305
    for grouped data 293–7
    for ungrouped data 292–3
Medicare levy 47
mixed numbers 4
mode 280, 300, 303, 305
    for grouped data 300–3
    for ungrouped data 300
Morgan Gallup Poll 377–8
moving average method 581–90
Müller-Lyer illusion 213
multi-stage sampling 378, 534
multiple bar charts 246, 247
multiple discounts *see* chain discounts

nominal interest rate 87
non-random samples *see* judgment samples
non-sampling errors 210, 383
normal probability distribution curve 305, 429–31
    areas under the curve 431–4
        table of 434–46, 461
    standardised normal curve 434
numbers, signed 11–12
numerator 3

ogives 247, 259–61
open-ended interval 252
optical illusions 213–15
Orbison illusion 215
ordered pairs 167
ordinate 167
Organisation for Economic Cooperation and Development (OECD) 205, 537
origin, point of 168, 574

Paasche formula 528–31, 537, 539
parameters 199, 372
Pearson's coefficient of variation 352
percentages 25–6, 208–9
    business application 28–60
    conversion to fractions/decimals 26–7
    use in tables 233
perpetuity 118
personal income tax 47–8
pictograms 233, 237–40
pie charts 240–3, 244
Ponzo illusion 213

population 199, 371, 372
portfolio investment 89
present value 108–10
    calculation of interest rate 110–12
present value at compound interest table 625–9
present value of an annuity table 647–61
price indexes 514, 515–18, 522–3, 532–4
primary data 200, 205, 206
principal 87
probability 391–4
    addition rule 399–402
    Bayes' Theorem 418–21
    Central Limit Theorem 447–55
    conditional 410, 418–21
    defined 394–5
    dependent events 411–17
    independent events 402–3
    joint 397–8
    marginal 397–8
    multiplication rule 402–3, 411–17
    mutually exclusive (disjoint) events 400
    non-exclusive (joint) events 400–2
    subtraction rule 395–7
    total 393–4
    tree diagrams 392–3, 404, 405–6, 414–16, 419
    Venn diagrams 395, 400, 401, 407–9, 411
probability distribution 201
probability sampling 373–4
    best-selling book lists 211
product moment formula 473
profit 53–60, 183
profit and loss 53–60
profit margin 53
profitability ratios 75–7
proper fractions 4
proportions 81–5

quantity indexes 514, 518–19
quantity relatives 518
quartile deviation 323, 329
    of grouped data 325–8
    of ungrouped data 323–5
questionnaires 380
    mailed 381
    personal interview 380–1
    telephone 382
quota sampling 373, 379

random number sample 374–5
random sampling *see* probability sampling
range 321–2
rank correlation coefficient 480–4
ratio scale *see* semi-logarithmic scale
ratios 70–4
   acid test 80
   current working capital 80
   efficiency 78–9
   financial 75
   liquidity 79
   profitability 75–7
real interest rate 87
real terms 209–10
reducing balance (diminishing value)
     depreciation method 143, 148–9
   depreciation schedule 150–2
   formula 153–6
regression analysis 201, 465, 486–8
   forecasting 488–94
regression line 465, 467, 473, 486, 489, 491, 493, 494, 496
repeating (non-terminating) decimal 9
Reserve Bank of Australia (RBA) 89, 204–5, 220
residual value 143
Retail Traders' Association 205
return on sales 76–7
return on total assets 75–6
rounding off 23–4
Roy Morgan Research Centre 205, 377

sales tax 48–51
Sales Tax Act 1992 49
sample 199, 371, 372
sample point 392
sample size 372–3
sample space 392–3
sample statistics 372
sampling 201, 210–12, 369–74
   cluster 377–8
   distribution of the mean 446–55
   multi-stage 378, 534
   quota 379
   simple random 374–5
   stratified 376–7
   systematic 375–6
   with replacement 403
   without replacement 405, 413–14

sampling errors 210, 378, 382, 542
Sander's parallelogram 214
scales 265–6
scatter diagrams 465–71, 489, 491
scientific notation 15–16
seasonal adjustment techniques 218
seasonal factors, in statistics 218, 219–20
seasonal index 594–610
   as deseasonalising tool 607–10
   as forecasting tool 607, 610
seasonal variation 562, 564–5, 594
   ratio-to-moving average 597–607
   simple average method 594–7, 606
secondary data 200, 205–6
secular trend analysis 567–8
   freehand graphic method 568–70, 581
   least-squares method 572–80, 581
   moving average method 581–90
   semi-average method 570–2, 573, 581
securities 36
semi-average method 570–2, 573, 581
semi-logarithmic scale 266, 267–8
sets 392, 407
signed numbers 11–12
significant figures 22
simple aggregate (composite) indexes 522–6
simple fractions 4
simple indexes 515
   price 515–18
   quantity 518–19
   value 519–21
simple interest 87–90, 103
   calculation 90–5, 104–8
   flat vs effective rate 95–7
simple linear regression analysis 486–8
simple relative indexes *see* simple indexes
simultaneous equations, graphic method 176–82
sinking fund 125–8
skewness 305
   coefficient of 366–7
   negative 305, 306
   positive 305, 306
Small Business Index 515
smoothing constant 591
Spearman rank correlation 480–1
spurious correlation 479
stamp duty 51–2

standard deviation 338, 350
   for a sample 338
   for grouped data (population) 342–4
   for ungrouped data (population) 338–41
   shorthand method of calculation 341–2, 344–9
standard error of estimate 382–3, 473, 496–500
standard error of the mean 447
standard form *see* scientific notation
standardised normal curve 434
statistical distortion 218–19
statistical inference 199–200, 372
statistics 198–9
   abuse of 207–21, 229
   descriptive 199, 200
      see also visual presentation of data
   inferential 199–200, 372
   meaningless (faked) 221
   methods 200–1
   seasonal factors 218, 219–20
   uses 201–3
stock on hand 78–9
stock turnover 78
straight-line (prime-cost) depreciation method 143, 144–7
stratification 376–7
sum-of-the-digits depreciation method 143
surveys
   phone-in (radio) 211
   telephone 382
symbols, in statistics 372
systematic sampling 375–6

tabular presentation 231–3, 234–7
taxation
   company tax 46–7
   personal income tax 47–8
   sales tax 48–51
   stamp duty 51–2
telephone surveys 382
television audience monitoring 211–12
terminating decimal 9
terms, definition of 213
time series 201, 561–2, 567
   cyclical fluctuations 562, 563–4, 581
   exponential smoothing 590–2
   irregular (residual) variations 562, 565–6

   seasonal variations 562, 564–5, 594–610
   secular trend 562–3, 567–8
total costs 183
total depreciation 142
trade discounts 42
tree diagrams 392–3, 404, 405–6, 414–16, 419
turning point, of graph 188
two-axis graphing *see* line graphs

United Nations 205
units-of-production depreciation method 143, 157–9
universal set *see* sample space

validity, of data 204, 205, 210
value indexes 514, 519–21
variable costs 183
variables 464–5
   bivariate 465
   dependent 465, 469, 496
   independent 465, 496
variation, coefficient of 352–4
Venn diagrams 395, 400, 401, 407–9, 411
visual presentation of data 227–9
   bar charts 243–7, 248
   checklist 270–1
   frequency distribution 247–8, 250–61
   line graphs 261–5
   pictograms 233, 237–40
   pie charts 240–3, 244
   principles 229–30
   scales 265–6, 267, 268
   selectivity 229
   tables 231–3, 234–7

weighted composite (aggregate) indexes 515
   average of price relative 532–7
   average of quantity relative 539–40
   price 527–31
   quantity 537–9
   value 540–1
weighted mean 283–4
weighting 534–5, 542
whole numbers *see* integers
World Bank 205
written-down value 142–3

z score 434–5
Zollner illusion 214